Advances in Intelligent Systems and Computing

Volume 1277

The series "Advances in Intelligent Systems and Computing" contains publications on theory, applications, and design methods of Intelligent Systems and Intelligent Computing. Virtually all disciplines such as engineering, natural sciences, computer and information science, ICT, economics, business, e-commerce, environment, healthcare, life science are covered. The list of topics spans all the areas of modern intelligent systems and computing such as: computational intelligence, soft computing including neural networks, fuzzy systems, evolutionary computing and the fusion of these paradigms, social intelligence, ambient intelligence, computational neuroscience, artificial life, virtual worlds and society, cognitive science and systems, Perception and Vision, DNA and immune based systems, self-organizing and adaptive systems, e-Learning and teaching, human-centered and human-centric computing, recommender systems, intelligent control, robotics and mechatronics including human-machine teaming, knowledge-based paradigms, learning paradigms, machine ethics, intelligent data analysis, knowledge management, intelligent agents, intelligent decision making and support, intelligent network security, trust management, interactive entertainment, Web intelligence and multimedia.

The publications within "Advances in Intelligent Systems and Computing" are primarily proceedings of important conferences, symposia and congresses. They cover significant recent developments in the field, both of a foundational and applicable character. An important characteristic feature of the series is the short publication time and world-wide distribution. This permits a rapid and broad dissemination of research results.

Indexed by SCOPUS, DBLP, EI Compendex, INSPEC, WTI Frankfurt eG, zbMATH, Japanese Science and Technology Agency (JST), SCImago.

All books published in the series are submitted for consideration in Web of Science.

More information about this series at http://www.springer.com/series/11156

Miguel Botto-Tobar · Marcelo Zambrano Vizuete ·
Angela Díaz Cadena
Editors

Innovation and Research

A Driving Force for Socio-Econo-Technological Development

Springer

Editors
Miguel Botto-Tobar
Eindhoven University of Technology
Eindhoven, Noord-Brabant, The Netherlands

Marcelo Zambrano Vizuete
Universidad Técnica del Norte
Ibarra, Ecuador

Angela Díaz Cadena
University of Valencia
Valencia, Valencia, Spain

ISSN 2194-5357 ISSN 2194-5365 (electronic)
Advances in Intelligent Systems and Computing
ISBN 978-3-030-60466-0 ISBN 978-3-030-60467-7 (eBook)
https://doi.org/10.1007/978-3-030-60467-7

This Springer imprint is published by the registered company Springer Nature Switzerland AG
The registered company address is: Gewerbestrasse 11, 6330 Cham, Switzerland

Preface

The First International Conference of Research and Innovation-CI3 2020 was held virtually on June 18–20, 2020, receiving hundreds of participants. It aimed to disseminate the research project results that are being carried out in different higher education institutions, research centers, and the business sector. Furthermore, CI3 2020 was jointly supported and co-organized by the most relevant ecuadorian institutes: Instituto Superior Tecnológico Rumiñahui (ISTER), Libertad, Bolivariano, Vida Nueva, Espíritu Santo, Sudamericano Loja, Central Técnico; and sponsored by the Universidad Nacional Mayor de San Marcos (Peru), the Federal University of Goias (Brazil) and Hostos–Community University of New York (USA), and GDEON.

The CI3 main objective is to promote the development of research activities in the HEIs and relationship between the productive and academic sectors of Ecuador, contributing to the National Development Plan "Toda una vida 2017–2021" fulfillment.

"Research as a pillar of higher education and business improvement" was the conference motto and hinted at how research, technological innovation, and academia must be related to the productive sector to leverage social and business development.

Given the global crisis caused by COVID-19, the conference was held as virtual in the ISTER Virtual Campus as the main venue. The lectures were transmitted through computer tools such as Zoom and Facebook Live.

The content of this volume is related to the following subjects:

- Technological Trends
- Electronics
- Software
- Computational Intelligence and Information Systems
- Communications
- Security
- E-learning, E-Government, and E-business

Organization

General Chair

Marcelo Zambrano V.	Instituto Tecnológico Superior Rumiñahui, Ecuador

Editorial Committee

Marcelo Zambrano V.	Instituto Tecnológico Superior Rumiñahui, Ecuador
Miguel Botto-Tobar	Eindhoven University of Technology, The Netherlands
Ángela Díaz Cadena	Universidad de Valencia, Spain

Publication Chairs

Marcelo Zambrano V.	Instituto Tecnológico Superior Rumiñahui, Ecuador
Miguel Botto-Tobar	Eindhoven University of Technology, The Netherlands

Organizing Committee

Wladimir Paredes	Instituto Tecnológico Superior Rumiñahui, Ecuador
Marcelo Zambrano V.	Instituto Tecnológico Superior Rumiñahui, Ecuador
Miguel Botto-Tobar	Eindhoven University of Technology, The Netherlands
Luis Andrés Chavez	Instituto Tecnológico Superior Rumiñahui, Ecuador

Steering Committee

Ángel Ernesto Huerta
Carmita Suárez
Wladimir Paredes
José Luís Flores
Ana Marcela Cordero
Maritza Salazar Veloz
Wilfrido Robalino
Roberto Tolozano
Sandra Jarrín
Alicia Soto

Scientific Committee

Alonso Estrada
Ana Zambrano
Doris Macías
Martha Fernández
Norma Molina
Mariana Lima
Marina Polo
Lizbeth Ximena Suárez Morales
Elfio Pérez Figueiras
Tulio Ramírez
Leonardo Carvajal
José Luis Da Silva
Gustavo Hernández
Germán Aníbal Narváez Vásquez
Francisco Pérez
Javier Hingant
Fernando Carrera
Miguel Ángel Zúñiga
Santiago Vidal
Luis Tello
William Zamora
Silvana Gamboa
Martha Cecilia Paredes
Andrés Rosales
Álvaro Jiménez Sánchez
Vladimir Bonilla Venegas
Gustavo Scaglia
Patricio Cruz
Pedro Maldonado
Hugo Arcos
Fabián Pérez
Hólger Capa Santos
Aracely Yandún
Ismenia Araujo
Luz Marina Rodríguez
Maritza Salazar
Ángela Díaz Cadena
Alex Santamaría
Verónica Falconí
Isabel Cristina Meléndez Mogollón
Míriam Romero Saldarriaga
Rosa Rugel Rivas
Lucía Begnini
Ana Jacqueline Noblecilla Olaya
Eliana Acurio
Nataly Aracely Pozo Viera
Patricia Otero
Luz María Tobar Subía
Fernanda Tusa
María Páez
Estefanía Melisa Rodríguez Santos
Anabel Porte
Magda Francisca Cejas
Zila Isabel Estévez Fajardo
Martha Fernández Rodríguez
Josnel Martínez
Sergio Montes
Jhonny Barrera
Darwin Aguilar
Fabián Sáenz

Ángel Jaramillo
Roberto Camana
Juan Carlos Santillán Lima
Fredy Javier Landy Hurtado
Borys Hernán Culqui Culqui
Frank Angel Lemoine Quintero
Henry Cadena Poveda
Iván Iglesias Navarro
Juan Félix Ripalda Yánez
Ángel Freddy Rodríguez Torres
Pablo Andrés Benavides Bastidas
César Mayorga Abril
Cristian Paúl Fabara Rodríguez
Christian Vásquez Falcony
Nelson Granda Gutiérrez
Diego Iván Pilaquinga Abadiano
Fausto Valencia
Cosme Damián Mejía Echeverría
Pablo Velarde Rueda
Marco Herrera
Francisco Rosero
Diego Iván Cajamarca Carrasco
Patricio Arguello Mendoza
Darwin Raúl Noroña Salcedo
Óscar González
César Hernán Rodríguez Garavito
Alex Patricio Toapanta Guacapiña
Robinson Gabriel Guachi Guachi
Álvaro Javier Prado Romo
Washington Caraguay
Geovanny Marcelo Mendoza
Cristian Andrés Tasiguano Pozo
Eric Cuenca
Esteban Montúfar
Javier Martínez Gómez
Álvaro Manuel Quinche Suquilanda
Leonardo Chancusig Chisag

Contents

Communications

Mobile Application with Cloud-Based Computer Vision Capability for University Students' Library Services

Joe Llerena-Izquierdo(✉), Fernando Procel-Jupiter, and Alison Cunalema-Arana

Universidad Politécnica Salesiana, Guayaquil, Ecuador
jllerena@ups.edu.ec

Abstract. This paper presents the development of the smart device mobile application "Book's Recognition". The app recognizes the text of library book titles at the library of the Universidad Politécnica Salesiana in the city of Guayaquil, Ecuador. Through a service stored in Amazon Web Service (AWS), Mobil Vision's algorithms for text recognition, and Google's API on the Android platform, the app "Book's Recognition" allows its user to recognize the text of the title of a physical book in an innovative and effective way, showing the user basic information about the book in real time. The application can be offered as a service of the library. The purpose of this development is to awaken university student's interest about new and creative forms of intelligent investigation with resources from the university's main library, and furthermore to facilitate the investigative process by providing information on non-digitalized, and digitalized, books that hold valuable and relevant information for all generations. The mobile app can be downloaded the following website: https://github.com/seimus96/mobile_vision.

Keywords: Machine vision · Text recognition · University library services

1 Introduction

University students currently search for information in a traditional or modern way, with the use of an intelligent device or without it [1]. Due to rapid technological growth and the development of different procedures to access information in different professional fields [2], users have begun to adapt new ways of interaction, performing more efficient, friendly and effective search tasks. Regarding university library services, it is important that the access to the library's information be dynamic, safe and agile. With existing new technologies and the use of mobile devices, it is increasingly common for students to access information through digital platforms for their homework or investigations [3], and increasingly less common for them to use physical libraries [4]. Educational libraries, especially university libraries, maintain library services that utilize informatic tools with varying technology to search for books. For example, the Salesian Polytechnic University in Ecuador allows access to its book

M. Botto-Tobar et al. (Eds.): CI3 2020, AISC 1277, pp. 3–15, 2021.
https://doi.org/10.1007/978-3-030-60467-7_1

catalog using an integrated library management system (ALEPH)[1] (see Fig. 1), as a search engine. This system is developed by Green Data[2], a business based out of Barcelona, Spain, currently integrated through ALMA[3], and manager of Spanish academic libraries. Professors and students can access the service through their university portal[4], and enter into ALEPH through the web. Searches only show if the book, author or topic in question is available at each of the three libraries of the university located in the cities of Guayaquil, Quito or Cuenca.

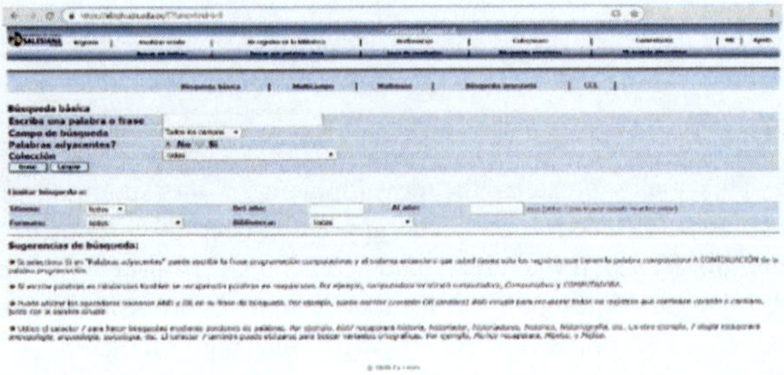

Fig. 1. The Salesian Polytechnic University of Ecuador's ALEPH system

The usual way university students (between 18 and 25 years old) look for a book upon arriving at the library is by finding the area with books related to the topic they are interested in and then browsing book by book (see Fig. 2).

Fig. 2. Usual way students look for books at the university library

[1] Available at https://aleph.ups.edu.ec/F?func=find-b-0.
[2] Available at https://www.greendata.es/.
[3] Available at https://www.exlibrisgroup.com/products/alma-library-services-platform/.
[4] On line catalogue available at https://www.ups.edu.ec/bibliotecas.

Finally, the process of loaning a book for reading or investigation is concluded when the loan is registered by the library assistant at the designated check-out desk (see Fig. 3).

Fig. 3. Images of the current loan system

1.1 Application Programming Interfaces Related to Real-Time Text Recognition

With the advances in the capture of information and the recognition of pattens, especially text recognition [5], new technologies have permitted the development of new services such as document analysis, access to devices via patterns, time savings in manual writing and minimization of errors in search engines [6]. With optical character recognition (OCR), a technique widely used with or without Internet connection, the text of an image when it is digitized is recognized or if it fails to be recognized during digitalization, while the user is writing in real time [7, 8]. An external library is required to implement OCR.

Table 1. Text recognition algorithms

Identifier	Name	Uses
SIFT	Scale invariant feature transform	[9, 10]
MSER	Maximally stable external region	[11, 12]
SURF	Speeded up robust features	[13, 14]
SWT	Stroke width transformation	[15]
EdgeBox	Edge boxes	[16, 17]
GLTR	Giant langauge model test room	[18]
Other	Other variations and types	[19, 20]

There are several algorithms for the recognition of text oriented images, each with different variations and uses (see Table 1). They are valuable for their ability to detect text and perform multiple tasks. Each algorithm has its own procedure to recognize text from images. Some of the algorithms recognize the main characteristics of an image, classifying them as static or movement, as is the case of the SIFT algorithm.

Google Cloud's API Vision offers accessible learning modules through API Rest. It assigns labels and classifies images into multiple categories, detects objects and faces, and reads handwritten or printed text. As both API vision and Flutter (a workspace to develop mobile apps) are Google tools, they have direct compatibility. Their use in this project is justified by their mutual integration which does not generate any conflict.

2 Methods and Materials

This study uses an empirical-analytic investigation method with a quantitative focus. The applied method is quasi-experimental using the technique of a survey of a random group from a specific sample.

From 2010–2019 the Salesian Polytechnic University of Ecuador has grown at the national level at a rate of 18.18%, (((recent value/previous value) – 1) x 100) (see Fig. 4). Currently it has three branches in three different cities: Cuenca (main branch), Quito and Guayaquil. This project is realized in the Guayaquil branch of the university. In Guayaquil the university has two campuses: "Centenario" y "María Auxiliadora". Campuses have approximately 7500 y 800 students, respectively (see Fig. 5).

	Total	Quito	Guayaquil	Cuenca
2010	22000	10000	7000	5000
2011	22500	10500	7100	4900
2012	23000	11000	7200	4800
2013	24500	12500	7300	4700
2014	24869	12700	7569	4600
2015	25000	12800	7700	4500
2016	25300	13100	7800	4400
2017	25600	13400	7900	4300
2018	25900	13600	8100	4200
2019	26000	13700	8300	4000

Fig. 4. Total number of students at the Salesian Polytechnic University of Ecuador

The university's main library in Guayaquil is located on the "Centenario" campus. It is a four-story building and provides library services to the students of the university in Guayaquil. The target group of this study are the students of this campus. The size of the population chosen for the study is 7500 students, with a confidence level of 95% and a margin of error of 4.69%, obtaining the sample size of 413 participants.

The "Book's Recognition" app is designed with Mobile Vision, a programming interface application (API) from Google. It uses optical character recognition technology in images in more than 50 languages and different types of archives[5], integrated into the services that the online platform offers. The API allows for the text recognition needed in physical libraries. It is installed in a mobile application, and it does not need

[5] Ver en: https://cloud.google.com/vision/?hl=es.

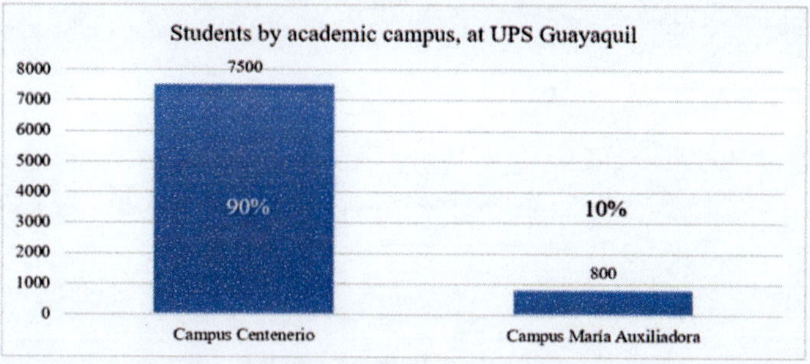

Fig. 5. Percentage of students per campus at the Guayaquil Branch of the Universidad Politécnica Salesiana.

mobile data to function. Text detection is performed in different languages, and the default language is Spanish. Text is recognized in segments by lines, words or blocks (see Fig. 6).

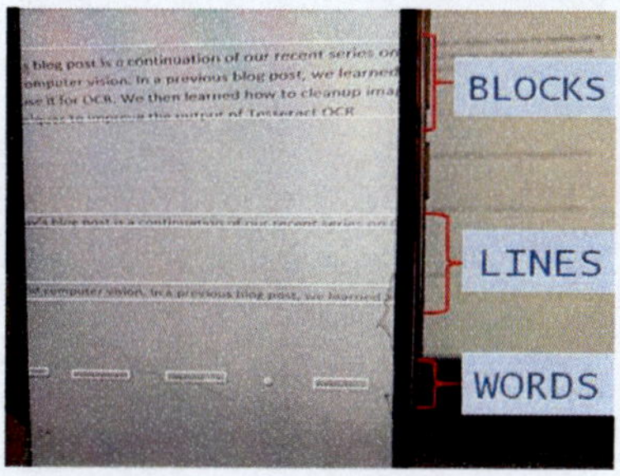

Fig. 6. Text recognition possibilities through API Mobile Vision

Flutter[6], an SDK made by Google to develop multiplatform mobile applications, was used to develop the application. Dart, an object-oriented programming language, is used via the descriptive programming method.

Dart is an AOT (*Ahead Of Time*), which allows Flutter to be fast and to personalize each structure. It is also *JIT* (Just in Time). To favor a more efficient development, the "*Hot Reload*" (see Fig. 7) method is used, which allows for instant compilation of changes made in the project. It also has the characteristic that, while updating a parameter of an object, the object does not lose its value.

[6] Disponible en: https://flutter-es.io/.

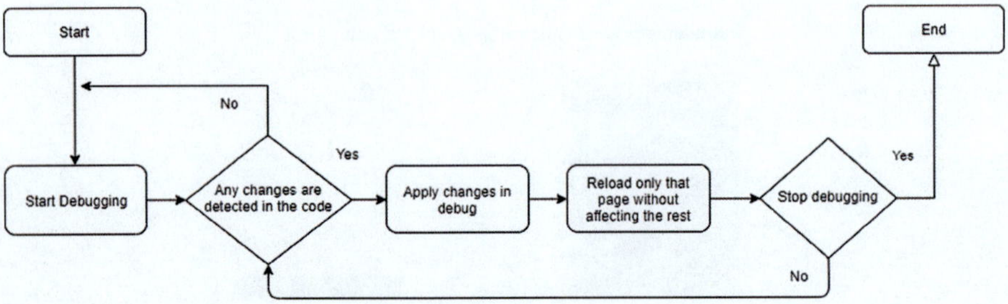

Fig. 7. *Hot Reload* function graph in *Flutter*

The project has four phases: the first phase runs API tests and algorithm analysis for real-time text recognition (see Fig. 8); the second phase develops the mobile application for Android mobile operating systems; the third phase develops the web service for the exchange and manipulation of the required data.

Fig. 8. Text detection tests using the recognition algorithm

In the fourth phase, tests are carried out in the physical space of the library with students of different years and majors (see Fig. 9). Finally, a survey is applied with an explanatory video[7], to measure the percentage of acceptance and library users' perception of the application.

Fig. 9. Testing the mobile app at the university library

[7] Explanation at: https://www.youtube.com/watch?v=eoDJLw63LEE&feature=youtu.be.

2.1 Development of the Mobile Application with Text Recognition

The mobile application "Book's Recognition", uses the *Flutter Mobile Vision* packet. The package enables the scanning of Bar Codes (including Quick Response codes or QR), the recognition of text and facial recognition. Its installation requires dependency on the package file (flutter_mobile_vision: ^0.1.3), through the command $ flutter pub get, in order to download the necessary complements. The projects principle file requires the line *import'package:flutter_mobile_vision/flutter_mobile_vision.dart';* to import the main file that makes the operation of the API possible in the application. Likewise FlutterMobileVision.start().then((x) => setState(() {})); initiates the use of the complement.

When the program is initiated by starting the application, the necessary permissions are requested which validate the types of cameras available on the device, as well as the resolutions of each camera. This allows the application to capture information about the resolutions of the integrated cameras (see Fig. 10a). In addition, lines of code are added to request the corresponding permission on the Android platform.

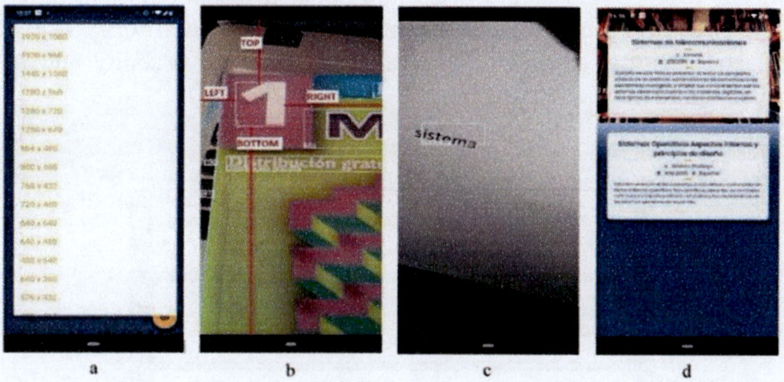

Fig. 10. a) Resolutions available in the device, b) positions of captures of text block, c) text detection and d) list of books with the selected word.

For the implementation in the mobile application, a friendly interface is used to present the texts captured by the camera. The values obtained from the captured text, with the exception of the value itself are the language and the positions (up, down, left, and right) in relation to the block of captured text (see Fig. 10b). A server was implemented in AWS, which contains the database, its administrator and the service stored for the information exchange. The information exchange service is developed in Hypertext Preprocessor (PHP) language searching for captured text in the database of book titles managed by MySQL. The entire text is captured, then the blank spaces on the sides of the image are eliminated. With the entirety of the text captured without the spaces, a search is run through the database table that contains the book titles (see Table 2). All incidents require the use of the phrase: tituloLibro LIKE '%'| |recognized text| |'%';.

Table 2. Data dictionary for the table: Tlibros, base

Column	Description	Mandatory/Key
id	Sequential book code with increment	*/OK
nombre	Original book title	*/no OK
descripcion	Short description of book contents	*/no OK
Autor	Author's name	*/no OK
imagen	URL of book's image	*/no OK
idioma	Original language of the book	*/no OK
fecha_publicacion	Book publication date	*/no OK
tema	General theme of the book	*/no OK

For example, if the recognized text is the word "system" (see Fig. 10c) the books that will be displayed will have the word "system" in their title, displayed in the list of information stored in the database (see Fig. 10d).

An improved version of the "Book's Recognition" application (see Fig. 11) allows the optimization of book searches at the time of text recognition that only contains valid information according to the standards described in ISO/IEC 25000: 2014 [21, 22], such as functional adequacy, performance efficiency, usability and reliability, which are taken into consideration.

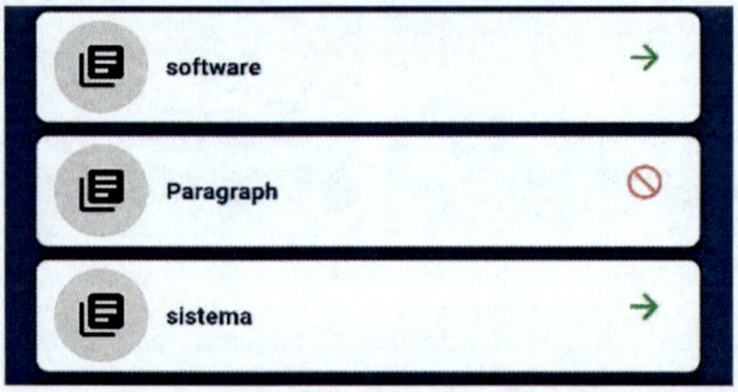

Fig. 11. Final version of the "Book's Recognition" application, which includes a functional improvement, more efficient search, improved usability and reliability

3 Results

The data obtained from the responses of the 413 participants surveyed to fulfil the objective of the investigation was tabulated. 67% of all participants indicated that they have had some experience with mobile applications with text recognition, allowing for high usability ratings by users who installed the application on their mobile devices. 33% of users reported that this was their first experience with text recognition applications.

83% of participants responded that new technologies, such as artificial intelligence algorithms, can generate new forms of behavior (see Fig. 12).

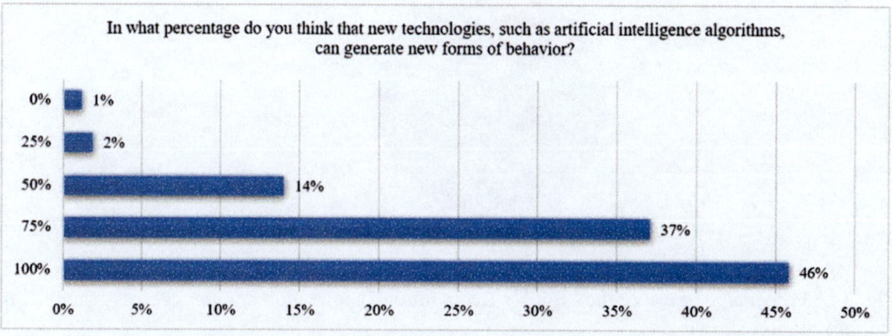

Fig. 12. Percentage of users who indicate that new forms of behavior may arise with the use of new technologies

On the other hand, 85% of participants agree that library services can be improved with information searches supported by mobile applications for specific purposes, such as locating information by selecting a non-digitized book within the physical library (see Fig. 13).

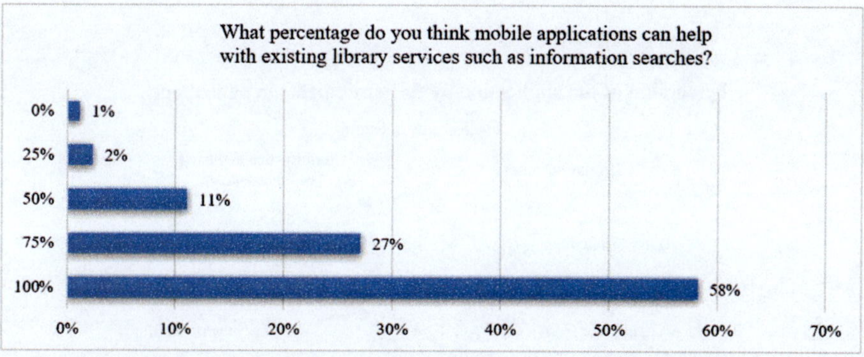

Fig. 13. Percentage of participants that support the use of mobile applications in library services

The "Book's Recognition" application is easy to use for 93% of those who participated in the study. On the other hand, 91% of participants were interested in the information provided by the application about the books recognized by the text recognition algorithm for digitized and non-digitized books in the University library (see Fig. 14).

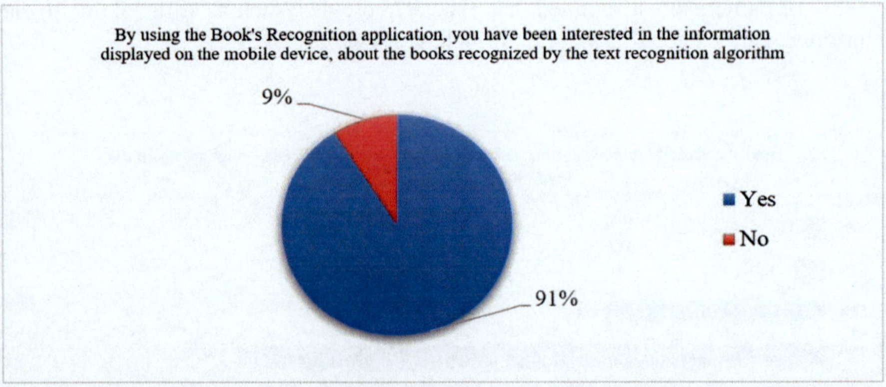

Fig. 14. Percentage of use of the "Book's Recognition" application and the interest generated by the information shown

It should be noted that participants' perception regarding the following characteristics: intelligibility in its use; performance in the use of the resources used by the application (memory and disk space); compatibility in the ease of interoperating the application; user friendliness; functional adequacy; attractive design; accessibility of the search feature; effective searches as well as its easy installation, exceed 70% acceptance as shown by the indicator "agree" and "totally agree". With this approval rating the "Book's Recognition" application achieves its objective among study participants (see Fig. 15).

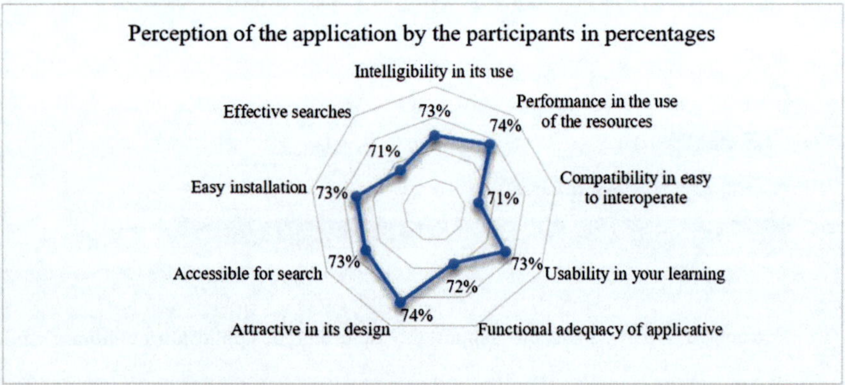

Fig. 15. Perception of "Book's Recognition" application by percentage

83% of study participants believe that the "Book's Recognition" application promotes university students' interest to perform searches from within the library for physical and digital books through new behavior with the use of a smart device.

Additionally, 96% would recommend the use of the application to educational institutions for its implementation and use. 80% of users recognize the application as innovative and affirm that it opens new opportunities for services using artificial intelligence, and that the technological development opens new possibilities not only for academic applications but also for administrative applications.

4 Discussion

The ways of using emerging technologies allows for the development of new behaviors but these behaviors are not definitive. A culture of research in young people is supported with tools capable of improving access times and location of information. This paper explores a novel inclusion of existing technology with the availability of the Google Mobile Vision API. It is clarified that there may be other possibilities of development to the problem posed in this document.

5 Conclusions

This paper presents the development of a mobile application on Android platforms, with computer vision capability developed with Google Mobile Vision API services. This technology awakens the interest of university students to search for information through new behaviors in the physical librarian spaces, behaviors that improve their research processes using digitized and physical books.

The levels of acceptance of the application exceed the expectations of the authors, especially due to the high percentage of users who value the application, the easy installation, the intuitive use of the app, the efficient search results, and the innovation in the academic world especially as a library service.

The project opens new possibilities to develop applications in other areas, not only in academia, but also in administrative areas and above all in information search management. The study highlights the interest of participants to use creative forms of intelligent searches and the application facilitates the research process with the use of books that have not been digitized but rather are available in the physical library of the university.

Acknowledgments. We thank the Salesian Polytechnic University; the team of people who work in the library of the university's Guayaquil branch for carrying out the project and testing the developed prototypes; the university students in academic period 55; the engineering majors for being critical in their responses to the development and implementation of the application which allowed for improvements and a new version of the application; and the GIEACI research group (https://gieaci.blog.ups.edu.ec/) for their support in the methodological-logical research process.

References

1. Kwasitsu, L., Chiu, A.M.: Mobile information behavior of Warner Pacific University students. Libr. Inf. Sci. Res. **41**, 139–150 (2019). https://doi.org/10.1016/j.lisr.2019.04.002
2. Llerena, J., Andina, M., Grijalva, J.: Mobile application to promote the Malecón 2000 tourism using augmented reality and geolocation. In: Proceedings - 3rd International Conference on Information Systems and Computer Science, INCISCOS 2018 (2018). https://doi.org/10.1109/INCISCOS.2018.00038
3. Izquierdo, J.L., Alfonso, M.R., Zambrano, M.A., Segovia, J.G.: Mobile application to encourage education in school chess students using augmented reality and m-learning. RISTI Rev. Iber. Sist. e Tecnol. Inf. **2019**, 120–133 (2019)
4. Meunier, B.: Library technology and innovation as a force for public good a case study from UCL library services. In: 2018 5th International Symposium on Emerging Trends and Technologies in Libraries and Information Services (ETTLIS), pp. 159–165. IEEE (2018). https://doi.org/10.1109/ETTLIS.2018.8485242
5. Ravagli, J., Ziran, Z., Marinai, S.: Text recognition and classification in floor plan images. In: 2019 International Conference on Document Analysis and Recognition Workshops (ICDARW), pp. 1–6. IEEE (2019). https://doi.org/10.1109/ICDARW.2019.00006
6. Ziran, Z., Marinai, S.: Object detection in floor plan images. In: Pancioni, L., Schwenker, F., Trentin, E. (eds.) ANNPR 2018. LNCS (LNAI), vol. 11081, pp. 383–394. Springer, Cham (2018). https://doi.org/10.1007/978-3-319-99978-4_30
7. Marne, M.G., Futane, P.R., Kolekar, S.B., Lakhadive, A.D., Marathe, S.K.: Identification of optimal optical character recognition (OCR) engine for proposed system. In: 2018 Fourth International Conference on Computing Communication Control and Automation (ICCUBEA), pp. 1–4. IEEE (2018). https://doi.org/10.1109/ICCUBEA.2018.8697487
8. Ozgen, A.C., Fasounaki, M., Ekenel, H.K.: Text detection in natural and computer-generated images. In: 2018 26th Signal Processing and Communications Applications Conference (SIU), pp. 1–4. IEEE (2018). https://doi.org/10.1109/SIU.2018.8404600
9. Hassan, A.K.A., Mahdi, B.S., Mohammed, A.A.: Iraqi journal of science. University of Baghdad, College of Science (2019)
10. Manasa Devi, M., Seetha, M., Viswanada Raju, S., Srinivasa Rao, D.: Detection and tracking of text from video using MSER and SIFT. In: Satapathy, S.C., Raju, K.S., Shyamala, K., Krishna, D.R., Favorskaya, M.N. (eds.) ICETE 2019. LAIS, vol. 4, pp. 719–727. Springer, Cham (2020). https://doi.org/10.1007/978-3-030-24318-0_82
11. Mehta, K., Patel, J., Dubey, N.: Text extraction from book cover using MSER. SSRN Electron. J. (2019). https://doi.org/10.2139/ssrn.3358207
12. Qin, S., Manduchi, R.: A fast and robust text spotter. In: 2016 IEEE Winter Conference on Applications of Computer Vision (WACV), pp. 1–8. IEEE (2016). https://doi.org/10.1109/WACV.2016.7477663
13. Sharma, M.K., Dhaka, V.S.: Segmentation of handwritten words using structured support vector machine. Pattern Anal. Appl. **23**(3), 1355–1367 (2019). https://doi.org/10.1007/s10044-019-00843-x
14. Proma, T.P., Hossan, M.Z., Amin, M.A.: Medicine recognition from colors and text. In: Proceedings of the 2019 3rd International Conference on Graphics and Signal Processing - ICGSP '19, pp. 39–43. ACM Press, New York (2019). https://doi.org/10.1145/3338472.3338484
15. Epshtein, B., Ofek, E., Wexler, Y.: Detecting text in natural scenes with stroke width transform. In: 2010 IEEE Computer Society Conference on Computer Vision and Pattern Recognition, pp. 2963–2970. IEEE (2010). https://doi.org/10.1109/CVPR.2010.5540041

16. NguyenVan, D., Lu, S., Tian, S., Ouarti, N., Mokhtari, M.: A pooling based scene text proposal technique for scene text reading in the wild. Pattern Recogn. **87**, 118–129 (2019). https://doi.org/10.1016/J.PATCOG.2018.10.012
17. Jaderberg, M., Simonyan, K.: Reading text in the wild with convolutional neural networks. Int. J. Comput. Vis. **116**(1), 1–20 (2016)
18. Gehrmann, S., Strobelt, H., Rush, A.M.: GLTR: Statistical Detection and Visualization of Generated Text, https://arxiv.org/abs/1906.04043 (2019)
19. Tang, Y., Wu, X.: Scene text detection and segmentation based on cascaded convolution neural networks. IEEE Trans. Image Process. **26**, 1509–1520 (2017). https://doi.org/10.1109/TIP.2017.2656474
20. He, T., Huang, W., Qiao, Y., Yao, J.: Text-attentional convolutional neural network for scene text detection. IEEE Trans. Image Process. **25**, 2529–2541 (2016). https://doi.org/10.1109/TIP.2016.2547588
21. Ramos, R.C.B., Villagran, N.V., Yoo, S.G., Quina, G.N.: Software quality assessment applied for the governmental organizations using ISO/IEC 25000. In: 2018 International Conference on eDemocracy & eGovernment (ICEDEG), pp. 311–316. IEEE (2018). https://doi.org/10.1109/ICEDEG.2018.8372327
22. ISO - ISO/IEC 25000:2014 - Systems and software engineering — Systems and software Quality Requirements and Evaluation (SQuaRE) — Guide to SQuaRE. https://www.iso.org/standard/64764.html. Accessed 05 Jan 2020

Algorithms for the Evolution for Electromagnetic Fields

Franyelit Suárez[1(✉)], Omar Flor[1], and Luis Rosales[2]

[1] Facultad de Ciencias e Ingeniería, Ingeniería Industrial, Universidad de Las Américas, Quito, Ecuador
franyelit.suarez@udla.edu.ec

[2] Universidad Experimental Politécnica Antonio José de Sucre, Puerto Ordaz, Estado Bolívar, Venezuela

Abstract. An algorithm is presented to solve non-linear wave equations and study the evolution characteristics of arbitrary spin fields in 3D. The eth formalism and the method of null characteristics allow discretization of differential equations. The numerical implementation of the method of null characteristics consists in evolving a scalar, electromagnetic or gravitational field on a family of null hypersurfaces in discrete sequence of time increments. While the eth formalism allows the discretization of angular coordinates. The code developed in Fortran 90 is stable and convergent in the second order and allowed the precise calculation of radiation for spin fields 0, 1 and 2 in the null infinity. The implemented algorithm is a first step to solve Maxwell's equations in the characteristic formulation.

Keywords: Characteristic formulation · Numerical methods · Radiation patterns

1 Introduction

The study of radiation is one of the problems that occupies a large part of the scientific community of modern and computational physics today. Gravitacional wave simulations [1–3], electromagnetic [4–6], and scalars [7–9] which correspond to spin fields two, one and zero respectively, are reported daily in scientific journals.

Many of these works are done from the numerical point of view and a determining factor is to find numerical solutions of equations of evolution of these fields. It is important to note that in the recent gravitational-waves detections, the simulations were the main for tune-up these detectors [10–12]. Electromagnetic fields contains a study of the wave equations, electric and magnetic fields, their propagation in a vacuum and other media. Simulations of scalar fields are justifiable, because the effects of gravitational waves are reproduced very well [13].

The numerical implementation of the null feature method consists of evolving a scalar, electromagnetic or gravitational field over a family of null hypersurfaces in a discrete sequence of time increments [14]. In particular, a stable and convergent three-dimensional code is developed based on the characteristic formulation of General Relativity have been development. This code requires data about the initial null

M. Botto-Tobar et al. (Eds.): CI3 2020, AISC 1277, pp. 16–27, 2021.
https://doi.org/10.1007/978-3-030-60467-7_2

hypersurface and boundary conditions on the tube world [15, 16] and let to find the waveform inte infinite null.

The code uses a second order finite difference evolution algorithm, based on delayed time steps on a three-dimensional null coordinate mesh. Another essential technology of the code developed is the eth module, which incorporates a computational version of the Newman-Penrose eth formalism [17].

The eth module is based on covering the topological sphere with two stereographic meshes, see Fig. 2, overlapping, corresponding to the northern and southern hemispheres. This allows to obtain precise expressions of second-order finite differences for angular derivatives. The eth calculation simplifies the equations, avoids singularities and is computationally efficient for the discretization of angular derivatives [18, 19].

2 Evolution Characteristics

The characteristic formulation is based on the characteristic initial value problem formulated by Bondi [20]. It has been implemented as a robust computational algorithm that allows to evolve empty spaces-times or with matter. Also, it has been applied to simulate the gravitational radiation emitted in space-times contained in black holes. The evolution is done through the space-time foliation by means of new cones, as shown in Fig. 1 [21].

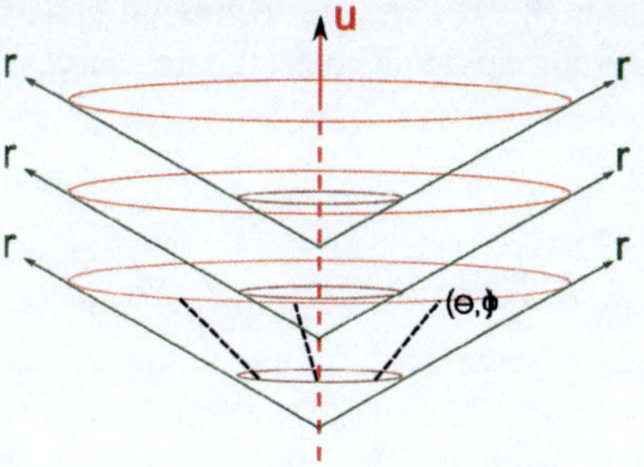

Fig. 1. Characteristic evolution of space time as a sequence of null cones open to the future.

2.1 Angular Derivatives

One way to use angular derivatives numerically is to change variables. For example, it has been used $y = -\cos\theta$ with excellent results to study gravitational radiation in axial symmetry and reflection. Another technique is to use eth formalism (es el símbolo ð). A version of this formalism has been used to discretize the angular operators that appear in the Einstein field equations for the Bondi-Sachs metric. It consists of writing the angular derivatives (on the sphere) that appear in the equations by means of stereographic projections in terms of rectangular coordinates (q, p). The null parallelogram method, together with the eth formalism, is known as the characteristic

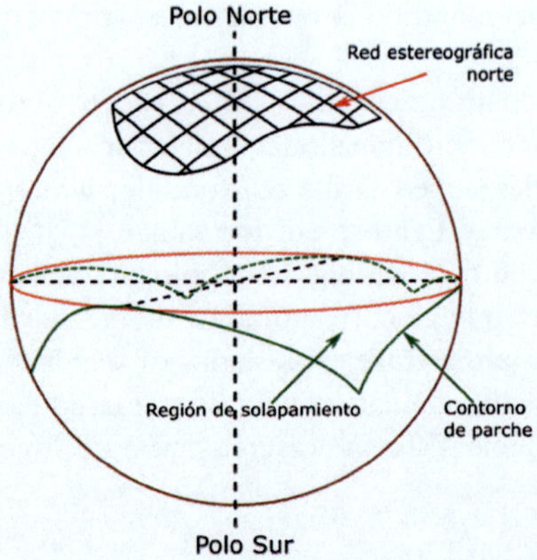

Fig. 2. Stereographic mesh showing a representation of the north patch based on the stereographic coordinates q, p. The dark line denotes the equator.

formulation of numerical relativity and has allowed simulating 3D gravitational radiation (three spatial and one temporal dimensions) with excellent astrophysical applications.

The analytical foundation of formalism is in pioneering works by Newman and Penrose in the 1960s [22]. In the 1990s, a discrete version of formalism was used to study gravitational radiation [8, 9, 12]. The technique consists of covering the topological sphere with two stereographic patches (square scarves, north and south) defined by this Eq. (1).

$$\zeta = q + ip \tag{1}$$

$$\begin{aligned} \zeta_N &= tg\left(\tfrac{\theta}{2}\right)e^{i\varphi} \quad \text{excluding } \theta = \pi \\ \zeta_S &= Cotg\left(\tfrac{\theta}{2}\right)e^{-i\varphi} \quad \text{excluding } \theta = 0 \end{aligned} \tag{2}$$

The Eq. 2 presents the patch for the north as for the south of the sphere, both patches are related in que Eq. 3 (Fig. 3).

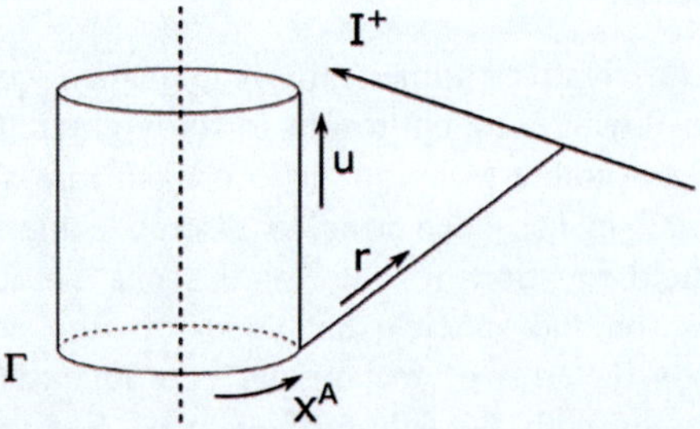

Fig. 3. The world tube and the coordinates are shown: temporal (u), radial (r) and the usual angles.

$$\zeta_S = \frac{1}{\zeta_N} \tag{3}$$

The eth formalism allows you to simulate fields with arbitrary spin. For example, gravitational, electromagnetic and scalar fields, which have spin 2, 1 and 0. Differential operators ð y $\bar{ð}$ (conjugate) they act on an arbitrary spin field in spherical coordinates θ, φ.

$$\begin{aligned} ð\Psi_S &= -(sen\theta)^S\left(\frac{\partial}{\partial\theta} + \frac{i}{sen\theta}\frac{\partial}{\partial\varphi}\right)(sen\theta)^{-S}\Psi_S \\ \bar{ð}\Psi_S &= -(sen\theta)^{-S}\left(\frac{\partial}{\partial\theta} - \frac{i}{sen\theta}\frac{\partial}{\partial\varphi}\right)(sen\theta)^{S}\Psi_S \end{aligned} \tag{4}$$

It is easy to demonstrate that the operator's action for zero spin fields is simply to derive the function from the coordinate indicated in the form (5) and the conjugate action (6).

$$ð\mathrm{F} = \partial_q F + i\partial_p F \tag{5}$$

$$\bar{ð}F = \partial_q F - i\partial_p F \tag{6}$$

where, F = F(u, r, q, p)

2.2 Null Characteristics Method

The null characteristics method allows to discretize the radial coordinate. Its main characteristic is the evolution through null cones open to the future and the use of compactification techniques, which allow them to be included the infinite in a finite mesh of points through the compactified coordinate in the form

$$x = \frac{r}{1+r} \tag{7}$$

Such that the range of evolution of the fields is between the boundary of the world tube and the infinite future null $0 \leq x \leq 1$.

Figure 4 shows the world tube where radiation propagates to infinity, each representing a hyper surface u = constant. The temporal and radial coordinates are shown, as well as the null parallelogram. The algorithm is based on knowing the fields in three points of the parallelogram and from these values determine them in the next point. This algorithm on a parallelogram is repeated until reaching infinity, then it is passed to the next cone (another hyper surface). Special attention requires the fields at r = 0 and r = infinity.z.

The sequence is shown for the times shown in the same graph in the hemisphere.

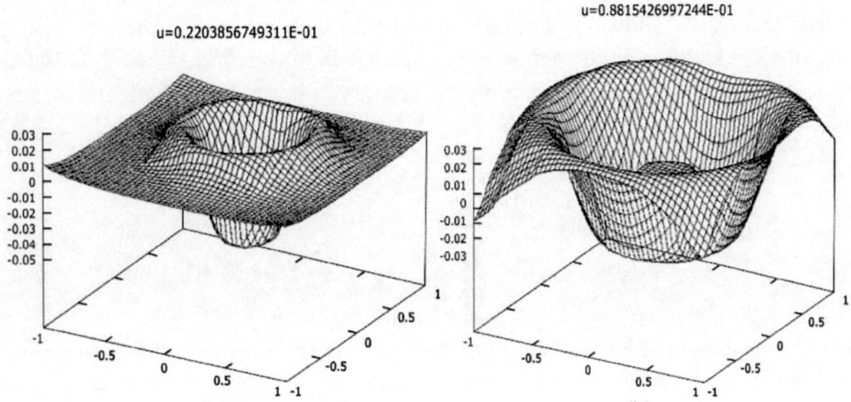

(a) (b)

Fig. 4. Radiation pattern for spin field two, and with s = 2 in the codes.

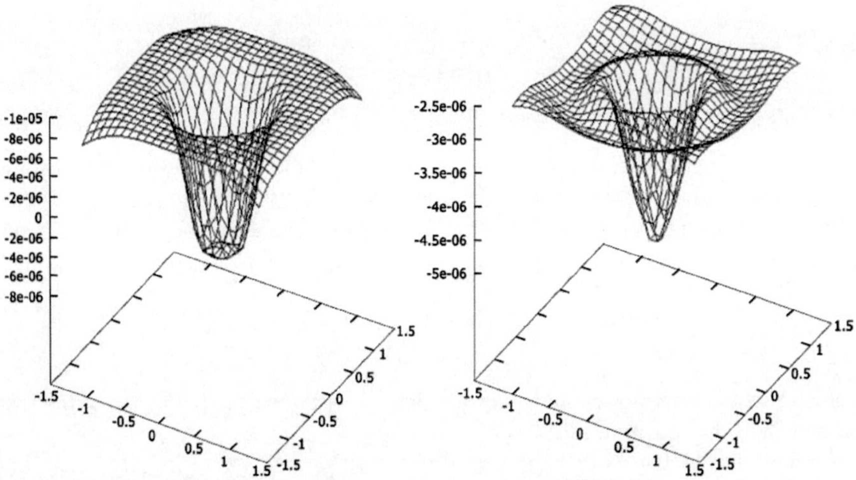

(a) (b)

Fig. 5. Radiation pattern for zero spin field, when s = 0 in the codes and can be used as code convergence. The sequence is shown for three times u = 1 and u = 10 for the southern hemisphere.

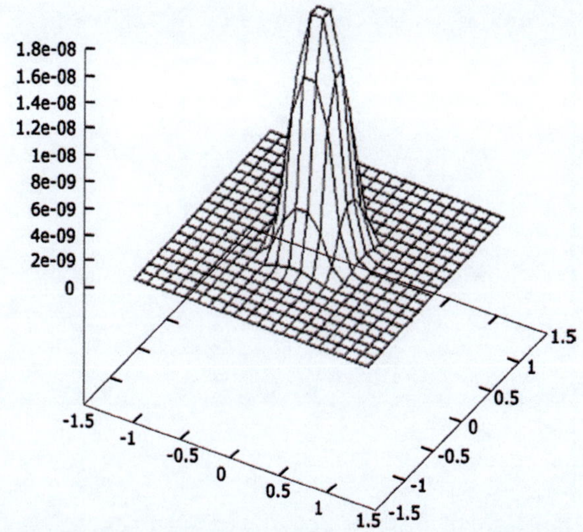

Fig. 6. Radiation pattern for spin field one when s = 0 in the codes and can be used as a convergence of the code when compared to previously obtained results. The sequence is shown for three times u = 3 and U = 8 in the southern hemisphere.

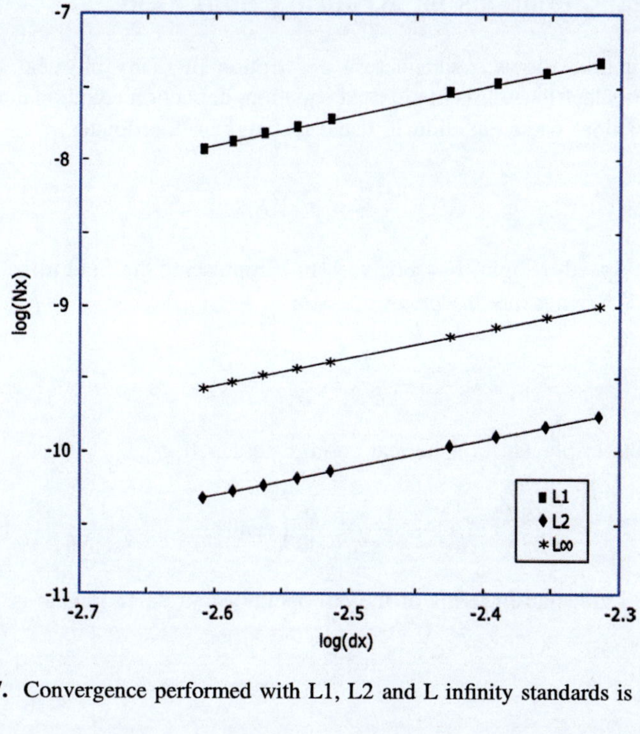

Fig. 7. Convergence performed with L1, L2 and L infinity standards is shown

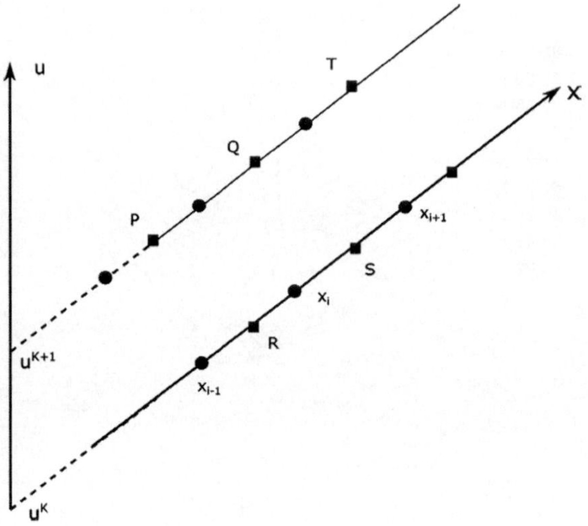

Fig. 8. Discretization of the radial coordinate x. The null parallelogram algorithm is shown. Once the fields at points P, R and S are known, the field at Q is determined. The parallelogram is repeated until reaching infinity. Then it goes to another hyper-surface.

3 Nonlinear Equations for Arbitrary Spin Fields

The wave equation allows a satisfactory description of many physical systems: the string equation, electric and magnetic field equation, deflection equation etc. In general, a three-dimensional wave equation in usual rectangular coordinates.

$$\frac{\partial^2 F}{\partial t^2} = c^2 \nabla^2 F \tag{8}$$

where c is the speed of light, F = F(t, x, y, z) It represents the field to be determined. The operator ∇^2 represents *laplacian* operator

$$\nabla^2 = \frac{\partial^2}{\partial x^2} + \frac{\partial^2}{\partial y^2} + \frac{\partial^2}{\partial z^2} \tag{9}$$

If you write the Laplacian in spherical coordinates. (r, θ, φ)

$$\nabla^2 = \frac{1}{r^2}\frac{\partial}{\partial r}\left(r^2\frac{\partial}{\partial r}\right) + \frac{1}{r^2 \mathrm{sen}\,\theta}\frac{\partial}{\partial \theta}\left(\mathrm{sen}\theta\frac{\partial}{\partial \theta}\right) + \frac{1}{r^2 \mathrm{sen}^2\theta}\frac{\partial^2}{\partial \phi^2} \tag{10}$$

Similarly, the Laplacian in terms of the eth operator and its conjugate is.

$$\eth\bar{\eth} F = 0 \tag{11}$$

Equivalently in stereographic coordinates (q, p) the Eq. (13) is written:

$$\frac{4}{(1+q^2+p^2)^2}\left(\frac{\partial^2 F}{\partial q^2}+\frac{\partial^2 F}{\partial p^2}\right)=0 \tag{12}$$

This last equation is very easy to discretize and solve numerically for finite differences. In any book of numerical analysis appears the way to discretize at the second order of approximation, as is our case.

For purposes of numerical implementation Eq. (12) can be written.

$$F,ur - F,rr + \eth\bar{\eth}F = 0 \tag{13}$$

where, F represents the field to be determined. The comma indicates derivatives with respect to the indicated coordinate. The spin of field F is determined by the spin of the initial data (to evolve). So that the action of the operator eth and its conjugate on F is done taking into account the spin of the field in the first hypersurface u = 0.

4 Results

Analytically the scalar wave Eq. (13) is integrated into the null parallelogram and in this way a discrete and exact expression is obtained. The codes consist of three main routines: one that includes the start (or start) at u = 0, gstart. Then a subroutine that calculates in all space time to infinity, gcol. Another routine, gscri, that allows you to program infinitely computationally. Other secondary: a subroutine that calculates eth derivatives, the routine where the fields are written for all time and all r, q and p and another routine where the initial data and boundary conditions are specified. All these routines are controlled by a main program (Fig. 9).

PSEUDOCÓDIGO

Find radiation for arbitrary spin fields

Write teh 3D wave equation

Write the angular part in terms of eth operators

Write the radial derivatives in terms of the compactified coordinate x

Define boundary conditions

Select the spin of the field to be determined (s = 0.1 or 2)

Specify the field in the initial hypersurface u = 0 and the boundary conditions

Find the fields at infinity for the northern and southern hemispheres

end

Fig. 9. Main program routines

The initial data and the boundary conditions

The boundary conditions are specified on the boundary of the world tube, and the initial data on the first hypersurface. So, the boundary conditions were:

- About the *world tube* Γ, F = 0.
- On the first hypersurface.

The initial field is specified in the form:

$$F(u=0, x, q, p) = \lambda(r - Ra)^4 (r - Rb)^4 \, {}^sYlm.. \tag{14}$$

where, ^{S}Ylm are the spherical harmonics of arbitrary spin s in stereographic coordinates en coordenadas estereográficas (q, p). They are used Ra = 3 y Rb = 6 for all code runs.

4.1 Numeric Implementation

The numerical evaluation of the angular momentum operator (11) requires the technique of finite difference in angular directions. The presence of singularities at the Laplacian operator's poles (or angular momentum) limits the range of steps in time in which the algorithm is stable. For this reason, two patches are adopted on a unitary sphere in terms of stereographic coordinates (12).

Analytically the scalar wave Eq. (13) is integrated in the null parallelogram and in this way a discrete and exact expression formulated in terms is obtained of an integral identity. The null parallelogram defined by points P, Q, R and S (see Fig. 8), is formed by two segments of incoming null geodesics (PQ, RS) that are on two successive cones of light separated separadosu, and two segments of Goedésicas null outgoing (RP, SQ), separated by Δr. The wave Eq. (13) in the null parallelogram is integrated into the form

$$\mathrm{FQ} = \mathrm{FP} + \mathrm{FS} - \mathrm{FR} + \int_{\Sigma} dudr \eth \bar{\eth} F / r^2 |_C \tag{15}$$

This equation is implemented in the code: the fields FQ, FP,FR y FS are the fields at the respective points and the angular part is integrated averaging in the center of the null parallelogram C.

The compactified radial coordinate is discretized as xi = i* x para i = 1, Nx. Stereographic coordinates are discretized qk = − 1 + (k − 3) q, pl = − 1+(l − 3) p, where, k and l are integers defined between −M y M.

The mesh used in these runs has dimensions: Np = 50, Nq = M y Nx = 60, where, Np, Nq y Nx represent the points in the angular and radial mesh, respectively.

The spacings in the angular mesh are:

$$\Delta p = \Delta q = \frac{1}{M - 5}$$

While for the radial mesh.

$$\Delta x = \frac{1}{Nx}$$

Finally, the codes are implemented in the compactified coordinate x. When passing the equations of the radial coordinate r to the compact one, it requires interpolations such that the value of the fields at points P, Q, R and S are calculated using a linear intergralation of Lagrange. The equation provides the numerical value of the field with a second order error. In the numerical calculation of all derivatives, a finite difference scheme was used in order of approximation.

Figure 5 shows the evolution in the null infinity for the time u = 1 and u = 6 of the radiation pattern for spin fields 2 in the northern hemisphere. Here the spherical harmonic selected was l = m = 2, for s = 2. Figure 6 shows the evolution in the null infinity for the time u = 2, 4, 6 in the southern hemisphere for spin fields 1, simulated with l = 2, m = 1 and s = 1 the spherical harmonic parameters. Figure 7 shows the evolution in the null infinity for the times u = 3, u = 5 and u = 15 in the southern hemisphere for zero spin fields. The spherical harmonic data l = 2, m = 1 and s = 0. The smoothness of the graphs is shown in total agreement with what is expected for zero spin fields. In this case, the algorithm to solve the Eq. (13) is the same to solve the differential equation for the electric potential in a quad-dimensional quadri space. Thus, the algorithm presented in this paper serves as a basis for numerically solving Maxwell's equations using the characteristic formulation.

4.2 Code Convergence

The set of possible analytical solutions of Eq. (13) is written:

$$F(u, r, q, p) = r1 + 1(u + 1)\,1 + 1\,(u + 2r + 1)\,1 + 1\,{}^{s}Yl, m \quad (16)$$

This solution is used to study the convergence of the code by comparing with the numerical solution evaluated at u = 0. This result guarantees that given an approximation of consistent finite difference stability is a necessary and sufficient condition to confirm the robustness of the code when incorporate spin field 0, 1 and 2. The convergence of the code was carried out with standards L1, L2 and L∞. The slopes of the lines in Fig. 6 are, respectively, 1.8, 2.1 and 1.7 in accordance with the expected order. The test consisted of varying the radial mesh from 40,….400 points and monitoring log (error) and logΔx, defining as the difference between the numerically calculated and the analytical field.

$$\text{error} = |\boldsymbol{Fnumeric} - \boldsymbol{Fanalitic}|$$

5 Conclusions

The study of spin fields 1 and 2 has been incorporated into the characteristic code and the stability and convergence of the code have been verified. In the case of zero spin, the algorithm is exactly the same to solve the equation for the electric potential numerically. As a first step to solve Maxwell's equations with the use of the null characteristics method and eth formalism.

The code developed to solve the wave equation for scalar fields of spin 1 is the preamble and then proceed to solve Maxwell's equations with the same techniques. Once the Maxwell equations have been solved numerically using the characteristic formulation, the numerical methods will be applied to real problems of Electromagnetic Theory and Electrical Engineering, such as the study of electromagnetic radiation emitted by a transmission line.

References

1. Advanced LIGO News, "LIGO Hanford's H1 Achieves Two-Hour Full Lock" (2015)
2. Abbott, B., Abbott, R., Abbott, T., Abernathy, M.: Observation of gravitational waves from a binary black hole merger. Phys. Rev. Lett. (2016). https://doi.org/10.1103/physrevlett.116.061102
3. Abbott, B., Abbott, R., Abbott, T., Abernathy, Y.M.: Properties of the binary black hole merger GW150914. Gener. Relativ. Quantum Cosmol. (2016)
4. Castelvecchi, D.: Gravitational-wave rumours in overdrive. Nature (2016). https://doi.org/10.1038/nature.2016.19161
5. Abbott, B., Abbott, R., Abbott, T., Abernathy, Y.M.: Astrophysical implications of the binary black hole merger GW150914. Astrophys. Journay Lett. **818**, 22 (2016)
6. Barreto, W., Silva, A., Gomez, R., Lehner, L.: The 3-dimensional Einstein-Klein-Gordon system in characteristic numerical relativity. Gener. Relativ. Quantum Cosmol. (2004). https://doi.org/10.1103/physrevd.71.064028
7. Bondi, H., Van der Bug, M., Metzner, A.: Gravitational waves in general relativity, VII. Waves from axi-symmetric isolated system. Roy. Soc. **269**, 21–52 (1962)
8. Gomez, R., Papadopoulos, P., Winicour, Y.J.: Null cone evolution of axisymmetric vacuum space–times. In: AIP (1994)
9. Penrose, R.: Asymptotic properties of fields and space-times. Phys. Rev. Lett. **10**, 66 (1963)
10. Goudsmit, S.A.: Phys. Rev. Lett. **15** (1965)
11. Newman, E., Penrose, R.: 10 exact gravitationally-conserved quantities. Phys. Rev. Lett. **15**, 231 (1965)
12. Gomez, R., Lehner, L., Papandopoulos, P., Winicour, J.: The eth formalism in numerical relativity. Gener. Relativ. Quantum Cosmol. **14**, 977 (1997)
13. Gómez, R., Winicour, Y.J.: Asymptotics of gravitational collapse of scalar waves. In: AIP (1998)
14. Barreto, W., Gómez, R., Lehner, L., Winicour, J.: Gravitational instability of a kink. Phys. Rev. D **54**(6), 3834 (1996)
15. Penrose, R., Rindler, Y.W.: Two-Spinor Calculus and Relativistic Fields (1984)
16. Thomas, J.W.: Numerical Partial Differential Equations: Finite Difference Methods (1995)
17. Lehner, L.: Numerical relativity: a review (2001)
18. Classical and Quantum Gravity (1997)

19. Barreto, R.M.: Phys. Rev. D **54**(1) (1996)
20. Gubser, S.S.: Absorption of photons and fermions by black holes in four dimensions (1997)
21. Neuberger, H.: Vectorlike gauge theories with almost massless fermions on the lattice. Phys. Rev. D (1998)
22. Bhardwa, V., Mathur, D., Rajgara, F.: Formation of negative ions upon irradiation of molecules by intense laser fields. Phys. Rev. D (1998)
23. Phys. Rev. D **60**(1) (1999)
24. Giudice, G.F., Kolb, E.W., Riotto, A.: Largest temperature of the radiation era and its cosmological implications. Phys. Rev. D **64**, 023508 (2001)
25. Gómez, R.: Gravitational waveforms with controlled accuracy. Phys. Rev. D **64**, 024007 (2001)
26. Zlochower, Y., Gómez, R., Husa, S., Lehner, L.: Mode coupling in the nonlinear response of black holes. Phys. Rev. D **68**, 084014 (2003)
27. Bishop, N.T., Gómez, R., Lehner, L., Maharaj, M.: Characteristic initial data for a star orbiting a black hole. Phys. Rev. D **72**, 024002 (2005)
28. Rham, C.D.: Beyond the low energy approximation in braneworld cosmology. Phys. Rev. D **71**, 024015 (2005)
29. Gomez, R., Husa, S., Lehner, L., Winicour, J.: Gravitational waves from a fissioning white hole. Phys. Revies D **66**, 064019 (2002)
30. Gomez, R., Frittelli, S.: First-order quasilinear canonical representation of the characteristic formulation of the Einstein equations. Phys. Rev. D **68**, 084013 (2003)
31. Newman, E.T.: Note on the Bondi-Metzner-Sachs group. In: AIP (1966)
32. Goldberg, J., Macfarlane, A.: Spin-s Spherical Harmonics and ð. In: AIP (1967)
33. Class. Quantum Gravity **7** (1990)
34. **198**(10) (1997)
35. Schnetter, E., Hawley, S.H., Hawke, I.: Evolutions in 3D numerical relativity using fixed mesh refinement. Class. Quantum Gravity **21**, 1465 (2004)
36. Sperhake, U., Kelly, B., Laguna, P., Smith, K.L.: Black hole head-on collisions and gravitational waves with fixed mesh-refinement and dynamic singularity excision. Phys. Rev. D **71**, 124042 (2005)
37. Baker, J.G., Centrella, J., Choi, D.-I., Koppitz, M.: Getting a kick out of numerical relativity. Astrophys. J. Lett. **653**, L93 (2006)

5G Network Security for IoT Implementation: A Systematic Literature Review

Manuel Montaño-Blacio, Johana Briceño-Sarmiento, and Fernando Pesántez-Bravo(✉)

Instituto Superior Tecnológico Sudamericano, Miguel Riofrío, 156-26 Loja, Ecuador
afpesantez@tecnologicosudamericano.edu.ec

Abstract. Fifth generation (5G) wireless technologies satisfies the growing demand for the Internet of Things (IoT), however, IoT devices are vulnerable to security threats due to the simplicity of their hardware and communication protocols, which imply possible attacks and security challenges. In this work we propose to conduct a systematic review of literature that relates 5G technologies to the internet of things and approach the security that the 5G network for IoT must provide. The Torres – Carrión method is used, raising four research questions: a) information security services, b) types of attacks in 5G-IoT, c) security in the layers of the IoT network architecture, d) strategies for 5G-IoT network security. Semantic search criteria were applied, in the Scopus database, obtaining 23 articles from 18 journals, the main studies were collected, it is evident that the blockchain is an efficient security mechanism that merits further study, that the physical layer is the one that receives the most active and passive attacks, such as denial of service (DoS) that is studied by several authors, together with mechanisms, architectures, protocols and algorithms that provide the security services of a mobile network.

Keywords: Security · 5G network · IoT · Attacks

1 Introduction

The growing development of IoT in the last years has admitted a number of connections of devices and objects, the fifth generation of 5G mobile technology, is a fundamental pillar to satisfies the demand for new services and the massive deployment, with this, the security risks increase and the problems of vulnerability and attacks in the different layers of network become more evident. There is still no complete security framework applicable to the 5G-IoT network, studies and tests are carried out to validate some architecture, mechanism or algorithm that ensures the transmission of information; encryption reduces attacks against devices [1], the IoT architecture based on layers on models and security features recognizes possible attacks and, the analysis of the network layer proposes solutions for the IoT industry [2, 3].

Therefore, the objective of this work is to contribute to the knowledge about security research that is implemented for a 5G-IoT network, to publicize existing proposals, vulnerabilities and attacks [4] that in some way affect the integrity, reliability

M. Botto-Tobar et al. (Eds.): CI3 2020, AISC 1277, pp. 28–40, 2021.
https://doi.org/10.1007/978-3-030-60467-7_3

and availability of information [5], devices or network architecture. Consequently, a systematic literature review of the subject is carried out to determine research questions, as well as to identify future research.

In this systematic review of the literature we use the method proposed by Torres-Carrión [6], which divides the process into three parts: planning, review report and presentation of results. We found 70 studies on the security of the 5G-IoT network. No systematic literature reviews were found that specifically address security in the 5G and IoT network together. Next, four research questions are posed related to network security services, establishing the most frequent attacks, knowing in which layer more attacks are carried out and defining the strategies used to control the transmission of information.

To execute the planning of the search process, general and specific inclusion and exclusion criteria were established. Variables that include theoretical support, standard and indicators are determined to organize the answers to each research question. Through the article search process, 23 studies were obtained, which, through the use of the Mendeley bibliography manager, were organized and managed.

Each of the articles is analyzed to list the studies based on the indicators of the proposed variables and determine comparisons with previous work and future research. Finally, the results of the study are presented in tables, together with the argumentation of the answer to each research question, leaving the topic open for future investigations of systematic reviews of literature through this methodological adaptation.

2 Method

For the SLR, of the Systematic review of the literature proposed by Torres-Carrión [6] adapted from Kitchenham and Bacca is applied, it divides the process into three main phases, planning, review report and presentation of results, which are detailed in Table 1.

Table 1. Phases of the Torres-Carrión methodology

Planning	Review report	Results presentation
Identification of the need for the review Current security status in 5G Conceptual mindfact Semantic Search Structure Research questions Review Protocol Development Related Systematic Reviews Journal selection and Database Definition of analysis categories	Search ID Selection of primary studies Note of study quality Data extraction and tracking Synthesis and data monitoring	Results Conclusions

2.1 Planning

Current State of Security in 5G. The 5G network is considered as a technological evolution, increases coverage, capacity, transmission speed and also carries security risks; the high speed of the connection and the number of devices connected at the same time could generate security gaps at the provider level and of course final users; these gaps would allow denial of service (DDoS) attacks [7], vulnerabilities and privacy issues; it is necessary then to improve the authentication of devices, the integrity of the data and the confidentiality of the information.

Research Questions. The Internet of Things (IoT) is changing communications and services today, new paradigms have emerged as they move from a conventional industry to an intelligent industry, having its massive development with the implementation of the 5g network. However, due to the heterogeneous environment in 5G networks and the nature of transmission of radio propagation, the guarantee of privacy security, authentication, authorization, control of access to devices and preservation of privacy is a challenging task. We are interested in 5G-IoT security [8] based on security factors such as: integrity, authentication, confidentiality and availability, so we consider the following research questions:

- RQ1 – Of the services that describe information security, which ones apply in the 5G-IoT network?
- RQ2 – In the 5G-IoT network, what attacks can occur?
- RQ3 – Of the network architecture layers, which one has worked the most in the 5G-IoT network security level?
- RQ4 – What strategies have been used in joint investigations for security in the 5G-IoT network?

Conceptual Mindfact. According to the Torres – Carrión methodology, the *conceptual mindfact* is carried out, which "allows the researcher to focus his attention on the real theoretical context of the research" [6].

In Fig. 1, the conceptual mindfact is presented that detailed schematically the conceptual structure of security in the 5G-IoT network; concept that derives from the areas of mobile communications and fifth generation networks (5G) [9]. Subclasses include integrity, authentication, reliability and availability [5, 10]. The 5G-IoT network security features that are of interest to the study are evident on the left side of the mind: architecture, protocols, algorithms, mechanisms and internet of things (IoT). Other fields of science, which are part of security, that will not be studied are: infrastructure and mobile telephony.

Development of Review Protocols

- General criteria
  - Studies that involve security in 5G-IoT networks.
  - Publications of the last 5 years, from 2016 to 2020.
  - The Scopus database is considered for the article search

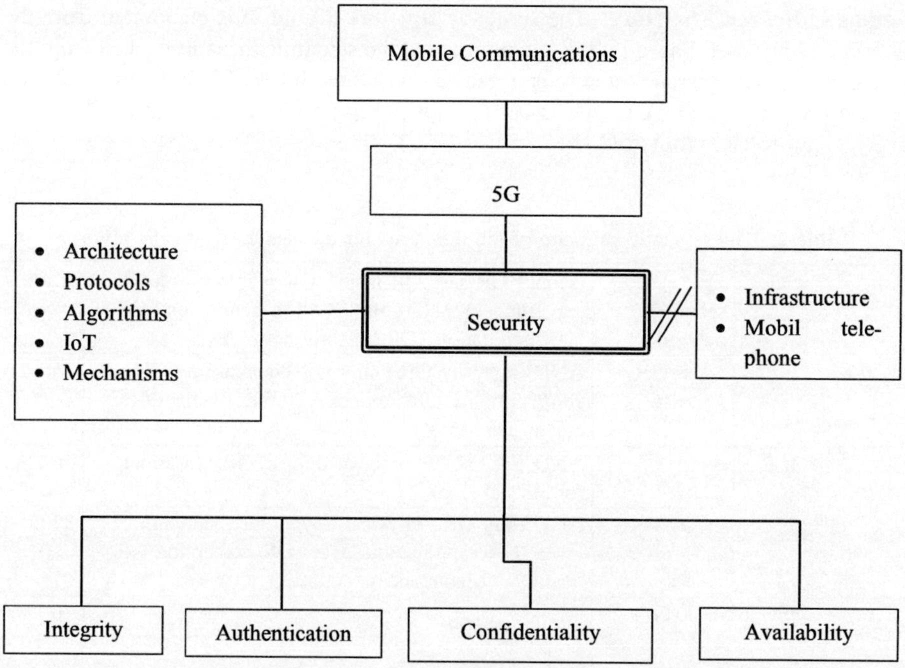

Fig. 1. Conceptual mindfact about the safety of 5G-IoT

- Specific criteria
  - For the search of services that describe the security of the information that are applied for 5G-IoT networks in RQ1, the ISO 27001 standard was considered.
  - To know what types of attacks 5G-IoT networks can present, according to RQ2, variables such as active and passive attacks have been used.
  - According to the architecture layers network of RQ3, is was considered physic layer, network and application.
  - It is necessary to know if there are joint investigations that propose security solutions in 5G-IoT networks, in order to verify whether a strategy that guarantees a secure network can be used, according to what is expressed in RQ4.

- Exclusion criteria
  - Studies in which reference is made to security in infrastructure and mobile telephony.
  - Only studies published in journals will be taken into account.

Semantic Search Structure. The semantic structure (Table 2) is elaborated from the conceptual mindset, based on the synonymy and the scientific thesaurus taken from the website [11]. The search structure is organized into three levels: 5G (L1), IoT (L2) and review protocols (L3). A search is carried out grouping L1 and L2 and results in 70 articles that refer to topic 5G-IoT.

Table 2. The semantic structure of the thesaurus for the search of specific articles.

L1	5G	5g security networks	(secur* OR safe OR insur* OR preser-vation OR surveillance) AND (5g OR "fifth generation" OR "last generation" OR "mobile networks 5g")	2168
		+ service security	AND (integrity OR auth* OR disponibility OR confidentiality OR vulnerability OR attacks OR amen* OR threat)	750
L2	IoT	+ IoT	AND (iot OR "Internet of things" OR "In-ternet for all" OR ioe)	78428
		+ layer	AND (phy OR "Physical Layer" OR "network layer" OR "application layer" OR perception OR mac OR "Media access control")	3728
[L1 AND L2]	[5G AND IoT]			70
L3	PROTOCOLS			
	Year		2016–2020	70
	Subject area		Computer Science	61
	Document Type		Article	23
	Languaje		English	23
	Quality criteria		Detailed reading of each article, to establish its relationship with the research area	19

Following the proposed methodology four revision protocols are applied, it starts considering the last 5 years, then it is limited to the area of computer science, selecting only articles in English, since they represent results of the relevant investigations on the problem raised. Finally, a detailed reading of each article is made to establish its relationship with security in 5G-IoT networks.

Related Systematic Reviews. The systematic review of related works is essential to achieve an original and especially useful contribution to the scientific community. A search was made in the Scopus database (Table 3) using the Syntax of database. In the works that were found, they did not fully answer the questions proposed in this investigation; in some cases analyzed, only part of the question is addressed and not within the context of this work, the general script was used with filtering oriented to a "review", "slr", "meta-analysis" or "survey".

Table 3. Recent studies of reviews of the literature on security in 5G-IoT networks.

Study	Analysis	Articles reviewed
[2]	The literature review, only at one point focuses on security and privacy for IoT, provides industry solutions from 3GPP. Related studies focus on the network layer to ensure that IoT traffic is delivered securely and efficiently between MTC devices (machine type communication) and mobile networks	>100
[1]	In this article, the authors study what concerns the applications, protocols, vulnerabilities and security of IoT networks They have made a systematic review of literature in which they have included analysis of articles related to IoT during the period 2015–2017 and conclude that the medical application is the most recurrent, in terms of attacks, they argue that the one that goes against The device is the most common and that as a tool to counteract these, encryption is used. By relating this article to the work realized, it can be said that the analysis developed focuses only on IoT, it is quite clear, but does not involve the implementation in 5G networks	111
[3]	Analyze the IoT architecture based on layered models and security features supported by security requirements such as the CIA triad, confidentiality, integrity and availability and other requirements such as accounting, auditability, non-repudiation, and privacy. It recognizes the possible security attacks of an IoT network in the four main layers, in which it generally proposes countermeasures to detect some security risks but not concrete solutions	57

Journal Selection and Database

Database. The database selected for the search is Scopus, the same one that has prestige in different fields of science; tools were used that contributed to obtaining 70 articles that refer to security in 5G-IoT networks; after that a selection is made based on the specified criteria.

List of Journals. The 23 articles found belong to 18 journals. The most relevant journal is IEEE Internet of Things, by its metrics it is located in quartile Q1 with an impact factor of 1.4; reflects a value of 129.36 and according to Google academic, has an h5 index of 70, refer to Table 4. It publishes articles on the latest advances, as well as review articles, on the various aspects of IoT. Topics include the architecture of the IoT system, the technologies that allow IoT, IoT communication and network protocols, such as network coding, IoT services and applications [12].

Definition of Analysis Categories. This section defines the categories of analysis for each research question, which facilitate grouping the articles according to the criteria that allow obtaining a response systematically to the questions proposed in 2.1.

Table 4. List of journals organized by category according to SJR 2018.

Ord	Journals name	Nro. Papers	SJR		h5 Google	Value
			IF	Cuartil		
SJR Science Edition						
1	IEEE Internet of Things Journal	4	1,4	Q1	70	129,360
2	IEEE Journal on Selected Areas in Communications	1	2,29	Q1	90	68,013
3	IEEE Transactions on Industrial Informatics Transactions	1	1,68	Q1	86	47,678
4	Journal of Network and Computer Applications	1	0,9	Q1	70	20,790
5	Future Generation Computer Systems	1	0,84	Q1	73	20,236
6	IEEE Access	1	0,61	Q1	89	17,916
7	Cognitive Computation	1	1,06	Q1	36	12,593
8	Computers and Security	1	0,67	Q1	50	11,055
9	Electronics Switzerland	2	0,46	Q1	1	0,304

- RQ1 – Of the services that describe information security, which ones apply in the 5G-IoT network? variables of ISO 27001 are considered.
  - Categories of the ISO 27001 standard (Information security): integrity, availability, confidentiality and authentication [13].
- RQ2 – In the 5G-IoT networks, what attacks can occur? The variables considered are passive and active attacks.
  - Passive attack categories: release of message contents and traffic analysis [4].
  - Categories active attacks: masquerade, replay, modification of messages and denial of service [4].
- RQ3 – Of the network architecture layers, which one has worked the most in the 5G-IoT network security level? IoT network architecture variables are considered.
  - IoT architecture categories: Physical layer, network and application [14].
- RQ4 – What strategies have been used in joint investigations for security in the 5GIoT network? It is considered as variable security strategies.
  - The categories that are taken into account for the analysis are: architectures, mechanisms, protocols and algorithms [15].

3 Review Report

3.1 Of the Services that Describe Information Security, Which ones Apply in 5G-IoT Network?

From the services that describe information security, studies show that the security of the 5G-IoT network is linked to the services proposed in the ISO 27001 standard, which focuses on confidentiality, authentication, integrity and availability to achieve information security, 45% of the articles reviewed maintain that confidentiality is the most sensitive part of these networks and propose some strategies (RQ4) to guarantee this service, 35% have parallel to confidentiality, authentication as one of the critical services that gives authorized users access to a network [16], regardless of the integrity and availability of the complement for a secure network scheme (Table 5).

Table 5. Items that had the ISO 27001 standard.

RQ1	From the services that describe information security, are they considered 5G-IoT network?		*f*
ISO 27001	Integrity	[17–19]	3
	Disponibility	[18]	1
	Confidentiality	[17–25]	9
	Authentication	[16–18, 23, 25–27]	7

3.2 RQ2 – In the 5G-IoT Network, What Attacks can Occur?

The attacks present in a network, Stallings [4] classify them into passive and active attacks; in the reviewed articles it is mentioned that in the 5G-IoT network, the most frequent attacks in this type of networks are, denial of service (DoS), replay, masquerade and traffic analysis. In [20] it presents a routing attack due to the dynamic infrastructure of the IoT network, analyzes and proposes measures to counteract the sinkhole and selective forwarding. An important effort is the analysis in [28, 29] to avoid the denial attack of service (DoS). In [18], it mentions a considerable contribution in defining possible security attacks and services based on the new service requirements and 5G network uses (Table 6).

Table 6. Articles that refer to passive and active attacks.

RQ2	In the 5G-IoT network, which attcks can occurs?		*f*
Passive attacks	Release of message contents	[26, 30]	2
	Traffic analysis	[23–25]	3
Active attacks	Masquerade	[16, 23, 28]	3
	Replay	[20, 22, 23, 30]	4
	Modification of messages	[25, 30]	2
	Denial of service	[20, 23, 28–30]	5

3.3 RQ3 – Of the Layers of Network Architecture, Which has Worked More in the Level of Network Security 5G-IoT?

Traditionally network security has been implemented in the upper layers, however, with the development of new types of services to support a large number of users and connected devices, IoT networks evolve and massify, making the physical layer (PHY) a more attractive target for hackers. The articles analyzed show that the physical layer (PHY) in the 5G-IoT network is the most considered to develop security strategies, in [31], it is mentioned that the study of physical capacity has received a growing interest and analyzes possible safety techniques; PHY presents security breaches that cause problems when applying techniques such as multiple entry, multiple exit (MIMO) [24], In addition, it is expressed that problems may occur due to dynamic infrastructure, protocols and heterogeneity of mobile objects [20]. Other authors have focused their study on improving the safety of physical capacity, proposing new strategies (Table 7).

Table 7. Articles that mention vulnerabilities in network layers.

RQ3	Of the layers of network architecture, which has worked more in the level of network security 5G-IoT?		*f*
Network architecture	Physical	[18, 21, 22, 24, 31, 32]	6
	Network	[16, 17, 20]	3
	Application	[16, 29]	2

3.4 RQ4 – What Strategies have been USED in Joint Investigations for Security in 5G-IoT Networks?

Security in the 5g-IoT network is a crucial point, several authors have proposed different strategies to ensure that the information that is transmitted has confidentiality, availability, integrity and of course it reaches the user who requires it; solutions such as blockchain [33] applied to IoT are proposed to solve problems of interoperability, privacy, security, traceability and network reliability through a distributed environment; as a mechanism is proposes crowdsourcing to mitigate local and remote attacks; The authors also present authentication systems, encryption and algorithms that complement the security frameworks (Table 8).

Table 8. Articles that analyze and/or propose network security strategies.

RQ4	What strategies have been used in joint investigations for security in 5G-IoT networks?		*f*
Security strategies	Architecture	[18, 23, 29, 33]	4
	Mechanisms	[18, 20, 22, 26, 27, 29–31, 34]	9
	Protocols	[16, 23, 25–27, 30]	6
	Algorithms	[17, 20, 23, 24, 30, 33]	6

4 Discussion

To ensure the communication of IoT devices in a 5G network, [17–25] it maintains that confidentiality must be taken into account to guarantee that the data is available only for users who are authorized and do not have interference or are heard by others not authorized; in [16–18, 23, 25–27] it is stated that authentication is one of the information security services that allow the data delivered to be genuine like devices or applications, these services together with the availability [18] that guarantees that the data and devices are available in real-time, and the integrity [17–19] that means that the data is not manipulated by intentional or unintentional interference during the delivery of Information on the network are critical aspects that must be considered when establishing a security framework for a 5G-IoT network.

Attacks on network security are analyzed by each of the layers, in the physical or perception layer [18, 21, 22, 24, 31, 32] responsible for collecting data from devices such as sensors or activators, there are attacks such as modification of messages in which malicious code or false data can be injected that allow the attacker to access the functions of the IoT devices and also send false data to users or applications, directly false information and therefore erroneous IoT services; In the replay attack [20, 22, 23, 30] in IoT, the attacker uses a device to send the final element true identification information, and to make it recognized as another component of the network.

The network layer is responsible for routing information to the destination and its security is focused on the availability of resources; In this capacity, there are attacks such as masquerade [16, 23, 28] which is an attack that fraudulently takes the identity of an authorized user of a computer system using stolen passwords or logins to access information. From an IoT device; the DoS (denial of service) attack [20, 23, 28–30] that denies the availability of IoT services due to the massive transmission of information to the red; in [20] they present a routing attack, the sinkhole in which a node like the one that transmits the information is passed and the neighboring nodes recognize it, sending data, breaking confidentiality and facilitating additional attacks such as DoS; traffic analysis [23–25] in which the attacker intercepts the messages sent, placing itself between the communication, also, the release message [26, 30] the attacker can read and capture the message, but cannot modify it.

The application layer [16, 29] is the one that stays the services for the end-user and its security is focused on software attacks, among which are the attack that involves the end-user to obtain their passwords, passwords through emails or web pages, malicious virus that obtain or modify confidential data and malicious scripts that spoil the functions of the IoT network.

To mitigate the attacks that occur in the 5G-IoT network, solutions are proposed in [16–18, 20, 23–27, 29–31, 33, 34] focused on architectures, protocols, mechanisms and algorithms, among the most relevant to solve security problems are the blockchain [33] that allows red interoperability, privacy, traceability and reliability through a distributed environment and, as a mechanism, propose crowdsourcing [22] to mitigate local and remote attacks; The authors also present authentication, encryption and algorithm systems that complement the security frameworks.

5 Conclusions and future work

Security in the 5G-IoT network is a research paradigm, some studies analyze network security, others the IoT security requirements; Few related studies refer to the 5G-IoT network within a security framework, which guarantees authentication, availability, and confidentiality [5], therefore, this article provides an analysis of information security services, the possible attacks that They present in the layers of IoT architecture (perception, network, and application), as well as frameworks for the secure implementation of IoT, among which blockchain technology [33] is effective in stopping denial of service (DoS) attacks [29], Because it is the most recurrent due to the number of interconnected devices in the network, crowdsourcing [22] is proposed to mitigate local and remote attacks.

The physical layer has been the one that presents the most attacks of different types [21, 24, 31], these include passive and active attacks and, of course, related to the information security services analyzed. Finally, the strategies proposed in the articles are based on architectures, mechanisms, protocols, and algorithms that are aimed at providing security to IoT in 5G, allowing offering greater speed and efficiency to control IoT devices. It is important to delve into several of the proposed strategies and continue adapting some already known in the 5G-IoT network, achieving more reliable security frameworks, devices without risks and information without interruptions and intruders, it would be important to take advantage of artificial intelligence technologies to combat the network security issues.

References

1. Amiruddin, A., Ratna, A.A.P., Sari, R.F.: Systematic review of internet of things security. Int. J. Commun. Networks Inf. Secur. **11**, 248–255 (2019)
2. Zhang, S., Wang, Y., Zhou, W.: Towards secure 5G networks: a survey. Comput. Networks. **162**, 106871 (2019). https://doi.org/10.1016/j.comnet.2019.106871
3. Gafurov, K., Chung, T.M.: Comprehensive survey on internet of things, architecture, security aspects, applications, related technologies, economic perspective, and future directions. J. Inf. Process. Syst. **15**, 797–819 (2019). https://doi.org/10.3745/JIPS.03.0125
4. Stallings, W.: Cryptography and Network Security Principles and practice, 5th edn. Prentice Hall, Upper Saddle River (2011)
5. Calder, A.: Information Security based on ISO 27001/ISO 27002. van Haren Publishing, Hertogenbosch (2009)
6. Torres-Carrion, P.V., Gonzalez-Gonzalez, C.S., Aciar, S., Rodriguez-Morales, G.: Methodology for systematic literature review applied to engineering and education. In: IEEE Global Engineering Education Conference, EDUCON, pp. 1364–1373. IEEE Computer Society (2018)
7. Gurusamy, D., Deva Priya, M., Yibgeta, B., Bekalu, A.: DDoS risk in 5G enabled iot and solutions. Int. J. Eng. Adv. Technol. **8**, 1574–1578 (2019)
8. Li, S., Xu, L. Da, Zhao, S.: 5G Internet of Things: A survey (2018). https://linkinghub.elsevier.com/retrieve/pii/S2452414X18300037
9. Liyanage, M., Ahmad, I., Abro, A.B., Gurtov, A., Ylianttila, M.: A Comprehensive Guide to 5G Security. Wiley, Hoboken (2018)

10. Iso.org: ISO/IEC 30141:2018(en), Internet of Things (loT)—Reference Architecture. https://www.iso.org/obp/ui/es/#iso:std:iso-iec:30141:ed-1:v1:en
11. thesaurus.com: Thesaurus. https://www.thesaurus.com
12. Wang, H.: IEEE Internet Things J. https://ieeexplore.ieee.org/xpl/aboutJournal.jsp?punumber=6488907
13. Romero Castro, M.I., Figueroa Morán, G.L., Vera Navarrete, D.S., Álava Cruzatty, J.E., Parrales Anzúles, G.R., Álava Mero, C.J., Murillo Quimiz, Á.L., Castillo Merino, M.A.: Introducción a la seguridad informática y el análisis de vulnerabilidades (2018)
14. Chaudhuri, A.: Internet of Things, for Things, and by Things. CRC Press, Boca Raton (2018)
15. UIT: La seguridad de las telecomunicaciones y las tecnologías de la información. 136 (2006)
16. Ouaissa, M., Ouaissa, M., Rhattoy, A.: An efficient and secure authentication and key agreement protocol of LTE mobile network for an IoT system. Int. J. Intell. Eng. Syst. **12**, 212–222 (2019). https://doi.org/10.22266/ijies2019.0831.20
17. Heigl, M., Doerr, L., Tiefnig, N., Fiala, D., Schramm, M.: A resource-preserving self-regulating Uncoupled MAC algorithm to be applied in incident detection. Comput. Secur. **85**, 270–287 (2019). https://doi.org/10.1016/j.cose.2019.05.010
18. Fang, D., Qian, Y., Hu, R.Q.: Security for 5G mobile wireless networks. IEEE Access. **6**, 4850–4874 (2017). https://doi.org/10.1109/ACCESS.2017.2779146
19. Militano, L., Orsino, A., Araniti, G., Iera, A.: NB-IoT for D2D-enhanced content uploading with social trustworthiness in 5G systems. Fut. Internet. **9**, 31 (2017). https://doi.org/10.3390/fi9030031
20. Santos, A.L., Cervantes, C.A.V., Nogueira, M., Kantarci, B.: Clustering and reliability-driven mitigation of routing attacks in massive IoT systems. J. Internet Serv. Appl. **10**, 18 (2019). https://doi.org/10.1186/s13174-019-0117-8
21. Wang, N., Wang, P., Alipour-Fanid, A., Jiao, L., Zeng, K.: Physical-layer security of 5G wireless networks for IoT: challenges and opportunities. IEEE Internet Things J. **6**, 8169–8181 (2019). https://doi.org/10.1109/JIOT.2019.2927379
22. Nieto, A., Acien, A., Fernandez, G.: Crowdsourcing analysis in 5G IoT: cybersecurity threats and mitigation. Mob. Networks Appl. **24**, 881–889 (2019). https://doi.org/10.1007/s11036-018-1146-4
23. Cao, J., Yu, P., Ma, M., Gao, W.: Fast authentication and data transfer scheme for massive NB-IoT devices in 3GPP 5G network. IEEE Internet Things J. **6**, 1561–1575 (2019). https://doi.org/10.1109/JIOT.2018.2846803
24. Xu, L., Chen, J., Liu, M., Wang, X.: Active eavesdropping detection based on large-dimensional random matrix theory for massive MIMO-enabled IoT. Electronics **8**, 146 (2019). https://doi.org/10.3390/electronics8020146
25. Sharma, V., You, I., Leu, F.Y., Atiquzzaman, M.: Secure and efficient protocol for fast handover in 5G mobile Xhaul networks. J. Netw. Comput. Appl. **102**, 38–57 (2018). https://doi.org/10.1016/j.jnca.2017.11.004
26. Arul, R., Raja, G., Almagrabi, A.O., Alkatheiri, M.S., Chauhdary, S.H., Bashir, A.K.: A quantum-safe key hierarchy and dynamic security association for LTE/SAE in 5G scenario. IEEE Trans. Ind. Informatics. **16**, 681–690 (2020). https://doi.org/10.1109/TII.2019.2949354
27. Huang, X., Craig, P., Lin, H., Yan, Z.: SecIoT: a security framework for the Internet of Things. Secur. Commun. Networks. **9**, 3083–3094 (2016). https://doi.org/10.1002/sec.1259
28. Safkhani, M., Shariat, M.: Implementation of secret disclosure attack against two IoT lightweight authentication protocols. J. Supercomput. **74**, 6220–6235 (2018). https://doi.org/10.1007/s11227-018-2538-8

29. Salva-Garcia, P., Alcaraz-Calero, J.M., Wang, Q., Bernabe, J.B., Skarmeta, A.: 5G NB-IoT: efficient network traffic filtering for multitenant IoT cellular networks. Secur. Commun. Netw. **2018**, 1–21 (2018). https://doi.org/10.1155/2018/9291506
30. Fu, Y., Yan, Z., Cao, J., Koné, O., Cao, X.: An automata based intrusion detection method for Internet of Things. Mob. Inf. Syst. **2017**, 1–13 (2017). https://doi.org/10.1155/2017/1750637
31. Zhang, S., Xu, X., Peng, J., Huang, K., Li, Z.: Physical layer security in massive internet of things: delay and security analysis. IET Commun. **13**, 93–98 (2019). https://doi.org/10.1049/iet-com.2018.5570
32. Memon, M.L., Saxena, N., Roy, A., Shin, D.R.: Backscatter communications: inception of the battery-free era—a comprehensive survey. Electronics **8**, 2–129 (2019). https://doi.org/10.3390/electronics8020129
33. Dai, H.N., Zheng, Z., Zhang, Y.: Blockchain for Internet of Things: a survey. IEEE Internet Things J. **6**, 8076–8094 (2019). https://doi.org/10.1109/JIOT.2019.2920987
34. Ni, J., Lin, X., Shen, X.S.: Efficient and secure service-oriented authentication supporting network slicing for 5G-enabled IoT. IEEE J. Sel. Areas Commun. **36**, 644–657 (2018). https://doi.org/10.1109/JSAC.2018.2815418

Validation of Dynamic Model for Communication Networks in Electric Vehicles

Diego Rojas(✉) and Efrén Fernandez

University of Azuay, Cuenca, Azuay 010107, Ecuador
drojas@uazuay.edu.ec

Abstract. Since the first implementation of electronic modules in conventional vehicles, communication between them was essential, CAN (Controller Area Network) was created to the communication between these modules. Advances in CAN have made it the most used communication protocol in electric vehicles. The communication networks present in electric vehicles demand high efficiency and control for the exchange of information between electronic modules. The use of various active, passive and electric traction safety systems generates a constant exchange of information and data that are important for the proper functioning of the electric vehicle. This paper presents the validation of a dynamic model for communication networks in electric vehicles. The model allows to analyze the frames of CAN and obtain the identifiers of the traction and speed modules to obtain a tool for validation during a fault situation. The main objective is the monitoring of the communication network to analyze possible failures that can be generated during operation. For the development of the model, the communication network configuration of a Kia Soul Electric vehicle is analyzed using Simulink-Network Vehicle Toolbox, LabVIEW-Ni-Xnet and the data acquisition card NI-9862.

Keywords: CAN · Network · LabVIEW · Simulink · NI-Xnet

1 Introduction

In recent years there has been a considerable increase in electronic components in automotive systems, forcing manufacturers to introduce greater advances in safety, reliability and comfort. Countless applications in these sectors demand more complex control modules and communication between them is paramount. Due to this breakthrough, new communication protocols have been developed, such as high-speed CAN, which is the most used in modern times, the X-by-wire, Flex Ray, Most [1].

In the new vehicles there are a large number of control modules, sensors and actuators, for this reason, the complexity of the communication network is increased. Due to this large increase in components, the following studies have been developed:

The article presented in [2], applies the rules of the SAE J1939 protocol to a pure electric bus with reliable results, but is limited to the issues of failures that may occur in the transmission of data and does not contemplate failures in the malfunction of the vehicle components.

M. Botto-Tobar et al. (Eds.): CI3 2020, AISC 1277, pp. 41–52, 2021.
https://doi.org/10.1007/978-3-030-60467-7_4

In [3], the fault codes of a basic scanner are emulated based on the module identifiers without examining the CAN data.

In the paper [4], they are based on fault codes generated by RPM and velocity modules and complement them with the codes predetermined by the ODB system without analyzing the data bits provided by the CAN protocol.

In the article [5], analyses hardware and communication with the SAE J1939 protocol and rules for the transmission of data and fault codes, but not analyses the detection of failures in certain damages that does not contemplate the rules of the established protocol.

Based on the analysis of the literature review, the development of a simulation application for fault detection is proposed through the use of the CAN protocol.

This paper contributes to the implementation of a dynamic CAN network model in an electric car with the purpose of verifying failures based on IDs, the article is organized in the following sections. Section 2 presents the methodology for developing the dynamic model, Sect. 3 presents the implementation of the proposed model, Sect. 4 shows the validation and verification and Sect. 5, the results and the conclusions.

The number of electronic modules implemented in an electric vehicle exceeds the modules that a conventional vehicle. An analysis of the CAN data of the Kia Soul EV show 67 ID's while a Kia Rio Xcite show only 8 ID's. The difference in the number of control modules is significant, so that the network topology for the communication components in the electric vehicle must be effective.

The CAN network in the electric vehicle is a high speed parallel configuration [1]. It is divided into four parts: DriveCAN, PowerCAN, ChassisCAN and DashCAN. DriveCAN has the MCU (Motor Controller Units) and the VCU (Vehicle Center Controller Units). PowerCAN contains the power system, which contains a BMS (Battery Management System), the CCS (Charger Controller System) and the VCU. ChassisCAN focuses on the control of the chassis system, such as EPS (Electronic Power Steering System) and the EHB (Electronic Hydraulic Brake System) DashCAN send the system messages through the electronic board.

The VCU is the most important controller in electric vehicle, it receives messages from the all modules, and send indications to the actuators. The four parts of CAN network have different baud rate [6].

In electric car, different types of networks will be found, either of different transmission speeds or different protocols. Communication between these types of networks is essential for driver comfort and safety. These topologies share data through a GATEWAY that allows communication between the modules found in these networks [7].

2 Dynamic Model Methodology

The analysis of the CAN data obtained in the Kia Soul 2015 electric vehicle have two models, the first one called on the go, for the validation of the dynamic model, and a second, laboratory, where the data will be analyzed without the vehicle.

In Fig. 1 the dynamic model is observed in which the CAN protocol data is obtained through the NI-9862 card, these data, for the laboratory model, are processed with the NI-XNET application and saved in a txt text file, this file is compiled into an excel file of csv extension to sort the data and its subsequent analysis in the applications developed in LabVIEW, in the case of Simulink, this file must be transformed into a table with MATLAB programming. In the case of the on the go model, the data is analyzed directly from the card to the applications in Simulink and LabVIEW.

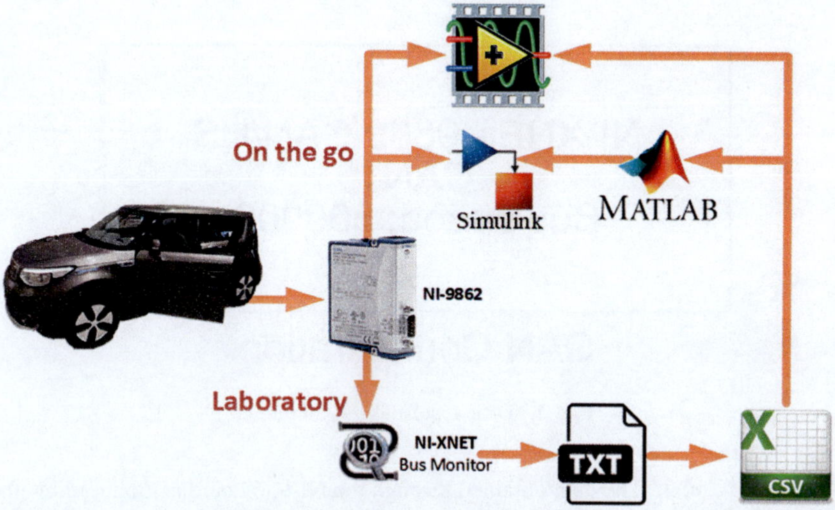

Fig. 1. Dynamic model methodology.

2.1 On the Go Model

This methodology is used for the analysis of CAN data directly in the vehicle. The configuration of the NI-9862 data acquisition card is directly in the programming of LabVIEW and Simulink. Figure 2 shows the configuration of the acquisition data through the CAN1 interface. In this figure you can see in the first instance the candb module which sends information from the CAN database and the interface module which, in this case, would be the NI-9862 acquisition card. These two configurations complement the NI-XNET Frame in Stream library, in charge of obtaining the data in

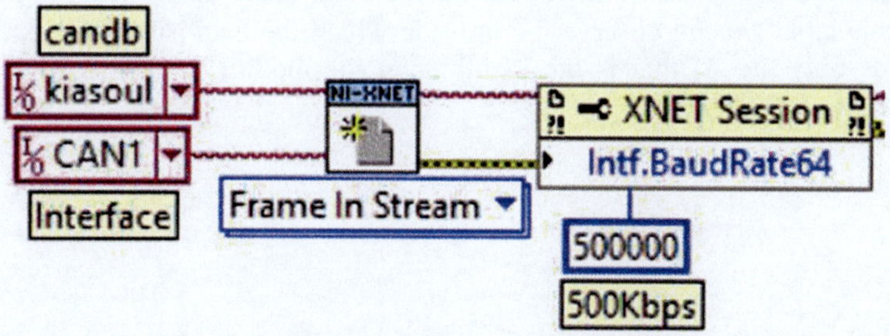

Fig. 2. Data acquisition in LabVIEW.

sequence of the CAN protocol of the electric vehicle. Finally, the XNET Session module, with the INTF.BaudRate64 configuration, gives you a value of 500000 to establish a data transmission rate of 500 Kbps.

Figure 3 shows the Simulink CAN configuration block with the same characteristics.

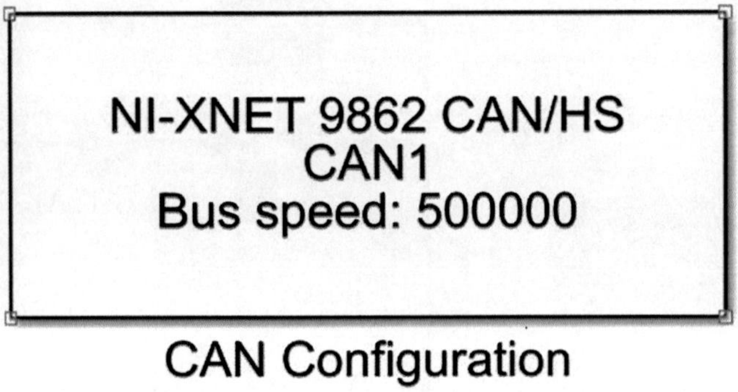

Fig. 3. Data acquisition in Simulink.

Laboratory Model. The data obtained through the NI-9862 card is monitored with the NI-XNET Bus Monitor application, which allows us to save all data frames in a text file of.txt extension. From Excel, the text file is imported, generating a spreadsheet delimited by commas whit.csv extension for direct analysis in LabVIEW, on the contrary, for use in Simulink, it is necessary to create a structure compatible with the block CAN Replay. The CAN Replay block is responsible for sending the data frame continuously to a virtual channel that will act as the data acquisition card. For the creation of the structure, the programming in MATLAB is necessary, this creates a.mat extension file that contains the compatible structure. Table 1 shows the fields that make up the structure used in Simulink. This table is composed of the fields: Data with size of 8 bytes and a uint8 class, Error with size of 1 byte and a uint8 class, Extended with size of 1 byte and a uint8 class, ID with size of 1 byte and a uint32 class, Length with 1 byte size and a uint8 class, Remote with 1 byte size and a uint8 class and Timestamp with 1 byte size and a double class.

The Table 2 shows the configuration of the comma-delimited file that LabVIEW uses. In this table we can observe the order in which the data is processed, taking as valid fields only the ID that is the identifier of the module and Payload that is the information to be analyzed.

Table 1. Table structs for Simulink.

Name	Size	Class	Value
Data	8 × 1	uint8	8 × # of samples taken
Error	1 × 1	uint8	1 × # of samples taken
Extended	1 × 1	uint8	1 × # of samples taken
ID	1 × 1	Uint32	1 × # of samples taken
Length	1 × 1	uint8	1 × # of samples taken
Remote	1 × 1	uint8	1 × # of samples taken
Timestamp	1 × 1	double	1 × # of samples taken

Table 2. CSV file struct for LabView.

Timestamp	ID	Payload	Length
0,001633	291	00 00 00 00 FE F2 F2 00	8
0,001875	153	00 00 02 FF FD 07 40 47	8
0,002119	160	20 29 01 02 00 00 00 0A	8
0,002369	164	00 08 00 00 00 00 06 0E	8
0,002860	1F1	00 00 00 00 00 00 00 00	8
0,003108	200	00 A8 2C 10 00 BB 3C 13	8
0,003350	201	81 00 00 00 00 00 00 81	8

3 Dynamic Model Implementation

3.1 Laboratory Analysis

To start the CAN data study, is necessary the configuration of NI-9862 data acquisition card through the NI-XNET Bus Monitor application, in this, the CAN1 interface is configured at a baud rate of 500 Kbaud and the checks in listed only and termination. At the end of the session, a.txt file is obtained with all the fields that make up the CAN data frame, from this file, a comma delimited document is generated to start a work in LabVIEW program.

For the analysis in LabVIEW, the data is handled as an array obtained from a spreadsheet delimited by ";", this array is transformed to a cluster to obtain the rows in Timestamp, ID and Data, in this way we can analyze the frames CAN to modify data acquisition time changing the wait in FOR structure. In order to analyze the data, changes of state in the operation of the vehicle are induced, we managed to identify the module that contains the speed, RPM and the electric traction data. The data corresponding to ID 4F2, in byte 1, corresponds to the speed on a 1:2 scale. The RPM module, whit ID 1F1 and a scale 1:2, we verify that there are two data groups, bytes 3, 4 and bytes 6, 7, for which we organize from MSB (Most Significant Bit) to LSB (Least Significant Bit) to obtain the value in decimal, all this programming can be seen in Fig. 4.

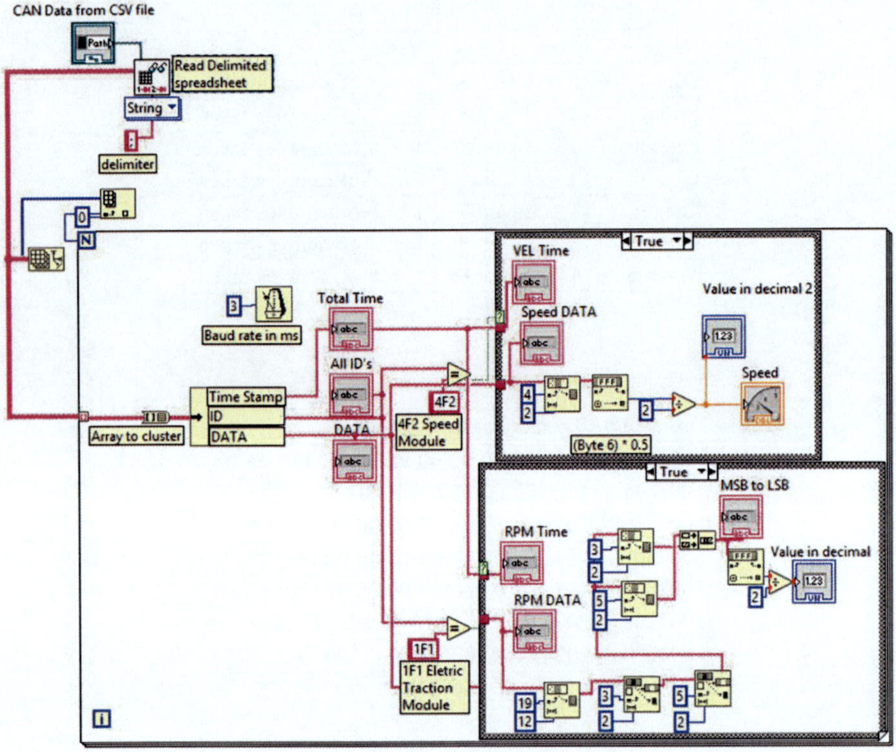

Fig. 4. LabVIEW program.

All CAN data is analyzed whit RMP and Speed to find the inverted input voltage ID, and this data is use to generate a fault detector.

In the data analysis of Simulink, a table compatible with the CAN Replay programming block was structured. A KiaSoulEV.mat file is generated with the programming in MATLAB.

In Simulink, a virtual channel is created for sending data to the CAN Receive blocks, which have the database of the RPM, Speed and Voltage modules (1F1 = 497, 4F2 = 1266, 524 = 1316), this program see in Fig. 5.

Simulink has the particularity of making the direct conversion of data according to the CANdb database, for this model, KiaSoulEV.dbc was created for electric vehicle Kia Soul with the two signals that we are analyzing. This database is configured in the CAN Unpack block within the Function-Call Subsystem.

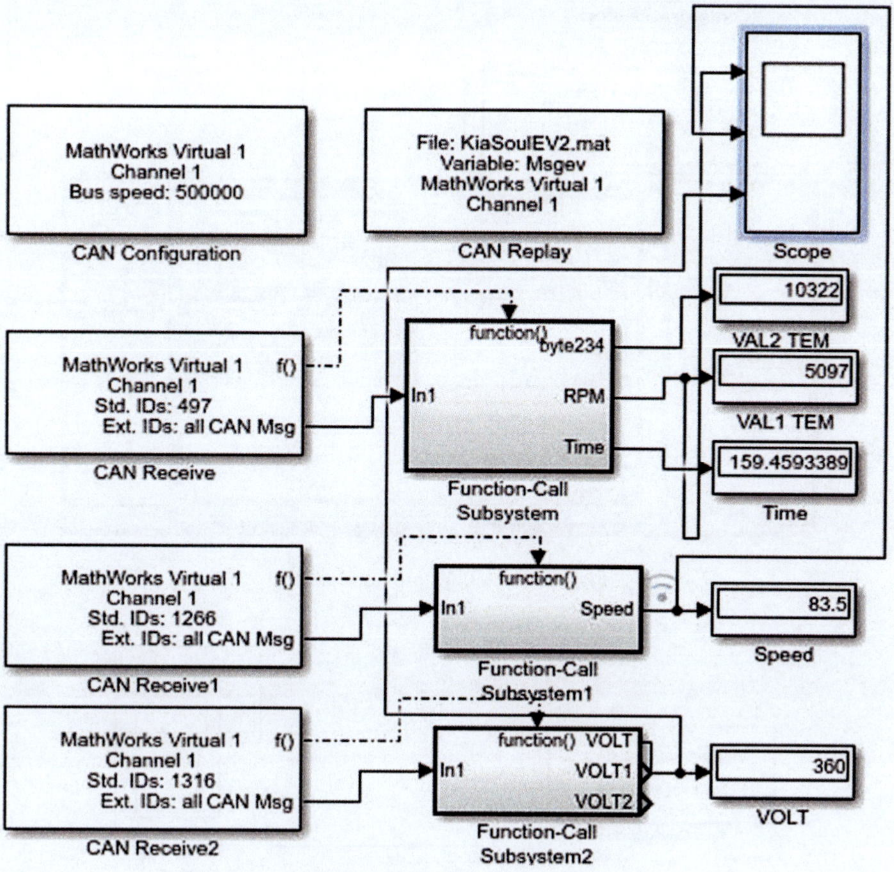

Fig. 5. Simulink program.

On the Go Analysis. The Software for analysis the CAN data frame in the laboratory, facilitate to understand the behavior of the vehicle, and thus develop the dynamic fault detection model. In the Fig. 6, show the programming in LabVIEW, where can see the configuration of the acquisition card and obtain the data with NI-XNET blocks that you have in the library.

In the Fig. 7, can see the simplicity of the model in Simulink, the use of Kia-SoulEV.dbc database in this programming is recommended. In this case, the NI-9862 data acquisition card is directly configured as the interface.

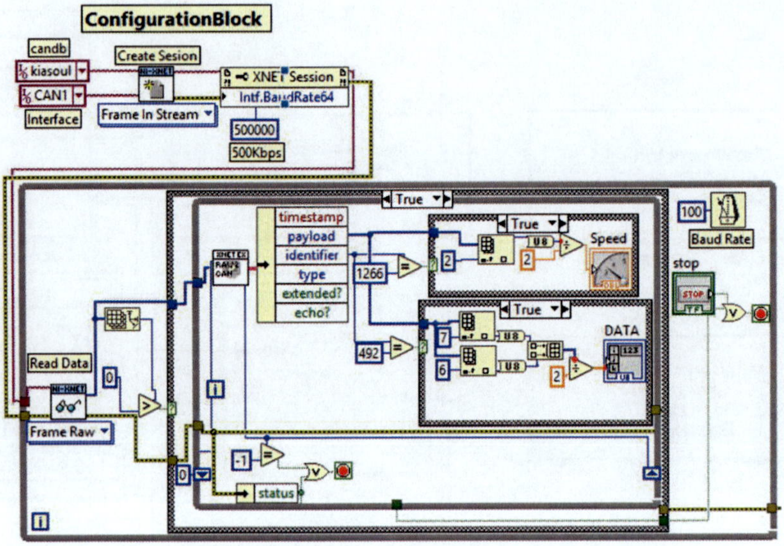

Fig. 6. LabVIEW program panel.

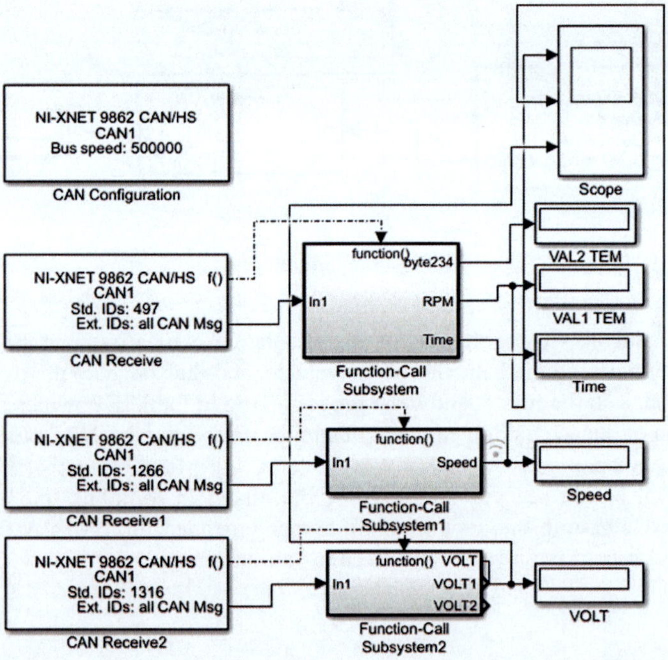

Fig. 7. Simulink program blocks.

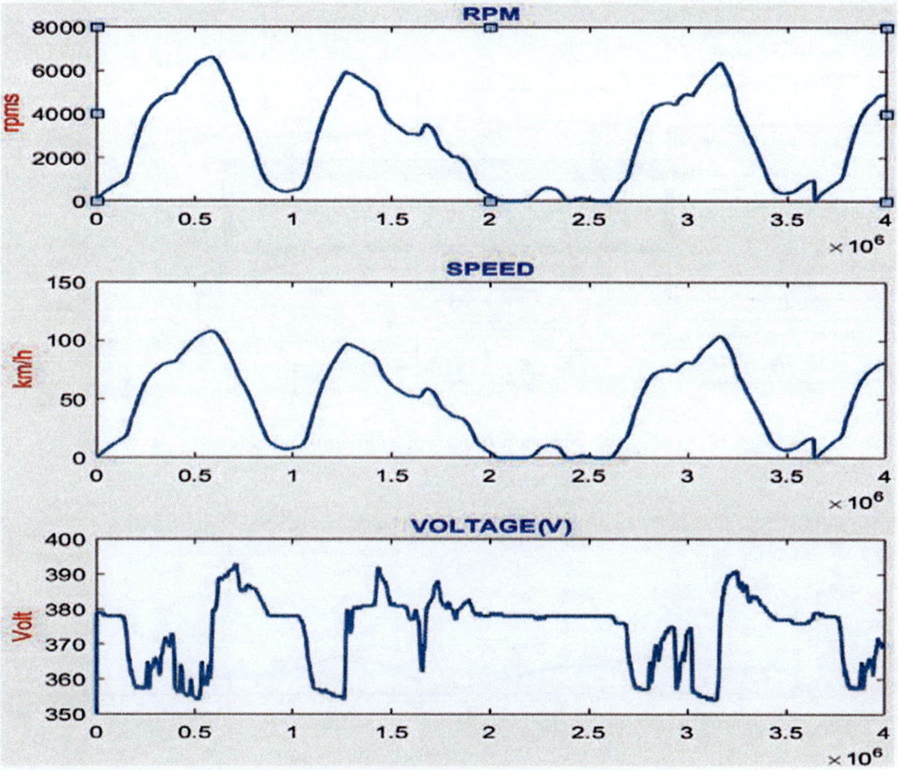

Fig. 8. Comparing data values.

4 Dynamic Model Validation

For validation, on the go model is used with the card connected to the vehicle, a fault is caused with a diagnostic tool connected in parallel to the data acquisition card.

The voltage is compared with the speed and RPM values (Fig. 8). Voltage varies between 354–390 V in normal operation. A sample of a short route has been taken, where we reach 100 km/h to verify that the voltage range 354–390 V is normal for the batteries function.

A visual indicator consisting of a warning box that appears in front of the program panel warns us of a possible high voltage. In Fig. 9, normal operation can be observed with a graphic speed indicator and numerical indicators of the acquired data.

For the failure, we have simulated a fault in the inverter's input voltage (Fig. 10) from a diagnostic tool, generating an abrupt voltage drop. In the Fig. 11, our software alerts to the real-time fault with an error box, while the graphical analysis verifies the voltage value.

A large number of tests were realized at different speeds and different driving times, giving reliable results at the time of fault detection. The dynamic model is designed for

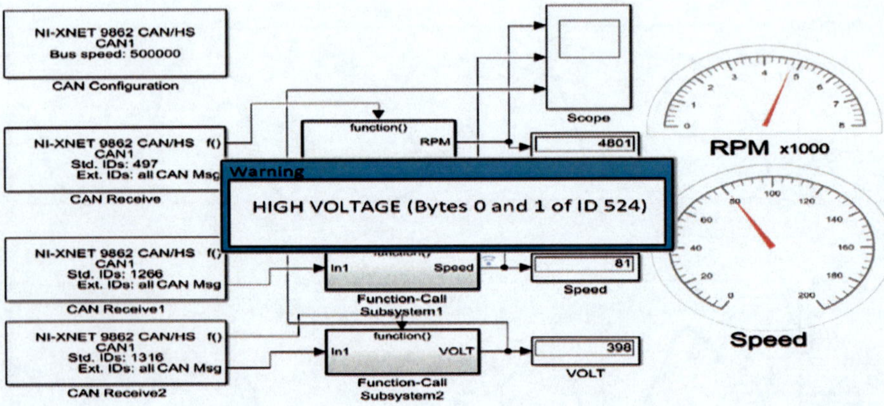

Fig. 9. Program blocks of validation model.

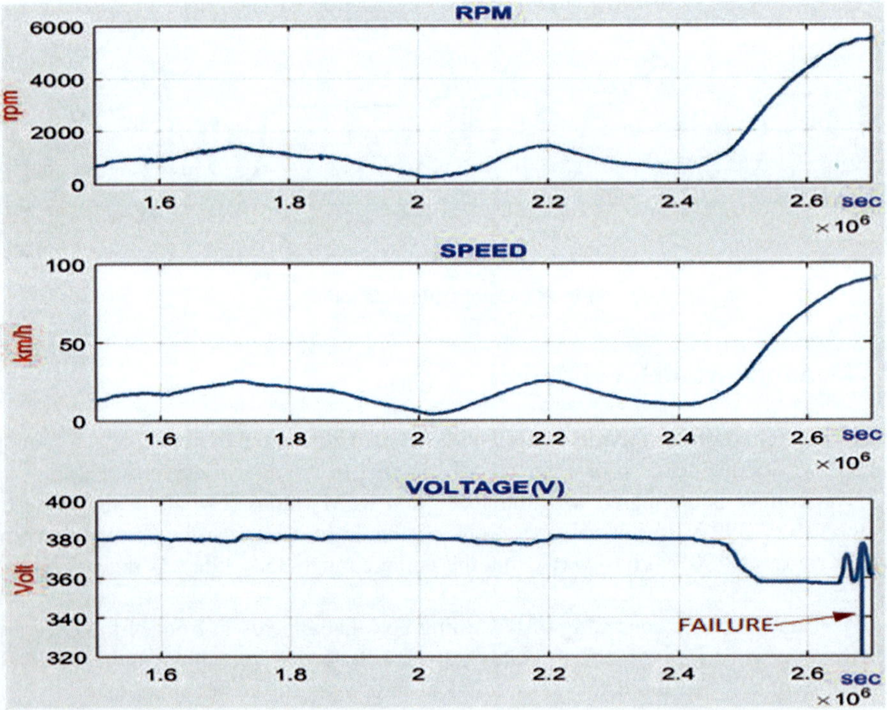

Fig. 10. Fault in the voltage.

the study of various identifiers by changing the reference values established in the relevant study with the tools developed in LabVIEW dedicated to the understanding of the data.

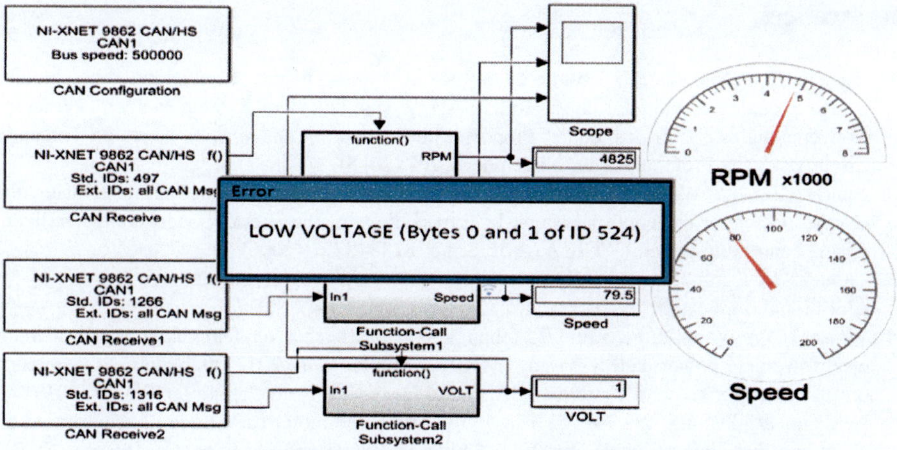

Fig. 11. Program blocks in error mode.

The tool developed in LabVIEW is used to analyze the CAN data frame of the file obtained in car tests. This is the reason why there is no model check with the vehicle running. However, the tool in LabVIEW, for the on the go model, also evaluates the vehicle while it is being operated.

The dynamic model is designed for the study of several identifiers that change the reference values established in it and with the tools developed in LabVIEW dedicated to the understanding of the data.

5 Conclusions

This article presents the validation of the dynamic model for communications in networks present in the Kia Soul Electric vehicle with the CAN protocol.

It is possible to obtain all the information provided by the modules that integrate a Kia Soul electric vehicle through the CAN communication protocol with the NI-9862 card for later analysis and the use of software that contains specialized libraries in the study of CAN networks. In this study, the Simulink and NI-XNET CAN libraries from LabVIEW were used.

A study of the data can be carried out directly in the car or in a laboratory with the tables and the generated database. We can detect in real time possible failures generated in the electric traction module, specifically through the input voltage of the inverter. The tool allows to validate errors and presents solutions to problems that normal diagnostic tools do not usually detect.

References

1. Esteban, E., Palomeque, F.: Estudio de la Red CAN y sus Diversas Evoluciones en Vehículos Convencionales e Estudio de la Red CAN y sus Diversas Evoluciones en Vehículos Convencionales e Híbridos, con el Propósito de Realizar un Diagnóstico Acertado Tomando Como Base sus Protocolos, no. November 2014 (2015)
2. Nan, J., Zai, L., Wang, Z., Wang, J.: Bus communication and control protocol using the electric passenger car control system. In: Proceedings of the World Congress on Intelligent Control and Automation (WCICA), vol. 2, pp. 8288–8291 (2006)
3. Camara, J., Cerqueira, T.A., da Silva, V.L.: Development of an automotive scanner for educational application. IEEE Lat. Am. Trans. **15**(1), 40–47 (2017)
4. Wang, L.Y., Wang, L.F., Liu, W., Zhang, Y.W.: Research on fault diagnosis system of electric vehicle power battery based on OBD technology. In: 2017 International Conference on Circuits, Devices and Systems, ICCDS 2017, vol. 2017, no. Janua, pp. 95–99 (2017)
5. Jianfeng, W., Dafang, W., Jie, X.: The design of electric motor car's body network based on CAN-bus distributed control. In: 2009 Chinese Control and Decision Conference, CCDC 2009, pp. 3712–3717 (2009)
6. Chen, X., Wu, J., Zhao, Y., Bai, H.T.: Design of CAN communication network with distributed control systems for pure electric vehicle. Adv. Mater. Res. **791–793**, 647–651 (2013)
7. Bustillo, J.M.: REDES Y MULTIPLEXADOS, pp. 1–22 (2019)

Access with Identification Technology by Radio Frequency for the Eloy Alfaro Higher Technological Institute

Darío Fernando Yépez Ponce[1,2](✉), Héctor Mauricio Yépez Ponce[2], and Edison Andrés Proaño Lapuerta[3]

[1] Instituto Superior Tecnológico Eloy Alfaro, Esmeraldas, Ecuador
dfyp1991@gmail.com
[2] CEO Ardutech.ec, Otavalo, Ecuador
[3] Instituto Superior Tecnológico Luis Tello, Esmeraldas, Ecuador

Abstract. Radio frequency identification (RFID) technology, performs the self-identification of cards or tags that store a unique identification code; this information is obtained wirelessly by a special reader. This type of technology in recent years has been adopted more frequently in both the industrial and domestic sectors due to its low cost and benefits compared to other similar technologies. The implementation of control systems with this technology generates great benefits such as increased security, generation of assistance reports and automatic access blocking. In this work a study of RFID technology is carried out exploring its main advantages and disadvantages compared to other self-identification technologies available in the market; Based on this study, an access control system with configurable, time-scalable, secure and low-cost RFID technology with free hardware and software for the Instituto Superior Tecnológico Eloy Alfaro (ISTEA) was created.

Keywords: RFID · RFID system · Access with RFID

1 Introduction

Existing professionals and future professionals must propose innovative and low-cost solutions to solve their problems; especially when starting to develop their own systems little by little the mentality of consumerism that currently governs ecuadorian society will change. Thus beginning to contribute from the higher education institutions (HEI) in the change of the productive matrix of the country. With the implementation of this system, the theoretical knowledge available in the different sources of information will be put into practice at the same time as it could be the beginning of a future venture with technology and tools that no commercial self- identification system has in the middle.

For [1], radio frequency identification (RFID) technology has had a great boom in recent years due to the relative reduction in market prices and the increase in its capabilities; An additional advantage of this technology is that it does not need physical contact such as inserting the card into a slot or waiting for it to be recognized optically;

M. Botto-Tobar et al. (Eds.): CI3 2020, AISC 1277, pp. 53–64, 2021.
https://doi.org/10.1007/978-3-030-60467-7_5

but only by approaching it at a certain distance from the reader, the card is validated by providing a virtual signature (unique identifiers of each card).

Currently, RFID technology is mainly used in the field of security, such as border crossings, identity credentials, vehicle control, identification of livestock, baggage control at airports and items for rent or loan, video clips, libraries, automotive industry, automation processes, agricultural sector; as well as in the retail market as an anti-theft device [2]. Over the years several technologies of self-identification have emerged and among its many applications, mention can be made of personnel access control [1, 3].

The need to carry out the study and subsequent implementation of an access control system for the ISTEA, arises as an alternative to solve several daily problems facing the institute such as: a) the institute does not have its own facilities, which is why it works in the physical facilities of the Eloy Alfaro school; These facilities are located at the back of the school with independent access. This access is used by the students of the school to enter and leave without any control, a situation that has generated conflicts between the authorities of both institutions; b) the institute does not have a concierge or guard to help control the entry and exit of people to its facilities, which is why teachers and students often have to wait for another teacher or student to open the door from the interior, during which time both have been victims of crime; and c) the insecurity present in the exterior of the institute due to its geographical location makes having the door open a counterproductive option in the afternoon and evening because only one third of the teaching staff remain and the institution has been stolen more than three times.

Taking into account the above, it is proposed to provide a technological solution to this problem through RFID technology. The general objective was: to implement a system that allows the control of access of authorities, teachers and students to the ISTEA, using radiofrequency identification technology and as specific objectives: a) determine the parameters and requirements to consider to implement a system that It involves RFID technology analyzing the advantages and disadvantages that this technology has compared to other self-identification solutions; b) design a configurable and scalable application so that it can adapt to different topologies through the use of free software; and c) implement a low-cost system through the use of free hardware for autonomous access to the institute's facilities.

With the implementation of this system, will it improve the entry and exit of administrators, teachers and students of the institute's facilities? And will the system to be implemented be safe and low cost? These are the hypotheses that were raised.

For the implementation of this system it is necessary to understand the operation of RFID technology and have human talent that has knowledge in electronics and programming.

To carry out the mentioned technological innovation, we proceeded to search and download information on the subject; after reviewing the content of articles, books and websites; Only those that were considered original and of importance for the development of the writing of this article were selected. A database was also created in which the data concerning the year of publication, authors, name and summary of each of the downloaded files were recorded in order to have a memory aid that allowed this project to be properly routed.

2 Self Identification Technologies

2.1 Self-identification Technologies

In this section an analysis is made of the most commercially used self-identification technologies and the components necessary to develop a project with these technologies.

Identification with Contact Memories

Contact memory buttons are a specific type of self-identification technology that requires physical contact with the button to read the label data. Contact memory has not been widely adopted as a self-identification solution because the three largest known systems of this technology are currently proprietary and if any of these is discontinued. Among its advantages we can mention: they are devices with multiple readings/writings and are very resistant, so they can be used in hostile environments [3].

Identification with Biometric Systems

This type of identification analyzes and/or measures physical characteristics, among the most used are: iris recognition, retinal reflection, hand geometry, facial geometry, hand or facial thermography, fingerprints and voice pattern. Biometric identification offers a significant advantage, given that, under this system, the person is explicitly identified, and not any credential or other object. Technology of this type is not applicable to solve certain problems because there are no systems that offer a reliability close to 100 percent and most of these systems have high costs [3].

Identification with Magnetic Cards

[3] defines that these systems are based on the reading of a magnetic strip. They use electromagnetic signals to record and encode information in a band that can be read by a machine for instant identification. The most widespread application is that of credit cards. Its main advantages are: access agility, unique identification, low cost and are not easily falsifiable; However, its continuous use deteriorates them physically as a result of friction at the time of reading and its information can be modified if it is approached to a relatively strong magnetic field, leaving it useless.

Identification with Barcode Cards

For many years it has been the technology most used by businesses to identify products for sale. This type of identification is carried out by coding the data in an image formed by combinations of bars and spaces. The images are read by special optical reading equipment through which the data can be communicated to a computer. In this type of technology, physical contact between the card and the reader is not necessary, however, there must be a line of sight between them [3]. This type of system is cheap, however, these cards are easily falsifiable or alterable, this being a great weakness for an access control system, which is why this type of technology is discarded for the application to be implemented.

Radio Frequency Identification

According to [3], the fundamental principle of RFID consists of two elements: a card or tag and an RFID reader. The reader interrogates the card or tag using a certain frequency and they answer remotely with the information it contains; The reader collects this information and sends it for processing to a control card.

2.2 Systems with RFID Technology

This section briefly describes the basic elements that an RFID system must have. The idea is to set the guidelines for everything that an account should have when you want to solve a problem by implementing a system with this type of technology.

Advantages of Radiofrequency Identification

The main advantages of RFID technology are:

1. Security. - It is a card that, due to its technological design, cannot be easily duplicated; each card has a different unique code.
2. Line of sight - It does not need a line of sight, so of all the self-identification systems it is the most agile and practical, for the reason, it does not need the card to be passed through a slot or in the correct direction.
3. Maintenance. - The readers are units without moving parts, which guarantees a correct operation without limit of use and without having to do some type of maintenance; They can also be installed outdoors without the weather, such as high and low ambient temperatures damaging it. The card or tag has no friction with the reader, so it does not wear out and its useful life is prolonged; This allows reuse. Proximity cards or tags come in several forms, the most widespread and standard is a fairly rigid rectangular-shaped plastic similar to a credential, which is prepared so that it can be customized by means of a print.
4. Rewritable - Some types of RFID cards or tags can be read and written multiple times.

Operating Frequencies of RFID Systems

[1] classify the frequencies of the RFID into four ranges:

1. Low frequency (9–135 kHz). - They have the disadvantage of a reading distance of a few centimeters and can read only one element at a time.

5. High frequency (13.56 MHz). - Coverage of stances from 1 cm to 1.5 m.
6. Ultra high frequency (0.3–1.2 GHz). - This range is used to have a greater distance between the card or tag and the reader, up to a distance of 4 m, depending on the manufacturer and the environment. These frequencies cannot penetrate metal or liquids unlike low frequencies.
7. Microwave (2.45–5.8 GHz). - The advantage of using such a wide range of frequencies is its resistance to the strong electromagnetic fields produced by electric motors; Therefore, these systems are used in automobile production lines. These cards or tags require more power and are more expensive, but it is possible to achieve readings at distances of up to 6 m.

According to [4], it states that there are three basic components in an RFID system: the card or tag, the reader and the host controller or equipment (Fig. 1).

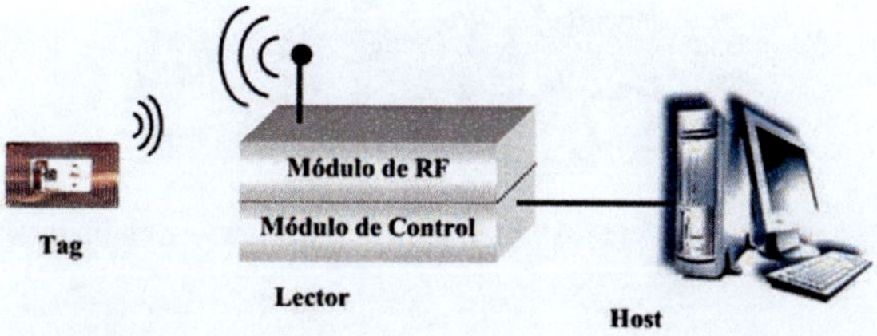

Fig. 1. Components of an RFID system. **Source:** Recovered from [3].

[3, 5] affirm that RFID technology can be divided mainly into two categories: a) passive systems; and b) active systems. In passive systems the card or tag does not incorporate a power source; while in active systems the card or tag incorporates a power source (batteries, batteries, others).

RFID Cards or Tags

They have a small circuit, integrated with a small antenna, capable of transmitting a unique serial number stored inside it to a reading device, in response to a request. The cards or tags have 2k bits of memory and do not allow simultaneous identification; since its manufacture, they are recorded with a unique identifier (UID) of 8 bytes that cannot be modified [appointment of flames]. The RFID cards that were used in the project are high frequency and read only. The UID provided by each card was used to link each user in the database with their respective UID (Fig. 2).

Fig. 2. RFID card. **Source:** Available in https://electronilab.co/tienda/tarjeta-de-proximidad-rfid-125kHz/

RFID Reader

It can only be read or read and write, it is composed of an antenna, an electronic radio frequency module and an electronic control module. This component allows you to read the cards or tags, and send the information obtained. The RFID card reader used in the project is the MFRC522 high frequency module, which is read only and communicates with an Arduino board (Fig. 3 and Table 1).

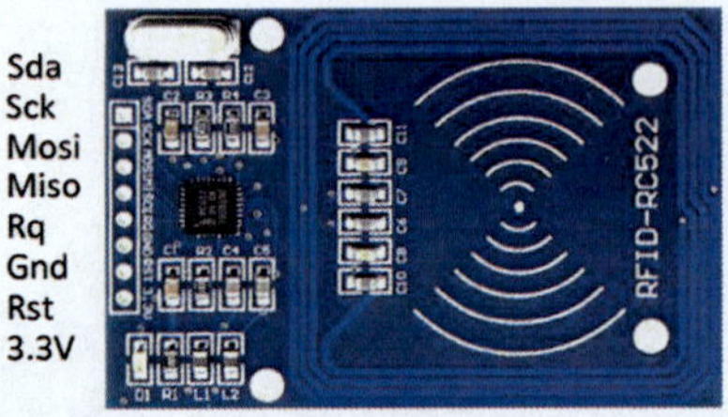

Fig. 3. MFRC522 reader. **Source:** Available in https://www.elabpeers.com/mfrc-522-rfid-reader.html

Table 1. Features of the MFRC522 reader.

Communication	BUS SPI, I2C y UART
Operating frequency	13,56 MHz
Reading distance	0 a 90 mm
Power consumption during writing	13–26 mA
Power consumption in standby	10–13 mA
Power consumption in sleep mode	Less than 80 uA
Operating voltage	3.3 V
EEPROM memory write cycles	More than de 100 000
Life of memory without power	Older than 10 years

Source: [6].

Host Controller or Computer

It is usually a PC, which runs a database and some control software. This component is the system orchestrator, because the software that allows the operation of the system is executed. The Arduino Mega 2560 card was used for the project. The Arduino card is responsible for receiving information from the reader, processing it and activating the actuator (electromagnetic lock) (Fig. 4 and Table 2).

Fig. 4. Arduino MEGA2560 development card. **Source:** Recovered form https://store.arduino.cc/usa/mega-2560-r3

Table 2. Characteristics of the Arduino Mega 2560 board.

Microcontroller	ATmega2560
Operating voltage	5 V
Input voltage	7 a 12 V
Digital input/output pins	54
PWM outputs	15
Analog inputs	16
Current for each input/output pin	40 mA
Flash memory	256 KB
SRAM memory	8 KB
Life of memory without power	Older than 10 years
EEPROM memory	4 KB
Clock speed	16 MHz

Source: [7].

3 Implementation

3.1 Algorithm

The algorithm developed was carried out in the Arduino IDE software, which is free and provided by the same company. The algorithm used is shown below:

```
#include <SPI.h>              //SPI protocol library.
#include <MFRC522.h>          //RFID module library.

#define RST_PIN 9             //Pin 9 to reset the MFRC522.
#define SS_PIN 53             //Pin 53 for the SS (SDA) of the MFRC522.
#define control_door 2        // Pin 2 to energize and de-energize the door.

int control = 0;              //Variable for access control.

byte Present_UID[4];          //Vector that will read the read UID.

//Assignment of each UID to a user.
byte User1[4] = {0x6A,0x59,0x11,0x0F};   //Mildred Azucena Murillo Benitez

MFRC522 mfrc522(SS_PIN, RST_PIN);        //Create the object for the MFRC522.

//Configuration
void setup () {
 Serial.begin(115200);                   //Speed of serial communication.
 SPI.begin();                            //Start SPI communication.
 mfrc522.PCD_Init();                     //Start the RFID reader MFRC522.
 Serial.println("Access Control:");      //Print message serially.
 pinMode(control_door, OUTPUT);          //Pin declared as output.
 digitalWrite(control_door, LOW);        //Pin disabled.
}

//Infinite cycle
void loop() {
 if(mfrc522.PICC_IsNewCardPresent()){        //Check if there is any UID present.
  if(mfrc522.PICC_ReadCardSerial()){         //Get the read UID.
   Serial.print(F("Card UID:"));             //Print message seriously.
    for(byte i=0;i<mfrc522.uid.size;i++){    //Get the size of the UID.
     Serial.print(mfrc522.uid.uidByte[i] < 0x10 ? " 0" : " ");   //Print UID.
     Serial.print(mfrc522.uid.uidByte[i], HEX);                  //Print UID.
     Present_UID[i]=mfrc522.uid.uidByte[i];     //Store the read UID in the dressing
room.
    }
   //Check the read UIDs with those stored to determine if it is valid.
```

```
      if(compareArray(ActualUID,Usuario1)){
        Serial.println("Bienvenida Mildred Murillo");  //Print message seriously.
        door();                                        //Call a sub function.
      }
      mfrc522.PICC_HaltA();        //End the current reading and enter standby mode.
    }
  }
}

//Sub function to compare the data of the vector read with the storage.
boolean compareArray(byte array1[],byte array2[]){
  for(byte i=0;i<mfrc522.uid.size;i++){
    if(array1[i] != array2[i])
    return(false);
  }
  return(true);
}

//Sub function to activate / deactivate the door.
void door(){
  Serial.println("Access granted ...");
  digitalWrite(control_door, HIGH);
  delay(10000);
  digitalWrite(control_door, LOW);
}
```

3.2 Hardware

In Fig. 5 the connection used in the implementation of the system is shown, the only change that must be taken into account is that instead of the motor an electromagnetic lock must be placed.

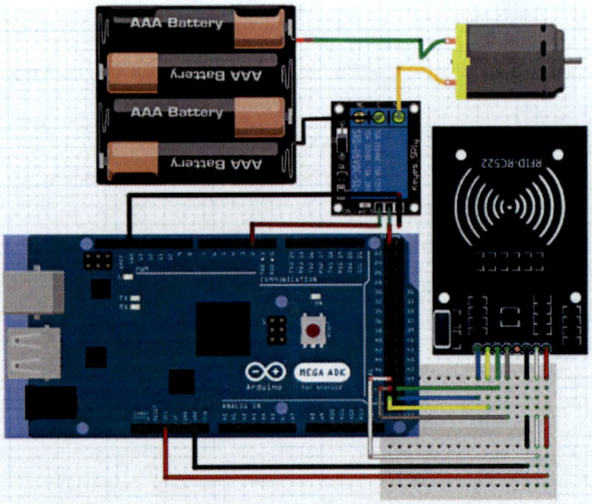

Fig. 5. Connection scheme developed in Fritzing. **Source:** Own.

4 Results

Once implemented the system for six months, it was conducted a survey among a sample of 61 teachers, students and administrators of the ISTEA to measure the impact of technological development implemented. A then, the most relevant results are displayed.

¿ Do you believe that when the automated access control system with RFID technology is implemented in the ISTEA, the safety of students, teachers and administrators has improved? (Table 3 and Fig. 6).

Table 3. Safety of students and teachers.

Options	Frequency	Percentage (%)
Yes	53	86,9%
No	7	11,5%
Other	1	1,6%
TOTAL	**61**	**100%**

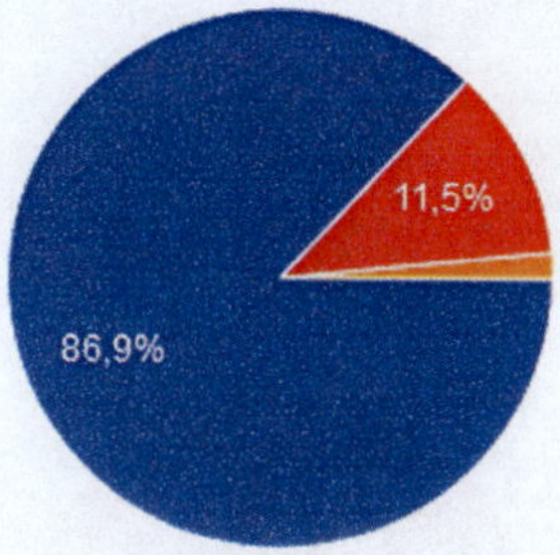

Fig. 6. Safety of students and teachers.

Value from 1 to 5, 5 being very safe and 1 very insecure. ¿How secure do you think is the automated access control system with RFID technology to the ISTEA? (Table 4 and Fig. 7).

Table 4. Security assessment of the RFID system.

Options	Frequency	Percentage (%)
Very insecure	0	0%
Insecure	6	9,8%
Normal	10	16,4%
Sure	24	39,3%
Very sure	21	34,4%
TOTAL	**61**	**100%**

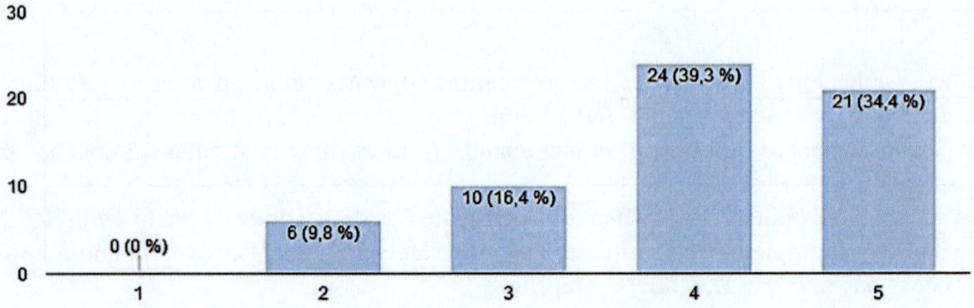

Fig. 7. Security assessment of the RFID system.

¿Which of the following factors do you consider to be the most relevant in the automated access control system with RFID technology? (Table 5 and Fig. 8).

Table 5. Most relevant aspect of the RFID system.

Options	Frequency	Percentage (%)
Security	33	54,1%
Technology	15	24,6%
Cost	3	4,9%
Quality	10	16,4%
TOTAL	**61**	**100%**

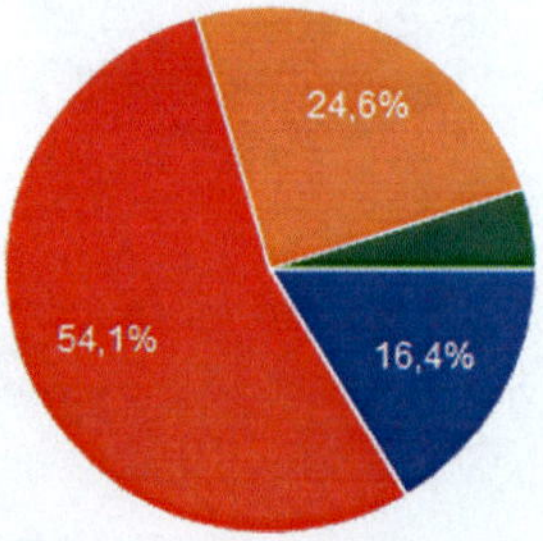

Fig. 8. Most relevant aspect of the RFID system.

¿What improvement would you propose to the automated access control system with RFID technology? If your answer is another, please specify (Table 6 and Fig. 9).

The technological project developed with open software and hardware, allows access control to the institute's facilities autonomously; above all it is a safe and scalable system of low cost. The estimated cost of the implemented system is $ 50; while commercial systems are over $ 100. In the presented article, the electronic diagram and the programming used are indicated so that the implemented system can be easily replicated and improved.

Table 6. Proposed improvements for the RFID system by the user.

Options	Frequency	Percentage (%)
Functionality	33	54,1%
Presentation	14	23,0%
Hydraulic arm	1	1,6%
Other technologies	1	1,6%
Automation	12	19,7%
TOTAL	**61**	**100%**

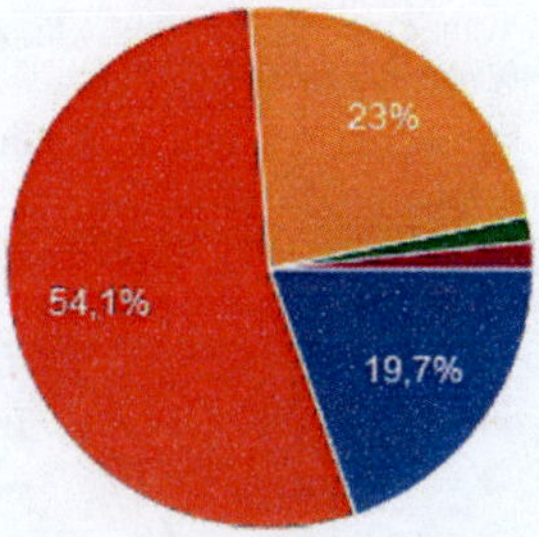

Fig. 9. Proposed improvements for the RFID system by the user.

As future work, a graphical interface will be implemented in the same way, developed with free software, to enter and eliminate users.

5 Conclusions

- With the open software and hardware platform, it is possible to develop novel systems at a low cost and with a high-level programming language.
- With the implementation of the access system with RFID technology, the safety of students, faculty and staff to increase, because the door no longer remains open for free access.
- More than 70% of surveyed users rate the system as safe, because it is very difficult to clone the access card.
- The 54.1% of the surveyed users state that the functionality of the system should be improved because it is difficult to remove a card or add a new one.

6 Recommendations

- Avoid placing the MFRC522 reader surrounded by metal, as there will be interference and readings cannot be taken properly; which would leave the system unusable.
- RFID cards must be 1.5 cm away from the reader so they can be read.

- Store the RFID cards at a suitable temperature between 15.5 and 95 degrees Celsius.
- Exposure to electrostatic discharge must be limited to RFID cards, as they can affect their performance, even render them unusable.

References

1. Overview, T.: RFID: a technical overview and its application to technology, no. June (2005)
2. Garfinkel, S.L., Juels, A., Pappu, R.: RFID privacy: an overview. IEEE Secur. Priv. **3**(3), 34–43 (2005)
3. Solangi, U.S., Memon, T.D., Noonari, A.S., Ansari, O.A.: An intelligent vehicular traffic signal control system with state flow chart design and FPGA prototyping. Mehran Univ. Res. J. Eng. Technol. **36**(2), 343–352 (2017)
4. Hunt, V.D., Puglia, A., Puglia, M.: RFID-A Guide to Radio Frequency Identification. Wiley, Hoboken (2006)
5. Phillips, T., Karygiannis, T., Kuhn, R.: Security standards for the RFID market. IEEE Secur. Priv. **3**(6), 85–89 (2005)
6. Ic, C.R.:1. Introduction (2007)
7. Consumption, U.P.: Microcontroller with Bytes In-System Programmable Flash ATmega640/V ATmega1280/V ATmega1281/V ATmega2560/V ATmega2561/V Preliminary

Computational Intelligence and Information Systems

Artificial Intelligence in Neuroeducation: The Influence of Emotions in the Learning Science

Yuliana Jiménez[1,5(✉)], Oscar Vivanco[2], Darwin Castillo[1], Pablo Torres[3], and Marco Jiménez[4]

[1] Departamento de Química y Ciencias Exactas, Sección de Fisicoquímica y Matemáticas, Universidad Técnica Particular de Loja (UTPL), San Cayetano Alto S/N, 1101608 Loja, Ecuador
{ydjimenez, dpcastillo}@utpl.edu.ec

[2] Departamento de Ciencias Biológicas, Sección de Biotecnología y Producción, Universidad Técnica Particular de Loja (UTPL), San Cayetano Alto S/N, 1101608 Loja, Ecuador
oavivanco@utpl.edu.ec

[3] Departamento de Ciencias de la Computación, Universidad Técnica Particular de Loja (UTPL), San Cayetano Alto S/N, 1101608 Loja, Ecuador
pvtorres@utpl.edu.ec

[4] Departamento de Psicología, Universidad Técnica Particular de Loja (UTPL), San Cayetano Alto S/N, 1101608 Loja, Ecuador
majimenez@utpl.edu.ec

[5] Instituto de Instrumentacion para la Imagen Molecular I3M, Universitat Politécnica de Valencia, Valencia, Spain

Abstract. Teaching is considered an emotional practice, in which emotions play a vital role in the cognitive and efficient learning processes. In this sense, the principal aim of this work is to present an artificial and psychological vision to analyze the facial expressions in higher education students. The proposed methodology is based on an emotional study during the math, physics, and biology evaluation process, applying a psychological test and face recording video under controlled conditions. The results indicate that the negative moods can influence, generating a low percentage in the accuracy of the answers during the evaluation. In conclusion, this reveals that it is necessary to provide new educational models that involve emotional development and a good teacher attitude. Finally, it is important to recognize that to enhance educational processes such as learning, it is necessary including innovative and emerging technologies, such as artificial intelligence methods.

Keywords: Artificial intelligence · Emotions · Learning · Face recognition · Neuroeducation

1 Introduction

Didactic science has tried to find rational explanations for teaching- learning experimental science processes. One proof of this is the constant educational innovation [1, 2], which is being carried out at the university level to improve these teaching processes.

M. Botto-Tobar et al. (Eds.): CI3 2020, AISC 1277, pp. 67–77, 2021.
https://doi.org/10.1007/978-3-030-60467-7_6

Innovation applied to higher education is intrinsically linked with development of new technologies. Likewise, advances in artificial intelligence (AI) open new challenges for teaching and learning fields in higher education, as well as the innovation, implementation programs and structurally changes on internal architecture of institutions of higher education [3].

Moursund and Popenici [3, 4] define to AI as area of scientific research in computing systems that can engage in humans apply through learning, adapting, synthesizing, self-correction and the use high complex information for data processing tasks. Thus, AI is both one of the most innovative and emerging technologies out there, e.g., research on face recognition from video sources that has intensified in recent years [5], in health-care systems (e.g., psychological analysis), image classification [6], object detection [7], voice recognition using machine learning and deep learning algorithms.

According to Poria [8], the emotion is intrinsic to human activities. He proposed an LSTM-based model multimodal sentiment analysis, which involves the identification of sentiments from videos and that enables us to capture contextual information from their surroundings in the video, thus aiding the classification processes.

It is widely documented that emotions and sciences learning has been investigated [9] using AI tools. On the other hand, is necessary to know that for a consolidated knowledge in our memory something fundamental is necessary: associated it with emotion. The learning generated in daily life or in a classroom that is associated with feelings-whether positive (joy or pride) or negative (fear or sadness) are those that will remain in our memory [10]. In the same way Logatt [10] mention that the aggressive or stressful educational spaces make it difficult to concentrate and most of the information provided in the classes will disappear. Instead, a motivating environment will not only predispose a better way of studying but also that knowledge could last much longer in our memory.

Likewise, in two experiments Brand, Reimer, and Opwis [11] show that negative mood impairs transfer effects and learning.

For example, imagine our student taking an exam: He feels nervous (effective); worried about failure (cognitive); he experiences an increase in cardiovascular (physiological) activation; manifest impulses to escape from the situation (motivation); and an anxious (expressive) facial expression on his face.

All of those issues have motivated the development of face recognition algorithms that draw from a large amount of information provided by videos. The facial expression can be analyzed using AI algorithms such as Machine learning [12, 13] that includes software able to recognize patterns and make predictions.

For the previous reasons, to get a better teaching and learning process in higher education this project pretends to carry out an emotional analysis, through AI face recognition algorithms in university students. In such a way, our main objectives: (1) Determine the emotional expression of students through digital assessment, and (2) Analyze the emotional climate impact in the learning process.

2 Methodology

2.1 Research Design

The study describes the analyses of emotional expression in various sciences and degrees such as Biochemistry, Chemical Engineering, Nursing, Telecommunications, and Business Administration.

The following procedure describes the flow chart of the techniques and tools used to carry out this teaching innovation project (Fig. 1).

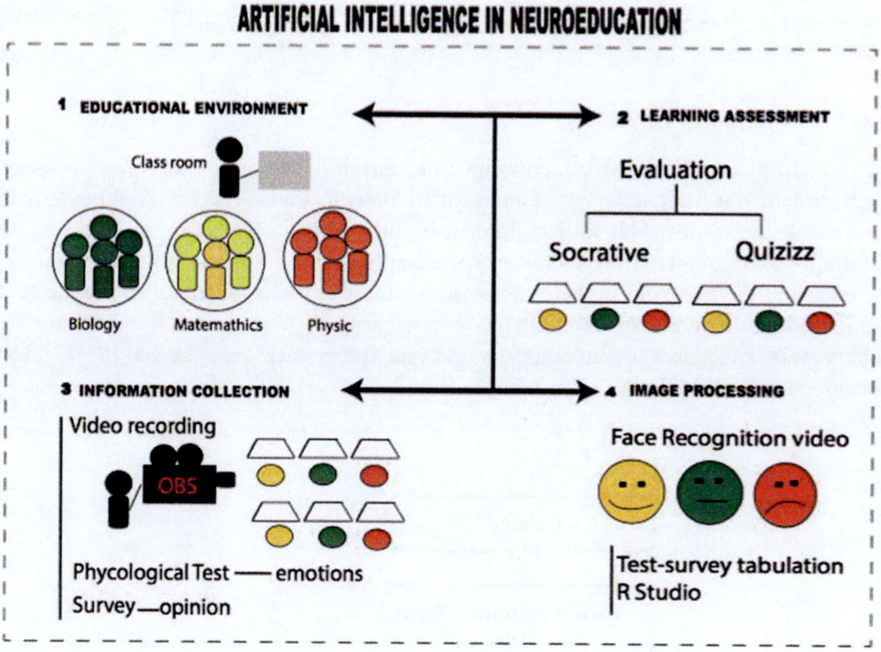

Fig. 1. Flow chart, describes the four phases applied to determine the emotional expression and emotional climate in the learning process of different subjects.

2.2 Participants

The study was carried out in one salon of Mathematics, Physics, and Biology classes at the Universidad Técnica Particular de Loja. The 61 participants were between 19 and 23 years old and selected as a non- probabilistic sample.

2.3 Measures

The students were evaluated according to the subject by using a multiple-choice knowledge test. These assessments were carefully designed through TICS such as

Quizziz and Socrative [14, 15]. A second test was applied to collect opinions regarding the teacher's ability to manage and generate a positive learning environment in the classroom in a survey. All facial expressions of students were captured using the OBS Studio software [16].

2.4 Procedure

The main resource was the face student's video recording during the evaluation process; different expressions were captured using the OBS Studio software. As a quantitative methodology, a psychological test and a survey were applied. The test determined emotions in the student, as well as the survey, collected opinions regarding the teacher's ability to manage and generate a positive learning environment in the classroom.

2.5 Information Processing

Previous to this research the students signed a consent letter for the use of information. Each student was assigned a computer with all analysis tools. The first trial began with the knowledge test of each subject applied by the Quizizz tool. Then, was carried out the online psychological test and survey were applied for collected opinions regarding the teacher's environment in the classroom. All these activities were registered through OBS Studio software. We use a database consisting of 61 facial motion videos. The gallery was subjected to an emotion analysis using face recognition [5, 17] tool developed in Visual Studio (C#) refer to Fig. 2.

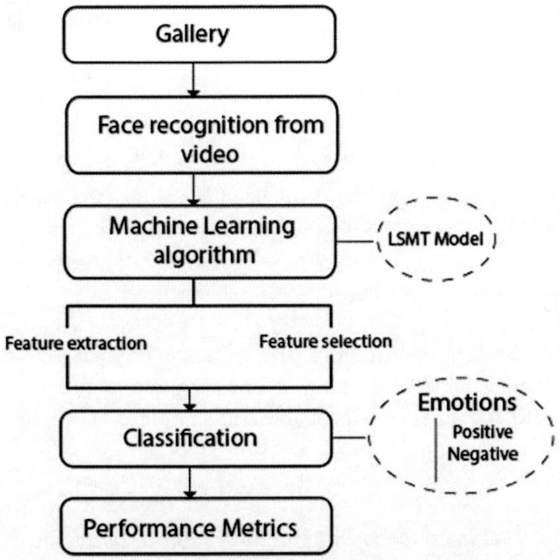

Fig. 2. The flowchart of face recognition procedure.

Images of students are converted to templates, which are then used for facial recognition. The Machine learning LSTM (Long Short Term Memory) method interprets features and extracted them from one or more images of an individual to create that individuals template and classify them.

Then, the facial expressions were captures during the evaluation process and classified according to attributes through image emotion recognition; detecting expressions such as happiness, satisfaction, surprise, contempt, fear, sadness, anger, and disgust.

2.6 Data Analysis

The data analysis was performed with R software version 1.2.5001 – © Sept 19, 2019 RStudio, Inc. The variables emotions frequencies (x) and emotion face video (y) were tested for homogeneity of variances and normality using the Levene and Shapiro-Wilk tests, respectively. Significant differences were expressed as $P < 0.05$ to both data sets. The data are expressed as the mean ± standard deviation (SD). Finally, the psychological test and survey were analyses through the averages.

3 Results

3.1 The Evaluation Through Digital Assessment

After processing Quizizz and Socrative test scores the averages are displayed in the following table.

Table 1. The Quiziz percentages summary obtained from the student's evaluation to know their performance in arithmetic, physics and biology

Students	Career	Partial evaluation (%) Correct Answers (CA)
21	Electronics and telecommunications, Administration Bio-chemistry and chemistry engineering	Math, 51%
20	Biology	Physics, 45%
20	Nursing	Biology, 58%

3.2 Emotional Video Expression in Recorder Phase

After the information processing using a face recognition algorithm (see Fig. 3a–c) the results show evidence about the greater presence of positive facial expression: happiness and surprise, followed by negative expressions: disgust, fear, and contempt (Table 2). Even though positive emotions predominated in the study, the negative feelings influenced in a low answer's accuracy (Table 1).

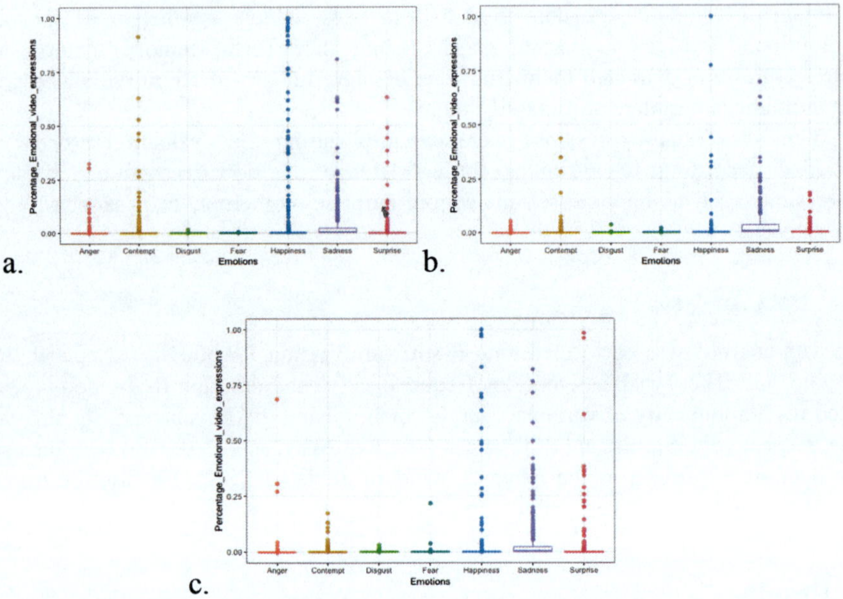

Fig. 3. Emotional expressions displayed by students during the evaluation process in: a. Math assessment. b. Physic assessment and c. Biology assessment.

The results analysis of emotional video expressions shows that the data for Math, Physic and Biology don't have a normal distribution. The Kruskal-Wallis rank sum test show statistical differences $P < 0.05$ between all analyzed groups by subject. The Pairwise comparisons using Wilcoxon rank sum test showed for Math anger, happiness and surprise as significant different. Besides in Physic, anger, happiness, surprise and contempt were significant. Biology happiness, anger, contempt, fear and disgust were significant.

Table 2. The relevant face emotions video results during Math, Physical and Biology evaluation with a sample of 21 students in each subject.

Emotions	Video recorder					
	Math		Physic		Biology	
	Mean	±SD	Mean	± SD	Mean	± SD
Happiness	49		18	12	30	14
Surprise	6	4	3	1.7	16	8
Sadness	27	7	36	8.5	30	9
Fear	0.9	0.05	2	0.1	0.9	0.1
Contempt	9	5	5	3.4	2	0.1
Anger	2	1.9	12	5	3	0.3
Disgust	0.7	0.08	1	0.19	0.4	0.2

3.3 Emotion Test in the Learning Process

Furthermore, the physicological test shows happiness and fear as pre-dominant emotions; with similar scores are surprise and sadness. Like- wise, negative emotions were displayed as fear, anger and disgust.

Concerning the biology subject, it is observed that the higher scores are related to sadness and surprise emotions; with lower scores registered to fear and happiness. (Table 3).

Thus, comparing the math video results with the psychological test, we can highlight that in both cases the positive emotions with most frequency were surprise and happiness, predominating mostly against negative emotions such as contempt and sadness (Table 2).

Further, with respect to physic video and psychological test, it was observed that in both cases the emotion with greater frequency was happiness, followed by sadness and surprising feelings with a lower percentage (refer Table 3).

Finally, comparing Biology and Physics results, it is noted that in both cases the emotion with greater frequency was sadness, while happiness and surprise evidence a lower percentage (see Table 3).

Table 3. The relevant emotions test percentage (x)* results applied in Math, Physic and Biology subjects with a sample of 21 students in each subject. * 0% none; 0% < x < 50% some; x > 50%most.

Psychological test (%)			
Emotions	Math	Physic	Biology
Happiness	23	45	10
Surprise	18	10	20
Sadness	0	10	30
Fear	0	30	15
Contempt	6	6	0
Anger	2	3	2
Disgust	3	2	2

3.4 Survey Opinion About Learning Environment in the Classroom

Further, 76.2%, 81%, and 63.6% indicate that the Mathematics, Physics, and Biology teachers love their work and listen to their students, as well as 66.7%, 57.1%, 77.3% consider that teachers demonstrate confidence in their student. Although a good percentage 61.9% of physic students perceive a motivating environment, only 47.6% and 36.4% perceive a pleasant and stimulating environment in the Mathematics and Biology classroom (See Fig. 4).

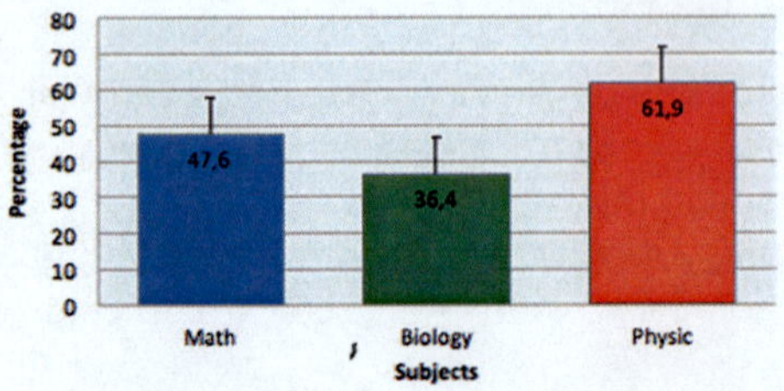

Fig. 4. Survey opinion percentage on emotional climate in classroom

Besides, the 71.4%, 61.9%, and, 50% indicate that the Mathematics, Physics and Biology teachers always use appropriate strategies and learning resources, as well as the 81%, 52.4%, and, 40.9%, are agreed in that the teachers promote continuous participation activities and they use adequate tools to measure their knowledge in 81%, 57.1%, and 40.9% respectively.

4 Discussion

4.1 Evaluation Through Digital Assessment

The importance of emotional education in learning sciences is an essential issue in the 21st Century, to achieve the expected learning in the students. In this work, we highlight several key research challenges associated with emotions and their influence on science learning [18, 19]. This type of educational approach seeks to combine the ability to comprehend and detect the student's specific emotions by new technologies like AI, during the assessment context and then to use these results to enhance the teaching-learning process [3, 4].

The evaluation test scores displayed low percentages of less than 58%, especially in physics. Even although positive moods prevailed, the negative emotions could influence decreasing right answers accuracy during the math, physics, and biology assessment. It was contrasting with Valiente [19] research, who demonstrated that the emotions may influence what students retain on a specific learning task.

4.2 Emotional Expression Evaluation Through Video

Evidence for such emotion comes from questionnaires, but also coding and analyzing expressions of emotion in a videotaped activity.

In this context, facial expression recognition [17] has been gaining popularity among AI researchers, also has become increasingly important for many applications in education (understanding student emotions) [20–22].

Likewise, Taskirar [23] proposed a face recognition method using a dynamic feature from expressive smile videos. In this study, the authors describe the biometric systems to measure and analyze the physical and behavioral characteristics of individuals.

As in math, the most frequent emotions were positive (surprise and happiness), predominating against negative emotions (contempt and sadness). However, in Biology and Physics, the emotion with greater frequency was the sadness. Likewise, according our statically analysis the emotions are related to subject. Indeed the pairwise comparisons between emotions show that different emotions could be expressed in each subject.

Comparing it with other related contexts as Borrego 2013 and Mellado 2014 [24, 25] determined that the emotions related to the contents of Biology and Geology are very positive; on the other hand, the memory of their emotions towards the contents of Physics and Chemistry in secondary school is mostly negative strong like fear, tension or despair. Among the positive ones, interest and curiosity stand out. This is going to be a constant that alerts us that in Physics and Chemistry we have a serious problem in the secondary stage.

On the other hand, in Science and Mathematics specialties, the causes of negative emotions are mainly attributed to the teachers and the content of the subject, while the causes of positive emotions are attributed to themselves as students and to problem-solving and laboratory activities [24].

In such way, the experimental results (psychological test and facial expressions video) displayed that emotions were very similar in both analyses.

4.3 Emotional Climate Impact in the Learning Process

The survey results indicate that a low student percentage perceive a pleasant and stimulating environment in mathematics (47.6%) and biology (36.4%) classroom. However, these scores were exceeding by the physics class with a motivating and better-punctuated environment (61.9%). Consequently, it is considered that a good teacher's attitude helps to combat negative emotional states such as stress, depression, fear, and sadness, contributing to better knowledge assimilation, ensuring positive educational environments that even allow increasing the approval of students' ratings [24].

Likewise, this reveals that it still has a lot of work to do, as actually, it is necessary to generate positive learning environments to improve the capacity students learning, as well as provide them with a better educational trying to reduce the anxiety [26] caused by the scientific study.

As indicated Garnica 2017 [27] about emotionally stable people, they are innovative, proactive, with a great responsibility sense and projection, who see challenges as great challenges and learning opportunities, with the ability to self-regulate their behavior, without falling into conformism.

5 Conclusion

Overall, we summarize that an effective emotion-shift recognition model and a right psychological test can aid to identify negative emotions and their influence in the precision of the answers during a learning assessment.

Thus, we believe that addressing each of the challenges outlined in this paper will not only enhance the understanding of the emotions but also will yield significant academic performance improvement. Also, it is considered important the teacher role as a neuro-educator, focusing on being empathetic, maintaining an emotional balance, generating good communication inside and outside the classroom.

Finally, as future work, we plan to apply a machine learning algorithm in students with depressive emotions by functional Magnetic Resonance Imaging (fMRI) and to extend these results in the learning processes.

Acknowledgment. The authors are gratefully with the Academic Vice-Rector of UTPL, also to MSc. Cesar Granda and MSc. Máximo Moreira for the support and contribution in this project.

References

1. Jiménez, Y., Castillo, D.: Educación de calidad mediante la estrategia Design Thinking. In: Edunovatic 2017. Conference Proceedings, pp. 472–481. Adaya Press (2018)
2. Jiménez, Y., Bautista, E., Carrillo, I., Castillo, D., Feijoo, D., Vivanco, O.: Simulation technologies to strengthen teaching-learning skills in Biochemistry, Nursing and Medical students. Revista Latinoamericana de Políticas y Administración de la educación (2018)
3. Popenici, S.A.D., Kerr, S.: Exploring the impact of artificial intelligence on teaching and learning in higher education. RPTEL **12**(1), 1–13 (2017). https://doi.org/10.1186/s41039-017-0062-8
4. Moursund, D.: Brief Introduction to Educational Implications of Artificial Intelligence. University of Oregon (2006)
5. Barr, J.R., Bowyer, K.W., Flynn, P.J., Biswas, S.: Face recognition from video: a review. Int. J. Pattern Recogn. Artif. Intell. **26**(05), 1266002 (2012). https://doi.org/10.1142/s0218001412660024
6. He, K., Zhang, X., Ren, S., Sun, J. Deep residual learning for image recognition. In: Proceedings of the IEEE Conference on Computer Vision and Pattern Recognition, pp. 770–778 (2016)
7. Redmon, J., Divvala, S., Girshick, R., Farhadi, A. You only look once: unified, real-time object detection. In: Proceedings of the IEEE Conference on Computer Vision and Pattern Recognition, pp. 779–788 (2016)
8. Poria, S., et al.: Emotion recognition in conversation: research challenges, datasets, and recent advances. 1905.02947 (2019)
9. Eligio, U.X.: An overview of the growth and trends of current research on emotions and mathematics. In: Understanding Emotions in Mathematical Thinking and Learning, pp. 3–41 (2017)
10. Logatt, C.A.: Cómo influyen las emociones en el aprendizaje (2016)
11. Brand, S., Reimer, T., Opwis, K.: How do we learn in a negative mood? Effects of a negative mood on transfer and learning. Learn. Instr. **17**(1), 1–16 (2007)
12. Guo, G., Zhang, N.: A survey on deep learning based face recognition. Comput. Vis. Image Underst. **189**, 102805 (2019). https://doi.org/10.1016/j.cviu.2019.102805. ISSN 1077-3142
13. Ding, C., Tao, D.: Trunk-branch ensemble convolutional neural networks for video- based face recognition. IEEE Trans. Pattern Anal. Mach. Intell. **40**(4), 1002–1014 (2018)
14. Valencia, Y.: De qué manera las Emociones Académicas influyen en el Aprendizaje. Iberciencia. Lima-Perú (2015)

15. Ruiz, D.: QUIZIZZ en el aula: evaluar jugando (2019). https://intef.es/wp-content/uploads/2018/11/Quizizz-en-el-aula-Evaluar-jugando.pdf. Accedido 1 Oct 2019
16. Bailey, H.: The OBS Project Contributors. Open Broadcasting Software (2017). https://www.obsproject.org/
17. Yi, D., Lei, Z., Liao, S., Li, S.Z.: Learning face representation from scratch. arXiv preprint arXiv:1411.7923 (2014)
18. Tyng, C.M., Amin, H.U., Saad, M.N.M., Malik, A.S.: The influences of emotion on learning and memory. Front. Psychol. **8**, 1454 (2017). https://doi.org/10.3389/fpsyg.2017.01454
19. Valiente, C., Swanson, J., Eisenberg, N.: Linking students' emotions and academic achievement: When and why emotions matter. Child. Dev. Perspect. **6**, 129–135 (2012). https://doi.org/10.1111/j.1750-8606.2011.00192.x
20. Colchester, K., Hagras, H., Alghazzawi, D., Aldabbagh, G.: A survey of artificial intelligence techniques employed for adaptive educational systems within e-learning platforms. J. Artif. Intell. Soft Comput. Res. **7**(1), 47–64 (2017). https://doi.org/10.1515/jaiscr-2017-0004
21. Grivokostopoulou, F., Perikos, I., Hatzilygeroudis, I.: An educational system for learning search algorithms and automatically assessing student performance. Int. J. Artif. Intell. Educ. **27**(1), 207–240 (2016). https://doi.org/10.1007/s40593-016-0116-x
22. Conati, C., et al.: AI in Education needs interpretable machine learning: Lessons from Open Learner Modelling 1807.00154 (2018)
23. Taskirar, M., Killioglu, M., Kahraman, N., Erdem, C.E.: Face recognition using dynamic features extracted from smile videos. In: 2019 IEEE International Symposium on Innovations in Intelligent SysTems and Applications (INISTA), Sofia, Bulgaria, pp. 1–6 (2019)
24. Borrego, E.C., Cortés, A.B.B., Mero, M.B., Jiménez, V.M.: Las emociones sobre la enseñanza-aprendizaje de las ciencias y las matemáticas de futures profesores de Secundaria. Revista EUREKA sobre enseñanza y divulgación de las ciencias, 514–532 (2013)
25. Mellado, V., Blanco, J.L., Borrachero, A.B., Cárdenas, J.A.: Las emociones en la enseñanza y el aprendizaje de las ciencias y las matemáticas (2014). https://doi.org/10.5565/rev/ensciencias.1478
26. Chang, H., Beilock, S.L.: The math anxiety-math performance link and its relation to individual and environmental factors: a review of current behavioral and psychophysiological research. Curr. Opin. Behav. Sci. **10**, 33–38 (2016)
27. Garnica, E.L., Brun, A.R., Martínez, G.C., García, V.M.Z., Vázquez, I.I.L.: Educación basada en emociones. In: *XIKUA* Boletín Científico de la Escuela Superior de Tlahuelilpan, vol. 5, no. 10 (2017)

Intelligent and Autonomous Guidance Through a Geometric Model for Conventional Vehicles

Danny Zea[1(✉)], Alex Toapanta[2], and Víctor Herrera Pérez[1]

[1] Facultad de Informática y Electrónica, Escuela Superior Politécnica de Chimborazo, Riobamba, Ecuador
danny.zea@espoch.edu.ec

[2] Instituto Superior Tecnológico Rumiñahui, Sangolqui, Ecuador

Abstract. Cyber-physical systems (CPS) in the automobile industry are facing major challenges related to the use and validation of these CPS, which entails high costs in the implementation and training tests in the physical world, thus limiting research. Therefore, there is a need to shorten the validation times of these CPS with the use of 3D simulation software. This research article proposes to simulate a CPS in the simulation software Webots, with the aim of emulating the autonomous movement of conventional vehicles by integrating a GPS sensor and a compass sensor which provide information on location and orientation, these data are used for the implementation of a geometric model by vectors, the same one that is developed in a controller that allows to take actions on the vehicles in the simulation software in order to emulate an urban traffic. Finally, a series of configurations have been made to evaluate the geometric model, managing to maintain the default speed of 94.194% with curves greater than 90 degrees. In addition, the validation of this system in a real environment through the instrumentation in land vehicles is drawn as future lines.

Keywords: Autonomous vehicle · Artificial intelligence · Cyber-physical systems · Geometric model · IoT

1 Introduction

The facilities in modern life are possible thanks to the great development of the evolution of information and communication technologies. Currently, almost all integrated systems present in our daily lives have communication capabilities and some level of automation or artificial intelligence. The confluence of cyber and physical spaces allowed traditional embedded systems to become cyber-physical systems (CPS), which are characterized by a coordinated integration between computing and physical processes through networks [1]. Modern CPS are present in advanced large-scale engineering systems such as avionics, health, transportation, automation and intelligent networks [2].

One of the cyber-physical systems (CPS) that has had an astonishing development is road transport [3], where the need arises to build safer and more reliable vehicles that can solve mobility problems in an autonomous and intelligent way principally related

M. Botto-Tobar et al. (Eds.): CI3 2020, AISC 1277, pp. 78–93, 2021.
https://doi.org/10.1007/978-3-030-60467-7_7

to the transport of goods and services [4]. The market in the automobile industry has grown in such a way that it has brought traffic congestions and problems in vehicular flow, especially in the urban sector. In the previous century, the vehicle was developed from a purely physical system based on the laws of mechanics and chemistry until becoming a sophisticated Cyber Physical System (CPS), that is composed of electronic components and control systems created to improve the performance and security [5]. However, most conventional vehicles have sensors that specifically monitor the infrastructure of the vehicle itself [5]; but these vehicles are also able to gather information about the physical world by integrating sensors, which can contribute to the localized detection of objects, considerably improving the reliability of conventional vehicles [6]. Some of the conventional modern vehicles are incorporated with external sensors such as proximity sensors that prevent the collision of vehicles and the run-over of passersby, although this helps reduce accidents, it has not provided adequate monitoring and control that allow to avoid traffic accidents.

One of the emerging scenarios of cyberphysical systems (CPS), are the autonomous vehicles, the same ones that to be able to transit, require the execution of complex tasks, as well as high yields [7]. The use of autonomous trains, for example, improves the efficiency of operation and the reduction of costs and infrastructure; however, on secondary tracks it is necessary to add sensors and communication systems that integrate a better and more efficient cyberphysical system (CPS) [8].

Cyberphysical systems (CPS) in the automobile industry on a large scale make it possible to improve road safety, traffic efficiency and human coexistence [9]; this is made possible not only with the connectivity between vehicles, but also with the integration of sensors in the infrastructure that surrounds each of the vehicles. In this way, all the information provided by the infrastructure can be used without the need for a vehicle to be in the environment of the generation of this information. All of this leads to complex automotive cyberphysical systems, even more, when modern vehicles integrate up to 100 Electronic Control Units (ECU) [10]. Today, research into cyberphysical systems (CPS) in the automotive industry is limited due to the high costs of the implementation and testing of monitoring and control systems in the physical world. Even instrumentation and operation require qualified personnel [11]. On the other hand, there is the security and reliability of data in a cyberphysical system (CPS) in the automotive industry [9], especially when there are threats that violate the interconnectivity of vehicles with the infrastructure that surrounds it. Due to this close relationship between security and protection in automotive CPS, a new approach to research has emerged in recent years [12, 13]. A possible solution to validate the CPS is the use of a simulation software, thus allowing to democratize the investigation in the improvement of vehicular flow [14]. The most commonly used methods for the autonomous displacement of land vehicles in a CPS using a geometric trajectory are Pure-Pursuit, vector search and Stanley [15]. However, the application of these methods tends to cause problems in steep curves and even more, at high speeds [15, 16]. This project simulates a CPS that emulates an urban environment, which includes conventional vehicles that circulate autonomously on a pre-established route by means of an automatic guide that is defined by a geometric model by vectors, which is evaluated based on the response in the steering angle of the land vehicle.

2 Description of the CPS Simulation Environment

The CPS simulation environment consists of the emulation of an ecosystem in the Webots R2019a simulation software, in which modeled and programmed vehicles interact through a controller. Webots' own libraries developed in the controller interact and control the vehicles through the sensory information received from and to the vehicle.

The emulation of the environment was implemented in Webots software. The model developed for the CPS simulation consists of two parts. On the one hand, there is the implementation of the 3D traffic scenario together with the vehicles. On the other hand, there is the controller that is developed in the C/C++ language in the Webots programming interface. The controller was also integrated in the interface of Webots software.

Then, in the controller, the information received by the virtual sensors implemented in the vehicles is processed (see Fig. 1). The interaction between the 3D traffic scenario and the controller is done in real time. The main objective of the processing of sensory information of the vehicles is the automatic guidance of them in a pre-established route.

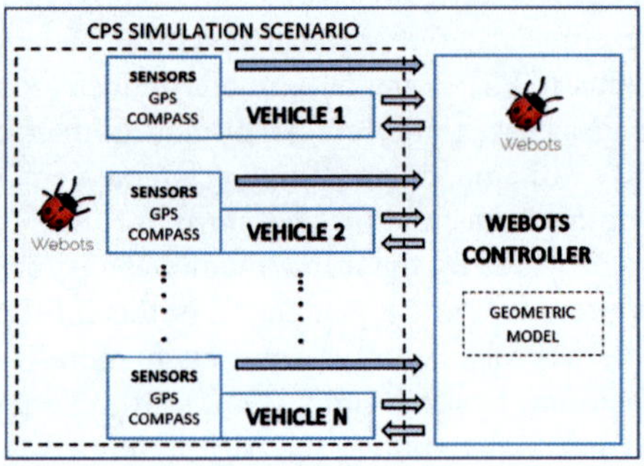

Fig. 1. General architecture of the geometric model for autonomous movement of vehicles.

2.1 3D Traffic Scenario

The 3D traffic scenario is a CPS, which was proposed to validate the automatic vehicle guidance. In Fig. 2 (a), a 3D simulation scenario is shown, which can be composed of several elements that allow virtual representation of a real urban environment. Figure 2 (b) shows the vehicles that represent the nodes and which are integrated by several virtual sensors such as the Global Positioning System (GPS) and a Compass. The representation of these virtual sensors in Webots software is depicted in Fig. 2 (c). These virtual sensors generate information that will be managed by the controller, which defines the necessary actions to control the autonomous operation of accelerator, brake and steering wheel.

The modeling of the 3D traffic scenario with the integration of conventional vehicles is a key factor to validate, for instance, the autonomous displacement of virtual nodes (vehicles) along a pre-established route. The aim of this configuration is to

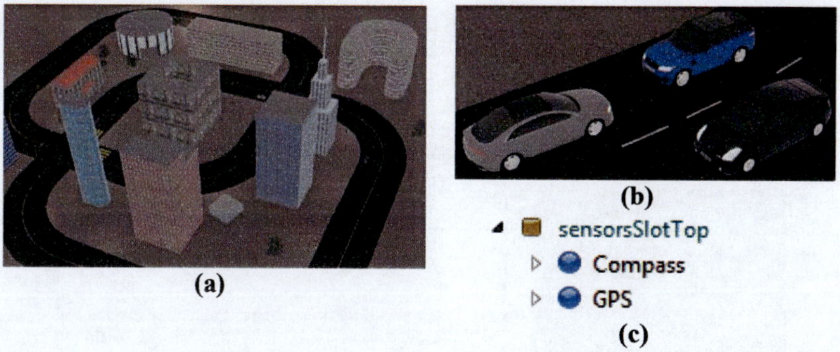

Fig. 2. CPS traffic scenario in Webots. (a) Aerial view of the simulation scenario, (b) Vehicle model and (c) Virtual sensors configuration.

design and customize the virtual CPS to represent the real world operation and help to the implementation of the validated autonomous control on a future physical scenario.

2.2 Controller

The controller was coded in Webots programming interface by using C/C++ programming language. The libraries and tools for applications in the automotive industry included in Webots were used in the controller. The aim of the controller is to be able of commanding autonomous movements of virtual nodes. The virtual nodes are in charge of sending the data referring to their position and orientation through the sensory information (e.g. GPS and compass). The controller can be implemented in several virtual nodes, however, is important to highlight that each vehicle takes a different route. Therefore, an independent controller for each virtual node is necessary. It is important to note that the controller of each vehicle varies in the coordinates that are defined for the autonomous route of each virtual nodes.

One of the considerations that must be taken into account is that the current routes and position cannot coincide, this in order to avoid the collision between the vehicles. Considering that the autonomous vehicles will be simulated under pre-defined routes, specific points were established in the 3D simulation scenario, which will be followed and matched by the controller (using the GPS information) in order to follow the route.

Once the last GPS coordinate has been reached, the vehicle returns to its first GPS position, continuously repeating this process.

When the vehicle reaches a coordinate, it seeks to orient itself with respect to the following coordinate by using the sensory information generated in the compass. The flowchart of the procedure described above is illustrated in Fig. 3.

The implementation of a geometric model in the controller allows obtaining an autonomous movement, avoiding the need of manual control of each vehicle. The aim of to be able of implementing more complex CPS with a close to real world validation. Besides, the geometrical model improves and facilitates the configuration of new route scenarios in real time.

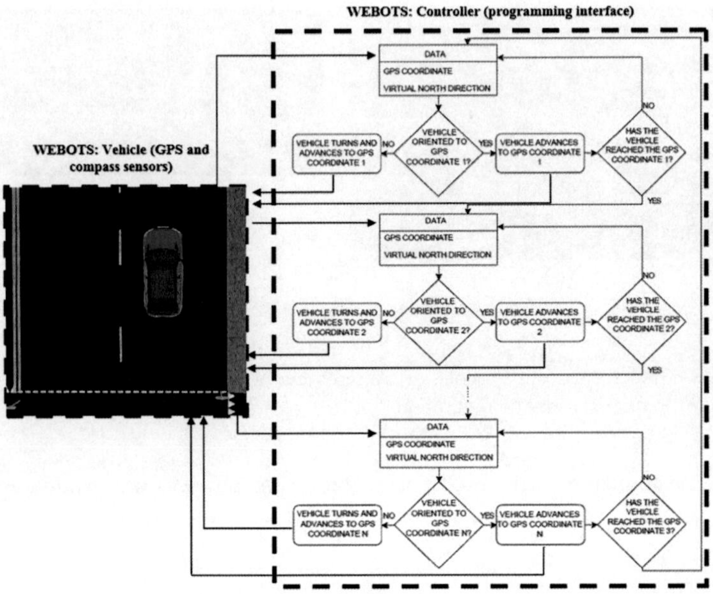

Fig. 3. Flowchart of the geometric model for autonomous vehicle guidance.

In this work, automatic guidance is possible by obtaining data from the GPS and the compass, which will be managed in the controller. As first step, the GPS coordinates that determine where the vehicle will circulate are established. These coordinates are defined as follows:

$$V\_P\_G[P\_GPS] = \{\mathrm{P\_GPS[X]},\ \mathrm{P\_GPS[Z]}\} \tag{1}$$

where V_P_GPS represents the vector of GPS coordinates; P_GPS the index or position of each element in the vector; P_GPS [X] and P_GPS [Z] are the coordinates of each GPS point, these coordinates are established at strategic points that allow making a closed circuit where the vehicle will drive. Then, a target speed for the vehicle is configured for each point of the vector (2) by taking into account the speed limitation and characteristics of the road.

$$\mathrm{V\_V\_GPS[V\_GPS]} = \{\mathrm{P\_GPS\_V[X]}, \mathrm{P\_GPS\_V[Z]}\} \tag{2}$$

where V_V_GPS is the real-time GPS vector of the vehicle; V_GPS is the number or position of each vector; P_GPS_V [X], P_GPS_V [Z] are the GPS coordinates of the vehicle. Subsequently, the vehicle orientation is obtained:

$$\mathrm{V\_V\_COMPASS[V\_COMPASS]} = \{\mathrm{V\_COMPASS[X]}, \mathrm{V\_COMPASS[Z]}\} \tag{3}$$

where, V_V_COMPASS is the real-time compass orientation vector of the vehicle; V_COMPASS the number or position of each vector; V_COMPASS [X],

V_COMPASS [Z] are the points of the vector on the orientation of the compass of the vehicle with respect to the north established in the 3D simulation scenario. In the following vector, we get the difference between the GPS positions vector and the vehicle GPS position vector:

$$\mathrm{V\_RESULT\_GPS[RESULT\_GPS]} = \mathrm{V\_P\_GPS[P\_GPS]} - \mathrm{V\_V\_GPS[V\_GPS]} \tag{4}$$

Then, we obtain the vector module V_RESULT_GPS [RESULT_GPS] in the following variable:

$$\mathrm{M\_V\_RESULT\_GPS} = \sqrt[2]{(V\_RESULT\_GPS[X])^2 + (V\_V\_GPS[Z])^2} \tag{5}$$

where, M_ V_RESULT_GPS is the distance between the vector of GPS positions and the vector of GPS positions of the vehicle. The vehicle advances with the established speed until this distance is equal to 0. The orientation of the vehicle is important; allowing it to move in the right direction, for this reason, obtaining the angle at each moment allows knowing the orientation of the vehicle with the following equation:

$$\begin{aligned} ANGLE\_VEHICLE = {} & tan^{-1}\left[\frac{V\_RESULT\_GPS\,[Z]}{V\_RESULT\_GPS\,[X]}\right] \\ & - tan^{-1}\left[\frac{V\_COMPASS\_GPS\,[Z]}{V\_COMPASS\_GPS\,[X]}\right] \end{aligned} \tag{6}$$

where, ANGLE_VEHICLE is the difference between the angle of the direction of the vehicle and the angle of the direction of the compass with respect to the north of the 3D scenario. If, the result of this difference is less than zero, a π rad is added; in case of being greater than zero, the same value is assigned. The final value is assigned as appropriate in the following variable:

$$(ANGLE\_VEHICLE < 0) \rightarrow BETA = ANGLE\_VEHICLE + \pi \tag{7}$$

$$(ANGLE\_VEHICLE > 0) \rightarrow BETA = ANGLE\_VEHICLE \tag{8}$$

$$BETA\_VEHICLE = BETA - \pi \tag{9}$$

where, BETA_VEHICLE represents the final angle that acts on the wheels of the vehicle allowing it to steer itself; a positive angle allows it to turn right and a negative angle allows it to turn left. This process described above is repeated successively both to solve the angle as well as to calculate the speed, until reaching the appropriate orientation and decreasing the distance to the next GPS position vector. When the last GPS position vector is reached, the vehicle is oriented and directed to the first GPS position vector, repeating this process successively. It is important to note that the established final angle and speed are values that are used by functions of WEBOTS that allow controlling the vehicle. Similarly, the values received from the GPS and

COMPASS sensors are made possible by using different functions that allow interacting with sensors installed in the vehicle.

The implemented geometric model, without a doubt, allows to simulate the circulation of a vehicle in a pre-established circuit, which means emulating the control of the vehicle by means of a particular driver.

3 Case Study: Autonomous Guidance of Vehicles

The particular 3D scenario is defined to evaluate and validate the automatic guidance of conventional vehicles. The scenario emulates an existing configuration of a virtual environment previously developed in WEBOTS called "city night" which is composed of two circular avenues that intersect twice. There are also other implemented elements such as: traffic lights, buildings, horizontal and vertical signaling that simulate an urban traffic environment and several vehicles incorporated with sensors (virtual mobile nodes) (see Fig. 4).

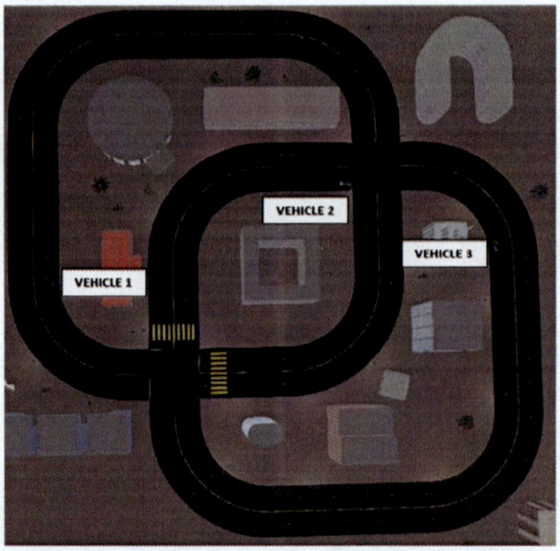

Fig. 4. Aerial view of the simulation scenario of the CPS.

In this case study, the NEO-M8T module is modeled for the GPS sensor and the HMC5883L module for the compass sensor. Table 1 and 2 show the main configuration parameters of the sensors applied in the corresponding virtual elements in WEBOTS. The outputs of the GPS model are the global coordinates of the location of the vehicle and the outputs of the compass model is a vector that indicates the direction with respect to the virtual north.

Table 1. GPS sensor model configuration: Specifications of the virtual CPS sensor.

Specifications/Inputs	NEO-M8T
Sensitivity (Tracking and Navigation)	−166 dBm
Horizontal position accuracy	2.0 m
Course accuracy	0.3°
Speed accuracy	0.05 m/s
Maximum navigation update rate	10 Hz

Table 2. Compass sensor model configuration: specifications of the virtual CPS sensor.

Specifications/Inputs	NEO-M8T
Resolution	5 milli-gauss
Measurement period	6 ms

The virtual sensors mentioned above are incorporated in each one of the vehicles included in the 3D simulation scenario. The routes for this particular case should not be crossed, in order to allow the proper flow of each vehicle and avoid collisions among them. The routes defined for each vehicle are depicted in Fig. 5.

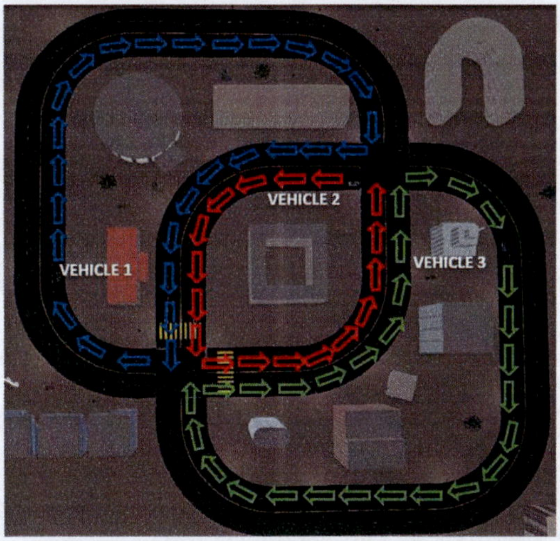

Fig. 5. Defined routes for each vehicle in the CPS simulation scenario.

3.1 Experimental Configuration

The simulation scenario was configured with three fully autonomous vehicles (distributed mobile nodes) with the corresponding sensors on board. The analysis consists of a data processing algorithm from the outputs of the GPS and compass sensors that allow the controllability of the vehicle to orient and direct it to each GPS coordinate of the pre-designed route.

In this scenario, three steps are considered for the autonomous vehicle operation: to obtain the absolute distance between the vehicle and the next GPS coordinate to be reached, the orientation of the vehicle with respect to the virtual north and, the angle resulting from the orientation of the vehicle with respect to GPS coordinates to be achieved. If the resulting angle is less than 0, π is added and if it is greater than 0, this same resulting angle is assigned.

Table 3. Set of data generated in the first configuration (C1) of the geometric model in WEBOTS.

C/V	C1	
	25 km/h (SA)	
	T (s)	DW
TOYOTA	39.744	620
LINCOLN	20.864	325
RANGE ROVER	39.744	620

The complete set of information for this first configuration (C1) contains 620 data generated in WEBOTS (DW) with a simulation time (T (s)) of 39.74 s and a constant speed (Vk) of 25 km/h.

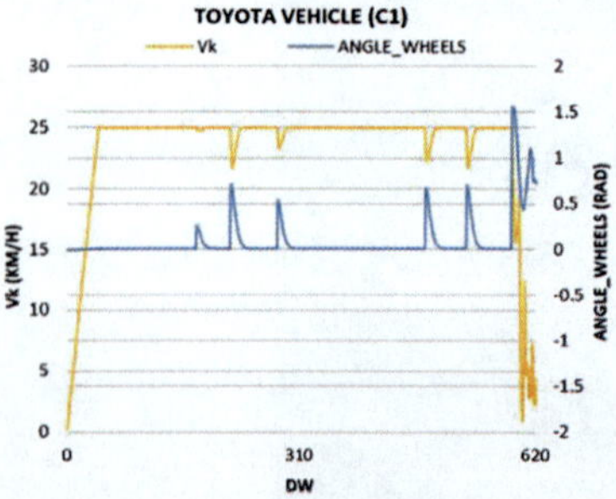

(a)

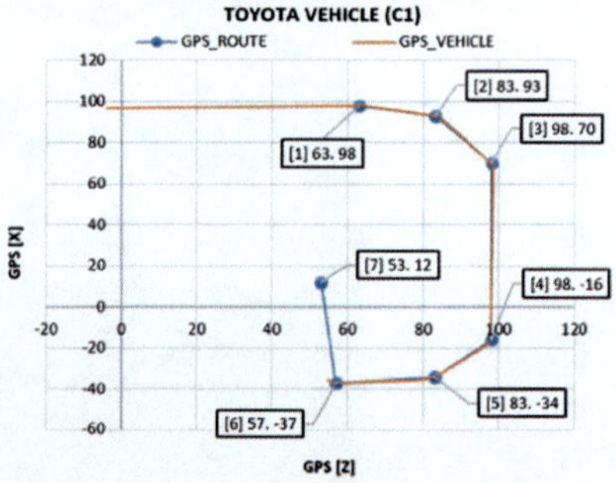

(b)

Fig. 6. Results of the experimental configuration of the geometric model in the Toyota vehicle. (a) Constant speed and angle on the wheels. (b) Route of the vehicle.

The generated data (DW) varies from vehicle to vehicle, because up to that data there is an overturn of the vehicles when trying to turn at intersections with an angle of 90 degrees. In addition, abrupt changes in the orientation of the front wheels are caused when the vehicle tries to orient itself towards the next GPS coordinate, which results in a reduction of constant speed and especially causing the vehicle to lose stability (see Fig. 6, Fig. 7 and Fig. 8).

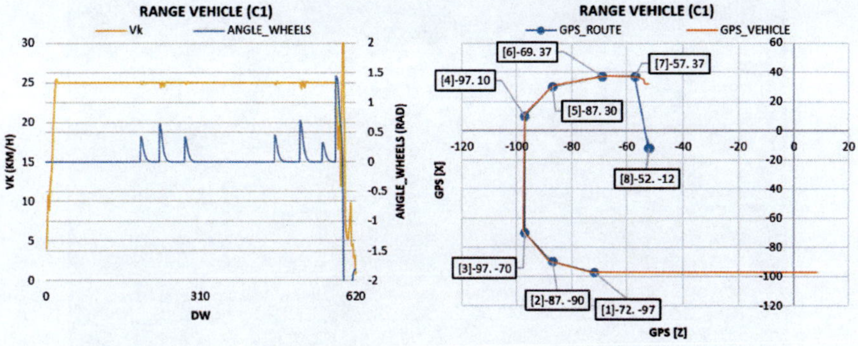

(a) (b)

Fig. 7. Results of the experimental configuration of the geometric model in the Range vehicle. (a) Constant speed and angle on the wheels. (b) Route of the vehicle.

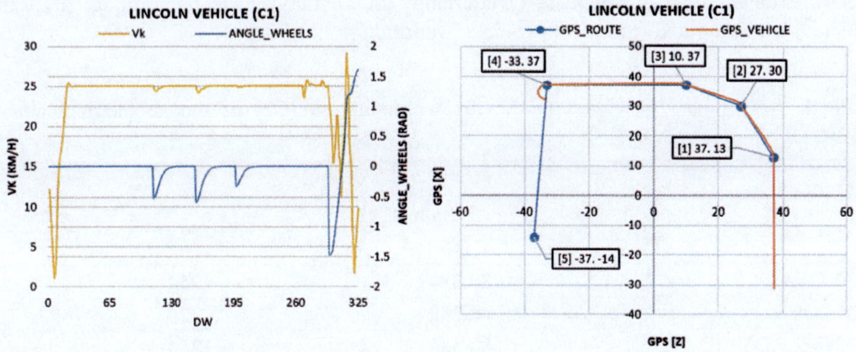

(a) (b)

Fig. 8. Results of the experimental configuration of the geometric model in the Lincoln vehicle. (a) Constant speed and angle on the wheels. (b) Route of the vehicle.

The performed configuration (C1) does not allow the complete route of the vehicles, because the Toyota vehicle travelled to the sixth coordinate, the Range vehicle travelled to the seventh coordinate and the Lincoln vehicle to the fourth coordinate. Figure 9 shows the critical points where an overturning can occur.

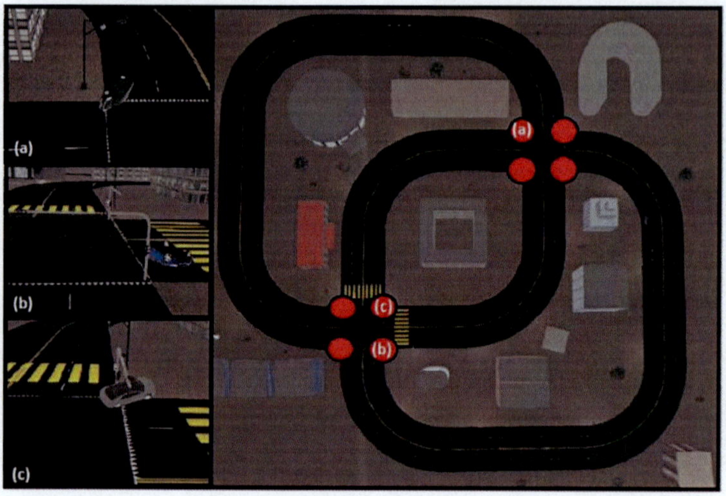

Fig. 9. Critical Points for Vehicle Overturning. (a) Toyota vehicle overturning, (b) Range vehicle overturning and (c) Lincoln vehicle overturning.

Table 4. Set of data generated in the first configuration (C2) of the geometric model in WEBOTS.

C/V	C2	
	25 km/h (SA)	
	T (s)	DW
TOYOTA	80.384	1254
LINCOLN	46.208	720
RANGE ROVER	80.384	1254

3.2 Testing of the Geometric Model

As already mentioned, the vehicle continues the route as long as the angle between one GPS point reached and the next one is less than 75°, but if this angle is exceeded the vehicle attempts to curve and a rollover occurs. For this reason, it is important to control the response angle. The angle admitted in WEBOTS library accepts values in the range [$-\pi/2$ to $\pi/2$]. Therefore, when the resulting angle value is out of range, the maximum allowed value ($\pm\pi/2$) is assigned. On the other hand, when the difference between the generated angle and the next angle is less than $-\pi/10$, then $\pi/10$ is subtracted, whereas if it is greater than $\pi/10$, then $\pi/10$ is added.

Based on the above principles, a second configuration (C2) is defined, to provide more stability during the vehicle turns and maintain a constant speed. This configuration is presented in Table 4.

Similar to the C1 configuration, the number of data generated (DW) varies from one vehicle to another, and the results of the operation is depicted in Fig. 10, Fig. 11 and Fig. 12 for the three analyzed vehicles.

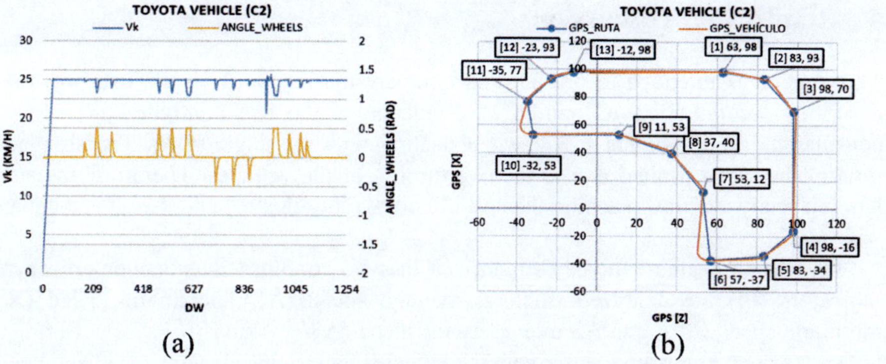

Fig. 10. Training results of the second configuration (C2) of the geometric model in the Toyota vehicle. (a) Constant speed and angle on the wheels. (b) Complete route of the Toyota vehicle.

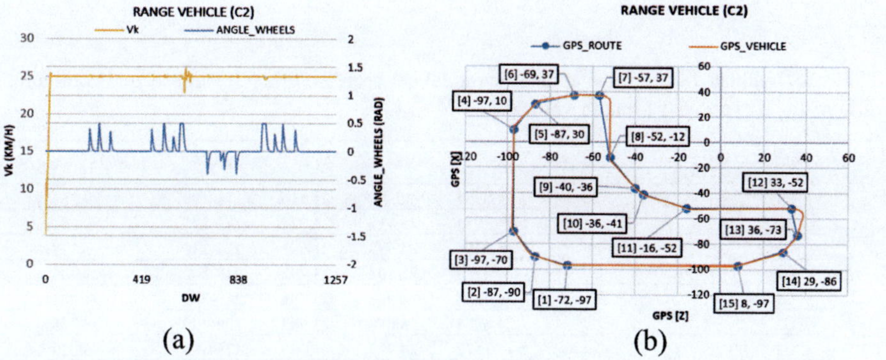

Fig. 11. Training results of the second configuration (C2) of the geometric model in the Lincoln vehicle. (a) Constant speed and angle on the wheels. (b) Complete route of the Lincoln vehicle.

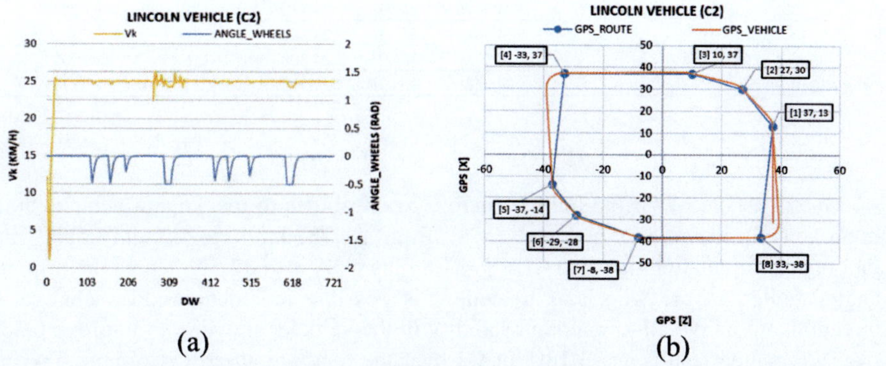

Fig. 12. Training results of the second configuration (C2) of the geometric model in the Range vehicle. (a) Constant speed and angle on the wheels. (b) Complete route of the Range vehicle.

4 Results and Discussion

This section is intended to present and compare the results from the two previously described configuration (C1 and C2). Evidence of the two configurations shows a considerable improvement in C2 when dealing with the angle (CA). The simulation time of the C1 is limited due to the overturning of the vehicles. Therefore, the simulation is analyzed until reaching the specific point (coordinates) and then it is compared with C2.

In order to evaluate the performance of the two configurations, some performance indicators (PI) are analyzed, such as average speed (AS), maximum speed (XS), minimum speed (MS) and AS over constant speed (AS/(25 km/h)). These performance indicators are taken from the moment the vehicle changes its constant speed and begins to orient itself from an initial GPS coordinate (CI) to the next reached GPS coordinate (CA), that is, since there is a considerable change in the angle of the wheels. Because the Toyota and the Range vehicle follow similar route, only two vehicles are evaluated and compared. The obtained results are shown in Table 5.

Table 5. Evaluated data of the geometric model between the first (C1) and second configuration (C2) of the Toyota and Lincoln vehicles.

S/PI		CI		CA		AS	XS	MS	DW	[AS/(25km/h)] %
		[X]	[Z]	[X]	[Z]					
TOYOTA	C1	98	63	93	83	24.8954	25.07	24.716	171-190	99.581%
	C2					24.8954	25.113	24.723	171-191	99.581%
	C1	93	83	70	98	24.0496	24.997	21.681	217-238	96.199%
	C2					24.4262	25.103	23.297	218-242	97.705%
	C1	70	98	-16	98	24.4657	24.996	23.298	280-289	97.863%
	C2					24.6863	25.088	23.943	280-299	98.745%
	C1	-16	98	-34	83	24.1572	24.997	22.249	474-495	96.629%
	C2					24.4577	25.114	23.300	476-499	97.831%
	C1	-34	83	-37	57	24.0744	24.997	21.724	528-549	96.298%
	C2					24.4580	25.085	23.295	529-553	97.832%
LINCOLN	C1	13	37	30	27	24.8070	24.998	24.316	105-126	99.228%
	C2					24.8841	25.051	24.612	106-128	99.536%
	C1	30	27	37	10	24.7476	24.998	24.116	150-171	98.991%
	C2					24.8615	25.015	24.506	151-174	99.446%
	C1	37	10	37	-33	24.9386	25.021	24.800	192-210	99.755%
	C2					24.9676	25.028	24.915	193-211	99.871%

One of the most visible points where C2 contributes to the Toyota vehicle, is the improvement in minimum speed (MS) from 21,724 km/h to 23.295 km/h when looking for orientation between CI = [−34.83] and CA = [−37.57], where the response angle applied to the wheels is treated, it is possible to reduce sudden changes in direction and above all to maintain stability in the vehicle, allowing to improve 6.28% over C1, which results as 93.18% in C2 over the constant speed (See Table 5).

In Fig. 13, it is observed that in C2, VX and VM are very close to constant speed, which contributes to a smooth movement without sudden changes in the direction of the vehicle (see Fig. 13).

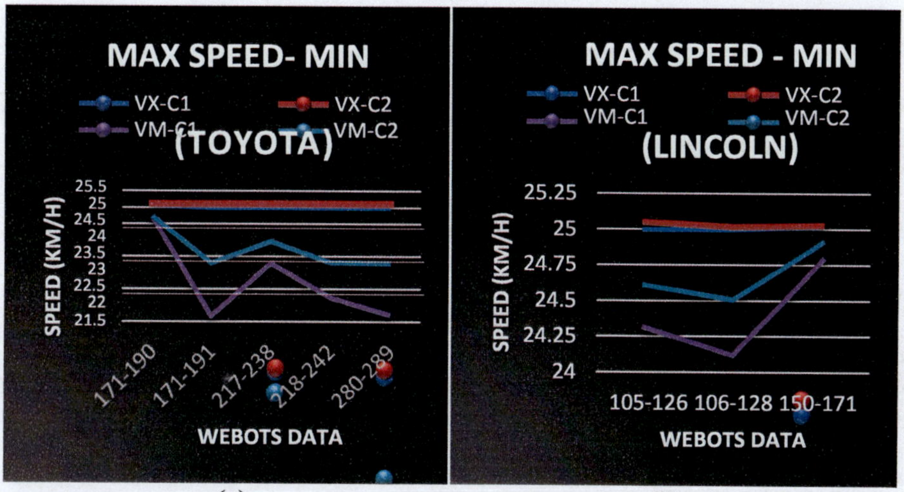

(a) (b)

Fig. 13. Evaluation of the geometric model between the first (C1) and second configuration (C2). (a) Toyota vehicle and (b) Lincoln vehicle.

By examining Fig. 13, data is obtained on the average speed in the total distance of travel (VPT) of the Toyota vehicle, where for the C1 is 22,734 km/h, while in the C2 it is achieved to reach 24,548 km/h, obtaining an improvement of 7.256% over the C1, while the (VPT) in the C2 of the Lincoln vehicle only achieves an improvement of 1.068% over the C1, this is because there are no sharp curves in the displacement (see Table 6).

Table 6. Average speed on the total distance of travel of the Toyota and Lincoln vehicles.

V/VPT		VPT (km/h)	VPT (%)
		25 km/h	(%)
TOYOTA	C1	22.734	90.934
	C2	24.548	94.194
LINCOLN	C1	24.411	97.643
	C2	24.677	98.711

It is evident that C2 allows simulating an autonomous movement along a pre-established route, since on the one hand, it improves the minimum speed reached by the vehicle, and on the other hand it prevents the vehicle from tipping over in pronounced 90-degree curves. All this means that C2 efficiently emulates an autonomous displacement of a vehicle and from this geometric model to achieve the simulation of urban traffic in a more real and complex CPS.

5 Conclusions

The design and implementation of a geometric model that allows autonomous guidance through a pre-established route of conventional vehicles in order to emulate urban traffic in a 3D simulation environment was proposed and implemented in this document. The proposal was developed by integrating a GPS sensor and a compass sensor in each vehicle, which provide information on the location and orientation of the vehicle. The outputs of the virtual sensors were used as inputs to the geometric model which was implemented in the controller. Based on this algorithm, different actions are taken on the modeled vehicle in the 3D simulation scenario. Two configurations were considered to evaluate the behavior of the proposal.

In the first configuration (C1) there is an abrupt change in the front wheels of the vehicles, which causes a decrease in the constant speed, loss of stability and a rollover when the vehicle tries to turn towards an intersection greater than 90° degrees. In the second configuration (C2), greater stability of the vehicle was obtained and it is possible to drive over routes with curves greater than 90°. The two configurations were evaluated up to the point before the overturning through several performance indicators, among them, the minimum speed reached when the vehicle seeks to orient itself to the next GPS coordinate.

In the Lincoln vehicle, three coordinates were analyzed before overturning, maintaining an average speed of 97.643% in the first configuration, while in the second configuration an average 98.711% over the constant speed. On the other hand, in the Toyota vehicle it was possible to analyze five coordinates before the overturning occurs; having an average speed of 90,934% in the first configuration and 98,711% in the second configuration. Based on the results, the proposed approach (C2) is more efficient to simulate the autonomous movement of vehicles by a predefined route.

Finally, automatic guidance through the geometric model for conventional vehicles will be integrated and validated in real driving environments through electromechanical instrumentation in a 2015 Toyota Yaris vehicle with automatic transmission.

References

1. Park, K.-J., Zheng, R., Liu, X.: Cyber-physical systems: milestones and research challenges. Comput. Commun. **36**(1), 1–7 (2012)
2. Villalonga Jaén, A., Castaño Romero, F., Haber Guerra, R., Beruvides López, G., Arenas, J.: El control de sistemas ciberfísicos industriales. Revisión y primera aproximación (2018)
3. Work, D., Bayen, A., Jacobson, Q.: Automotive cyber physical systems in the context of human mobility. National Workshop on High-Confidence Automotive Cyber-Physical Systems, September, pp. 3–5 (2008)

4. Krasniqi, X., Hajrizi, E.: Use of IoT technology to drive the automotive industry from connected to full autonomous vehicles. IFAC-PapersOnLine **49**(29), 269–274 (2016)
5. Hegde, R., Mishra, G., Gurumurthy, K.S.: An insight into the hardware and software complexity of ECUs in vehicles. In: Advances in Computing and Information Technology, pp. 99–106 (2011)
6. Castaño, F., Beruvides, G., Villalonga, A., Haber, R.E.: Self-tuning method for increased obstacle detection reliability based on internet of things LiDAR sensor models. Sensors (Switzerland) **18**(5), 1–16 (2018)
7. Dai, X., Chang, W., Zhao, S., Burns, A.: A dual-mode strategy for performance-maximisation and resource-efficient CPS design. ACM Trans. Embed. Comput. Syst. **18**(5s) (2019)
8. Hofbauer, D., Schmittner, C., Brandstetter, M., Tauber, M.: Autonomous CPS mobility securely designed. In: International Symposium on a World of Wireless, Mobile and Multimedia Networks, WoWMoM 2019 (2019)
9. Schmittner, C., Ma, Z., Schoitsch, E., Gruber, T.: A case study of FMVEA and CHASSIS as safety and security co-analysis method for automotive cyber-physical systems. In: CPSS 2015 - Proceedings 1st ACM Workshop on Cyber-Physical System Security, Part ASIACCS 2015, pp. 69–80 (2015)
10. Abelein, U., Lochner, H., Hahn, D., Straube, S.: Complexity, quality and robustness - the challenges of tomorrow's automotive electronics, pp. 870–871 (2013)
11. Zhang, Z., Eyisi, E., Koutsoukos, X., Porter, J., Karsai, G., Sztipanovits, J.: A co-simulation framework for design of time-triggered automotive cyber physical systems. Simul. Model. Pract. Theory **43**, 16–33 (2014)
12. Schneider, D., Armengaud, E., Schoitsch, E.: Towards trust assurance and certification in cyber-physical systems. In: Lecture Notes in Computer Science (including Subseries Lecture Notes in Artificial Intelligence and Lecture Notes in Bioinformatics), LNCS, vol. 8696, pp. 180–191 (2014)
13. Banerjee, A., Venkatasubramanian, K.K., Mukherjee, T., Gupta, S.K.S.: Ensuring safety, security, and sustainability of mission-critical cyber-physical systems. Proc. IEEE **100**(1), 283–299 (2012)
14. Zea, D.J.: Sensores IoT para aplicaciones de conducción autónoma. Simulación y evaluación. Sensores IoT para aplicaciones de conducción autónoma. Simulación y evaluación (2018). http://oa.upm.es/52628/1/
15. Parkl, M.-W., Lee, S.-W., Han, W.-Y.: Development of lateral control system for autonomous vehicle based on adaptive pure pursuit algorithm, ICCAS, pp. 369–372 (2011)
16. Snider, J.M.: Automatic steering methods for autonomous automobile path tracking, Work, February, pp. 1–78 (2009)

Integration of Artificial Intelligence as a Tool for an Online Education Model

William Villegas-Ch[1(✉)] and Xavier Palacios-Pacheco[2]

[1] Universidad de Las Américas, Av. de los Granados E12-41 y Colimes esq., Quito, Ecuador
william.villegas@udla.edu.ec

[2] Universidad Internacional del Ecuador, Av. Simón Bolívar y Av. Jorge Fernández, Quito, Ecuador

Abstract. Currently, universities base all their efforts on improving student learning. To achieve this, they improve their educational methods and work in training the teaching staff in new techniques that transform traditional learning into active learning. Active learning seeks to make the student the main actor of his own education, with this, the student is free to learn what he owes and does so in a time appropriate to his needs. This need grows when the evolution of education conforms to online education models. Online education, although presented as an accessible option for all sectors of society has several complications such as high dropout and low learning rates. New technologies act as ideal assistants to solve these problems, because, through an application, they have the ability to interact with people and improve learning processes. Artificial intelligence interacts with users and simulates a person. These tools allow improving management in educational processes. This work aims to apply a Chatbot applied to a learning management system in the online education model.

Keywords: Artificial intelligence · Chatbot · Learning · LMS

1 Introduction

Online education models break the paradigms and problems of the modality of face-to-face education such as time and geography. Online education solves these problems and meets the needs of students, where everyone has their own way and time of learning. However, online education must be organized with an adequate methodology where student learning is guaranteed. For which, the inclusion of information technologies (IT) in online education is mandatory and makes the student the protagonist of their own learning, in addition to generating an active education. The implementation of IT and active education has a great connection, since these become the appropriate tools for the use of resources and the development of activities in a student-based learning system [1]. IT is presented as virtual assistants in online education, its objective is to provide ideal environments to improve the understanding of the resources available to the student. Its integration in these models generates a positive environment in the interaction of the students and a greater motivation towards learning topics.

M. Botto-Tobar et al. (Eds.): CI3 2020, AISC 1277, pp. 94–105, 2021.
https://doi.org/10.1007/978-3-030-60467-7_8

IT contributes to the strengthening of educational platforms and academic management of universities. This is possible through computer systems and tools that identify variables, make decisions and recommend activities [2]. Generating a system that meets these characteristics requires a robust architecture, focused on the discovery of knowledge. To do this, a data analysis and processing model must be established. This treatment ranges from the acquisition of the data, regardless of its type and format. The tools that have the capacity to process and analyze data are the Business Intelligence (BI) or Big Data [3] platforms.

The implementation of a data analysis platform in a university depends largely on the volume and type of data, and each institution defines the granularity of the analysis based on the questions it wishes to answer. Once knowledge of the data has been obtained, it is necessary to add value to these results, this is done by education experts who take actions that improve learning. In this work, the interpretation of results and decision-making that allow students to obtain personalized learning are delegated to artificial intelligence (AI) [4]. To fulfill this objective, it is necessary that there is a protocol that allows the coexistence between data analysis, an AI model and the methodology proposed by the tutors that is embedded within the resources and activities presented to the students [5]. A platform that integrates these components requires a process that identifies and evaluates numerous variables, through data analysis, data that students generate from interaction with the academic systems of each university [6]. By identifying the variables, they are processed and the needs of the students are established. The results are transferred and processed in an academic Chatbot that uses AI for decision-making based on data learning and interaction with students. The decision-making carried out by the Chatbot is executed through the recommendation of activities. Each recommended activity aims to improve student understanding, in addition these are aligned to the way each one learns, as a result learning is improved in an online education model. The entire process must be embedded in a single system, which improves the potential of using IT transparently for the user. The universities that offer online education models generally make use of a learning management system (LMS), this work uses this resource as a link for the integration of data analysis and AI, in addition, which is a familiar system for students [7].

The work is distributed in the following sections, in Sect. 2 the concepts used for the implementation of the method are collected, Sect. 3 develops the method, Sect. 4 conducts the discussion based on the results obtained, Sect. 5 presents the conclusions.

2 Theoretical Foundation

2.1 Online Education

The development of Information and Communication Technologies (ICT) allow the generation of new educational models in which all people can access education independently of factors such as time or where they are [8]. Online education allows the development of educational programs that have the web as the setting for teaching and learning.

An online education model allows to establish a dialogue or a learning experience without the need to coincide in a physical space in a certain time. The possibility of creating digital environments with a variety of resources and activities designed for students, allows for autonomous learning. IT provides a number of resources and tools that are used by tutors to generate feedback on each activity or even allow for asynchronous meetings where students can answer questions or reaffirm their knowledge.

2.2 Business Intelligence

The term BI is used to refer to the technologies, tools, practices, and applications that organizations use to collect, integrate, analyze, and present their data in a way that generates insightful, actionable information [9]. Information gathered through leveraging services and software to transform intelligence reporting and business tactical decisions. BI serves to support and enable companies to make smart decisions based on tactical information. BI systems analyze all aspects of an organization. Compare your strengths, weaknesses, and capabilities with market conditions [10]. This allows organizations a competitive advantage. Furthermore, this tool gives companies the possibility-dad to increase their opportunities in the market.

2.3 Chatbot

A Chatbot is a tool that allows you to have automated conversations with users and generate leads, resolve user concerns or provide a service [11]. Chatbot's generally use AI to simulate a natural language conversation. The AI not only provides this software with an improvement in their processing capabilities and understanding of conversations, it also offers the possibility of analyzing the feeling of said conversation, which means that the Chatbot is not only used to automate pre-established responses and responses, but to generate valuable reputation reports, sentiment analysis and engagement with brands through the conversations that users have with them [12]. The AI-developed and machine learning algorithms allow Chatbot's to be able to learn. They can intuit the habits and understand the tastes and preferences of the users [13].

3 Method

For the development of the method that allows an AI system to recommend activities in an online education model, it consists of stages such as, the identification of variables, the analysis of data, the design of the recommender system, etc. In Fig. 1, the architecture that integrates AI in an LMS is presented, being this the most important resource for the educational model where the research is carried out [2].

The proposed architecture is applied in related stages, where it is essential that each one of them fulfill its functions to guarantee the results and give value to the learning process.

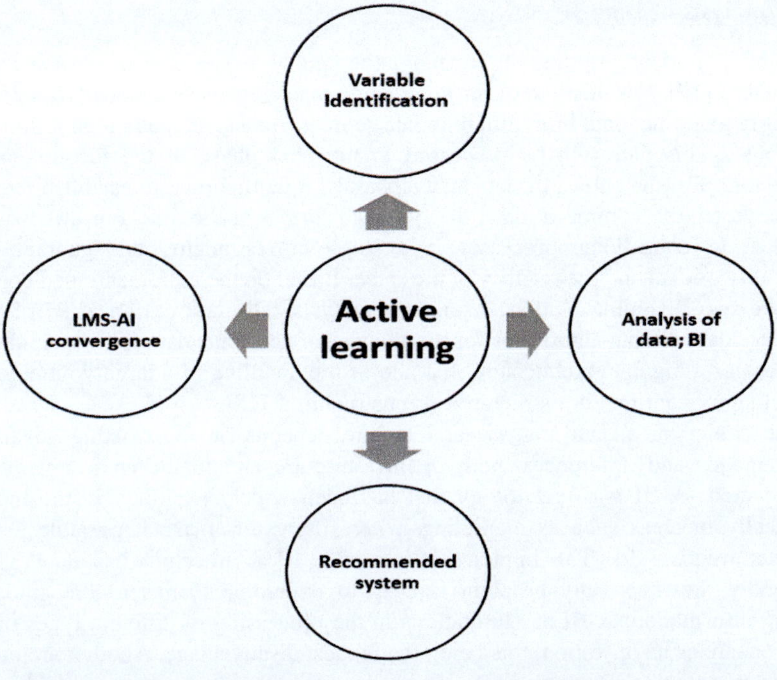

Fig. 1. Architecture of a comprehensive system for the development of an active learning environment.

3.1 Variable Identification

The proposed model is divided into stages that build an environment based on the needs of the students to guarantee the effectiveness of an online education model. The first stage is the identification of variables that affect student performance and learning. For this, a robust architecture is created where the technical, administrative and academic areas participate. The inclusion of these areas allows identifying the variables that cause problems in student learning. The analysis of the variables even allows us to detect deeper problems, such as university dropout or low academic effectiveness. These variables respond to the model proposed by Bean that identifies the academic, psychosocial and environmental variables that cause attrition syndrome [14].

Academic variables identify student performance and integration and make up the repositories that store student academic information. Psychosocial variables include peer interaction and interaction with teachers, as well as the inclusion of the student in the online educational model [15]. This model includes information from the LMS as the main repository of the university that participates in this study for the development of activities [16]. The environmental variables are parameterized according to funding, external social relations, transfer opportunities and the interaction that the student registers in the LMS and the use of ICT.

3.2 Analysis of Data

To identify student progress in learning, the model makes use of a data analysis architecture [17]. For this, a tool for knowledge management is included, this tool has the ability to extract data from different data sources, transform it and load it into a data warehouse. This data will be processed in the next phase of the recommendation architecture. For the choice of data analysis tools, it is important to establish the needs to be covered, the volume of data, the type and format of the data. For this work it is important to consolidate processes in a single infrastructure, this guarantees the availability of data and the security of the same. Based on the aforementioned concepts, there are two favorable platforms for data analysis. The first option is BI platforms which use data mining algorithms for the analysis of academic data [3], a second option is the use of a big data architecture that allows the handling of a large volume of data without significant that are structured or unstructured [18].

The choice of a data analysis architecture depends on the existing advantages, disadvantages and differences, both in infrastructure and in the environment to be implemented. A BI is a technology that has been widely exploited in the industry, specifically in areas such as marketing, where its results make it possible to know customer trends [19]. The implementation of a BI architecture has its degree of complexity, however, any problem is easy to overcome thanks to the amount of existing information. A BI has limitations in the processing of different types of data, which penalizes its operation, this being the biggest disadvantage. Another architecture that has taken great emphasis is Big Data, these architectures allow processing large volumes of data in very short times, in addition, the use of this technology in services where data is not structured, has made Big data the ideal architecture for integration with any technology [20]. A factor considered as a disadvantage is the complexity of its implementation and the high knowledge required for the development of new searches and applications.

This research focuses on the management of data that comes from a single system, specifically the university's LMS. This feature makes the analysis handle only structured data, which has a direct impact on the choice of the tool to be used, in addition to the functionality and agility of implementation. These parameters are easily covered by a BI aligned to Knowledge Discovery in Knowledge Discovery in Databases (KDD) [21]. Figure 2, presents the phases of a KDD model included in the BI architecture. The process begins with the selection of the university's sources, these sources are exclusively part of applications that manage a database in SQL and certain applications such as the LMS use a MySQL connection [22, 23]. There is a certain amount of data found in other formats such as spreadsheets or plain text files. All these sources are considered within the selection phase for which use is made of extraction, transformation and loading tools extract, transform and load (ETL) [24].

In the preprocessing stage, a prior review of data quality is carried out, this includes the elimination of incomplete, inconsistent or repeated data [25]. The transformation stage establishes a single format in all the data extracted from the sources, occasionally, this allows setting additional values of date/time type. The inclusion of these data allows the generation of projections or history, for the educational model this information is necessary to know the trends in students in certain periods or educational

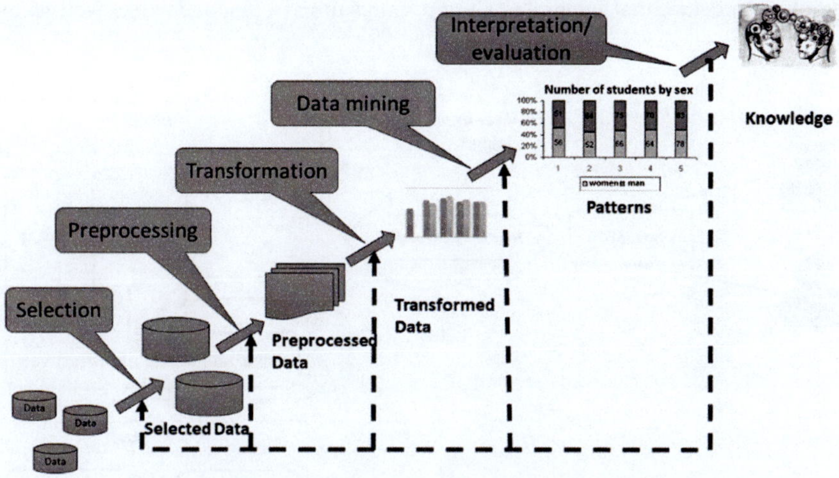

Fig. 2. KDD architecture applied to a BI environment [17].

cycles. The data transformed by the ETL is saved to a multidimensional repository known as the data warehouse. A data warehouse offers the analysis process the availability of the data and guarantees the quality of the data that is consumed in a next stage for its analysis. In the analysis, techniques such as data mining are used where various patterns are identified in the data generated by students through algorithms [26]. The identified patterns are classified in order to determine the students who share certain variables such as the degree of difficulty in the development of academic activities. The results of algorithms such as association, segmentation, classification, etc., are interpreted by experts to arrive at knowledge.

3.3 Recommended System

The process of recommending activities is entrusted to the AI, there are several AI systems that adjust to the needs of an online education model, established to improve learning [27]. Among the most common systems are Chatbot, expert systems, recognition systems, behavior-based intelligence, etc. For the good development of a model of recommendation of activities it is necessary to focus on a system that adjusts to the needs of the students and whose interaction contributes to the motivation of the students. The system that is aligned to these characteristics and that has more enhancement with the management and interaction with users is the Chatbot. Chatbot's are computer programs that allow you to establish a conversation with a person in natural language. This ability allows the proposed model to interact with students and become a virtual assistant who records their requirements, learns from them, and takes an action that improves learning. Chatbot's incorporate AI systems, therefore, they have the possibility to learn from the interaction with students over time [12].

Integrating a commonly used Chatbot, to improve the learning process of university students, depends on the parameters analyzed in previous phases. Figure 3 contains the

Chatbot architecture that integrates BI and its results as components, as well as other data sources.

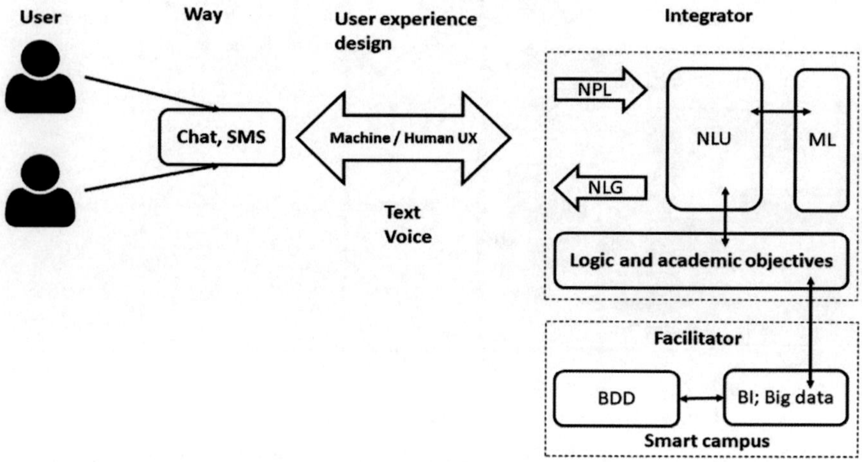

Fig. 3. Communication architecture between a Chatbot - Data analysis [28].

The architecture to be executed starts with the interaction of the user with the AI system, so that the interaction is friendly to the student, it is done through a chat or SMS. By having an autonomous system, the student is more participatory to the point that they make all kinds of queries to the Chatbot, making them a learning assistant. To achieve this, the system even feeds on the information obtained in the analysis of data on the methodology that teachers use within the classroom. These methodologies can be measured through the different activities and evaluation mechanisms developed by the students. These activities by university policies are developed with the use of LMS [28].

The Chatbot applied to learning have become the ideal assistant for teachers, since they can handle all kinds of information corresponding to the academic performance of students. The activities carried out by the students are centralized in the LMS, this is the main system that the student uses for the review of resources and development of academic activities [29].

Chatbot communication with the user is executed through an application-based communication channel, generally web and points towards an intermediate layer known as UX. Chatbot's handle two types of UX, interface and UX writing. The interface UX is in charge of defining how content is presented within the channel, for example, Facebook Messenger [30]. Communication with the user is in charge of the writing UX. This establishes the how, the bot communicates, for example, through texts, images, conversation flows, contexts and any tool that allows answering everything the user asks or answers.

The next component in the architecture is the integrator, which offers Totally Natural Language Processing (NLP) and is the AI part of the Chatbot. The NPL is responsible for acquiring, identifying and processing natural language, the result of which goes to another processing stage known as, natural language understanding (NLU). The NLU identifies the user's intention, it is a process that goes hand in hand with machine learning (ML). The Chatbot that is responsible for the management of educational management integrates pre-established rules of AI. In addition, it integrates a neural network that allows the Chatbot to learn, which implies that it is necessary to train the neural network so that it suggests activities that meet the needs of students [31].

In the facilitator component are the data that have been previously processed and analyzed by the BI, here, all the data sources that allow identifying the student and their academic performance are included. The facilitator communicates the academic part with the logical part, for example, the student starts her session in the LMS where there is a record of activities that need her attention. The Chatbot knows this information, and starts the interaction with a greeting and informs you about pending activities [32]. The information of the student, as well as, of the academic performance come from the analysis carried out by the BI. The Chatbot with this information initiates a second interaction with the student where he looks for the reasons why he has not been able to carry out these activities, the NLP and the NLU interpret the student's situation. Finally, with all the information, the IA recommends a specific type of activity in the LMS learning module.

3.4 Integration of LMS and Artificial Intelligence

To provide the greatest functionality to the complete architecture, it is important to create modules in the LMS, the objective is to integrate the AI and BI systems into a single repository from which to launch the different applications. The IA guarantees personalized experiences for each student by analyzing their behavior and performance in different subjects, interacts directly with students and identifying the way they learn [33]. The information obtained is used to understand the student's interest in relation to learning. This knowledge is used to improve the resources of each subject, in addition, adjustments can be made to the courses and adjustments, which is reflected in a less linear but more effective way to retain information. The AI makes the LMS respond to the individual needs of the user by intelligently considering them to the requests that each one makes the interest of the students. The AI system as it becomes available from more data being fed into processing, more AI information about individual student needs, making the learning platform a continuous improvement engine that grows alongside your students [34]. By having more data that the system processes, more artificial intelligence learn about the individual needs of the students, which turns the learning platform into a continuous improvement engine that grows alongside its students.

4 Discussion

In traditional models, monitoring depends on the evaluator's criteria, which means that there is a bias determined by the criteria of one or a group of people. In the online education models this generates inconveniences, since no greater interaction with the student. To solve these problems, it is necessary to carry out an analysis that covers a greater number of variables and the appropriate way to do it is through IT. Integrating data analytics and AI into LMS prevents learning development from being biased to personal judgment. The analysis is objective and seeks to work on the needs presented by the students, for this, the BI architecture uses an ETL that allows establishing connection chains with different databases.

The ETLs used as reference are SQL Integration Services and Pentaho Data Integration in its open-use version. ETLs allow any type of data source to be integrated into the analysis, processed, transformed and loaded into a data warehouse. This process ensures data quality and reliability in the following phases. Mining algorithms identify patterns in student data and allow them to be classified according to their strengths and weaknesses in learning.

By identifying the weaknesses of the students, value is added to the analysis and the results are sent to the AI system in search of knowledge. The difference from a traditional educational data mining model is that decision-making and actions in these models depend on the people in charge of learning. This being a weakness in the process, since the final decision depends on the criteria of a person. Furthermore, any educational monitoring and improvement model must generate immediate action, which is something that the proposed model guarantees, unlike traditional models. The proposed model contributes significantly to the learning process; the integration of an AI module allows decision-making in real time. The module has all the student information and the Chatbot learns and interacts with the user in a personalized way.

5 Conclusions

This work seeks to offer the student an assistant for their learning. For this, it integrates two technologies considered as emerging that allow the analysis of student performance, through the implementation of a BI that manages student data. These data contain information on the academic performance of each student, however, getting to know this data requires data processing. To do this, a KDD model is applied based on data mining algorithms that allow identifying patterns in students and classifying them in such a way that models can be created for a personalized education. The results of the analysis are used by the AI system through a Chatbot. The Chatbots, in addition to interacting with students creating an environment of trust and interest with users, allow obtaining information on topics that interest each student. This information allows creating an educational model attached to the way each student learns.

The academic follow-up is a very arduous task that consumes many resources of a university, this in an online education model has a greater impact, since the interaction with a teacher is minimal. This forces universities to seek support in IT, the use of emerging technologies provides innovative solutions that base their results on data analysis and exploit knowledge with AI systems such as Chatbot. In this way, each event, request or trend can be registered and used to improve learning. In addition, having a direct interaction with the student allows the system to feedback and learn from each user.

References

1. Kirkup, G., Kirkwood, A.: Information and communications technologies (ICT) in higher education teaching - a tale of gradualism rather than revolotion. In: Learning, Media and Technology, pp. 185–199 (2005). https://doi.org/10.1080/17439880500093810
2. Salcines, E.G., Morales, C.R., Ventura, S., Castro, C.D.: Sistema recomendador colaborativo usando minería de datos distribuida para la mejora continua de cursos e-learning. IEEE-RITA **3**, 19–30 (2008)
3. Bele, J.L., Bele, D., Pirnat, R., Rqþdulü, Q.: Business Intelligence in e-Learning, pp. 369–375. IEEE (2015)
4. Riquelme Santos, J.C., Ruiz, R., Gilbert, K., Asociación Española de Inteligencia Artificial: Inteligencia artificial : revista iberoamericana de inteligencia artificial. Asociación Española para la Inteligencia Artificial (AEPIA) (1997)
5. Valdez, A., Cortes, G., Castaneda, S., Vazquez, L., Medina, J., Haces, G.: Development and implementation of the balanced scorecard for a higher educational institution using business intelligence tools. Int. J. Adv. Comput. Sci. Appl. **8**, 164–170 (2017). https://doi.org/10.14569/ijacsa.2017.081022
6. Buenaño-Fernandez, D., Villegas-CH, W., Luján-Mora, S.: The use of tools of data mining to decision making in engineering education—a systematic mapping study. Comput. Appl. Eng. Educ., 1–15 (2019). https://doi.org/10.1002/cae.22100
7. Villegas-Ch, W., Palacios-Pacheco, X., Buenaño-Fernandez, D., Luján-Mora, S.: Comprehensive learning system based on the analysis of data and the recommendation of activities in a distance education environment. Int. J. Eng. Educ. **35**, 1316–1325 (2019)
8. Lee, S.J., Lee, H., Kim, T.T.: A study on the instructor role in dealing with mixed contents: how it affects learner satisfaction and retention in e-learning. Sustain. **10** (2018). https://doi.org/10.3390/su10030850
9. Nautilus, S.: The Data Warehousing Insitute
10. Combita Niño, H.A., Cómbita Niño, J.P., Morales Ortega, R.: Business intelligence governance framework in a university: Universidad de la costa case study. Int. J. Inf. Manage. 0–1 (2018). https://doi.org/10.1016/j.ijinfomgt.2018.11.012
11. Hill, J., Ford, W.R., Farreras, I.G.: Computers in human behavior real conversations with artificial intelligence: a comparison between human – human online conversations and human – Chatbot conversations. Comput. Hum. Behav. **49**, 245–250 (2015). https://doi.org/10.1016/j.chb.2015.02.026
12. Cahn, B.J.: CHATBOT: architecture, design, & development. In: University of Pennsylvania School of Engineering and Applied Science Department of Computer and Information Science, pp. 1–46 (2017)
13. Al-zubaide, H., Issa, A.A.: OntBot: ontology based ChatBot. In: International Symposium on Innovations in Information and Communications Technology, pp. 7–12 (2011)
14. Himmel, E.: Modelo de análisis de la deserción estudiantil en la educación superior. Calid. en la Educ. 0, 91 (2002). https://doi.org/10.31619/caledu.n17.409
15. Bean, J.P.: The application of a model of turnover in work organizations to the student attrition process. Rev. High. Educ. **6**, 129–148 (1983)

16. Bean, J.P.: Interaction effects based on class level in an explanatory model of college student dropout syndrome. Am. Educ. Res. J. **22**, 35–64 (1985). https://doi.org/10.3102/00028312022001035
17. Villegas-Ch, W., Luján-Mora, S.: Analysis of data mining techniques applied to LMS for personalized education. In: IEEE World Engineering Education Conference (EDUNINE), pp. 85–89 (2017). https://doi.org/10.1109/EDUNINE.2017.7918188
18. Villegas-Ch, W., Lujan-Mora, S., Buenano-Fernandez, D.: Application of a data mining method in to LMS for the improvement of engineering courses in networks. In: Chova, L.G., Martinez, A.L., Torres, I. (eds.) 10th International Conference of Education, Research and Innovation (Iceri2017), pp. 6374–6381. IATED-International Association Technology Education & Development, Lauri Volpi 6, Valenica, Burjassot, 46100, Spain (2017)
19. Riofrio, G., Encalada, E., Guaman, D., Aguilar, J.: Business intelligence applied to learning analytics in student-centered learning processes. In: Proceedings Latin American Computing Conference, pp. 1–10. IEEE, Arequipa (2015). https://doi.org/10.1109/CLEI.2015.7360045
20. Peng, M.Y.-P., Tuan, S.-H., Liu, F.-C.: Establishment of business intelligence and big data analysis for higher education. In: Proceedings of the International Conference on Business and Information Management. ACM. pp. 121–125 (2017). https://doi.org/10.1145/3134271.3134296
21. Fayyad, U., Gregory Piatetsky-Shapiro, P.S.: Knowledge discovery and data mining: towards a unifying framework. In: KDD, vol. 96, pp. 82–88 (1996)
22. Villegas-Ch, W., Luján-Mora, S., Buenaño-Fernandez, D.: Towards the integration of business intelligence tools applied to educational data mining. In: IEEE World Engineering Education Conference (EDUNINE), pp. 1–5. IEEE (2018). https://doi.org/10.1109/EDUNINE.2018.8450954
23. Comendador, B.E.V., Rabago, L.W., Tanguilig, B.T.: An educational model based on Knowledge Discovery in Databases (KDD) to predict learner's behavior using classification techniques. In: IEEE International Conference Signal Processing, Communication, Computing Conference Proceedings, pp. 1–6 (2016). https://doi.org/10.1109/ICSPCC.2016.7753623
24. Pan, B., Zhang, G., Qin, X.: Design and realization of an ETL method in business intelligence project. In: IEEE International Conference on Cloud Computing and Big Data Analysis, pp. 275–279 (2018). https://doi.org/10.1109/ICCCBDA.2018.8386526
25. Rodzi, N.A.H.M., Othman, M.S., Yusuf, L.M.: Significance of data integration and ETL in business intelligence framework for higher education. In: Proceedings International Conference on Science in Information Technology: Big Data Spectrum for Future Information Economy, pp. 181–186 (2016). https://doi.org/10.1109/ICSITech.2015.7407800
26. Villegas-Ch, W., Luján-Mora, S.: Análisis de las herramientas de minería de datos para la mejora del E-learning en Plataformas LMS. In: TIC actualizadas para una nueva docencia universitaria, pp. 761–774. McGraw-Hill (2016)
27. Villegas-Ch, W., Palacios-Pacheco, X., Luján-Mora, S.: Artificial intelligence as a support technique for university learning. In: IEEE World Conference on Engineering Education (EDUNINE), pp. 1–6 (2019). https://doi.org/10.1109/EDUNINE.2019.8875833
28. Villegas-ch, W., Arias-Navarrete, A., Palacios-pacheco, X.: Proposal of an architecture for the integration of a Chatbot with artificial intelligence in a smart campus for the improvement of learning. Sustain. **12**, 1500 (2020). https://doi.org/10.3390/su12041500
29. Rahman, A.M., Al Mamun, A., Islam, A.: Programming challenges of Chatbot: current and future prospective. In: 5th IEEE Region 10 Humanitarian Technology Conference 2017, R10-HTC 2017, pp. 75–78 (2018). https://doi.org/10.1109/R10-HTC.2017.8288910

30. Shukla, V.K., Verma, A.: Enhancing LMS experience through AIML Base and Retrieval Base Chatbot using R Language. In: International Conference on Automation, Computational and Technology Management, pp. 561–567. IEEE (2019). https://doi.org/10.1109/ICACTM.2019.8776684
31. Sharma, R., Patel, M.: Survey on Chatbot design techniques in speech conversation systems. Int. J. Adv. Comput. Sci. Appl. **5**, 37–46 (2018). https://doi.org/10.17148/iarjset.2018.596
32. Shawar, B.A., Atwell, E.S.: Using corpora in machine-learning Chatbot systems. Int. J. Corpus Linguist. **10**, 489–516 (2005). https://doi.org/10.1075/ijcl.10.4.06sha
33. Jayakody, J., Perera, I.: Enhancing competencies of less-able students to achieve learning outcomes: Learner aware tool support through Business intelligence. In: Proceedings 2016 IEEE International Conference on Teaching, Assessment, and Learning for Engineering, TALE 2016, pp. 154–160 (2017). https://doi.org/10.1109/TALE.2016.7851787
34. Science, N., Bishwabidyalay, G., Mamun, S. Al: review of integrated applications with AIML based Chatbot. In: 1st International Conference on Computer & Information Engineering, pp. 87–90 (2015). https://doi.org/10.1109/CCIE.2015.7399324

Augmented Reality as an Academic Training Experience in Higher Education

Wilma Gavilanes[1(✉)], Blanca Cuji[1], Oliver Toalombo[1], and Juan Carlos Fiallos[2]

[1] Technical University of Ambato, Ambato, Ecuador
wilmalgavilanesl@uta.edu.ec
[2] Escuela Superior Politécnica de Chimborazo, Riobamba, Ecuador

Abstract. Educational processes in general demand the use of digital resources in the classroom. Experimentation with real objects allows academic training to be more interactive and dynamic. In this sense, the incorporation of tools such as Augmented Reality in the classroom is reflecting a new way of teaching and learning.

The objective of this article was to apply the OARA methodology for the design of learning objects with Augmented Reality (AR) and through its use to determine the level of technological acceptance (TAM), validate the design of the final product and assess the academic performance of third semesters students of the Basic Education program of the Faculty of Human Sciences and Education of the Technical University of Ambato.

In this university teaching experience, the results showed that students value highly the use of content enriched with AR using their mobile devices, as well as the design of the final product was highly satisfactory and evidenced an improvement in student's academic performance.

Keywords: Academic training · Augmented reality · OARA methodology TAM model · Mobile devices · Teaching-learning

1 Introduction

In today's knowledge society the AR is presenting itself as a technology with strong possibilities that offer opportunities to innovate content and teaching methods to develop academic skills in the process of university education.

According to the latest Horizon reports [1] Edutrens and Hypercycle of the Gartner company, they present important advances on educational technologies such as: AR and VR that are having a strong effect on higher education, due to the degree of interest that its use awakens, also it increases motivation towards the cognitive contents presented to university students [2] all this along with the increasing universalization of mobile devices enabling access to content enriched with AR.

Augmented reality can be defined as an emerging technology that allows to combine physical and digital information in real time, allows to add layers of visual

M. Botto-Tobar et al. (Eds.): CI3 2020, AISC 1277, pp. 106–116, 2021.
https://doi.org/10.1007/978-3-030-60467-7_9

information about the real world around us, using technological devices such as tablets, smart phones or augmented reality glasses [3].

According to [4], it is necessary to implement new, more interactive tools that arouse the interest and motivation in students, keeping them from using traditional resources and awakening interest in discovery, where the student can visualize and interact with 3D objects that generate real elements through technology using mobile devices, projecting reality in the classroom [5].

Many experiences have been developed on the use of AR at different educational levels and different academic disciplines, with emphasis on the educational field, several authors [6–8] describe advantages and disadvantages of the use of AR such as.

- Increases motivation and interest of students
- Promotes the acquisition of investigative skills
- Promotion and development of critical thinking and problem solving capacity
- Promotes the development of laboratory practices, minimizing risks
- Teachers must develop academic skills in the management of applications for resources design with AR; there are currently several technological tools available and access-free, which enable the design of content with AR without having knowledge of programming easily and intuitively
- The devices with which they work may have some degree of difficulty in accessing the resource
- Resources with broad accessibility are needed

In this research shows the design of a resource enriched with AR applying the OARA methodology [9], which consists of 4 phases: Information Structure, Multimedia Design, Assembly and Publication and Validation. The work included 68 students of the Basic Education program of the Faculty of Human Sciences and Education of the Technical University of Ambato, an experimental, field methodology was applied with probabilistic sampling, applying surveys to collect data and subsequent analysis using the statistical software Spss, 2 surveys were applied, one of the TAM model, another of design validation, A test was also conducted at the beginning of the process and another at the end to determine the learning result achieved by the students.

2 Educational Experiences with Augmented Reality in Higher Education

At the European level [10] in the study "Mobile devices and augmented reality in the learning of university students" carried out in Spain with the aim of knowing the motivation degree that notes enriched with augmented reality arouse and determine how it influences academic performance; The design, production and evaluation of an OA was carried out, concluding that the students involved at the University of Seville maintain a positive acceptance when making use of this emerging technology using video as a teaching means, since factors such as attention, trust, relevance and satisfaction come around when students interact with different learning objects with AR, thus significantly improving academic performance. From this perspective it can be said that this technology enables learning and, at the same time, creates an innovative environment for both the teacher and the student.

Likewise [11] in this research the educational possibilities of the AR in university education were evidenced through innovative experiences with 117 students who take the subjects of Information Technology and Communication in Social Education at the Pablo de Olavide University, from Seville, (Spain). An open questionnaire was designed to examine in greater depth the functionality, limitations and formative possibilities of augmented reality by students who have used the same system in their formative processes. It was considered that AR is applicable mainly in childhood, also for educators and social workers. However, training and economic investment are required to ensure success in the classroom.

Similarly [12] analyze how AR offers great changes in the university field, in areas such as medicine, by developing four objects with 3D animations; manipulation and transfer of videos; concluding that medical students access pedagogical content in a simple way due to the availability of having a mobile device. On the other hand, it was found that gender does not influence the degree of technological acceptance, affecting positively motivation. However, some students have problems interacting with AR objects due to their technological unfamiliarity, but with the frequent use of this technology, these negative elements disappear.

Another AR innovation project is the Media Lab of the IlCE (Latin American Institute of Educational Communication), based in Mexico and has 14 member countries, all Latin American; one of its main objectives is to apply technology as a dynamic tool and educational innovation, generating projects to improve learning using technological resources [13].

In recent years, countries such as Colombia have begun to explore emerging technologies in their educational programs, such as the work of [14] "*Emerging technologies, challenge for Colombian higher education*" they conclude that by incorporating such technology, learning becomes collaborative, meaningful, multimodal and participatory promoting self-learning with the arrival of ICT in the academic field, generating substantial changes in the way knowledge is taught and transmitted, this has broken the paradigm of traditional classrooms through new digital tools, mobile and customized hardware applications, all of them supported by the latest communication and information technologies.

Similarly [15] at the National and Open University in Colombia based on a training of diploma courses, specializations, masters and doctorates supported in E-learning education through a program called Trainer of Trainers (PFF in Spanish), it implements a teacher training approach called TPACK, with technological use, contents and pedagogical knowledge totally related to the augmented reality being flexible and didactic, a good educational practice with ICT was found, helping to understand some of the learning difficulties. It was established that teachers assume resources based on AR with the aim of strengthening their teaching methods, it was also concluded that educational resources with AR motivate the development of logical, critical, scientific and experimental technological thinking.

In another South American country in Chiclayo, Peru, the University of San Martín de Porres (USMP) developed educational experiences that expose some interaction options of the AR in different fields such as: architecture, entertainment, academic training, art, medicine; specifically, AR is related to education, allowing students to interact with virtual objects as a reinforcement for the classroom [16].

And in Ecuador we will mention the AR project "Virtopsia", developed by the Private Technical University of Loja (UTPL), for the subject of Legal Medicine, as a tool to support the development of practices, since access to the forensic center is restricted and very crowded by the number of students, this project allows the possibility of simulating autopsies of real cases, in a didactic way from a mobile device and offline, coupling the content to the legal medicine specialization, this is a free-access application to any student or educational institution [17].

Finally, the works developed at the Technical University of Ambato, Faculty of Humanities and Education can be mentioned, they have worked with students of the Teaching Degree in computer science as OA designers with AR in different areas and subjects, to be used by students of the Basic Education, Tourism, Educational Psychology, student consumers of these OA using mobile devices, concluding that they improve motivation and academic performance of students [18, 19].

3 Methodology

The methodology used was experimental, quantitative, consisting of 4 phases, which are part of the OARA methodology [14], (see Fig. 1), it is an original proposal designed by the researcher, the first 3 phases were developed in the months of April-June 2019, The investigator, the teacher of the subject and the students who consume information worked together.

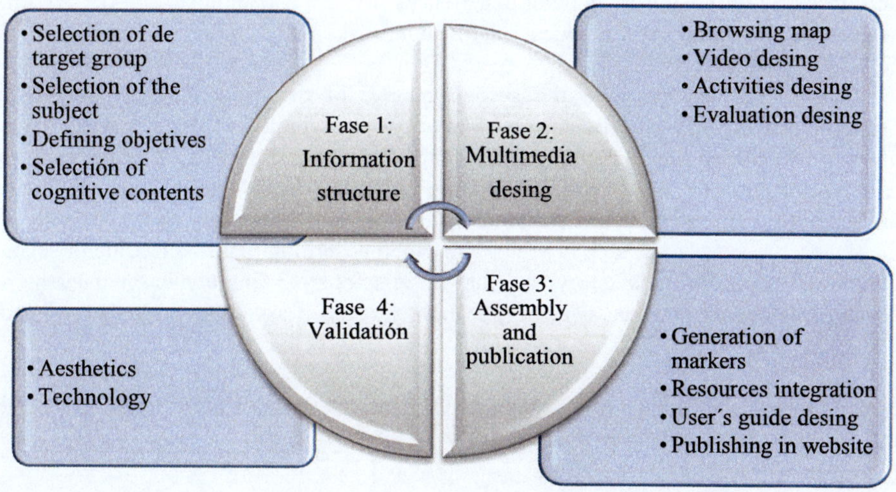

Fig. 1. Methodology for the design of OA with AR (OARA) Source: Prepared by the authors

Phase 1: Information structure, in this initial phase basic elements for the development of the learning object are detailed, in this case, the subject selected was Use of ICT tools corresponding to the cognitive contents of the first semester of the Basic Education program of the Faculty of Human Sciences and Education of the Technical University of Ambato. In Phase 2: Multi-media Design, all aspects necessary for the design and publication of multimedia content such as: videos, audios, images, icons, presentations, interactive activities and evaluations, a free software for the design of resources was used and images were taken from free sites that have a creative commons license.

For Phase 3: Assembly and Publication, the Unity 3D tool created by Unity Technologies and SDK Vuforia were used, these tools were considered for their robustness and their permanent use without access limitation, compared to others applications used with ZAPPAR, LAYAR or ROAR that have a maximum operating limit of the application.

For the generation of the markers or information activators, it was complemented with VUFORIA which is a free Qualcomm development kit for the creation of Augmented Reality compatible with the Unity 3D framework that allowed the elaboration, storage and recognition of the target image for displaying virtual information.

An easy access and navigation environment was designed for the dissemination of the application, in Wix, the site composed by the sections: 1.- Home (main page), 2.- Information (guide to download the mobile application), 3.- Objectives (prior and already acquired knowledge), 4.- Bookmarks (content activators), 5.- AR brochure (digital educational material), to know the site you can access the following: Access link: https://olivertoqui2552.wixsite.com/web2ar.

Phase 4 Validation: Two instruments were designed; a file with the TAM model introduced by [20] and used in other experiences [21, 22] and another "ad hoc" record, to evaluate the design of the educational resource taken as a reference by [23]; additionally 2 reagent tests.

This phase took place in the second quarter of the academic period (July - September-2019), the experiment took 8 weeks over a period of 4 h per week. This process was carried out in collaboration with the teacher tutor of each course and the research teacher, the TAM model consisting of a total of 14 items was used, it was divided into 4 sections: Ease, Perceived Utility, Attitude, Intention for use in each of the stages, the answer for each item was presented according to a Likert scale of seven options (1 Highly unsatisfactory, 2 Quite unsatisfactory 3 Slightly unsatisfactory, 4 Neither satisfactory nor unsatisfactory, 5 Slightly satisfactory, 6 Quite satisfactory and 7 Highly satisfactory).

To evaluate the designed resource, an "ad-hoc" questionnaire divided into 4 categories was used, Quality of Content, Multimedia Design, Utility, Accessibility [9] with a Likert scale of seven options.

4 Results

The instruments proposed were validated by Cronbach's alpha coefficient of reliability. The values obtained in each of the categories are displayed, obtaining an average in the first case for Model TAM of 0, 98 (see Table 1), and in the second case of the resources designed, a value of 0.96 (see Table 2), these values were found to show the highest scale of the level of reliability of the instruments used.

Table 1. Cronbach's alpha categories of the TAM model

Ease of use	0.901
Perceived Profit	0.856
Attitude of use	0.928
Intension of use	0.862

Source: Prepared by the authors

Table 2. Cronbach's alpha design validation

Quality of the content	0,959
Multimedia Design	0, 952
Utility	0,943
Accessibility	0,878

Source: Prepared by the authors

The values show the students' perception in relation and standard deviations (see Table 3), which allow to determine the compliance of the model that indicates the intention of use and Attitude of use of a technology is determined by the variables perceived utility and ease of use, It is evidenced in this case that the easier the use of the application is and presents utility in the training process, this is motivating and interesting, allowing the students to explore innovative and attractive cognitive content, enabling the academic training process, this can be referenced in other similar works [8, 23].

Using SPSS 23.0 was used, carrying out a CrossChart test between categories of the TAM model and used coefficient of Kendall, showing that, at the junction of variables Ease - Attitude a value of 0.68 was obtained; between Ease -Intension 0, 62; Perceived Profit-Attitude 0.65; Perceived Utility-Intent to Use 0.66, showing that there is an average and strong correlation between these groups of pairs of variables, allowing to demonstrate compliance with the Model of Technologic acceptance TAM where ease and benefit of the resource impact directly in the intention and attitude of using of technology (Table 4).

Table 3. Mean and standard deviations by category Model TAM

Easy to use	Mean	Standard deviation
1. How do you think the degree of usability of the app was?	6,3382	, 90785
2. The use of this application in your class was…?	6,2794	, 87836
3. Did you manage this resource?	6,1324	1,03526
4. From your point of view, how would you rate the accessibility of the resource with OA that we have presented to you?	6,1471	, 95049
Perceived profit	Mean	Standard deviation
5. How did you like the usefulness of the contents?	6,1912	, 88533
6. Does this help me solve my tasks in a way?	5,8382	1,21692
7. What was your level of learning?	6,0294	, 91375
8. Interactive activities and evaluations developed, how did you like them?	6,3235	, 87146
Attitude of Use	Mean	Standard deviation
9. When using this resource your level of motivation was…?	6,2941	, 88197
10. I would like to use the OA in class again if I had opportunity	6,1765	, 96105
11.I would like to use OA to learn other topics	6,1176	1,00044
Usage intent	Mean	Standard deviation
12. To what extent do you consider that the use of OA makes learning more interesting?	6,1912	, 90203
13. How did you like using your mobile device to operate the OA?	6,1471	1,02600
14. To what extent the use of OA in class would be beneficial?	6,2059	, 89039

Source: Prepared by the authors

Table 4. Cross charts categories model TAM

	Value	Approximate significance
Ordinal by ordinal Tau-c de Ken dall	, 686	, 000
Ordinal by ordinal Tau-c de Ken dall	, 626	, 000
Ordinal by ordinal Tau-c de Ken dall	, 652	, 000
Ordinal by ordinal Tau-c de Kendall	, 663	, 000
N of valid cases	68	

Source: Prepared by the authors

Concerning the analysis of the resource designed, the questionnaire has 4 sections related to: Quality Content, Multimedia design, usefulness, Accessibility. - It could be evidenced that most of the students think that the designed resource was highly satisfactory, moderately satisfactory and mostly satisfactory and in very few cases Unsatisfactory, (see Fig. 2).

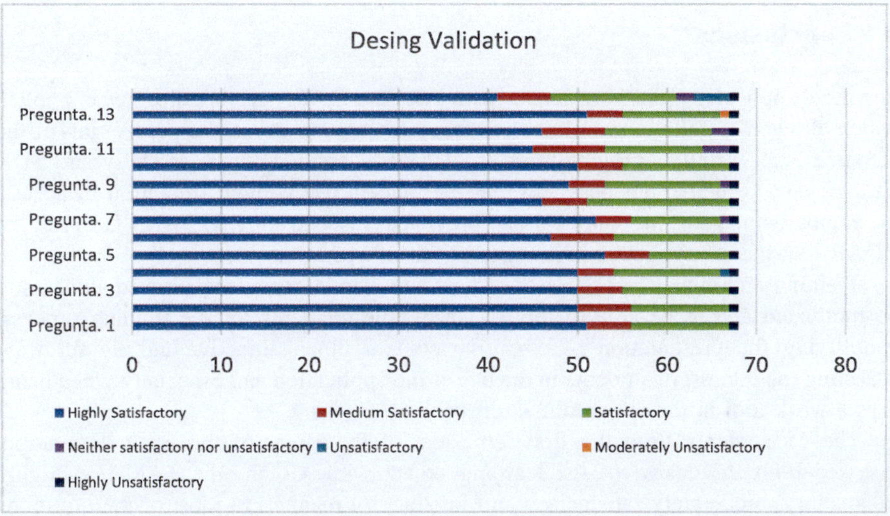

Fig. 2. Design validation

With the data of the qualifications obtained, the SPSS 23.0 software was used to perform the Wilcoxon hypothesis validation test whose results show, with a significance of less than 1% (99% reliability level) that the pre-test qualifications are significantly different from the values of the post-test.

Therefore, there is an improvement in learning using content enriched with AR, as evidenced by the values in Tables 5 and 6.

Table 5. PostTest and PreTest data

		N	Average range	Sum of ranks
Postest - Pretest	Negative ranges	8[a]	19,25	154,00
	Positive ranges	46[b]	28,93	1331,00
	Draws	14[c]		
	Total	68		

[a]Postest < Pretest
[b]Postest > Pretest
[c]Postest = Pretest

Table 6. Wilconxon level of significance

	Postest - Pretest
Z	$-5{,}124^{b}$
Sig. asymptotic (bilateral)	,000

[b]Postest > Pretest

5 Conclusions

For the design of the learning object with AR, the OARA methodology was applied which allowed to follow an ordered sequence of steps to get to the publication of the resource successfully, programming tools were applied such as Unity and SDK Vuforia due to their robustness and their permanent use without limitation of access, in comparison with the applications previously used with ZAPPAR, LAYAR or ROAR that has an application maximum limit of operation.

Training and induction processes of new educational resources allowed students to be motivated and to be able to rely on accessible elements for the training process, simplifying the presentation of cognitive content in an attractive and playful way, attracting the interest of students in the use of the application and especially when using it as a work tool in the classroom assertively.

The 68 students from the first semesters of the career of the Basic Education, assessed both the design of the learning object with AR and the TAM in a highly satisfactory, moderately satisfactory and satisfactory manner; in most of the questions posed to determine the relevance of the designed educational resource.

Both accessibility and functionality of the resource were relevant, all students and teachers could access from anywhere and anytime to the designed resource by downloading the application from the website, allowing access in an easy and intuitive way, since the designed resource was new and unknown by the students.

With the data statistics obtained it was evident that students have a high degree of interest and motivation for the devices by using mobile as part of the teaching-learning process, the ease of use and usefulness of the content allow them to activate both intention and attitude of use for technological resources oriented to two specific contents, thus complying with what is determined in the TAM model.

Applying a pretest and posttest it was possible to evidence that the learning results were superior after having used the designed resource, the students went through a process of previous training, later they went through a process of experimentation using the object enriched with AR using mobile devices, this allowed the students to strengthen the learning and to evidence better results in their academic performance.

The experimentation developed was highly valued by the entire educational community, as future work, it is intended for teachers to be trained in tool management for OA design with AR.

References

1. Johnson, L., Adams Becker, S., Cummins, M., Estrada, V., Freeman, A., Hall, C.: NMC Informe Horizon 2016 Edición Superior de Educación (2016)
2. Cabero Almenara, J., Fernández Robles, B.: Las tecnologías digitales emergentes entran en la Universidad: RA y RV. RIED. Rev. Iberoam. Educ. a Distancia **21**(2), 119 (2018)
3. Badilla Quesada, M., Sandoval Poveda, A.M.: Realidad aumentada como tecnología aplicada a la educación superior: Una experiencia en desarrollo. Innovaciones Educ. **17**(23), 41–50 (2015)
4. Fabregat Gesa, R.: Combinando la realidad aumentada con las plataformas de e-elearning adaptativas. Enl@ce Rev. Venez. Inf. Tecnol. y Conoc. **9**(2), 69–78 (2012)
5. Cabero-Almenara, J., Llorente-Cejudo, C., Gutiérrez-Castillo, J.J.: Evaluación por y desde los usuarios: objetos de aprendizaje con Realidad aumentada. Rev. Educ. a Distancia **0**(53) (2017)
6. Akçayır, M., Akçayır, G.: Advantages and challenges associated with augmented reality for education: a systematic review of the literature. Educ. Res. Rev. **20**, 1–11 (2017)
7. Cabero Almenara, J., Fernadéz Robles, B., Marín Díaz, V.: Dispositivos móviles y realidad aumentada en el aprendizaje del alumnado universitario **20**(2), 167–185 (2017)
8. Kellems, R.O., Cacciatore, G., Osborne, K.: Using an augmented reality-based teaching strategy to teach mathematics to secondary students with disabilities. Career Dev. Transit. Except. Individ. **42**(4), 253–258 (2019)
9. Gavilanes Lopez, W., Abasolo, M.J., Cuji Chacha, B.R., Aguirre Sailema, L.: Technological acceptance model (TAM) using augmented reality in University Learning Scenarios. In: 2019 14th Iberian Conference on Information Systems and Technologies (CISTI), pp. 1–6 (2019)
10. Cabero Almenara, V.J., Fernández Robles, B., Díaz, M.: Dispositivos móviles y realidad aumentada en el aprendizaje del alumnado universitario. Rev. Iberoam. Educ. a Distancia **20**, 167–185 (2017)
11. Cabero Almenara, J., Vázquez Cano, E., López Meneses, E.: Uso de la Realidad Aumentada como Recurso Didáctico en la Enseñanza Universitaria **11**(1), 25–34 (2018)
12. Barroso, J., Almenara, J.C., Moreno, A.M.: La utilización de objetos de aprendizaje en Realidad Aumentada en la enseñanza de la medicina. Innoeduca. Int. J. Technol. Educ. Innov. **2**(2), 77–83 (2016)
13. Barroso Osuna, J., Cabero-Almenara, J., Ortiz, R.V.: USO EDUCATIVO DE LA RA: EXPERIENCIAS EN ESPAÑA Y MÉXICO. Enseñanza Teach. Rev. Interuniv. Didáctica **36**(2), 7–29 (2018)
14. Márquez Díaz, J.E.: Tecnologías emergentes, reto para la educación Superior Colombiana. Ingeniare **23** (2017)
15. Pedraza, G., Esperanza, C., Plata, A., Flor, O., Cárdenas, L., García, M.: La experiencia de Realidad Aumentada (RA) en la formación del profesorado en la Universidad Nacional Abierta y a Distancia UNAD Colombia. Píxel-Bit. Rev. Medios y Educ. **51**, 111–131 (2017)
16. Cárdenas Ruiz, H.A., Mesa Jiménez, F.Y., Suarez Barón, M.J.: Realidad aumentada (RA) aplicaciones y desafíos para su uso en el aula de clase. Dialnet **35**, 137–148 (2018)
17. Samaniego-Franco, J.B., Agila-Palacios, M.V., Jara-Roa, D.I., Sarango-Lapo, C.P.: Realidad Aumentada como recurso de apoyo en el proceso enseñanza-aprendizaje en Medicina Legal. Rev. Ibérica Sist. e Tecnol. Informação; Lousada **E17**, 972–984 (2018)
18. Sanchez Guerrero, J., Salazar Mera, J., Gavilanes Lopez, W., Sanchez Reinoso, R., Tamami Davila, C.: Use of augmented reality AR in university environments. In: 2018 International Conference on eDemocracy & eGovernment (ICEDEG), pp. 291–297 (2018)

19. Gavilanes López, W.: Percepción de los estudiantes en relación al uso de la realidad aumentada en los procesos pedagógicos. Rev. Publicando 4(11(1)), 388–397 (2017)
20. Davis, F.D.: Perceived usefulness, perceived ease of use, and user acceptance of information technology. MIS Q. **13**(3), 319–340 (1989)
21. Cabero Almenara, J., Pérez Díez de los Ríos, J.L.: Validación del modelo TAM de adopción de la Realidad Aumentada mediante ecuaciones estructurales. Estud. sobre Educ. 34(0), 129–153 (2018)
22. Bárbara, F.: Aplicación del Modelo de Aceptación Tecnológica (TAM) al uso de la realidad aumentada en estudios universitarios de educación primaria. En R. Roig. Tecnol. innovación e Investig. en los procesos enseñanza-aprendizaje. Barcelona Octaedro (3) (2016)
23. Garay, U., Tejada, E., Maíz, I.: Valoración de objetos educativos enriquecidos con realidad aumentada: una experiencia con almnado de máster universitario. Píxel-Bit Rev. Medios y Educ. **50**, 19–31 (2017)

Machine Monitoring Based on Cyberphysical Systems for Industry 4.0

Toapanta Alex[1(✉)], Zea Danny[2], Tasiguano Cristian[1], Vera María[1], and Paspuel Carlos[1]

[1] Instituto Superior Tecnológico Rumiñahui, Sangolquí, Ecuador
alex.toapanta@ister.edu.ec
[2] Escuela Superior Politécnica de Chimborazo, Riobamba, Ecuador

Abstract. This article presents the design, and implementation of an architecture based on cyber-physical systems (CPS), which are mechanisms controlled and/or monitored by computational algorithms and closely integrated with the internet, to supervise variables acquired from a machine with computer numerical control (CNC) and a manipulator robot. This architecture uses low-cost components and is fully implemented on a Raspberry Pi 2 Model B computer. Furthermore, the architecture uses a topology of multiple client-server programmed on Python language for an appropriate global visualization of the variables acquired from the connected machines. A user interface (HMI) was also developed for the real-time visualization of the key machine indicators selected according to the user criteria. Moreover, the architecture was tested in an industrial part machining process and the stability of the data acquisition system was validated based on the machine condition information. Therefore, the architecture has sufficient robustness and capability to be used in other manufacturing environments. Finally, the presented system also becomes a useful tool in the decision-making process since it uses continuously updated information, statistics and trend history of each machine that composes the CPS.

Keywords: Condition monitoring · Knowledge modeling and knowledge-based systems · Intelligent manufacturing systems

1 Introduction

Industrial processes involve different elements for the start-up of manufacturing, being necessary to have all the information that allows determining behavioral patterns, to develop techniques for making decisions accurately, through continuous monitoring and in real-time of the process variables [1]. Thus emerging, the need to implement new technologies, which will transform the overall production of the formation of fully integrated, automated and optimized processes.

For this, a tool is required that allows a total interconnection between the machines of the industry to be able to acquire, store, visualize and manage the relevant information of any manufacturing process through complete digitization of it, known as Industry 4.0. This concept is associated with the fourth industrial revolution, which is considered a new level of digitization, organization, and control of manufacturing

M. Botto-Tobar et al. (Eds.): CI3 2020, AISC 1277, pp. 117–127, 2021.
https://doi.org/10.1007/978-3-030-60467-7_10

methods to obtain reconfigurable, reusable, as well as autonomous and intelligent manufacturing systems [2]. Industry 4.0 is based on multiple technologies (IoT, CPS, Big Data, Collaborative Robotics, cloud computing, additive manufacturing, among others), some of these already implemented and others in the process of development, such as are the so-called Cyber-Physical Systems (CPS) [3].

Cyber-Physical Systems (CPS) constitute physical and engineering systems with the ability to interact with the physical world, in which operations are fully integrated, supervised and controlled by a computational nucleus [4]. In this way, what is sought through the implementation of a CPS is to provide physical components or objects (equipment, machinery, objects, etc.) with computing, communication and storage capabilities (cybernetic system), to convert them into intelligent objects, being able in this way to work together forming distributed and autonomous ecosystems [5]. Within this framework, CPS can relate to physical objects to monitor, store and control information available of any manufacturing process to promote learning and evolution capacities [6]. CPS-based monitoring systems are essential to generate the best and fastest results in terms of the use of resources, machines, profitability, and safety of manufacturing processes, that is, achieving a productive unit offering the possibility to understand in a detailed way the behavior and the dynamics of the processes [7].

In particular, the manufacturing industry is implementing CPS-based condition monitoring (CM) systems, since they allow identifying the fluctuations and significant variations of the variables and signals provided by the sensors, with the intention of detecting, preventing and correcting possible failures or breakdowns [8]. In industrial processes, the most widely reported CM applications focus on: vibration analysis and diagnosis, force analysis, axis advance velocity analysis, lubricant analysis, acoustic emission analysis; infrared thermography, ultrasound tests, analysis of the state of the motor and consumption, among others [9].

In this document, a multiplatform state monitoring (CM) architecture is proposed, using cyber-physical systems (CPS), focused on the variables of forwarding speed, axis positions, vibrations and electrical parameters of a CNC machine tool and a manipulator robot. Likewise, generating a human-machine interface that allows not only to visualize the relevant variables of the manufacturing process but also shows a historical graph of the behavior of variables selected by the operator and the visualization of statistical data, management of events and alarms, storage and generation of reports at the factory level.

The article is organized into five sections, as follows: Sect. 2 provides the details of the system architecture; Sect. 3 explains the development of the human-machine interface; Sect. 4 shows and discusses the results obtained. Finally, in Sect. 5, the conclusions of this work are indicated.

2 CPS System Architecture Implemented

The scheme of the implemented CPS architecture is illustrated in Fig. 1. The CPS architecture of the system is distributed in two fundamental nodes that are in charge of establishing the continuous flow of information between the present levels. The nodes contain the communication protocols necessary to interact with the different physical devices to have a system that is compatible with any type of machine, breaking the paradigm that each machine has its communication protocol [10, 11]. On the one hand, there is the node located at the field level, which is responsible for acquiring the information of the PLC of the machine tool CNC and the manipulator robot, as well as sensors signals deployed in each machine (acoustic emissions, vibrations, etc.). The primary function of this node is to guarantee communication and data flow between the field level made up of sensors, actuators and internal variables of the machines and the control level made up of the computer in charge of the data capture for further processing and storage. Data sent by the different machines are received on the computer via TCP-IP communication. On the other hand, we have the node compound by an embedded computer Raspberry Pi 3, which constitutes the central node in charge of receipt information of each node of the field level (CNC and Robot Manipulator machine), and it constitutes an intermediate communication node between the physical part and subsequently to the Internet. Internally, these communication nodes are composed of algorithms, both for extracting information and for its subsequent treatment and storage, achieving the visualization of the data by the operator locally, and sending information to the cloud.

In general, the central node serves for the processing and sending of process information emitted by the local nodes to the cloud, allowing the monitoring and supervision of all events and alarms that occur at the plant level. In the first instance, if an error or failure (bad configuration, alarms, etc.) occurs in any of the local nodes, an exchange of messages is generated, alerting of the event produced and immediately notifying the user about recommendations or immediate actions that must be executed. Secondly, the

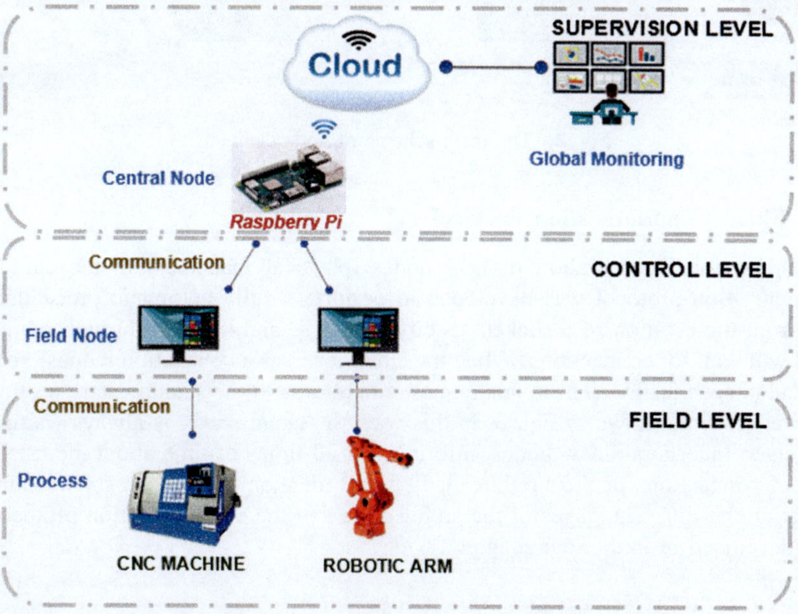

Fig. 1. CPS architecture implemented.

system offers the possibility of storing those events or alarms in order to apply algorithms based on machine learning (neural networks, artificial networks, grouping, etc.), to implement control systems with self-monitoring capabilities and allow estimating or predicting possible damage to machine components, improving the planning of preventive maintenance, reducing downtime due to production stops [12, 13].

The connectivity between the nodes present in the CPS architecture is controlled by the users, that is, the user decides whether to monitor all the machines at once or just one by one [14].

In Fig. 2, the most important elements of the node are shown at each level of the process. The field node contains the configuration of all communication protocols needed to interact with different physical devices, as well as the data processing process for sending the information to the next level. On the other hand, the central node is in charge of collecting all the information sent by the field nodes to subsequently apply information processing algorithms (statistics, fast Fourier transform, etc.), in order to obtain the most important characteristics that serve to model and estimate the current state of the process.

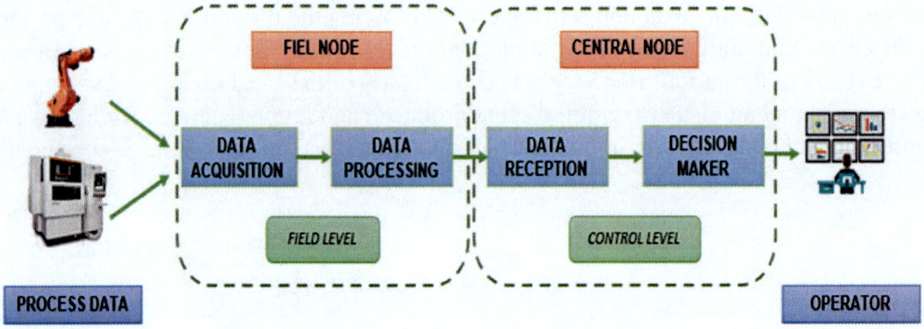

Fig. 2. The main scheme of node elements.

2.1 Cloud Communication Protocol

To achieve the incorporation of new nodes (physical machines) in the cloud, the communication protocol was developed to perform a fully automatic procedure that goes from the creation of a socket, specifying the IP and the port through which the cloud will wait for connections. When the cloud receives a connection request, redirect a thread to specifically address this request and returns to the standby state waiting for more requests, as shown in Fig. 3. In this way, the cloud server is always waiting for requests to incorporate new nodes, informing at all times of time about the establishment of connections or their failure. It is worth mentioning that the software implemented in the different stages of the architecture and the communication protocol has been developed in Python language.

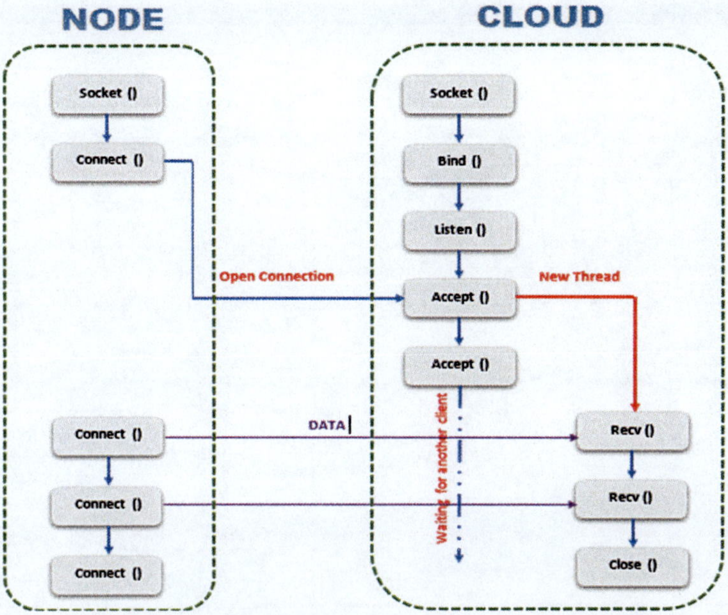

Fig. 3. Communication protocol operation.

The machine parameters that can be monitored are the temperature, the speed of advance of the axes, the axes position, vibrations, acoustic emissions and electrical parameters of each motor. These representative variables are obtained from multiple local sensors integrated into each machine, as well as from the information provided by the machine itself.

3 Human – Machine Interface

The information acquired from the physical processes through the implemented architecture is available in real-time, both for visualization and for calculating values that establish the dispersion percentage of selected variables. Also, the operation of different process state indicators and trends of the main variables that have been chosen by the operator can be visualized through the implemented human-machine interface. These modeling elements are used to estimate the current behavior of the machine or machines.

The developed interface can also generate and store reports of the values of the variables involved. The stored data can be exported to standard file formats (TXT, HML), is freely accessible to any user or operator of the machine at the factory level. In Fig. 4, the implemented human-machine interface is presented. This interface is multiplatform, which means that it can be applied to any type of industrial machine.

The historical trends graphics show the behavior in real-time of the variable selected by the operator (axis advance speed, axis positions, engine power, vibrations

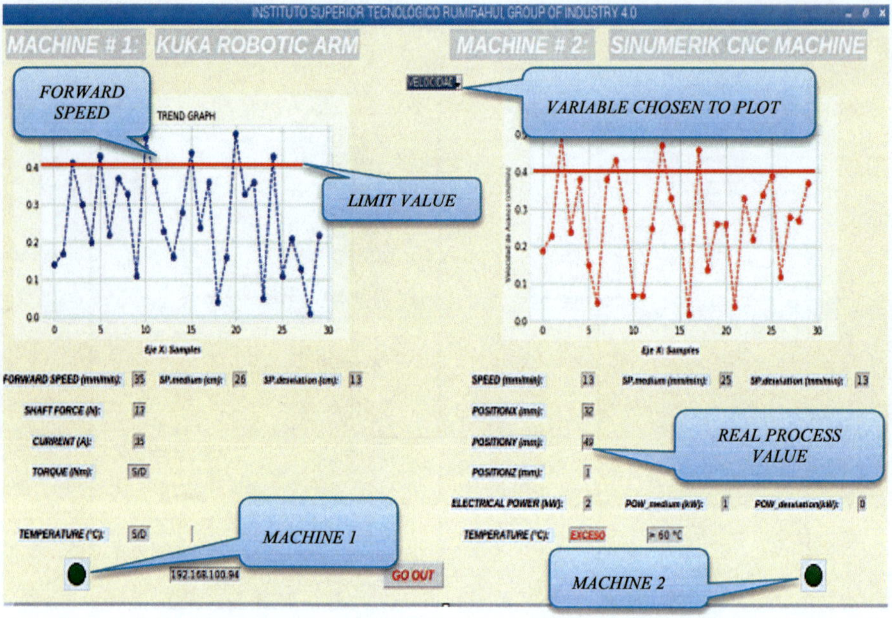

Fig. 4. Visual - analytical interface for variable tracking.

and strength on the axis), considering the limit value presented by the physical variable of the process. All points above the limit value are considered states critical presented in the manufacturing process for different reasons (hard materials, complex designs, bad condition at the tips of the shaft, machine motors overload, excessive vibration, noise extreme, etc.). These states are stored and subsequently evaluated to implement future control algorithms to predict damage and generate maintenance events autonomously. Besides, variables indicators allow you to visualize the actual values of the variables under the monitoring process. These indicators consist of two columns, which show the name of the variables with the respective units and the current value, being permanently updated according to the parameter assigned time or sampling frequency, also managing the possible alarms that may occur in the process development. These changes are represented by a color change of the displayed variable setting the black to "normal" and red to indicate "alarm". In case of a change in the variable value that exceeds the limit preset, returning to black when the value of the variable is below the limit, indicating that it is operating normally.

Monitoring and analysis of the process are equally evident, through reports anomalies generated and stored in a TXT file type, which registers the IP address of the machine, the type of event occurred and the moment (day-hour) that it happened. Additionally, a report is generated when the equipment overheats, which consists of the event produced, with the respective identifier and moment (hour), as can be seen in Fig. 5.

It is important to note that two records are produced, the first when the equipment experiences overheating and the second when it returns to normal temperature, in order to know the exact time that the machine was operating outside of normal values, information vital to predict fall will in the future.

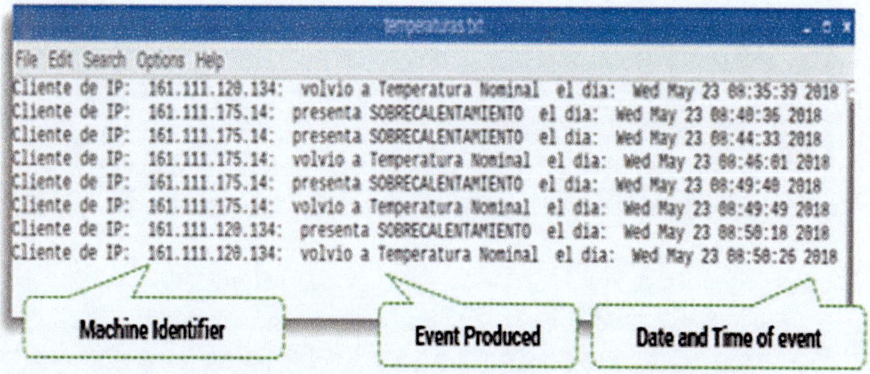

Fig. 5. Report of events generated in a part machining process.

4 Analysis of Results

In this section, the visual analysis of the result of the representative variables related to the state of the machine is presented to show the behavior of the CNC machine and the manipulator Robot in a process of machining parts used in the automotive industry (transmission components). From the experimental data obtained and stored from the machining process, statistical calculations are carried out, such as the standard deviation of a certain variable, which will serve as a guide to establish the behavior of these variables in the machining process.

In Fig. 6, the behavior of the axis feed speed variable is represented, for which 3000 data from the machining process were used in this case. As one can see, certain monitored values are outside the permissible range established by the machine manufacturer, which means that the machine works in unusual conditions in certain periods. This is mainly due to the hardness of the material with the one that is working, as well as the final form that needs to be obtained in the material. Through these values, future damage to the machine can be predicted.

In Fig. 7, the standard deviation of the data obtained from the variable speed of advance of the axis is shown, using which it is possible to estimate the wear of the components involved in the manufacturing process. Similarly, it can be seen that the peaks that are in the upper part of the limit line correspond to values outside the range established by the manufacturer. Peaks that correspond to the excess speed of advance of the axis, due to the presence of soft parts in the material used in the machining process. It is worth mentioning that Fig. 6 and Fig. 7 corresponds to the statistical analysis of the variable forward speed of the axis.

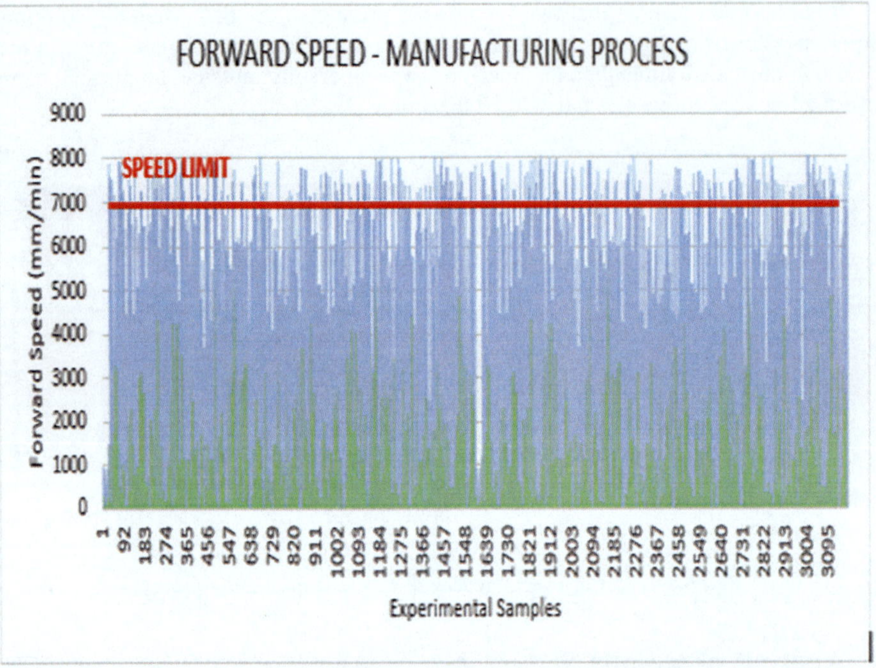

Fig. 6. Visual - analytical interface for monitoring the axis advance speed variable.

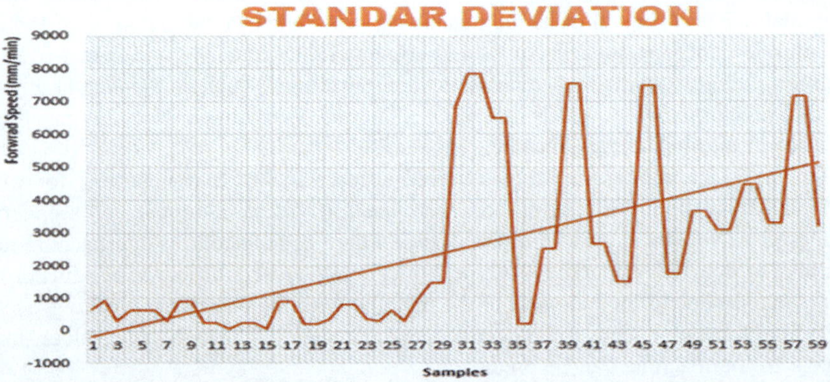

Fig. 7. Statistics of the behavior of variables in a machining process.

Similarly, the experimental analysis of the monitoring result (500 data, 2 min) of the variable acoustic emission in the shaft bearings are presented, applied to the same process of machining parts for the automotive industry. Measurement of the acoustic emission generated by the bearings is an effective way to detect the state of degradation of the bearings, even at low speeds. In Fig. 8, the waveform of the acoustic signal emitted by a bearing without any defect is shown, in which the normal operation of the bearing can be seen without the presence of any type of transients, remaining practically constant in a certain value.

However, in Fig. 9, the waveform emitted by a defective bearing is shown. The presence of transients (peaks of short duration) can be observed. These transients correspond to wear on the bearings, which can be caused by the applied load, poor lubrication, oxidation, excess speed, among others. Such variations in both amplitude and time will be used later to know the real nature of the defect in the bearing and to be able to predict the remaining lifetime of the bearing.

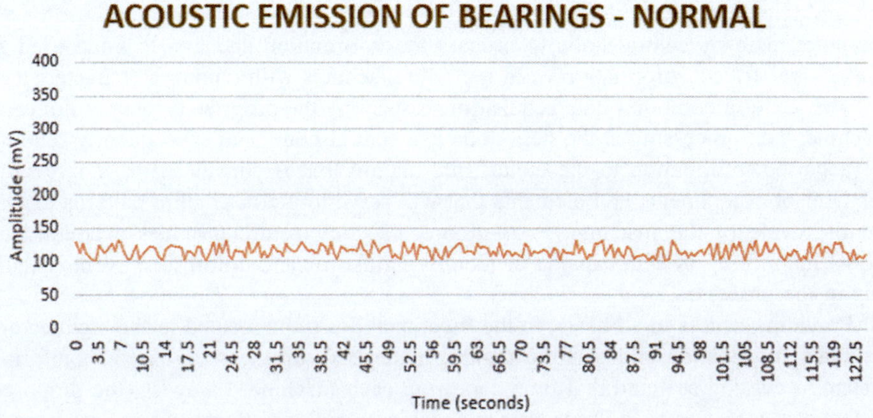

Fig. 8. Acoustic signal waveform - bearing in good condition.

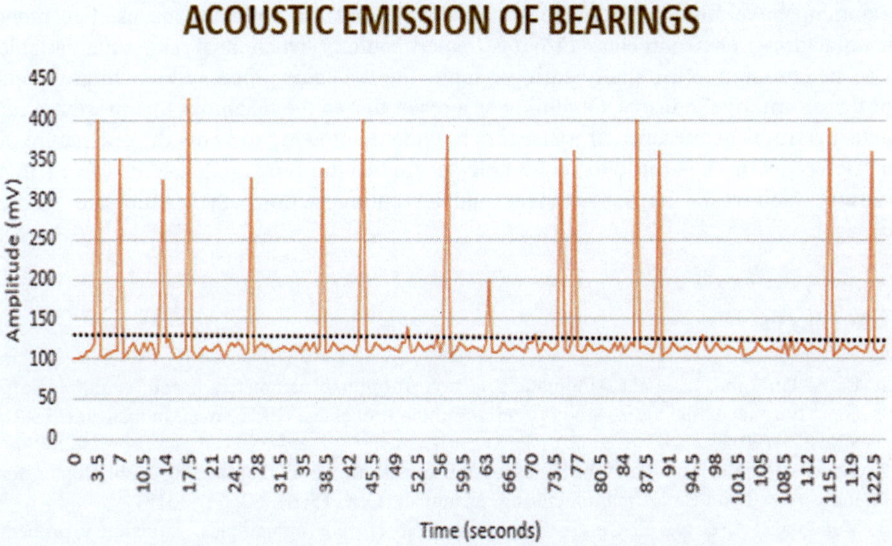

Fig. 9. Acoustic signal waveform - bearing in bad condition.

Obtaining and graphical representation of the different variables of the machining process serves as the basis for designing and developing predictive control models that anticipate mechanical failures, extend the remaining useful life of manufacturing systems or parts, facilitating their replacement, and take actions that anticipate future damage.

5 Conclusions

This document presents the general design and implementation of a monitoring architecture using cyber-physical systems to know the status of a CNC machine tool and a Manipulating Robot. The proposal integrates a solution that includes a low-cost computer, using a central node to process the information and a field node used to interact and collect information from multiple machines with common characteristics.

The solution combines data acquisition, respecting the proprietary protocol of each machine, data processing at the field node and central node, and subsequently sending the acquired information to the cloud; Information that is considered to monitor the behavior of components and elements that will serve to predict failure patterns during the life cycle of the machine. Furthermore, the architecture provides a continuous global monitoring system capable of identifying alarms and unforeseen events occurring in the processes.

Providing access to all users, to the functionalities implemented in the architecture to obtain information and perform control, if desired, on the state-based monitoring system, accessing real and real-time data from each machine. Likewise, the proposed system was subjected to multiple tests in a real industrial manufacturing environment in the GAMHE group laboratory, certifying correct operation in complex machining processes. These results demonstrate that both the design, implementation, and programming have sufficient robustness, efficiency, and the ability to be used in other manufacturing environments. Proving, experimentally when analyzing the variable acoustic emission in the shaft bearings, applied to the same process of machining parts for the automotive industry. Obtaining as a result that in the machines that present wear in the bearings the presence of transistors is evident, allowing to know the real nature of the defect in the bearing and to be able to predict the remaining useful life of that bearing, facilitating its replacement and executing actions that anticipate future damages.

References

1. La Fé-Perdomo, I., et al.: Automatic selection of optimal parameters based on simple soft-computing methods: a case study of micromilling processes. IEEE Trans. Industr. Inf. **15**(2), 800–811 (2019)
2. Ahuett, H., Kurfess, T.: A brief discussion on the trends of habilitating technologies for Industry 4.0 and Smart manufacturing. Manufact. Lett. **15**(B), 60–63 (2018)
3. Villalonga, J., et al.: El Control de sistemas ciberfísicos industriales. Revisión y primera aproximación. En: XXXIX Jornadas de Automática **1**, 1–6 (2018)

4. Geissbauer, R., Vedso, J., Schrauf, S.: Industry 4.0: building the digital enterprise. Glob. Ind. 4.0 Surv. **83**, 1–8 (2016)
5. Lee, E., Seshia, S.: Introduction to Embedded Systems: A Cyber-Physical Systems Approach, 2nd edn. MIT Press, Cambridge (2016)
6. O'Donovan, P., et al.: A fog computing industrial cyber-physical system for embedded low-latency machine learning Industry 4.0 applications. Manuf. Lett. **15**(B), 139–142 (2018)
7. Mittal, S., et al.: Smart manufacturing: characteristics, technologies and enabling factors'. Proc. Inst. Mech. Eng. Part B J. Eng. Manuf. **233**(5), 1342–1361 (2019)
8. Liu, C., Jiang, P.: A cyber-physical system architecture in shop floor for intelligent manufacturing. Procedia CIRP **56**, 372–377 (2016)
9. Mourtzis, D., et al.: A cloud-based approach for maintenance of machine tools and equipment based on shop-floor monitoring. Procedia CIRP **41**, 655–660 (2016)
10. Villalonga, A., et al.: Condition-based monitoring architecture for CNC machine tools based on global knowledge. IFAC-PapersOnLine **51**(11), 200–204 (2018)
11. Lima, J., Massino, S.: Manual de Detección de Fallas de una Línea Piloto de Producción de Quesos basado en Conocimiento Experto. Información tecnológica **19**(3), 65–74 (2008)
12. Larrinaga, F., et al.: Analysis of technological architectures for the new paradigm of the industry 4.0. DYNA **94**(3), 267–271 (2019)
13. Villalonga, A., et al.: Monitorización inteligente del estado de rodamientos basada en técnicas de aprendizaje automático. En XL Jornadas de Automática, 234–241 (2019)
14. Ali, Y., Rahman, R., Hamzah, R.: Acoustic emission signal analysis and artificial intelligence techniques in machine condition monitoring and fault diagnosis: a review. Jurnal Teknologi **69**(2), 121–126 (2014)

E-learning, E-Government and E-business

Collaborative Work in the Development of Assessments on a Moodle Learning Platform with ExamView

Roberto López-Chila(✉), Joe Llerena-Izquierdo, and Nicolas Sumba-Nacipucha

Universidad Politécnica Salesiana, Guayaquil, Ecuador
{rlopezch, jllerena, nsumba}@ups.edu.ec

Abstract. This document presents the research work focused on the development of learning evaluations by teachers at the Salesian Polytechnic University (Guayaquil, Ecuador), in the Centenario campus, in the traditional way and with the use of the ExamView program. It is also identified that the collaborative work in the realization of questions can be shared in a feasible way by using a format known by the user, with or without expertise in the use of professional programs or virtual environments. VLE platforms, although they have modules for editing questions for evaluations through questionnaires, the interaction interfaces lack of familiarity when compared to a common text editor. This work shows that when using the ExamView program, teachers improve their effectiveness in the development of evaluations, improve the technological skills of the teacher when preparing the evaluation and sharing it with their colleagues, motivate the frequent development of them about that modality and allow adequate conditions for execution in the VLE.

Keywords: Technology enhanced learning · Moodle-based E-learning · B-learning

1 Introduction

The inclusion of information technologies in the educational field, allows the use of evaluative models to measure student learning [1], especially in universities, educational platforms such as virtual learning environments (VLE) allow the use of secure formats for the elaboration of the evaluations [2], in addition the traditional resource that is the physical document is still used. Existing platforms [3, 4], present their own environment for creating and editing evaluations through interfaces designed from their best development interpretation but not friendly compared to a professional text editor.

Different authors present learning evaluation methodologies with the use of tools, [5–7], and techniques that are coupled on VLE, [8–10] and supported with resources Digital and/or traditional resources innovate processes to achieve learning achievements in the form of indicators.

In the Salesian Polytechnic University of Ecuador, formed by three offices in three cities of the country, in the city of Cuenca, the headquarters are located, in the city of

M. Botto-Tobar et al. (Eds.): CI3 2020, AISC 1277, pp. 131–141, 2021.
https://doi.org/10.1007/978-3-030-60467-7_11

Quito and Guayaquil the two-remaining headquarters. It has its VLE on the Moodle platform in Amazon Web Services (AWS) that provides its access services, to 25,000 students in the face to face classes and distance modalities, to the year 2019.

The number of teachers has increased due to the increase in students at different locations, around 1000 teachers. With this, the number of physical resources has also increased, looking for new efficient ways of using paper at the time of evaluations. This work is carried out at the headquarters of the city of Guayaquil where there are currently two campuses, "Centenario" and "María Auxiliadora", each with 7500 and 800 students respectively and 200 professors between full time and part time. The evaluations in an academic period, which consists of two partials each, are two; with paper exams under a specific format. In the same way, during the learning process, evaluation activities are carried out, at least two in each part, giving a total of four (or more, according to teacher planning or of the teaching cooperative faculty) throughout the period.

Existing Traditional and Digital Evaluation Resources
This document presents the research work focused on the development of learning evaluations by teachers at the Salesian Polytechnic University, Guayaquil headquarters, in the Centenario campus, in the traditional way and with the use of the ExamView program. It is also identified that the collaborative work in the realization of questions can be shared in a feasible way by using a format known by the user, with or without expertise in the use of professional programs or virtual environments. VLE platforms, although they have modules for editing questions for evaluations through questionnaires, the interaction interfaces lack of familiarity when compared to a common text editor. This work shows that when using the ExamView program, teachers improve their effectiveness in the development of evaluations, improve the technological skills of the teacher when preparing the evaluation and sharing it with their colleagues, motivate the frequent development of them about that modality and allow adequate conditions for execution in the VLE (Fig. 1).

Fig. 1. Comparative evaluation models, a) use of physical document, b) use of VLE

In the academic environment there are different versions of the professional word processor, with this, the familiarity of using the same editor allows recognize tools that allow the preparation of the final document. The versions of the professional editor are not a cause of impediment in the creation of the evaluation, but the teachers have used at least some existing version (see Fig. 2).

Fig. 2. Use of the text editor for the development of the evaluation in physics, traditional way by teachers in the preparation of the document

The university allows teachers in face-to-face and distance mode, to continuously access the VLE without restrictions, to develop their teaching-learning activities with the available environment (Moodle) and minimum tools in it for the preparation of the material required in their activities, and its proper configuration [11, 12].

The VLE, have their own tools for the generation of evaluations, for experienced users or trained in moodle, also allows the incorporation of other developed programs [13], in Table 1 the different file formats with textual content for the import of items or questions in the VLE, it is clear that it is understood that the same platform must be connected, accessed by the teacher, for the realization in mentioned environment. The existing formats for moodle, have a format with knowledge of the markup language with eXtensible Markup Language (xml) tags. ExamView allows a working environment on Word for windows.

Table 1. Import formats accepted on the Moodle platform.

Formats	Description	File format
Blackboard V6+	Blackboard format	Compress: Zip
ExamView	ExamView para MsWord format	Compress: Zip
Aiken	Simple text format	Text File
Absent word of	Simple text format	Text File
GIFT	Markup format	Text File
Moodle XML	Markup format	Text File
WebCT	Markup format	Text File
Incrusted answers (Cloze)	Markup format	Text File

A teacher elaborates on the program ExamView test generator, a bank of questions to generate the document in rich format (similar to a word document for windows) that allows its proper printing in case of physically requiring it or exporting it in a compressed format for import into the VLE.

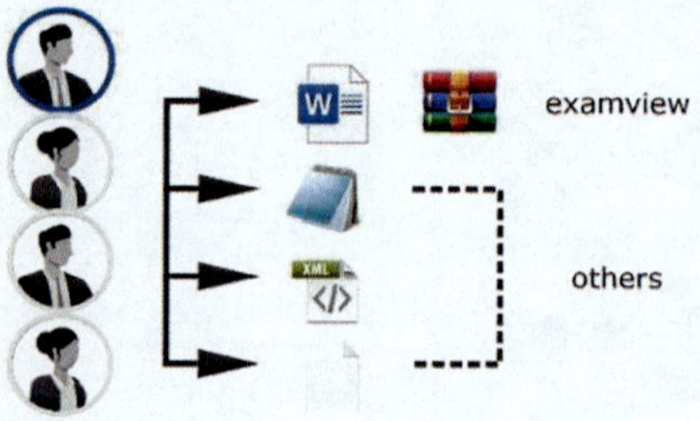

Fig. 3. ExamView format on Microsoft word and its export in compressed format

While the other formats require that the teacher knows the structure of marks similar to the hypertext language (html) or special characters as well as appropriate separations, spaces or line breaks. The program used in this work has a format like the professional text editor commonly used by most university teachers (see Fig. 3).

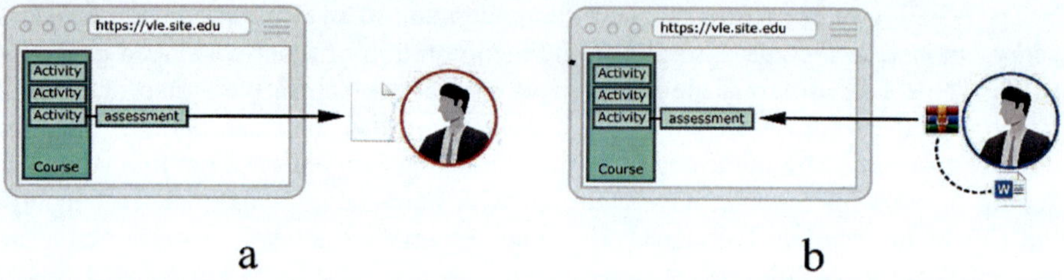

Fig. 4. a) Use of the platform for the preparation of the evaluation, b) use of the text editor for the preparation of the evaluation and upload of the export file to the VLE

The use of the VLE environment limits the teacher's freedom of work, having to access the platform with its authentication and with the use of existing tools (see Fig. 4a), which lack of familiarity by users, they require more time in the design and construction of the evaluation. The program used for research allows teachers to prepare evaluations of a workspace similar to the text editor and provides a compressed file with the same information in valid format for the VLE (see Fig. 4b).

2 Methods and Materials

A hypothetical-deductive methodology with a mixed approach is developed, a bibliographic investigation is initially carried out that uses the documentary technique to gather information on the use and benefits of technological tools for the elaboration.

Preparation of questionnaires. Then two data collection techniques are performed to establish inferences, in the first one the field technique is used through student surveys, to collect the experiences when conducting evaluations with the use of the virtual learning platform and without it, in the second one uses interviews with expert teachers, to gather the experiences when developing the exercise bank for an evaluation in the learning platform. Finally, the hypothesis is verified by inductive procedures from the data obtained, the conclusions of this work are defined by inferences proven experimentally by a deductive process.

The survey technique is applied to six courses, of the first cycles, of the engineering degrees of the Salesian Polytechnic University (Guayaquil, Ecuador), campus Centenario campus, each course with 40 students, a total population of 240 students. With a margin of error of 5% and a confidence level of 99%, giving the number of participating students of 177. The Google forms tool is used for data collection. Finally, the answers are tabulated according to three criteria according to the experience in the evaluation models (see Fig. 5). These criteria are effectiveness in the development of the evaluation, motivation for development and conditions for execution. Three teachers are experts in the use of the ExamView program, from the six courses.

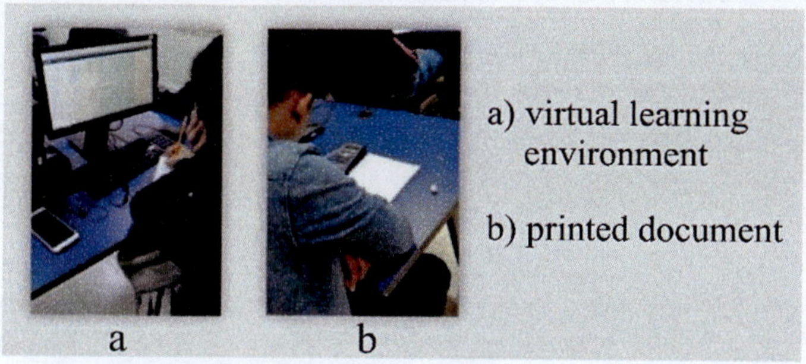

Fig. 5. Evaluation models applicable at the Salesian Polytechnic University (Guayaquil, Ecuador), a) using the VLE and b) using the physical document.

3 Results

Thirty characteristics were analyzed (see Fig. 6) gathered into three comparative criteria between one model and the other (see Fig. 5). The criteria of efficacy, motivation and conditions for the use of one model or the other were chosen from the interviews with the experts to classify the characteristics of each model.

In the criterion of effectiveness for model A, it is evaluated whether it allows the teacher to assess adequately the knowledge of students, if it allows the student to clarify what they have learned in a specific subject, if it demands to the teacher a clear wording in the form, if it requires to value the complexity of the statement vs. the number of questions, and if model A gives relevance to the way in that the question is presented (see Fig. 7).

		1	2	3	4	5
	EFFECTIVENESS	Totally in disagreement	Somewhat disagree	Neither agree nor disagree	Somewhat agree	Totally in agreement
	Model A					
1	It allows the teacher to properly evaluate the knowledge in the student.	3%	3%	12%	25%	56%
2	It allows the student to clarify what they learned in a specific topic.	1%	5%	14%	27%	54%
3	It requires the teacher a clear wording in the way of asking.	1%	2%	10%	34%	53%
4	It requires assessing the complexity of the statement vs. the number of questions.	1%	4%	15%	33%	47%
5	It gives relevance to the way in which the question is presented.	1%	2%	16%	30%	51%
	Model B					
6	It allows the teacher to properly evaluate the knowledge in the student.	5%	6%	16%	23%	51%
7	It allows the student to clarify what they learned in a specific topic.	4%	6%	21%	28%	41%
8	It facilitates the student to clearly understand the wording of the question.	3%	8%	20%	31%	38%
9	The complexity of the statement is adjusted vs. the number of questions.	3%	9%	23%	27%	38%
10	Note the relevance to the way in which the question is presented.	3%	8%	20%	27%	41%
	MOTIVATION	Totally in disagreement	Somewhat disagree	Neither agree nor disagree	Somewhat agree	Totally in agreement
	Model A					
11	It allows the teacher to shorten the delivery time of the grade.	3%	3%	6%	22%	65%
12	It allows the student to obtain their result in real time.	1%	3%	6%	19%	72%
13	The teacher becomes familiar with the construction of the questions with an easy-to-use tool	2%	2%	10%	31%	55%
14	establishes interest in conducting the evaluation.	2%	3%	9%	31%	55%
15	exercises the autonomy of the student to carry out the evaluation.	2%	5%	12%	25%	55%
	Model B					
16	It allows the teacher to shorten the delivery time of the grade.	11%	15%	18%	27%	29%
17	It allows the student to obtain their result in real time.	14%	19%	21%	18%	29%
18	It gives the teacher the possibility to carry out several evaluation models.	5%	11%	26%	25%	32%
19	establishes interest in conducting the evaluation.	5%	8%	29%	27%	31%
20	The teacher uses a high cost for its elaboration.	6%	10%	25%	27%	33%
	TERMS	Totally in disagreement	Somewhat disagree	Neither agree nor disagree	Somewhat agree	Totally in agreement
	Model A					
21	It allows the teacher to use less physical resources (paper, reproductions, etc.).	5%	2%	6%	12%	75%
22	It allows the student to use fewer resources (purchase sheets, etc.).	4%	1%	4%	23%	68%
23	It requires the teacher to know tools for the elaboration of the content of the question.	2%	2%	8%	23%	65%
24	It requires the expertise of the teacher when calculating the development time of the	2%	3%	8%	31%	56%
25	It facilitates to the teacher the availability of a team for the elaboration	2%	2%	9%	27%	60%
	Model B					
26	It allows the teacher different skills in the elaboration of questions that model A.	2%	7%	24%	31%	37%
27	It allows the student a different perception to the question than model A.	2%	6%	26%	33%	33%
28	It facilitates the student to understand the question in model A.	4%	6%	27%	29%	34%
29	It reflects the adequate time required by the student to complete the question.	4%	8%	24%	32%	32%
30	It provides adequate space for the student.	4%	7%	21%	31%	37%

Fig. 6. Characteristics evaluated by the participants

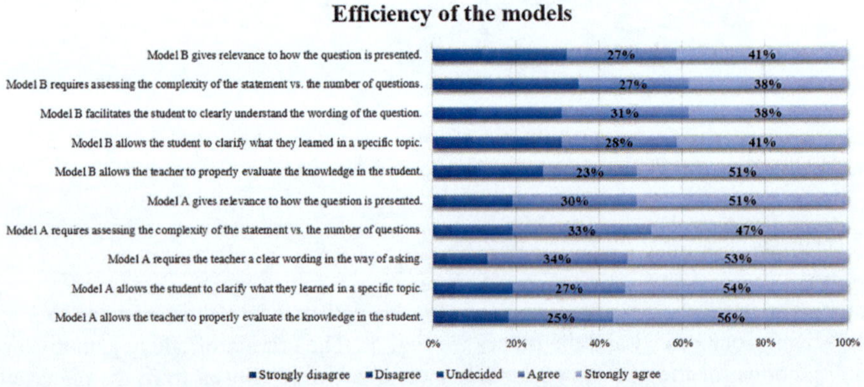

Fig. 7. Relevant percentage of the effectiveness of the models

For model B, it is evaluated, if it allows the teacher to properly evaluate the knowledge in the student, if it allows the student to clarify what they have learned in a specific topic, if it facilitates the student to clearly understand the wording of the question, if adjust the complexity of the statement vs. number of questions and, if in model B, gives relevance in how the question is presented.

In the motivation criterion for model A, it is evaluated whether it allows the teacher to shorten the delivery time of the score, if it allows the student to obtain their result in real-time, if the teacher becomes familiar with the construction of the questions with a easy-to-use tool (ExamView), if establishes interest in the realization of the evaluation, and if model A, exercises the student autonomy to carry out the evaluation (see Fig. 8).

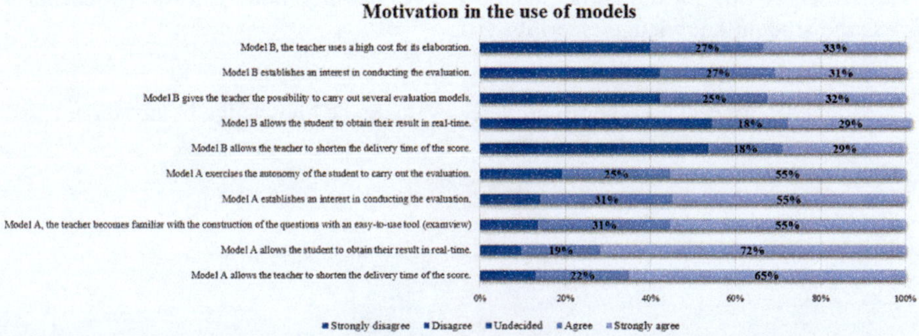

Fig. 8. Relevant percentage of motivation in the use of the models

For model B, it is evaluated, if it allows the teacher to shorten the delivery time of the qualification, it allows the student to obtain its result in real time, if it gives the teacher the possibility to perform several evaluation models, if interest is established in carrying out the evaluation, and if in model B, the teacher uses a high cost for its elaboration.

In the criteria of conditions for model A, it is evaluated if it allows the teacher to use less physical resources (paper, reproductions, etc.), if it allows the student to use fewer resources (to acquire sheets, etc.), if it requires the teacher to know the tools for the elaboration of the content of the question, if it requires the expertise of the teacher when calculating the time of development of the question, and if model A, facilitates the teacher the availability of a computer equipment for the preparation of the questionnaire (see Fig. 9).

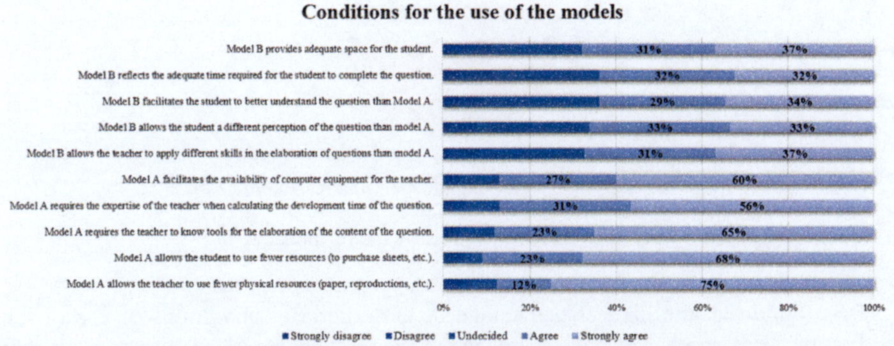

Fig. 9. Relevant percentage of the conditions for the use of the models

For model B, it is evaluated, if it allows the teacher to apply different skills in the elaboration of questions than model A, if it allows the student a perception different from the question that model A, if it facilitates the student a better understanding of the question than model A, if it reflects the adequate time required by the student for the realization of the question, and if model B, provides adequate space for the student.

It is shown that Model A has that the highest percentage in the criteria of effectiveness, where 56% of the participants perceive that this model allows the teacher to assess the student knowledge (see Fig. 10).

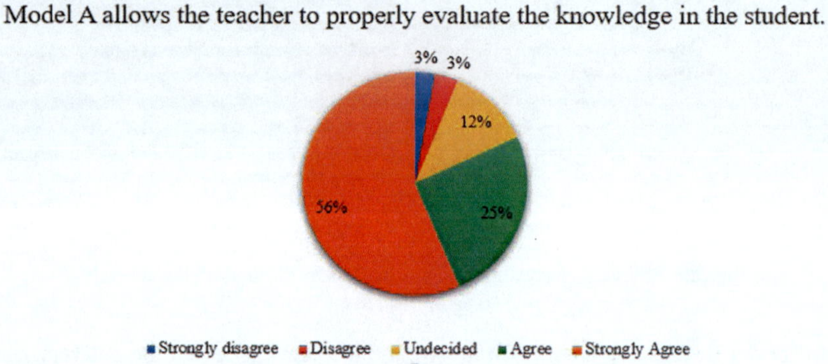

Fig. 10. Efficiency percentage of model A

Model A has the highest percentage in the motivation criteria of the models, participants with 72% perceive that said model allows the student to obtain their evaluation result in a short and real-time (see Fig. 11).

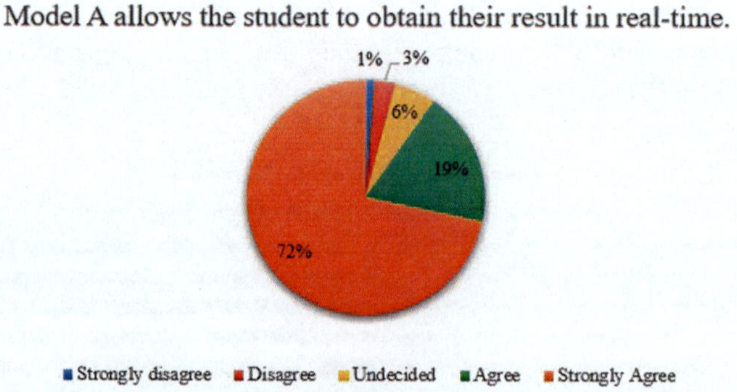

Fig. 11. Motivation percentage of model A

It is found that the highest percentage in the criteria of conditions of use of the models, model A reaches a 75% acceptance of the participants in the characteristic that this model allows the teacher to use less physical resources such as paper use, reproductions, staples etc., and especially the time required to do it (see Fig. 12).

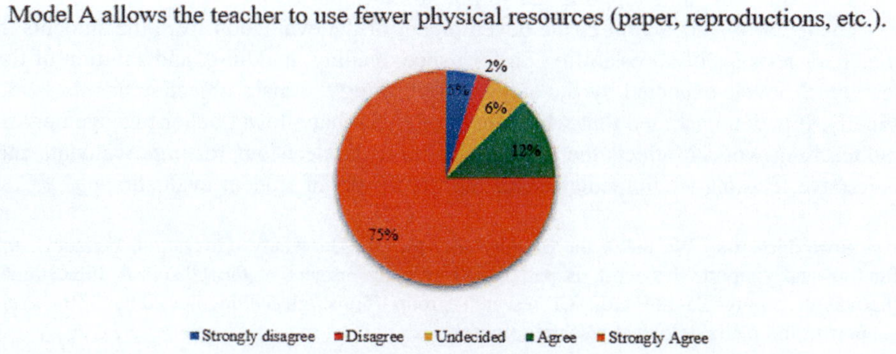

Fig. 12. Percentage of conditions of model A

4 Discussion

It is indicated in this work, that when there are different applications and text editors for the realization of a physical document, the performance, and benefits of a specific program strengthens the management by the teacher, with or without technical skills in the management and expertise of VLEs or of said text editor programs to achieve the elaboration of the evaluation activity, without making use of the paper and in a collaborative way, reducing the time in the elaboration and execution of the same, which is the proposed objective.

5 Conclusion

A collaborative work proposal is presented that demonstrates the effective use of the application ExamView in the development of learning evaluations at the Salesiana Polytechnic University in Guayaquil, Ecuador.

It is evident that the application allows the realization of evaluation structures on the text editor commonly used by teachers and generates a valid export file for import into the VLE, with this the teacher has two formats for application, through the Traditional form printed on paper, which is not required, and digitally in the VLE for proper configuration and automatic execution.

It is concluded that the lack of preparation in the use of technologies affects the teaching work in the preparation of evaluation tests for students, the traditional use of a test affects the time of teaching work and to some extent the indicators of achievement in student learning. The information on these tests remains on paper and must be tabulated manually, causing loss of time and mistakes as it is not registered in a repository, automatically or immediately after the execution of the evaluation.

The use of the evaluation development tool, ExamView, considerably reduces the preparation time of the evaluation in two ways: (1) a test structure for physical reproduction is generated, if required and (2) the evaluation qualification it is generated thanks to the VLE that supports the format to be imported.

The level of satisfaction in the development of the evaluation from the students in the characteristics of accessibility, performance, reading, handling, and solution of the test, reach levels expected by the authors confirming the main objective of this work. Finally, it is demonstrated that when applying tools that allow a better management of the teaching work, it affects the time it takes to make decisions for improvements and corrective learning by immediately having the results of student evaluations.

Acknowledgments. We thank the Salesian Polytechnic University (Guayaquil, Ecuador), for funding and support, this work is part of the research project of the GITICEA Educational Innovation group. To the GIEACI research group (https://gieaci.blog.ups.edu.ec/) for their support in the methodological research process.

References

1. Triyana, I.G.N., Sri Ratmini, N.K., Mandra, I.W., Ruscitadewi, N.W., Wira Adi Armaeni, K., Etika Adnyani, N.P.: The use of moodle-based E-learning in evaluating students' learning. J. Penjaminan Mutu. **5**, 165 (2019). https://doi.org/10.25078/jpm.v5i2.1089
2. Bañeres, D., Noguera, I., Elena Rodríguez, M., Guerrero-Roldán, A.: Using an intelligent tutoring system with plagiarism detection to enhance e-assessment, 5 September (2019). https://doi.org/10.1007/978-3-319-98557-2_33
3. Kulkarni, P.V., Rai, S., Kale, R.: Recommender system in elearning: a survey. In: Proceeding of International Conference on Computational Science and Applications, pp. 119–126 (2020)
4. Cifuentes-Vicente, P.: Uso de la herramienta Moodle por los alumnos en las modalidades presencial y semipresencial. revistas.upsa.es (2019)
5. Llerena Izquierdo, J.F.: El Uso De Grabaciones Por Video Como Recurso De Evaluación De Conocimientos De Aprendizajes. In: El Uso De Grabaciones Por Video Como Recurso De Evaluación De Conocimientos De Aprendizajes. Editorial Abya-Yala (2018)
6. López, R.: El uso del video como herramienta para la enseñanza y aprendizaje de la matemática en los primeros niveles de educación superior. Mem. Académica CITIS, pp. 117–128 (2015)
7. Troncoso-Pantoja, C.A., Díaz-Aedo, F., Amaya-Placencia, J.P., Pincheira-Aguilera, S.: Elaboración de videos didácticos: un espacio para el aprendizaje activo. FEM Rev. la Fund. Educ. Médica **22**, 91–92 (2019)
8. Francés, J., Bleda, S., Calzado Estepa, E.M., Martínez Guardiola, F.J., Heredia-Avalos, S., Hernández Prados, A., Hidalgo Otamendi, A., Vera Guarinos, J., Yebra Calleja, M.S.: Análisis y aplicación de nuevas metodologías docentes basadas en clase invertida y gamificación a través de Moodle (2019)
9. Peña, B.: Análisis del aprendizaje a través de la huella en Moodle: aplicación en la asignatura de Termodinámica Técnica. In: Libro de Actas IN-RED 2019: V Congreso de Innovación Edicativa y Docencia en Red, pp. 315–328. Editorial Universitat Politècnica de València, València (2019). https://doi.org/10.4995/INRED2019.2019.10415

10. Estrada, J.C., Nacipucha, N.S., Chila, R.L.: El uso de los códigos QR: una herramienta alternativa en la tecnología educacional. Rev. Publicando. **5**, 83–106 (2018)
11. Suleri, J.I., Suleri, A.J.: Comparing virtual learning, classical classroom learning and blended learning. Eur. J. Sustain. Dev. Res. **3**, 8 (2019)
12. Corona, M.C., Corona, F.C., Corona, A.C.: 4.6 Aprendizaje con plataforma Moodle, en modalidades E-learning y B-learning en Educación Normal. INNOVACIÓN EN EL PROCESO Aprendiz **1**, 447 (2019)
13. Ramirez Herrera, S., et al.: Integración de Opensimulator y Moodle, para la evaluación de actividades desarrolladas en entornos virtuales tridimensionales (2019)

Gamification as an Educational Strategy to Strengthen Cognitive Abilities of Mathematics in School Children

Ligia Jácome-Amores[1,2(✉)], Wimper Rivera Freire[1,2], and Richard Sánchez Sánchez[1,2]

[1] Facultad de Ingeniería y Tecnologías de la Información y Comunicación (FITIC), Universidad Tecnológica Indoamerica, Av. Manuelita Sáenz y Agramonte, 180103 Ambato, Ecuador
ligiajacome@uti.edu.ec
[2] Unidad de Posgrado-Maestría en Educación, Universidad Tecnológica Indoamerica, Av. Manuelita Sáenz y Agramonte, 180103 Ambato, Ecuador

Abstract. This article presents a study on the usefulness of M@TILDA, a gamification tool, developed to strengthen children's learning in the area of mathematics, which sometimes turns out to be one of the most complicated when it comes to developing the learning in the children of second and third of basic. The methodology used was based on three phases: In the first, an exploratory investigation was carried out through surveys and interviews not structured with authorities and teachers of educational centers in Tungurahua (Ecuador). In the second, the technological tool was developed, with a dynamic difficulty adaptation mechanism (DDA), which allows ergonomic and challenging behavior. Finally, in the third, a quasi-experimental study was carried out in which 25 children of school age participated. The quantitative results showed significant statistical differences in performance levels, so the teachers stated that M@THILDA is a potential tool to develop learning strategies and that it would support their teaching work in the classroom.

Keywords: Gamification · Mathematical learning · TIC

1 Introduction

This article presents a study on the usefulness of M@TILDA, a gamification tool, developed to strengthen children's learning in the area of mathematics, which sometimes turns out to be one of the most complicated when it comes to developing the learning of second graders. The methodology used was based on three phases: In the first, an exploratory investigation was carried out through surveys and not structured interviews with authorities and teachers of educational centers in Tungurahua (Ecuador). In the second, the technological tool was developed, with a dynamic difficulty adaptation mechanism (DDA), which allows ergonomic and challenging behavior. Finally, in the third, a quasi-experimental study was executed to make 25 children of school age participate. The quantitative results showed significant statistical differences in performance levels, so the teachers stated that M@THILDA is a potential tool to

M. Botto-Tobar et al. (Eds.): CI3 2020, AISC 1277, pp. 142–150, 2021.
https://doi.org/10.1007/978-3-030-60467-7_12

develop learning strategies and that it would support their teaching work in the classroom.

In the field of research, there have been many studies with positive results when intervening in learning with serious games and other gamification applications; such is the case of NumberWorlds [3], and studies [4, 13, 15]; that emphasize the development of mathematical skills, through recreational activities, applying board games that enhance the relationship between counting and quantity, between number and space, mathematical operations, notions of greater, less than; in a variety of situations, number names and graphical representation of numerical magnitude, and accuracy, by placing numbers on the number line and in numerical comparison operations. On the other hand, the works carried out in [2, 5] where studies of relevance were developed, creating educational software with augmented reality, where they propose with a virtual board to teach regular young children, numerical representations.

All these studies have contributed in some way to cognitive development in the area of mathematics, in children of regular education in different countries. However, in Ecuador there are few studies carried out in this area. And those that have been analyzed by the research team, limited because, during use, they always present the same scenes and the same activities to be performed; What turned out to the boy after the first manipulations boring and little challenging. In addition, several of them are very basic and with few possibilities of play, many of them do not have degrees of complexity or different levels of difficulty, nor do they show a dynamic adaptation in the games so that they help the child to fall behind according to the domain that berry showing. This is why we have seen the opportunity to create a gamification tool (M@THILDA) to help dynamically strengthen the learning of basic mathematics in children who are entering the school stage in an always entertaining and challenging interface thanks to the mechanism and adaptive flow of the game with the implementation of a dynamic adaptation algorithm (DDA).

2 Related Works

In order to support the scientific bases of this study, some parameters for bibliographic research were defined, with these parameters it was possible to find various documents They were exported to a specific collection of research papers that showed significant results in learning by applying gamification tools at different educational levels. Such is the case of the study carried out in [6], where the experience of students using a gamified activity that encourages the graphical representation of mathematical functions is explored and described. On the other hand, we have the research where a proposal for the gamification of the video game "Clash Royale" is presented for students of the third year of Compulsory Secondary Education in the province of Jaén. In the studio, a total of six sessions were developed, consisting of cooperative activities and games, which make use of various elements of said video game, with the aim of making the teaching-learning process more stimulating and motivating for students [7].

On the other hand, the project where a didactic strategy was designed, for the development of skills in the approach and resolution of arithmetic problems, in rural educational institutions with a multigrade modality, supported by gamification and

Polya methods, for Cycle II degrees of the Las Pavas and Valle de Tenjo Educational Institutions; with significant results in the training sessions [8].

More over this study investigated how five gamification user types may relate to six mainly used online learning activities in a distance online bachelor's and master's class in educational science through the use of a systematic approach. A total of 86 students participated in the questionnaire in a cross-sectional study. The findings showed average agreement shares for all five gamification user types. The correlations revealed that the six online learning activities were at least significantly connected to one of the five gamification user types, and both person-centered and environment-centered perspectives were displayed [9].

The study where an experiment was carried out for four months with 40 university students from first year programming courses. Students were randomly assigned to one of two versions of the programming learning environment: a gamified version made up of rankings, points and badges, and the original non-gamified version. The results of the contribution show that gamification affected users in different ways depending on their personality traits. These results indicate that the effect of gamification depends on the specific characteristics of the users [10].

In the same way, research where I try to increase student participation in learning through the gamification technique in difficult subjects such as programming language courses, through elements of play in activities; The results showed an increase in the score of the elements of the game and a good effect on the participation of the students [11]. Finally, in researching the potential of gamification to promote knowledge retention using an action research approach, where the results of a longitudinal study that included 617 secondary and tertiary education students conducted over a two-year period. Various workshop designs incorporating numerous gamification elements were compared with non-gamified workshop designs, tested and refined over time. Improved workshop designs led to higher levels of knowledge retention that exceeded reference values in the educational literature [12].

3 Method

The method used is outlined in three phases: first, interviewing authorities and some teachers working in different Tungurahua Province schools in the first phase, defining the learning method used and characterizing the tool to be established in accordance with the technical requirements to be taken into account.

3.1 Preliminar Research

An exploratory enquiry focused on the regular education school that would participate in this research, and 25 children in their early stages of schooling. Then, through a survey applied to 4 teachers of second and third grade of basic, and the observations made by the researchers during the field visits, it was possible to know the method that was applied in the classroom: the guide the government's guide text was used and the teacher's own didactic material was prepared, and that basically consists of tangible objects such as cards that illustrated representations and mathematical symbols, cardboard figures, beads, and abacus. During class development, teachers instructed children about the mathematical activities and operations they should perform. In certain cases, the children could not concentrate and follow the instructions given, and the type of teaching material they handled complicated their skills development. This required the teacher to work individually with each of the children, but this was an obstacle because of the number of children in their care making it even more difficult to advance in the contents and the development of the teaching work in general.

When the researchers asked the teachers if they used any digital medium in teaching, they mentioned that, although all the teachers have a computer, and the educational centers have some computer labs that they could access a few hours and some days to the week, they did not know how to integrate this technological resource in their educational work, and the few digital tools they knew, were not mathematics or did not fit the educational needs of children. When teachers were proposed to make a digital tool that helps children learn math by playing, they were very interested, as this was alternative teaching and a strategy to teach new knowledge in mathematics, one of the areas that at the discretion of the teachers was the most conflictive in this educational level.

Finally, with the teacher's guidance, a letter was written to parents seeking their permission to test the tool with their children; the organizational and technological viability for applying the tool was also checked, with the help of the teachers, and the interpretation that it should have.

3.2 M@THILDA'S Implementation

This playful learning tool was developed at Uniti, an IDE for video game development. The M@THILDA'S interface is presented in a web environment and consists of a repository of learning objects, each of which carries a challenge or challenge that the child must face, and which has as learning units, the sense of the number, count, add, subtract, relate between major-minor than, and build basic mathematical operations by itself (where it combines the previous ones). In addition, each learning object has nine-level recreational activities linked to the correct learning unit; all this accompanied by visual and auditory elements, through Digital Signal Processing (PSD) and TTS voice synthesizer [1]. In Fig. 1, the architecture of the tool is shown.

It is important to note that the generation of a certain level within a learning object is random, taking into account all possible combinations of numbers to work, including those already used in the previously reached levels. In its architecture, it consists of an algorithm of adaptation of dynamic difficulty (DDA) of play and adaptive workflow (FAT), which allows the tool to always be challenging and sustainable for the child [12]. The DDA, which is based on the child's performance in the game, causes the child to change the difficulty without the child noticing [16]; and it consists in reducing or increasing the degree of difficulty between two related levels according to the domain that has reached the current level.

Each level ends, rewarding the child with 1, 2 or 3 stars, for their achievements in one level. If at the end of a level, you have obtained a minimum star, the next challenge will be easier to overcome; but, on the contrary, if at the end of a certain level, you have obtained the maximum number of stars, the next level will be a little more difficult to

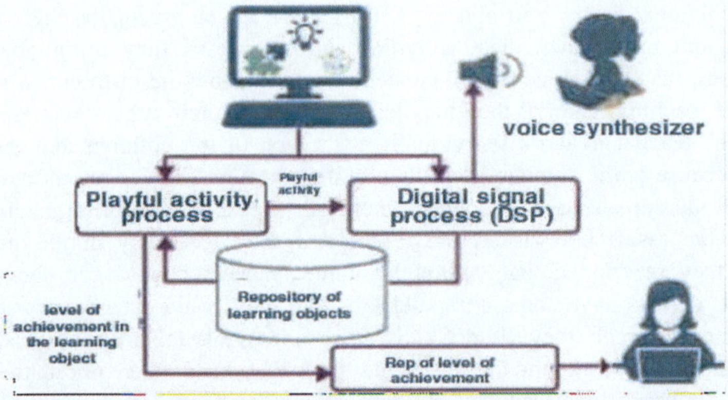

Fig. 1. M@THILDA'S architecture

reach, all this accompanied by flexibility in the child's workflow, as he makes his own decisions through the game, since he can decide on what learning object he wants to practice [4]. An extract of the code and the capture of the user interface of the developed tool are shown in Fig. 2.

```
if (button.frame < 4 && Juego.starsArray[button.levelNumber - 1] != 3)
{
    Juego.level = button.levelNumber;
    Util.reiniciarNivel();
    game.state.start('seleccionarObjeto');
} else {
    var buttonTween = game.add.tween(button);
    buttonTween.to({alpha: 0.5}, 20, Phaser.Easing.Cubic.None);
    buttonTween.to({alpha: 1}, 20, Phaser.Easing.Cubic.None);
    buttonTween.to({alpha: 0.5}, 20, Phaser.Easing.Cubic.None);
    buttonTween.to({alpha: 1}, 20, Phaser.Easing.Cubic.None);
    buttonTween.start();
}
```

Fig. 2. Extract from the ADD code and the M@THILDA interface

After its implementation, the program was validated by a pilot test by a teacher, a psychologist, and Emilita, a girl of third grade with whom we checked each of the learning items that M@THILDA brings. Observations and important events were recorded to correct and refine the first versions of the system. During the tests Emilita showed interest, concentration, and motivation with each of the learning activities; The teacher and the educational psychologist commented on how they could apply the tool as a mechanism for strengthening classroom learning.

4 Evaluating M@THILDA

The first prototype of the system was evaluated by means of a pilot test, a teacher of and two children of third grade. This initial pilot was carried out to verify that the game works correctly, it also served to improve the user interface according to the teacher's observations or the children's reactions when interacting with the gamification tool. Subsequently, it was proposed to carry out an experimental phase in a real setting and with a representative sample to determine if the system provided statistically significant evidence regarding the use of the conventional method in the classroom.

4.1 Participants

Twenty-five third-grade boys and girls from the regular education system participated in the study. A document was prepared for all the participants to request the consent of the parents. With the authorization of the director of the institution and the collaboration of the teachers, all the informed consents and demographic data of the participants were collected, which can be summarized in Table 1.

Table 1. Datos demográficos de los participantes

Niñas	Niños	Edad Cronologica M(SD)	Nivel escolaridad
17	8	7,20(0,74)	3ro básica

4.2 Experimentation Instrument

M@THILDA was used as experimentation instruments, which was installed in 5 desktop computers in the institution. In addition, there were 2 portable computers provided by the research team. In addition, the tool was also installed on the teachers' computers.

4.3 Process

Previous to the experimentation in the classroom, as a first activity the teachers were trained in the use of technology in general and in the application of the gamification tool in particular. Subsequently, the research team, together with the teachers, carried out various tests of the game's operation and an approach to the children. Figure 3 shows Emilita, a 7-year-old girl from third grade, using the conventional method (a) and M@ATILDE learning objects (b).

The assessment was performed over a span of 8 weeks, with the trained teachers and children familiar with using the program. On the one hand, the children used the traditional method interchangeably (teacher-prepared cards) and to the system implemented by another. Finally, an evaluation was applied: (A) with the traditional method, and (B) with the use of the tool. The children of each group, with the help of the teachers, carried out the activities indicated to them with the two learning methods;

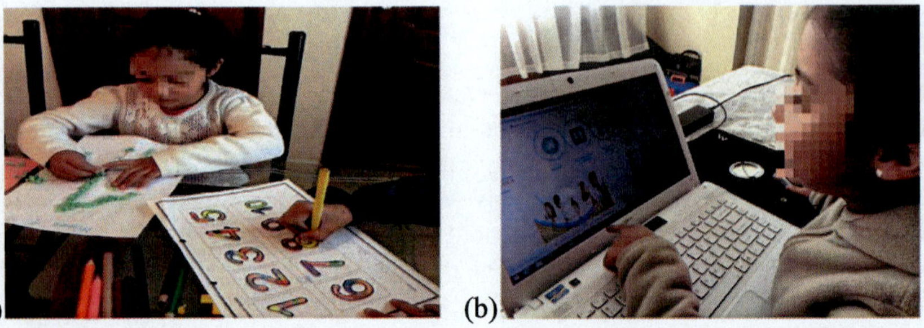

Fig. 3. Emilita's pictures: (a) conventional method cards, (b) M@THILDE's learning objects

Achievement levels, comments and relevant events such as interaction, motivation, and degree of concentration were recorded for qualitative analysis.

For the quantitative analysis of the data, it was found, in the first instance, that the data of the levels of achievement, with the two moments of the study (A/B), behaved normally; for this and the sample size, the Shapiro Wilks test ($p > 0.05$) was applied. Subsequently, the Student's parametric T test ($p < 0.05$) was applied with a 95% confidence level, to determine whether or not there were significant statistical differences between the levels of achievement, at these two moments.

5 Results

After the experimentation, the results showed that the achievement rates increased by 18.50% with the use of the recreational tool; about the rate of achievement with the conventional method. Determine if there were significant statistical differences between achievement levels when using the fun approach compared to achievement levels using the traditional method. First, the Shapiro-Wilks test was applied, showing the following results: The data of achievement levels, using the conventional method, were ($p = 0.500$), and using the serious game, of ($p = 0.160$); which indicates that the data of the levels of achievement with the two methods if they came from standardized samples.

Subsequently, with the results of normal data; Student's parametric T test was applied to analyze the statistical differences when comparing the achievement levels of the experimentation; showing as a result: ($t = -8.76$, $gl = 25$, $p = 0.000$). So it could be concluded that, with the use of M@THILDA in the classroom, the yields measured by the levels of achievement in children; if they were superior, compared to the use of the conventional method, since there were significant statistical differences, with 95% confidence and a level of significance of 0.

6 Conclusions

Creation of a gamification tool with visual and auditory elements in a versatile and clear user interface helped the child retain his/her interest in learning mathematics throughout the experiment, which strengthened the child's engagement and focus. The evaluation of the tool in the classroom allowed us to identify its usefulness in aspects of interaction and learning. This shows the results evidencing the remarkable M@THILDA'S contribution in children concerning motivation and degree of concentration and their levels of achievement, which on average improved significantly with 18.50%; versus the conventional method. The statistical model applied showed that in this study there were significant statistical differences in learning with the use of the tool versus the traditional method; with a significance level of ($p = 0.000$).

7 Future Work

This experience has allowed the identification of other latent needs and the development of cognitive and communication skills, such as communication and learning in children with dyslexia or dyscalculia. This is because in the institution there are children who present these conditions. There is also the question of whether the tool would be useful in children with Down Syndrome, since in the institution there are some students who also present this syndrome. Finally, it would be very interesting to investigate the development of learning strategies that teachers and specialists can create with the tool.

References

1. Asterisk: Text-To-Speech (TTS) y Automatic Speech Recognition (ASR). Disponible en (2009). http://www.wikiasterisk.com/index.php?title=TTS_y_ASR. Accessed 19 July 2017
2. Avila-Pesantez, D., et al.: Design of an augmented reality serious game for children with dyscalculia: a case study. In: International Conference on Technology Trends, pp. 165–175. Springer, Cham (2018)
3. Griffin, S.: Number worlds: a research-based mathematics program for young children. Engaging Young Children in Mathematics: Standards for Early Childhood Mathematics Education, pp. 325–342. Lawrence Erlbaum, New Jersey (2004)
4. Ligia, J.A., Magaly, A.R.P., Patricio, S.S.R., Elizabeth, S.S.P.: Assistive technological tools to strengthen interaction, communication and learning in children with different abilities. In: International Conference on Information Technology & Systems, pp. 340–350. Springer, Cham (2020)
5. Ramani, G., Siegler, R.: Promoting broad and stable improvements in lowincome children's numerical knowledge through playing number board games. Child Dev. **79**(2), 375–394 (2008)
6. García, F.Y.H., Rangel, E.G.H., Mera, N.A.G.: Gamificación en la enseñanza de las matemáticas: una revisión sistemática. Telos **22**(1), 62–75 (2020)
7. Expósito-Escudero, S.: GAMIFICACIÓN DEL VIDEOJUEGO "CLASH ROYALE" EN ESTUDIANTES DE SECUNDARIA (2019)

8. Mahecha Moreno, H.P., Casallas Forero, L.F.: Uso de estrategia didáctica apoyada en la gamificación para el desarrollo de habilidades en el planteamiento y resolución de problemas aritméticos, en instituciones educativas rurales (2019)
9. Bovermann, K., Bastiaens, T.J.: Towards a motivational design? Connecting gamification user types and online learning activities. Res. Pract. Technol. Enhanc. Learn. **15**(1), 1 (2020)
10. Smiderle, R., Rigo, S.J., Marques, L.B., de Miranda Coelho, J.A.P., Jaques, P.A.: The impact of gamification on students' learning, engagement and behavior based on their personality traits. Smart Learn. Environ. **7**(1), 1–11 (2020)
11. Khaleel, F.L., Ashaari, N.S., Meriam, T.S., Wook, T., Ismail, A.: The study of gamification application architecture for programming language course. In: Proceedings of the 9th International Conference on Ubiquitous Information Management and Communication, pp. 1–5, January 2015
12. Putz, L.M., Hofbauer, F., Treiblmaier, H.: Can gamification help to improve education? Findings from a longitudinal study. Comput. Hum. Behav. **110**, 106392 (2020)
13. Stanitsas, M., Kirytopoulos, K., Vareilles, E.: Facilitating sustainability transition through serious games: a systematic literature review. J. Clean. Product. **208**, 924–936 (2018)
14. Tremblay, J., Bouchard, B., Bouzouane, A.: Adaptive game mechanics for learning purposes-making serious games playable and fun. CSEDU **2**, 465–470 (2010)
15. Whyte, J.C., Bull, R.: Number games, magnitude representation, and basic number skills in preschoolers. Dev. Psychol. **44**(2), 588 (2008)
16. Wilson, A.J., Dehaene, S., Dubois, O., Fayol, M.: Effects of an adaptive game intervention on accessing number sense in lowsocioeconomicstatus kindergarten children. Mind Brain Educ. **3**(4), 224–234 (2009)

A Didactic Model with Technology 4.0 for Ubiquitous Learning at the UNIANDES University of Ecuador

Gustavo Eduardo Fernández Villacrés[1(✉)],
Karina De Lourdes Serrano Paredes[2], Isabel Cristina Mesa Cano[2],
and Jorge Viteri Moya[1]

[1] Universidad Regional Autónoma de Los Andes, Ambato, Ecuador
ua.eduardofernandez@uniandes.edu.ec
[2] Universidad Católica de Cuenca, Cuenca, Ecuador

Abstract. This article describes the execution of the didactic innovation project proposed at Uniandes University in the city of Ambato in the Republic of Ecuador. This has as objective: propose a new educational model entirely supported by the called 4.0 technology to achieve ubiquitous learning in the educational institution. Specifically, it deals with the problems related to vocational training, not aimed at achieving competencies for efficient job performance in the future. Also, the conflictive situation is associated with the low level of technological management by teachers as well as total ignorance about technology 4.0. The research was developed among the students, and professors of the University, initially, a theoretical foundation was made related to didactic models, technology 4.0, ubiquitous learning. In the end, a techno-alternative didactic model is proposed, fully supported by 4.0 technology applied through computers, and mobile devices.

Keywords: Didactic · Technology 4.0 · Ubiquitous learning · Education

1 Introduction

The term "ubiquitous learning" is given to the training process in a new comprehensive educational background, for some Hispanic authors, is easier to talk about a ubiquitous formation or even universal training, everybody agrees that it is a set of training activities supported entirely in the technology, they are accessible in any place, and from any device, it can be said that the ubiquitous learning is, in fact, the version technological learning general. We must consider having universal learning; this should be able to develop in any transmission medium. If you analyze the situation from a strictly technical standpoint, it is necessary to evaluate their ability to use information, and communication technologies both from the perspective of the users' necessities as well as the capabilities that new technologies bring ubiquitous training [1].

The Universidad Regional Autónoma de Los Andes "UNIANDES", is a Higher Education Center, a private, and secular law entity, with legal status, and administrative, and financial autonomy, which offers a comprehensive education to its students,

M. Botto-Tobar et al. (Eds.): CI3 2020, AISC 1277, pp. 151–162, 2021.
https://doi.org/10.1007/978-3-030-60467-7_13

without distinction of sex, race, religion or politics; therefore, the admission of students depends on their intellectual capabilities. "UNIANDES" was created in compliance with Article 7 of the Law of Universities, and Polytechnical Schools of Ecuador. It is based on the report No. 01235 of October 10, 1996, issued by the National Council of Universities and Polytechnical Schools (CONUEP); on the Law of Creation of the University issued by the National Congress on January 9, 1997, and its publication in the Official Register No. 07 of February 20, 1997 [2].

Uniandes University degrees develop the professional training process in the modalities of Presential, Semi-Presential, and Distance Learning. The teachers deploy and sequence the learning process following a training logic in which the learning results are developed. One of the careers that have had a very good acceptance in the central area of the country is the Medical Sciences Career, due to several factors such as large laboratories, renowned doctors, and frequent disciplinary control, it should be mentioned that the Medical Sciences Career is within the Faculty of Medical Sciences [3].

Uniandes University degrees develop the professional training process in the modalities of Presential, Semi-Presential and Distance Learning. The teachers deploy and sequence the learning process following a training logic in which the learning results are developed. One of the careers that have had a very good acceptance in the central area of the country is the Medical Sciences Career, due to several factors such as large laboratories, renowned doctors, and frequent disciplinary control, it should be mentioned that the Medical Sciences Career is within the Faculty of Medical Sciences [3].

In several of the visits made to the mentioned Faculty, some academic and pedagogical deficiencies in the teaching of the different subjects taught could be appreciated. With this premise, the initial diagnosis was carried out with the students of the Systems Career on the use of Information Technologies. For that purpose, a quick research on the levels of technology use was developed as a support to the educational process carried out by each teacher. The results of this preliminary research ratified the following difficulties:

a. Teachers have a very low level of mastery, and knowledge of new technologies, for this reason, their application is limited. That is to say, the teachers rely very little on the Tics to strengthen the teaching-learning process;
b. Complementary to this, it can be stated that the didactic process that the teachers of the career are developing has a great deficiency because it is not supported by new didactic strategies based on computer technology. It should not be forgotten that most students are digital natives, and that the technological environment is a natural environment for them, and this advantage is being wasted.
c. Mobile devices are not used as teaching aids by teachers
d. The career in their practice forces the frequent use of simulators, these are mechanic but nowadays the new simulators are digital, this means that the student should have a pre-knowledge of 3D virtual platforms to easily adapt to this new type of simulators.

Based on these difficulties, it has been possible to formulate the educational problem in the following terms: How to improve the learning process in the Universidad Regional Autónoma de Los Andes?, as a possible solution to the problem arises the project of "Didactic Innovation", which has as general objective: To propose a new didactic model supported by the new technologies of the generation called 4.0 for the improvement of the ubiquitous learning in the Universidad Regional Autónoma de Los Andes. To achieve this general objective, the following specific objectives have been proposed:

- Develop a theoretical foundation on didactic models, 4.0 technology, and ubiquitous learning.
- Diagnose the didactic process in the Faculty of Medical Science of Uniandes.
- Elaborate the didactic model supported by technology 4.0

The development of the project began with the respective theoretical foundation described below:

Model: it is a representation of reality that implies a distancing from it. It is the conceptual, and symbolic representation and therefore it is indirect, that being necessarily schematic becomes a partial and selective representation of some aspect of that reality, focusing the attention on what it considers important and despising what is not. A model is identified with a kind of interpretative scheme that selects data from reality, structures them, deciding which aspects are important to know about the reality to which it refers. A model is a mediating scheme between reality and thought, between the world and science [4].

The concept of didactics arises from the Greek word "didaktiko" whose meaning relates to the term "I teach". In general, didactics is related to the field of practical teaching, therefore it is an educational activity [5]. The origin of university didactics can be dated back to the Greek period when Plato created the Academy for the search of knowledge through philosophy. This Greek philosopher also followed the path laid out by Socrates, which led to learning through questions from students and incorporated other activities such as the symposium and lessons. In the Middle Ages, there was a great ecclesiastical influence, and it was St. Augustine who introduced the didactics of reading a passage of a book of the Bible and then the respective commentary through the teacher [6]. At the beginning of the seventeenth century, John Paul II pointed to didactics as a universal method of teaching, which is why on some occasions didactics is considered a teaching technique. At the beginning of the 20th century, the German university incorporated terms such as university didactics, university pedagogy, and academic pedagogy [7].

University didactics: is a special didactics oriented to the forms of teaching in higher education but based on general didactics as well as specific ones to understand its object of study. This process that forges professionals for society in an integral way is carried out in a human relationship between the professor, and the student [8].

Didactic model: It is defined as a series of sequences and actions designed by the teacher with pedagogical intention. These are built from theoretical knowledge and experience in educational practice. The didactic model is "a set of principles of an educational nature, the result of academic knowledge and practical experience, which serve to define the educational objectives, and intend to guide the teaching-learning processes that take place in the classroom [9].

Didactic strategies: Learning strategies are procedures (sets of steps, operations, or skills) that a student employs in a conscious, controlled, and intentional manner as flexible tools for meaningful learning, and problem-solving. However, at present, it seems that curricula at different educational levels promote students who are highly dependent on the instructional situation, with little or no conceptual knowledge about different disciplinary issues, but with few tools or cognitive instruments that help them to face new situations, and learning in different areas by themselves, and which are useful in the most diverse situations [10].

Didactic strategies: Learning strategies are procedures (sets of steps, operations, or skills) that a student employs in a conscious, controlled, and intentional manner as flexible tools for meaningful learning and problem-solving. However, at present, it seems that curricula at different educational levels promote students who are highly dependent on the instructional situation, with little or no conceptual knowledge about different disciplinary issues, but with few tools or cognitive instruments that help them to face new situations and learning in different areas by themselves, and which are useful in the most diverse situations [11]. In this last approach, the integration of the cognitive and the affective, of the instructional and the educational, as essential psychological and pedagogical requirements, is revealed as a determining characteristic. This process has as its essential purpose to contribute to the integral formation of the student's personality, constituting a fundamental way to acquire the knowledge, procedures, norms of behavior, and values bequeathed by humanity [12].

ICTs in education: UNESCO states the following regarding the use of new technologies in the field of education: "Information and communication technologies (ICTs) can contribute to universal access to education, equality in instruction, the exercise of quality teaching and learning, and the professional development of teachers, as well as to more efficient management and administration of the education system [13].

E-learning: The European Commission of Education gives the following definition of e-learning, commonly known as e-learning: "Learning that is supported by new technologies and the Internet to improve the quality of learning by facilitating access to resources and services as well as remote exchanges and collaboration. It can also be defined as training that is provided through the use of the Internet and new technologies and therefore its distinction from traditional education focuses on the potential of ICT in the educational field [14].

E-learning makes it possible to create learner-centered learning environments. They are characterized by being interactive, efficient, and easily accessible and distributed. An E-learning scenario should consider aspects such as institutional, pedagogical, technological, interface design, evaluation, management, support, and usage ethics. In this way, e-learning tries to be a combination of interactive resources that generate support and structured learning activities [15].

B-learning: Blended learning commonly referred to as "B-learning" refers to the mixing of face-to-face and virtual educational processes at the same time. This means that face-to-face teaching is combined with no-presential technology. This educational approach solves problems that E-learning has, such as the lack of human interaction, and platform costs. It is widely used in the business environment because it allows for cost reduction [16].

M-learning: During these last years of the 21st century, the development of wireless data networks has been appreciated, which has allowed the connection of devices such as electronic tablets, and smartphones to the Internet and, based on this, to be able to access educational content at any time and place. M-learning can be defined as a combination of e-learning and mobile and wireless computing to provide learning experiences [17].

Also, UNESCO's definition of mobile learning should be mentioned: "Mobile learning involves the use of mobile technology, alone or in combination with any other type of information and communication technology (ICT), to facilitate learning at any time and in any place". Some research works report that mobile devices are used in education as mediators in the didactic process because these devices can be used to consult diverse educational materials, this would mean that their use should promote the development of skills involved in the learning task. They also point out that the devices helped them to promote their thinking skills and to cooperate with their peers; similarly, it is highlighted that mobile devices can increase students' motivation within the classroom. It is also mentioned that the use of mobile phones and tablets could encourage students to become interested in a subject and spend more time on it [18].

The main characteristics of mobile learning include Ubiquity, that is, it allows the development of the teaching and learning process at any time and place. In principle, it could be associated with any mobile technology, but in the educational field, there are three that stand out: smartphones, digital tablets, and phablets, a device that is the result of the hybridization of the first two mentioned above [19].

The term "Industry 4.0" was established by the German government with the support of German industries to describe the digitization of industrial systems and processes, and their interconnection through the Internet of things, to achieve greater flexibility and individualization of production processes. It is a vision of the factory of the future or intelligent factory. The digital transformation of industry and companies with the integration of new disruptive technologies such as Big Data, the cloud and cybersecurity, all framed in the intelligent cities that are producing the advent and deployment of the fourth industrial revolution [20].

Sociology has determined that the main attributes of society 4.0 are the following: Technician: Today's society has a strong dependence on technology.

- Fast: Immediacy has come to be considered a primary value; it is about getting there before anyone else, too often without time to reflect on whether or not it is worthwhile to make the journey.
- Hypermedia: Multiple channels provide information that is always linked to other related information and provide immediate access to new information proposals. Plain text has been replaced by hypertext.
- Informed: The levels of information received by multiple channels are proud.
- Relative: Fashions and the latest trends set the values. Traditional humanistic values are in a downward spiral.

Superficial: As a consequence of speed as a value, people do not usually go deep enough into the subjects and are moved by conclusions about the facts based on simple blunt phrases read on the net. Due to lack of time to properly attend to all the information they receive, they end up basing themselves on the headlines without going down to read the body of the news and, of course, without taking time to reflect and direct each topic as it should be [21].

The basic technologies of industry 4.0 are:

- Artificial intelligence applied to all processes such as artificial vision, information searches and more
- The internet of things
- Cloud Computing
- Augmented and virtual reality
- Robotics and 3d simulation
- Three-dimensional printing.
- The Big Data and the analysis of large volumes of information
- Cybersecurity
- The communication called 5G
- Web analytics for mobiles
- Drones and intelligent vehicles
- Intelligent geolocation [20]

Virtual reality consists of producing, using computer programming, an environment that appears to the user's eyes as real and immersed in it, using the equipment of suitable hardware (three-dimensional glasses and optionally gloves). Augmented reality is the vision that one has in a physical environment of the real world, through a technological device that adds additional information to that perceived by the human eye. Utilizing a device or a set of devices, virtual information is added to the physical information perceived by natural vision. Tangible physical elements are combined with virtual elements and create an augmented reality in real-time [22].

On the other hand, it can be pointed out that we live in an ecosystem in which technologies have changed everyday life and new methods of manufacturing, decision making, relating and learning have appeared. This new reality demands that training be completely redefined. A reform of the educational system is necessary. It must move within a framework of universal ethical duties, it must facilitate, without abandoning the identity aspects of human groups, the acquisition of a global, transcultural and cosmopolitan vision, and it must make it possible to understand the background of conflicts, challenges and, achievements of humanity. It is necessary to teach how to undertake; to develop the capacity to observe, analyze, reason and propose; working at the same time to help discover what each student likes and is most passionate about and to identify whether they have the necessary skills to do so. To this effect, training must be flexible, adaptive and must offer open programs based on advanced teaching methodologies. The main thing to teach will continue to be "learning to learn" but for the new type of working person, basic knowledge and education for work must be different, and must prioritize creativity and innovation [21].

The world has changed a great deal in recent years, due to the rapid expansion of digital technologies and the social changes that have resulted from this and which have changed the world of work in the process. It must be understood that we need to adopt new attitudes and skills, the methods and tools used in the education system must be readjusted. We need an educational system that, while assuming the importance of combining the humanities and the scientific method, adjusts in a symbiotic way to the present society and the demands of the future, understanding at the same time that studying and learning are the same thing. We must accept that the learning process is multi-factorial and ubiquitous. It is not only necessary to teach how to read, listen, experiment and observe. We must also teach to stimulated the will to overcome and the ambition to improve, to overcome challenges and achieve success with loyalty and respect for others. Competences that have been omitted in today's School, as if leading people to the best of themselves and making them unfold their entire potential were something negative [22].

1.1 Methods

A transversal study has been designed of the problem related to the almost null use of technology 4.0 as an element of didactic support, it was proposed to develop a qualitative-quantitative investigation, the same one that was carried out in the place where the symptoms of the problem could be observed, in this case, the Faculty of Medical Sciences of the University UNIANDES of Ambato.

The methods used are:

Analytical-synthetic: Applied to develop a theoretical framework based on the collection of information and its synthesis. This collection was done in the denominated primary and secondary sources.

Inductive -Deductive: To induce a general solution from a particular one.

The main technique defined for Research, is the survey, both to Teachers and students.

The instruments associated with the technique are the respective questionnaires.

The population involved in the problem is composed of 1900 students and 138 professors belonging to the Faculty of Medical Sciences.

Afterward, and based on the statistical formula for finite populations, we proceed to obtain the sample to be investigated. This sample was calculated as 337 involved, of which 250 were students and 87 teachers. After the surveys were carried out, the data were tabulated in the SPSS program from which the respective frequency tables could be constructed.

1.2 Results

The following are the results of the research carried out both to teachers and students of the Faculty of Medical Sciences (Tables 1 and 2).

Table 1. Results of the survey of 87 teachers

No	Questions	Yes	No
1	Do you use technology 2.0 as a teaching aid?	31	56
2	Do you think that technology is advancing rapidly?	62	25
3	Is your level of technological management medium-high?	41	46
4	Have you heard of 4.0 technology?	17	70
5	Have you used 4.0 technology as a teaching aid?	5	82
6	Do you think that today's education should be supported by all existing technologies?	61	26

Source: Researchers

Table 2. Results of the survey of 250 students

No	Questions	Yes	No
1	Do your teachers use technology 2.0 as a teaching aid?	51	199
2	Do your teachers use didactic strategies strongly supported by technology?	87	163
3	Are your teachers' classes not very innovative?	203	47
4	Do you know about technology 4.0?	61	189
5	Do you think it would be advisable to mix all technologies to improve the educational process?	178	72
6	Do you think that your teachers are up to date with technology	24	273

Source: Researchers

Analyzing the results of the survey to the teachers, it can be highlighted that 64% of them, do not use the technologies of the web 2.0, also 71% are aware that the technology is developing quickly. It is also relevant that 57% claim to have a medium to a high level of technological skills. On the other hand, 80% of the teachers have not heard about 4.0 technology while 94% have never used it. Finally, 70% of teachers believe that all technologies can be used simultaneously to generate universal learning. About the survey conducted with students, the results are: 80% say that teachers do not use technology 2.0 in their classes, 65% say that teachers do not apply innovative teaching strategies supported by technology. On the other hand, 76% do not know about technology 4.0, and 90% say that teachers are not updated in the handling of technology.

Proposal: The proposed didactic model is called **"techno-alternative"** and has hybrid characteristics obtained from both a technological and an alternative (research-based) didactic model. The proposed structure is based on the model defined by Cruz Hernández and Ana Guaráte in their book "Didactic models for learning situations and contexts" published in 2017 (Table 3).

Table 3. Structure of the proposed didactic model.

Techno-alternative didactic model for teaching Medicine
Theme and problem
Here we define in a general way the subject that will be dealt with in the didactic sequence. Here too, the problem to be solved by the students through research, revision or re-study of a topic is proposed
Objectives
Here the objectives or learning outcomes are defined: The objectives have to do with the development of the capacity of reasoning and the spirit of initiative to produce medical solutions and generation of new knowledge

(*continued*)

Table 3. (*continued*)

Application process	
Teacher's role Student's role	**Teacher's role Student's role**
Before class (Inverted classroom)	
Selects and organizes the programmatic content to be developed in the learning situations in which the problems are to be solved Choose the sources of information that students can consult for the solution of problems individually or through teams Estimates the previous contents of the students about the subject to be developed in the solution of problems	Orients the student previously on what the learning situation will be about. Prepares for class, reviews and revises previous contents, researches related to the topic, reads and/or uses materials previously or personally indicated by the teacher
During	
Home	
It explains the procedure and steps for carrying out the problem-solving work For the solution of the problem, it is suggested to follow the following principles of reasoning: • Make an objective summary of the case. • Sort the information. • Prioritize symptoms and signs according to their sensitivity, specificity, predictive value, relative importance in the physiopathology of the disease, potential severity, etc. • Group the symptoms and signs under control • Distinguish between "hard" and "soft" syndromes [23] • - Do not hypertrophy the diagnosis by creating artificial syndromes • - Go from symptom and sign to syndrome, nosology and etiology • - To have a holistic vision avoiding reductionism • - There are sick people, not diseases.y "blandos"	Also the diagnosis can be by comparison, intuition, hypothesis A motivational activity is also initiated in accordance with what is going to be developed and based on the previous knowledge that the student has Assumes a posture of attention regarding the procedure of how the solution of the problem will be carried out It assumes a posture of attention on the motivational activity that the teacher presents before the resolution of the problem by their previous knowledge
Development	
Orients and supervises each of the groups according to the problem situation under study and the approach they use to solve the problem	They develop the work in the group based on work teams and following the methodology presented for its achievement, in one or a combination of the approaches according to how the problem situation is presented and the orientations given by the teacher

(*continued*)

Table 3. (*continued*)

Close	
Clarifies doubts Reinforce knowledge through exposure	Analyze the scope of the problem-solving results, what did or didn't I achieve, what did or didn't I do?
AFTER (Evaluation)	
Evaluates the discussion held to solve the problem Evaluates their performance and the learning achieved by the students Proposes improvements for activities, if necessary, performs reinforcements through tutoring	Presents, orally and/or in writing, his or her views on the subject and how the work and its product are developed
Expected lessons	
Problem-solving learning, autonomous learning, discovery learning, task-based learning, innovative learning	

Source: Eduardo Fernández

1.3 Discussion

According to Garrel and Guilera, a school disconnected from society is not a good omen for the future, this concept is fully adapted to the process that we are analyzing, it is difficult to understand that teachers of the XXI century, are with low levels of technological management, therefore means that the training provided is not in line with the time and that, unfortunately, is not training professionals for the future, perhaps it is doing for the present with limited skills that will soon be competitive disadvantages in the labor market. For the author Joyanes, the training for a performance in the industry 4.0, must be revised, so much the basic one as the university one and very especially the professional one, as well as the associated one to the update of knowledge of the active workers. These concepts fit perfectly in the analyzed problem, it can be pointed out that the Uniandes University would commit a greasy mistake if it allowed that its professors are outdated, also it becomes very dangerous the fact that there is a low level of technological handling **because** present and the future is eminently technological. It is possible to analyze the fact that a series of new technologies are approaching that must be learned by students because they are the ones that will be used in the real work environment and consequently will be used by new professionals.

1.4 Conclusions

The following conclusions can be drawn from the results obtained:

- The level of technical management of the teachers is quite low, that is why technology has not been an important element within the didactic support. This has generated an image of obsolescence on the part of the teachers in front of the students, who have not expressed themselves on the level of knowledge, but it has been possible to appreciate a disagreement with the fact that in the middle of the 21st-century technology is used as an element of didactic support.

- As for the specific handling of technologies, it can be concluded that some aspects of the elements of the web 2.0 are known, and that several tools belonging to it are used sporadically. It must be understood that the web 2.0 has already been used for more than 15 years, and it is understandable that it is known, although of course not at the expected level.
- As far as technology 4.0 is concerned, it can be concluded that not much is known about it, that something has been heard about it but that its learning, and even worse its application is very distant.
- Specifically, in the didactic aspects, it can be pointed out that the proposed model is very adaptable to any career, although its initial definitions are oriented towards medicine. It is obvious that not all the technologies of the so-called industry 4.0 can be directly applied, but some of them such as augmented reality, virtual reality, cloud computing, and 3d printing can be immediately adapted. Collaborative robotics is going to take some time to be integrated into educational processes, but in the not too distant future, it will be part of it.

In this eminently technological age, ubiquity is a fundamental characteristic, the use of multiple educational platforms that can be accessed from different electronic devices, also allows education to be developed in a face-to-face, virtual, and mobile way to generate ubiquitous learning that takes place anywhere, and at any time. Technology 4.0 is considered ideal for achieving ubiquitous learning.

References

1. Fernández, E.: U-learning. El futuro esta aquí. Alfayomega, México (2010)
2. Hernández, M., Urrutia, J.: Didáctica transdisciplinar en la Universidad. Edunt, Trujillo (2016)
3. Alzate, M.: Enseñanza y didáctiva universitarai. Ecoe ediciones, Bogota (2009)
4. Larriva, F.: La investigación de los modelos didácticos, 11 December 2001. https://e-spacio.uned.es/fez/eserv/bibliuned:20427/investigacion_modelos.pdf. Accessed 23 Jan 2017
5. Bedoya, J.: Epistemología y pedagogía. Ecoediciones, Bogotá (2005)
6. García, J.: Fundamentos del aprendizaje. Trillas, México (2008)
7. Lemaitre, M.: La educación superior como parte del sistema educativo en América Latina y el Caribe. IESALC, Unesco, Caracas (2018)
8. Páez, E., Montero, M.: eat, Educación y pedagogía. Pasajes, encuentros y conversaciones. UPTC, Universidad Pedagógica Tecnológica de Colombia, Tunja (2014)
9. Salamea, D.: Universidades del siglo XXI. Tomos I y II. Ediciones Killari, Quito (2015)
10. Medina, A., Salvador, F.: eat, Didáctica general. Pearson, Prentice-Hall, Madrid (2009)
11. Santiváñez, V.: Diseño curricular a partir de competencias. Ediciones de la U, Bogotá (2013)
12. Hernández, M.: La gestión didáctica para la formaciónde de administradores. Edunt, Trujillo (2017)
13. Burgos, J., Lozano, A.: eat, Tecnolgía educativa y redes de aprendizaje. Trillas, México (2010)
14. López, M.: Aprendizaje, competencias y Tic. Pearson Educación (2013)

15. Ariza, R.: José y Ruiz Palmero, Julio, Competencias, TIC e innovación. Ediciones de la U, Bogotá (2011)
16. Paredes, M., Gutierres, I.: eat, Recursos educativosen red. Síntesis, Madrid (2015)
17. Hébuterne, S.: Android. guía de desarrollo de aplicaciones móviles. Eniediciones, Barcelona (2014)
18. Benavides, R.: La planificación de la educación superior. Publicaciones Sapienti, Pucesa, Ambato (2014)
19. Montero, R.: Android. Desarrollo de aplicaciones móviles. Ediciones de la U. Ra-ma, Bogotá (2013)
20. Joyanes, L.: Industria 4.0. La cuarta revolución industrial. Alfaomega, Madrid (2018)
21. Garrel, A., Guilera L.: La industria 4.0 en la sociedad digital. Marge Books, Valencia (2018)
22. Paredes, J., De la Herran, A.: eat, La práctica de la innovación educativa. Síntesis, Madrid (2011)
23. Diaz, F., Hernández, G.: eat, Estrategias docentes para un aprendizaje ubicuo. McGrawHill, México (2010)

Good ICT Practices for the Integral Development of Ecuadorian Universities

Wladimir Paredes-Parada[1(✉)], Franz Del Pozo[2], Silvia Elizabeth García González[2], and Calvin Ndea[3]

[1] Instituto Tecnológico Universitario Rumiñahui, Sangolqui 171103, Ecuador
wladimir.paredes@ister.edu.ec
[2] Universidad Central del Ecuador, Av. Universitaria, Quito 170129, Ecuador
[3] University Jomo Kenyatta, University in Juja, Juja, Kenya

Abstract. The proliferation of information and communication technologies in all spheres of society, and especially in the area of education, has posed major challenges for developing countries. Bridging the gap in access to ICTs between different strata of society has become a constant task at all levels. The objective of this paper is to analyze the digital divide that exists between students and teachers in Ecuador's university institutions. The research was of an exploratory type, with a field design. A questionnaire was applied to a sample of 133 teachers and 906 students from a universe of 1654 people, selected through a stratified probabilistic sampling. The result of the exploration carried out allowed to point out the findings from a descriptive investigation, supported, equally, in background and theoretical references. The results show an initial reduction of the gap in the use of basic technological tools and the need to continue deepening in the use of new more complex technologies such as Academic ERPs, as a management and integration core of good ICT practices to achieve a comprehensive and harmonious development between the teaching - learning and academic - administrative processes, in order to alleviate the computer deficiencies that are still apparent in institutional controls, and thus ensure that good practices are adopted by higher education institutions to guarantee their comprehensive development for the benefit of education and the society in which they are immersed.

Keywords: Good practices in ICTs · Academic ERP · Data model · Data analytics · Integral development education

1 Introduction

In the process of technological development, which has taken place over the last few decades, various events have emerged through which society has attempted to describe the individual and social positions taken to face the changes. One of these developments has been the idea of transforming society in its economic, political, cultural and other dimensions brought about by the so-called technological revolution. Another relevant aspect in this change has been the conformation of the industrial society,

M. Botto-Tobar et al. (Eds.): CI3 2020, AISC 1277, pp. 163–170, 2021.
https://doi.org/10.1007/978-3-030-60467-7_14

which is the term coined by sociology to explain the modernization of the social structures of humanity in the light of the scope of the industrial revolution.

The industrial society, in the same way, was gestating changes in its foundations evolving to new models where the optimization and controls of the social doing was replaced by the processing and handling of the information giving step to what it calls society of the information, concept created by [8], that basically "concluded that the number of people who were dedicated to handle and process information was greater than the employees who made tasks based on a physical effort". A conclusion that years later, [1] complemented by pointing out that the "main axis of this will be theoretical knowledge and warned that knowledge-based services would become the central structure of the new economy and an information-based society".

In this theoretical referential context, the challenges that, from the beginning, the inclusion of the information and communication technologies in society has projected are visualized; called to contribute with the progress and solution of problems and oriented to influence directly in the culture of the population through the production of the knowledge. The information society, supported by technological resources, makes communication and transmission of this knowledge possible, strengthening global exchange among peoples. Thus, the innovation of technologies merged with the communicational process, denote today the knowledge society, where the modes of transfer and exchange have modified, in different ways, the ways in which activities are developed in modern society.

Undoubtedly, in the knowledge society, the environments of knowledge generation and production have meant greater equality and freedom of access. Producing, consequently, a higher level of education and socio-economic development of people; granting a high level of ethical responsibility to educational institutions, reaffirming its value as a social good that consolidates and generates scientific knowledge.

However, under this model, society is supposed to reach high levels of equality that allow democratization and access to information and communication technologies, promoting digital inclusion from a global vision of the world, but the reality is that the levels of social inequality are high, as well as the indices of economic, cultural, educational development on a planetary scale. There arises a problem in the implementation of new technologies when considering the existing digital gap and, therefore, the underutilization and technological dependence of developing countries with respect to developed ones.

Within this framework, there are studies that deal with the existing digital gap between nations. In the case of Latin America, the Economic Commission for Latin America and the Caribbean - [2] stated that "in order to present and examine the elements that characterize the dissemination of the digital paradigm and its impact on the growth and productivity of the countries of Latin America and the Caribbean, it is necessary to analyze the dynamics of national and international digital divides faced by the countries of the region, identifying their main determinants: the level and distribution of income and the formal education of the population, variables that are closely linked. A fundamental proposal for public policymaking emerges from these analyses.

Based on these guidelines, various studies and activities have been carried out in Ecuador to improve the quality of higher education institutions (IES), in order to overcome the digital divide and allow for the appropriate incorporation of information and communication technologies into academia. The premise is that every university should take into account that there is a technological gap between students and teachers, and the adequate application of policies and good practices in university teachers and researchers will allow this gap to be increasingly reduced and thus strengthen the teaching-learning process in education institutions.

In the modern day and age, every institution of higher education should have a guide of good practices for the proper use and application of technology based on the basic knowledge of teachers and researchers, in order to strengthen their operative effectiveness in support of their core processes - teaching and learning and administration.

Most Ecuadorian higher education institutions make use of the Learning Management Systems (LMS). These systems are used to deliver virtualized teaching and learning for components to learners stationed across the country and globally. Moodle is an LMS, which enjoys great acceptance among the Higher Education Institutions (IES) of Ecuador for the support of the teaching-learning process.

It has been observed that in the Ecuadorian IES, access to permanent to broadband Internet and technically acceptable bandwidth for a fast phased adoption and the intensive use of ICT to support educational process is possible. This complemented by Wi-Fi access networks, adequate libraries, infrastructure and teachers with postgraduate and doctoral degrees, or in the process of obtaining these academic degrees, contribute to the creation of a quality ICT dependent educational environment.

However, with regard to the academic management support systems used to plan, execute and control the academic-administrative processes of the IES with the objective of proper functioning of institutional management, there are precarious technological delays that make these processes more difficult and expensive, rendering the Ecuadorian IES to adopt static and time barred processes that end up generating inconsistent and unreliable information, with the indisputable consequences that this situation generates.

2 Scope

The origins of the work with ICTs in Ecuador can be seen in a regional group that originated the idea of merging criteria and strategies. [10] describe the strategy of incorporating ICTs into initial teacher training through educational reforms that seek to expand coverage and improve quality in higher education institutions and teacher training institutes in the countries of the Andean Region: Bolivia, Colombia, Ecuador, Peru, and Venezuela; they present 23 experiences in four of the five countries mentioned, which formed part of the research work: "Teacher Training and ICTs, an approach to mapping in the Andean Region.

The study carried out by [4] on the introduction of ICTs at the National Autonomous University of Mexico (UNAM) served as a prelude; as well as those of [7], who postulated the importance of the integration of ICTs by teachers in higher education classrooms, their idea was to propose ten online platforms, free and easy to use, to promote information among students and teachers under an online methodology; this new pedagogical model would be based on the concept understood as blended-learning.

For their part, [6] presents the challenges faced by teachers in the use of ICTs in higher education institutions and the changes in teaching arising from an exploratory study carried out within the framework of a national programme developed in Greece. The study highlighted the need for policy change in education. [11], carries out a study focused on the use of ICTs, from the student factor and demonstrates that the new professional in training must adapt to the conditions of the constant evolution of new technologies, if he wants to become a problem solver in production and services.

For their part, [9] present the ethical vision in the use of ICTs in higher education, pointing out that academic failure has an enormous weight in teacher management, since teachers are resistant to change and the use of new technologies. In essence, they support the need to change educational policies and include technology, its prac-tice and use as part of the moral values of the teacher in teaching.

[12] are conducting a study at Victoria University (Australia) and show the difficulties arising from the same local, national and international ICT challenges and propose a range of software needed for teacher upgrading and academic management to enable the autonomous handling of writing, numerical calculations and computer information in different contexts and media. Sample Heading (Third Level). Only two levels of headings should be numbered. Lower level headings remain unnumbered; they are formatted as run-in headings.

3 Methodology

The work focused on the use of both logical and empirical methods, as the analysis approach took the mixed paradigm as a reference. The research efforts were directed to determine, in the country, the use of technological tools, the Internet and virtual education, as well as the use of updated resources in higher level institutions for the search and use of information.

Surveys were based on the work of [7] and addressed to teachers, administrators and students. Stratified probability sampling was used in the selected institutions of higher education.

4 Results

The results of the study determined that: the technological resources used by teachers are dependent on their age, as can be seen in the following Table 1:

Table 1. Use of electronic and computer equipment.

Edad	Total edad	Proyector	PC	Laptop	Tablet	Smart phone	Phablet	Pizarra digital	USB	OTRO
25–29	10	63,3	70	75,6	62,2	58,9	16,7	11,1	68,9	6,7
30–34	20	58	57,6	82,6	19,3	82,6	0	8,3	72,3	0
35–39	21	66,7	80	86,7	60	43,3	3,3	10	63,3	3,3
40–44	25	68,8	87,5	87,5	27,1	81,3	2,1	0	91,7	8,3
45–49	20	77,9	92,3	57,2	45,4	57,2	5,1	9,2	39,5	0
50–54	19	70,6	90,5	78,4	37,2	59,7	0	9,1	70,6	0
55–59	11	85,7	92,9	53,6	46,4	75	0	0	42,9	0
60	1	77,8	88,9	66,7	22,2	66,7	0	0	66,7	0
Promedios		**71,1**	**82,5**	**73,5**	**40,0**	**65,6**	**3,4**	**6,0**	**64,5**	**2,3**

Phablets and digital whiteboards are considered to be more technologically advanced instruments, but they are rarely used (Table 2).

Table 2. Use of virtual education technology tools.

Rango	Total edad	Moodle	Educativa	Blackboard	Efront	Ninguno	Otra
25–29	10	0	51,1	13,3	0	28,9	6,7
30–34	20	16,3	23,5	3	0	26,9	37,5
35–39	18	36,7	23,3	3,3	0	43,3	0
40–44	27	10,4	31,9	0	0	29,9	27,8
45–49	19	12,8	56,9	9,2	0	7,7	13,3
50–54	20	44,4	16,9	4,8	0	12,6	19,7
55–59	11	0	19,6	0	0	53,6	32,1
60 +	7	33,3	22,2	11,1	0	33,3	0
Promedios		**19,2**	**30,7**	**5,6**	**0,0**	**29,5**	**17,1**

In the analysis it can be observed that there is very little percentage of teachers who use virtual education systems, despite this being one of the main tools in current education (Table 3).

Table 3. Tools - Technological resources.

Rango	Total edad	Bibliotecas digitales	Catálogos digitales	Repositorios digitales	ISI WEB	Otros
25–29	10	93,3	42,2	34,4	17,8	0
30–34	20	84,5	28,8	14,4	40,2	0
35–39	18	93,3	30	20	20	6,7
40–44	27	84,7	31,9	18,8	4,2	4,2
45–49	19	88,2	66,2	50,3	19,5	40
50–54	20	84,4	70,6	36,6	15,6	0
55–59	11	73,2	32,1	39,3	51,8	0
60+	7	77,8	11,1	44,4	33,3	0
Promedios		**84,9**	**39,1**	**32,3**	**25,3**	**6,4**

In the analysis it can be seen that the great majority uses virtual libraries as a tool for their educational activities, although this seems to be more a requirement of higher education control bodies than a real motivation (Table 4).

Table 4. Tools - internet resources.

Rango	Total edad	Mailin list	Blog	Naveg. internet	Co-rreo	Wiki	Pizarra virtual	Skype	Hangout	Face	Twitter	Mess	Video confe	Apps phone	2nd life	Otro
25–29	10	00	0	69	71	29	11	41	29	30	42	24	46	11	29	6,7
30–34	20	7,2	11	81	77	36	10	10	9,1	47	28	19	13	55	10	0
35–39	18	17	17	83	90	50	17	33	6,7	30	27	13	33	17	6,7	3,3
40–44	27	19	15	44	65	25	15	13	13	29	15	17	17	26	8,3	4,2
45-49	19	2,6	43	25	79	57	2,6	12	43	21	64	12	48	39	50	0
50–54	20	3	38	66	60	46	16	4,8	22	38	52	35	45	14	9,5	0
55–59	11	7,1	39	79	73	46	25	7,1	7,1	34	29	34	39	32	7,1	0
60	7	11	11	56	78	67	22	11	33	11	67	44	0	11	22	0
Promedios		8,3	21,8	62,7	74,0	44,5	14,8	16,5	20,2	30,0	40,3	24,8	30,1	25,5	17,9	1,8

It can be observed that the most widely used tool in all age groups is email. One of the important facts is that a tool as well known as Facebook is little used. In general, social networks have low percentages, with Twitter being the tool most commonly used on average. Phone applications for educational purposes are used in a low percentage (Table 5).

Table 5. Tools - programs.

Rang	Total edad	Word	Power point	Base de datos	Excel	SPSS	Prog diseño	Prog diseñoWEB	Prog video edit	Prog PDFedit	Prog reprod	Prog proy	Lattes	Otros
25-29	10	86,7	71,1	41,1	75,6	11,1	6,7	17,8	11,1	42,2	11,1	11,1	0	0
30-34	20	91,7	88,6	15,5	85,6	4,2	10,2	25,8	14,4	87,5	10,2	14,4	7,2	0
35-39	18	93,3	86,7	23,3	83,3	10	6,7	13,3	10	60	6,7	13,3	6,7	0
40-44	27	91,7	82,6	16,7	71,5	2,1	8,3	2,1	6,3	31,3	6,3	2,1	4,2	8,3
45-49	19	97,4	85,6	7,7	74,9	2,6	21	2,6	14,4	73,8	51,8	9,2	0	33,3
50-54	20	97	84,4	13,9	74,9	19,7	4,8	15,6	27,5	67,1	37	27,5	3	3
55-59	11	92,9	92,9	14,3	73,2	25	19,6	7,1	0	48,2	21,4	12,5	7,1	0
60	7	88,9	77,8	11,1	66,7	0	0	11,1	11,1	22,2	0	0	0	0
Promedios		**92,4**	**83,7**	**17,9**	**75,7**	**9,3**	**9,7**	**11,9**	**11,8**	**54,0**	**18,1**	**11,3**	**3,5**	**5,6**

As can be seen in this table almost all use Word as a working tool, followed by Power-Point, Excel and Pdf editing, which means that the use of Microsoft desktop tools continues to dominate the academic sector.

It is important to note, the very low percentage of Lattes use, probably because it is a more complex tool. Following the same trend, it is observed that the SPSS tool, widely used in all types of research and statistical analysis, is little used.

According to [5], the adoption by IES of decision support systems that are adaptable to academic resource planning is essential. Higher education institutions are increasingly interested in using flexible business model approaches (ERPs) to solve their academic resource allocation problems. However, very few schools have opted for this approach, much more considered that the planning of the allocation of academic resources is not a well-structured, data supported decision-making process and therefore it is necessary to adopt academic systems that are flexible and easily adaptable through parameterization and builds an information model that supports near real-time decision making. In addition, many of the decision-makers are necessarily involved in the academic planning process and may take different perspectives on the importance of achieving different goals and objectives. Successful resource management solutions can be expected to vary considerably from one decision-maker to another, as the cognitive processes, perceptions and assessments of the individual are taken into account

5 Conclusions and Recommendations

The analyses carried out led to the following conclusions and recommendations:

All higher education institutions in the country need to train teachers in ICTs, taking advantage of the model proposed by UNESCO [3, 10] to improve teacher practices in teaching and learning.

IES should provide teachers with computer equipment that allows adequate functionality of office packages such as word processors, presentations and spreadsheets. These devices must necessarily have a USB connector among other connectivity features. The data analyzed show that teachers use them more frequently, so it is the responsibility of higher education institutions to provide this fundamental input for the development of their academic classes.

Another of the technological equipment in which teachers must have the necessary competence to be able to use them is the digital blackboards, since these serve to adapt the interactivity of the students with their professors. The non-existence of the digital blackboard limits the possibility of knowledge generation and the extension of research, since it would oblige to carry out operational activities that take a long time, both for the teacher and the student.

For virtual education, the teacher must be prepared to handle LMS platforms, according to the study the most common LMS is Moodle. The use of these virtual educational platforms is fundamental for all teachers of the different modalities, both for full-time, part-time and distance, since it is an aid for the management of the teaching-learning process in which both teachers and students are partners.

These competencies must be aligned with the creation of courses and their administrative management, as well as the use of various resources that allow an adequate virtual education.

The study determined that teachers and students are not using databases to generate research and knowledge, so a policy of continuous improvement in the application of the use and exploitation of academic databases should be carried out.

Teachers should be able to use social networks for educational purposes, due to the tendency of their use among students, which allows instantaneous communication and group management.

It is important that IES adopt academic ERPs to automate their processes, this will generate information to support institutional decision making, as well as facilitate the adoption of a culture based on processes with reliable information. Academic ERPs have a great advantage over traditional academic management systems as they adapt easily to institutional and regulatory changes in education, as well as to the different visions of academic planners in an environment of multiple conflicting objectives.

In the academic ERP market, there are generally expensive solutions such as Ellucian's Banner, Universitas' OCU, among others, which in most cases makes adoption by Ecuadorian IES prohibitive. However, the emergence of academic ERPs, such as ERP a1- academia, based on open architectures offer the possibility that education centers with low budgets may consider it a viable option.

References

1. Bell, D.: The coming of Post-Industrial Society A venture in social. Forecasting, Harmondsworth, Peregrine (1976)
2. CEPAL: Conferencia Ministerial Regional Preparatoria de América Latina y el Caribe para la Cumbre Mundial sobre la Sociedad de la Información. CEPAL. Bávaro, Punta Cana, República Dominicana (2003)
3. Enlaces: Estándares TIC para la formación inicial docente: Una propuesta en el contexto chileno. Enlaces. Chile (2008)
4. Fombona, J., Pascual, M.: Las tecnologías de la información y la comunicación en la docencia universitaria. Estudio de casos en la Universidad Nacional Autónoma de México. (UNAM). Educación XXI **14**(2), 79–110 (2015)
5. Franz, L.S., Lee, W.M., Van Horn, J.C.: An adaptive decision support system for academic resource planning. Decis. Sci. (12), 276–293 (1981)
6. Kalogiannakis, M.: Training with ICT for ICT from the trainee's perspective. A local ICT teacher training experience. Educ. Inf. Technol. **15**(1), 3–17 (2010)
7. Lareki, A., Martínez de Morentin, J.I., Amenabar, N.: Towards an efficient training of university faculty on ICTs. Comput. Educ. **54**(2), 491–497 (2010)
8. Machlup, F.: La producción y distribución de conocimiento en los Estados Unidos. N. J. Princ. Univ. Press, Princeton (1962)
9. Nordkvelle, Y.T., Olson, J.: Visions for ICT, ethics and the practice of teachers. Educ. Inf. Technol. **10**(1–2), 21–32 (2005)
10. Rozo, A., Prada Dussán, M.: Panorama de la formación inicial docente y TIC en la Región Andina. Revista Educación y Pedagogía **24**(62), 191–204 (2013)
11. Van Der Vyver, G.: The search for the adaptable ICT student. J. Inf. Technol. Educ. Res. **8** (1), 19–28 (2013)
12. Venables, A., Tan, G.: Measuring up to ICT teaching and learning standards. Issues Inf. Sci. Inf. Technol. **9**, 29–40 (2013)

Electronics

Material Selection, Simulation and Validation for Cop Coils High Voltage Spark Plug Boots Insulators

Eduardo Portilla[1], Juan Gabriel Espinosa Aguilar[1], Javier Martínez-Gómez[1,2(✉)], and Gustavo Moreno[1]

[1] Universidad Internacional SEK, Quito Albert Einstein s/n and 5th, Quito, Ecuador
javier.martinez@uisek.edu.ec
[2] Instituto de Investigación Geológico y Energético (IIGE), Quito, Ecuador

Abstract. The following research presents the study and selection of an alternative composite material to manufacture electrical insulation devices for the high voltage circuit of Otto cycle internal combustion engines, better known as gasoline engines. High voltage insulators are generally made of materials such as rubber or silicone, whose material, when subjected to high temperatures, suffers technical degradation, causing electric power leaks and engine power losses. The purpose of this study is to suggest a composite material, with which only these insulating bodies can be manufactured. To do this, multicriteria selection techniques are used while the results are validated by virtual simulation of thermal character and normed experimental tests.

Keywords: Automotive vehicle · Temperature · Composite material · Electric power · Electronic circuit

1 Introduction

The selection of materials is one of the most important stages when designing a product. From the point of view of engineering design, the selection of materials is the process that aims to identify the appropriate materials for manufacturing processes [1].

Multi-criteria decision making (MCDM) methods for selecting materials are considered one of the design strategies implemented to achieve product efficiency [2, 3]. Because each product is different, each one may require numerous functions that could not be satisfied using a single material for all of them [4, 5]. A design that incorporates the selection of multiple materials is a viable alternative to achieve the functional requirements of a product. According to [1, 6]. The implementation of multiple materials in the design of the product leads to a greater performance of the product in terms of functionality, manufacturing capacity, costs, and aesthetics. For Caliskan, H. [7] MCDM methods are based on the comparative study of candidate materials: since they are techniques that help defining the characteristics of what is sought, a list of candidate materials must be established in order to determine what the main function will be and later select a single one. According to Aly et al. [8] selecting an appropriate

M. Botto-Tobar et al. (Eds.): CI3 2020, AISC 1277, pp. 173–184, 2021.
https://doi.org/10.1007/978-3-030-60467-7_15

material to manufacture a product generally involves a comparative evaluation of several criteria and attributes of the material, which are important when offering it as a service. This means that the comparison between materials is key when determining a material as suitable. On the other hand, it is necessary to validate the results of the MCDM. The design simulation helps manufacturers verify and validate the intended use of a product in the development phase, as well as its manufacturing capacity [9, 10]. Generally, the term "simulation" is used as a generic term for computer-aided engineering (CAE) [11, 12]. Several design simulations approaches have become standard components of product development in different sectors and continue to gain in importance, as this economical, fast, affordable and easy-to-use design simulation software allows users to cope with new technologies and applications.

Simulation models are sets of mathematical equations that represent the response of the system in a physical domain of interest. The complexity of mathematics depends on the availability of the data and varies according to the application and the design stage [11]. These simulations can cover aspects such as structural behavior, acoustics, system dynamics, resistance to shocks, thermal and flow analysis, stress analysis, fuel economy, the development of controls and much more [12].

However, none of the mentioned articles uses the MCDM methods with laboratory simulation and experimentation to enhance the design phase and corroborate the results. The purpose of this paper was to perform a MCDM of a spark plug boot (SPB). After the MCDM selection process, a thermal origin simulation was performed using the NX Nastram Software in the thermal simulation module. Finally, the results of the simulation were verified by submitting several specimens of SPB constructed with the material selected and considered optimal to laboratory tests of thermal and dielectric nature.

2 Materials and Methods

The materials and methods include the operation of the piece, the properties of materials used in the study, the explanation of the MCDM, how simulation was performed, and the thermogravimetry test. Hereunder, each of them will be explained.

2.1 Piece Performance

Automotive ignition coils are an essential part of automotive engines. According to Skinner & Lovers [13] they are complex small devices that initiate combustion in an engine. They provide high voltage electric power (34 kV), or spark to ignite the stoichiometric air-fuel mixture.

These devices are known as independent coils (coil on plug, COP). COP coils are composed of two main parts: the body of the coil or high voltage transformer (HVT), which houses the electrical and electronic components responsible for providing high voltage, and the high-tension lead (HTL) that transports energy from the coil to the spark plug. It should be noted that the HTL is removable from the coil. The HTL is covered with an electrical insulator which prevents electrical energy from leaking out of this circuit known as spark plug boot (SPB). The SPB high-voltage insulation undergoes technological degradation of the material, due to the working conditions, causing failures such as excessive slack in the boot in the sector that shelters the upper part of the spark plug or fissures in the material. The result of this degradation is a leakage of electrical energy, which causes the combustion of the stoichiometric mixture in the cylinder where the degradation takes place to be incomplete or null, thus causing failures such as: power losses of the engine (25% approximately), excessive fuel consumption due to the compensations managed by the multi point fuel injection system (MPFI) and high pollutant emissions. For these reasons, it is necessary to replace the insulating SPB. However, in the Ecuadorian market, the spare part cannot be found, but rather the complete COP coil, i.e., the SPB and the HVT coil together. But whenever there is a problem that only affects the SPB, buying the complete COP coil is an unnecessary expense, which has strong economic and environmental consequences (carbon footprint). Therefore, it would be ideal to be able to independently find the SPB [13–19].

2.2 Materials

The choice of materials was made through the preparation of a list with seven polymeric materials, which were evaluated under the following preliminary criteria:

1. having dielectric, thermal, chemical and mechanical qualities, which are necessary to fulfill the indicated function;
2. being easily found in the Ecuadorian market;
3. having a competitive price.

A matrix where the candidate materials and the criteria that will be considered can be seen is built. The materials were assigned the letter M while the criteria were assigned the letter C, as can be seen in Table 1. Only the first six criteria will be considered since criteria 7 and 8, corresponding to the chemical qualities, have the same value in all materials. The materials to be studied are high-density polyethylene (HDPE), low-density polyethylene (LDPE), polystyrene (PS), polyvinyl chloride (PVC) Polyamide 6- Nylon PA6, copolymer of Ethylene-Tetrafluoroethylene ETFE - Teflon and Polypropylene PP. The choice is based on the references [13, 20–22] that dwell on their properties. In all the criteria, a higher is better, except in the elasticity module.

2.3 Multicriteria Selection Methods and Weighting Method

The multicriteria selection methods used are: VIKOR, PUGH, TOPSIS, PROMETHEE, DOMINIC and COPRAS. Mentioned methods need the criteria with which the selection will be made to have a weight or quantitative value, for which the weighting method known as STATISTICAL VARIATION was applied.

Method of Statistical Variation: It is an objective statistical method that gathers the values of all the variables of a preestablished decision matrix and concatenates them with each other numerically until finding the value that most closely matches the ideal. This will assign a higher weight to the numerically higher criterion [2].

Table 1. Selected materials and their properties

Materials (M)	Dielectric strength V × m (C1)	Work temperature (C2)	Price $ × Kg. (C3)	Coefficient of thermal expansion × $10^{-6}K^{-1}$ (C4)	Thermal conductivity $Wm^{-1}K^{-1}$ (C5)	Elasticity module GPa (C6)	Resistance to Hydrocarbons (C7)	Resistance to grease and oils (C8)
Polyethylene AD (M1)	22	120	4	100	0.33	0.3	Yes	Yes
Polyethylene BD (M2)	22	90	4	100	0.52	1.2	Yes	Yes
Polystyrene (M3)	20	95	4.25	70	0.17	1.65	Yes	Yes
PVC (M4)	14	75	4	75	0.25	4.0	Yes	Yes
Nylon (M5)	25	160	4	95	0.28	3.0	Yes	Yes
Teflpon (M6)	25	160	4.5	90	0.24	0.8	Yes	Yes
Polypropylene (M7)	30	120	5	100	0.22	1.5	Yes	Yes

VIKOR Method: VIKOR method is the MCDM method originally developed to solve problems of deciding alternatives and criteria. Assuming that compromise is acceptable for conflict resolution, the decision maker wants a solution that is the closest to the ideal, and the alternatives are evaluated according to all established criteria. VIKOR ranks alternatives and determines the solution named compromise that is the closest to the ideal [23].

PUGH Method: For [24], the PUGH method is used to evaluate options in a design process, categorizing and quantifying criteria, material, processes, characteristics, etc. It gives them certain importance in order to create a scale.

TOPSIS Method: The TOPSIS method is used to sort preferences by similarity to ideal solution. TOPSIS is a multiple criteria method to identify solutions from a finite set of alternatives. The basic principle of TOPSIS method is to choose the alternative that has the shortest distance from the positive ideal solution and the farthest distance from the negative ideal solution. An ideal solution is defined as a collection of scores or values for all attributes considered [25].

PROMETHEE II Method: The PROMETHEE method involves using a list of alternatives and building a series of scales, which can be positive or negative and net, according to the desired characteristics of a certain valuation process. The process is as follows [26].

DOMINIC Method: The Dominic procedure is a qualitative method of material selection, which considers weight factors in the selection criteria. The selection is made with matrices. In the rows, the criteria and the weight factors are included, while the columns contain the candidate materials, identified with a letter. The operative method is described in [27].

COPRAS Method: The COPRAS method is a decision-making method which combines the values of the criteria with the weights of the criteria. Priorities are thus obtained, which are relative, based on the qualities looked for in a material. These positive and negative priorities are evaluated in such a way that performance levels of each material option can be observed, that is, their measurement is a percentage. In this way, a classification of the materials according to their numerical performance can be achieved [2].

2.4 Spearman Correlation Coefficient

Once the results of the MCDM have been obtained, the existence of a correlation between them must be measured, for which a tool such as the Spearman Correlation is used. Values obtained with this method can score from −1.0 to 1.0, going through 0, interpreted as follows: if they approach 1.0, it means that there is a positive correlation; if they approach −0.1, there is a negative correlation; and if the value is 0, there is no correlation. To do this, the results of the MCDM must be analyzed in pairs, making combinations so that they all come together.

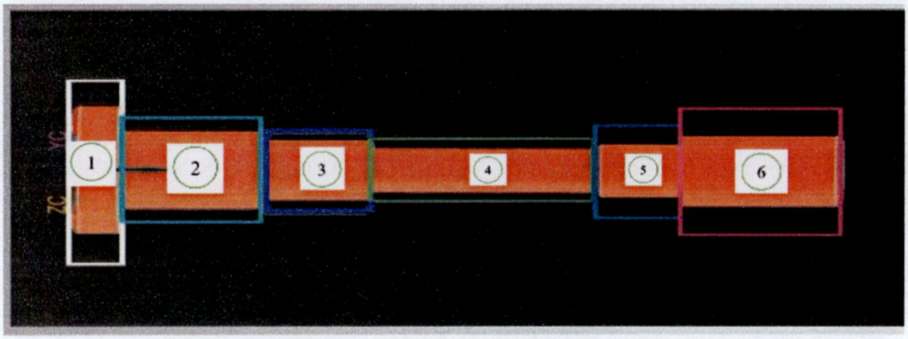

Fig. 1. SPB divided in 6 quadrants

2.5 Simulation

The simulation process shows us how effective could be the winner material in a virtual extreme work.

Procedure: 1. Open a new Fem and Sim environment; 2. Import a 3D CAD model; 3. Load the material from the software library; 4. Mesh the object 5. Set the thermal charge up; 6. Set the thermal restrictions up; 7. Run.

Thermal Simulation: This environment is known as Fem and Sim, which are the virtual places where the simulation will be carried out in the NX software and where a CAD model of the SPB will be loaded. The Fig. 1 shows a specimen of SPB divided in 6 quadrants in order to analyze the temperature effects in each one of them (Table 2).

Table 2. Border conditions for simulation

Material	Load (W)	Temperature °C	Coefficient of Thermal conductivity 23 °C ($V\ m^{-1}\ K^{-1}$)	Thermal exposure time From O sec. to 600 s	Computer time by simulation (Min.)
Polyethylene BD	30	160	0.3	Yes	10
Polyethylene AD	30	160	1.2	Yes	12
Polystyrene	30	160	1.65	Yes	14
PVC	30	160	4.0	Yes	15
Nylon	30	160	3.0	Yes	20
Teflon	30	160	0.8	Yes	17
Polypropylene	30	160	1.5	Yes	21

2.6 Thermogravimetry Test

The experimentation process tries to test the thermal characteristics of an SPB prototype built with the material with better results, carried to extreme values, beyond those that the material possesses.

The Thermogravimetric experimental process was carried out in the laboratory of the CIAP Polymer Applied Research Center of the National Polytechnic School of Ecuador. The test was carried out in accordance with the ASTM D3850-12 standard "Rapid Thermal Degradation of Solid Electrical Insulating Materials by Thermogravimetric Methods (TGA)". The test conditions are specified in Table 3.

Table 3. Test conditions

Factors	Characteristics/values
Laboratory:	CIAP/EPN
Norm:	ASTM D3850-12
Equipment:	Thermo balance
Gas:	Nitrogen
Gas flow:	50 ml/min
Crucible:	Platinum

3 Results and Discussion

The materials and methods include the results obtained in the MCDM, the results of the simulation in each of the materials, and the thermogravimetry test in the chosen material. Each will be explained below.

3.1 Results of the MCDM

In MCDM the winner material is which reaches the closest position to ZERO. A matrix where the candidate materials can be seen was constructed in Table 1. The next step was to apply the multicriteria methods in order to determine the most suitable material to manufacture the SPB insulators in a technical way. To be able to apply the multi-criteria methods, the criteria with which the selection was going to be carried out must be precisely taken into account: the dielectric strength, the working temperature, the coefficient of thermal expansion, the thermal conductivity, the modulus of elasticity, the resistance to hydrocarbons, the resistance to fats and oils, and the price. Once the criteria were determined, each criterion was to be weighted, that is, it had to be determined which of the criteria was the most important in this selection of materials. According to [2] the weighting factors express the relative importance of each criterion.

3.2 Results of the Statistical Variation Method

Figure 2 shows a statistical comparison of the results, where the best material is the one that holds the first position of the overall ranking the highest number of times. Nylon obtained the first place four times, and the second place twice.

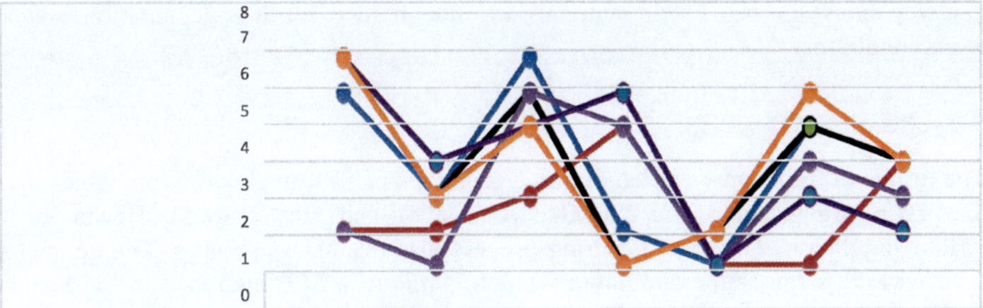

	Polietileno BD	Polietileno AD	Poliestireno	PVC	Nylon	Teflón	Polipropileno
VIKOR	6	3	7	2	1	5	4
PUGH	2	2	3	5	1	1	4
TOPSIS	7	3	6	1	2	5	4
PROMETHEE	2	1	6	5	1	4	3
DOMINIC	7	4	5	6	1	3	2
COPRAS	7	3	5	5	2	6	4

Fig. 2. Comparison of MCDM results

The results of the correlation of results of the MCDM methods appear in Table 4. It is observed that the VIKOR method has a correlation of 0.92 with TOPSIS and 0.85 with COPRAS, which indicates a great correlation. However, it has a low correlation with the PUGH method. In general, the PUGH method has correlations below 0.55.

Table 4. Spearman correlation in MCDM results.

	VIKOR	PUGH	TOPSIS	PROMETHEE	DOMINIC
PUGH	0.071	–	–	–	–
TOPSIS	0.92	−0.17	–	–	–
PROMETHEE	0.5	0.54	0.2	–	–
DOMINIC	0.46	0.42	0.46	0.38	–
COPRAS	0.85	−0.3	0.85	0.85	0.28

3.3 Results of Simulation

In the figure number three it is possible to see every single material, in the hardest conditions of work, with a security factor of 1,6, which means that, the temperature is around sixty percent higher than the regular work temperature. The color scale determines how warm are the several sections of the specimen of SPB. Being the red one the highest temperature (160 °C) and the de blue one the lowest temperature (60 °C) with an orange and yellow scale in between.

The heating points were setting up in the closest areas to the combustion chamber, that is why the warmest quadrants are the number 5 and 6. The materials that reach the red color show that these materials are getting very close to the thermal degradation. In that way the Nylon and Teflon materials are enough away for their degradation in such work conditions.

3.4 Results of the TGA

The results of TGA were carried out on a prototype of SPB made of nylon, which was the best resulting material of the MCDM selection and showed its ideal behavior in virtual simulation. The manufacturing process was manual machining. The geometry used was the same as the simulation, which is that of a SPB equipped in the Suzuki G16B engine used by Grand Vitara vehicles. The results obtained after the TGA thermogravimetric test, to a prototype built with Polyamide 6 (nylon), can be observed graphically in Fig. 14, where the curve of thermal degradation of the material starts from 400 °C approximately, evidencing that the SPB built with nylon will perfectly support the 1630 °C that it will have to face inserted in the cylinder head of an internal combustion engine (Figs. 3 and 4).

Table 5 shows the exact values thrown by the TGA, which warn of a degradation calculated as SPB weight loss from 420 °C.

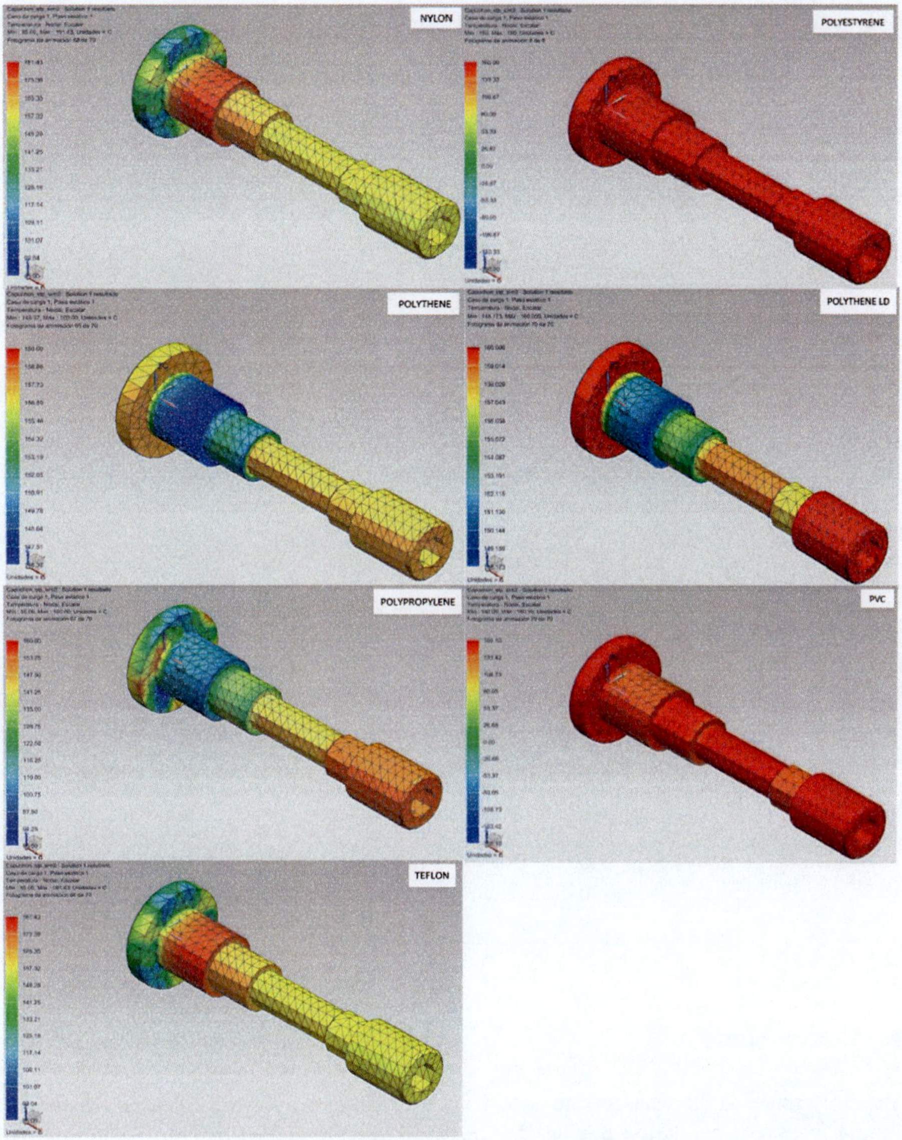

Fig. 3. Comparison of SIMULATION results

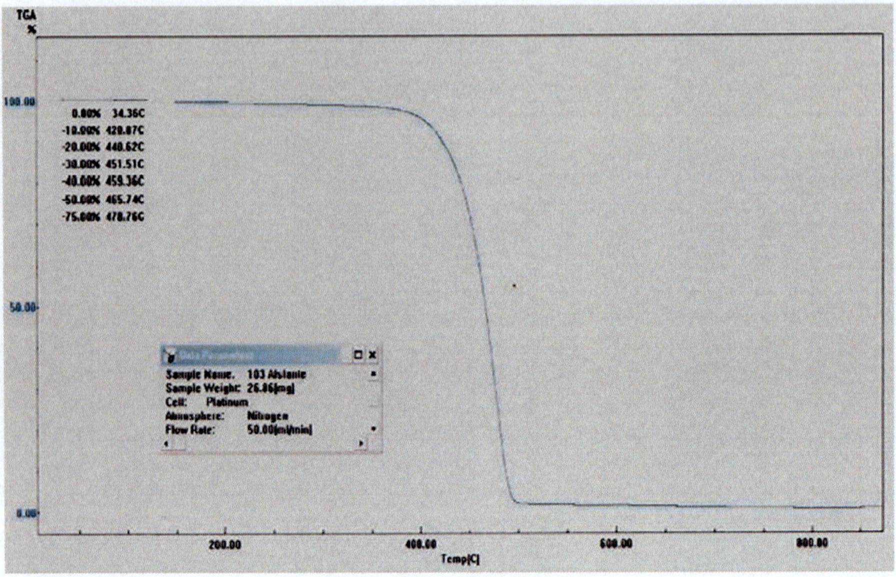

Fig. 4. Results of TGA

Table 5. Thermal degradation values of the SPB constructed with Nylon in the TGA

Lost weight (%)	Temperature
10	420.9
20	440.6
30	451.5
40	459.4
50	465.7
75	478.8

4 Conclusions

The present research was able to select an alternative composite material, to manufacture electrical insulation devices for the high voltage circuit of Otto cycle internal combustion engines. The selection of the material was made using three selection procedures: the application of multicriteria methods, virtual simulation, and experimentation. The use of these three methodologies helped each one validates the previous one and allowed establishing real limits.

It was determined that, of the non-traditional materials used to manufacture SPB studied in this research, the material suggested for constructing high-voltage insulators for COP coils is nylon. It was determined that 4 of the 6 multicriteria selection methods used in this research place nylon as the first option while the other two place it as a second option, representing 66.7% acceptance of nylon as the ideal material to manufacture SPB, according to the MCDM.

Through the virtual simulation process, it was determined that in the same environment where the edge conditions are the same, the thermal behavior of the materials varies significantly, since the quadrants show behavior alterations according to the material. It was established that, in order to comply with the dielectric isolation process of the SPB constructed with nylon, the dimensions are determinant, since the nylon has a low modulus of elasticity in comparison with the silicone. In this way, to be coupled to the geometry of the spark plug and perform the ideal insulation, it needs to have a minimum clearance.

Acknowledgement. This research takes part of the project P121819, Parque de Energias Renovables founded by Universidad International SEK.

References

1. Ashby, M.F., Cebon, D.: Materials selection in mechanical design. Le Journal de Physique IV **3**(C7), C7–C1 (1993)
2. Villacreses, G., Gaona, G., Martínez-Gómez, J., Jijón, D.J.: Wind farms suitability location using geographical information system (GIS), based on multi-criteria decision making (MCDM) methods: the case of continental Ecuador. Renewable Energy **109**, 275–286 (2017)
3. Villacís, S., Martínez, J., Riofrío, A.J., Carrión, D.F., Orozco, M.A., Vaca, D.: Energy efficiency analysis of different materials for cookware commonly used in induction cookers. Energy Procedia **75**, 925–930 (2015)
4. Gaona, D., Urresta, E., Marínez, J., Guerrón, G.: Medium-temperature phase-change materials thermal characterization by the T-History method and differential scanning calorimetry. Exp. Heat Transf. **30**(5), 463–474 (2017)
5. Kastillo, J.P., Martínez, J., Riofrio, A.J., Villacis, S.P., Orozco, M.A.: Computational fluid dynamic analysis of olive oil in different induction pots. In: 1st Pan-American Congress on Computational Mechanics–PANACM 2015, pp. 729–741 (2015)
6. Aldás, P.S.D., Constante, J., Tapia, G.C., Martínez-Gómez, J.: Monohull ship hydrodynamic simulation using CFD. Int. J. Math. Oper. Res. **15**(4), 417–433 (2019)
7. Caliskan, H., Kursuncu, B., Kurbanglu, C., Güven, S.: Material selection for the tool holder working under hard milling. Mater. Des. **45**, 473–479 (2013)
8. Aly, M.F., Hamza, K.T., Farag, M.M.: A materials selection procedure for sandwiched beams. Mater. Des. **1980–2015**(56), 219–226 (2014)
9. Godoy-Vaca, L., Almaguer, M., Martínez-Gómez, J., Lobato, A., Palme, M.: Analysis of solar chimneys in different climate zones-case of social housing in Ecuador. IOP Conf. Ser. Mater. Sci. Eng. **245**(7), 072045 (2017)
10. Gomes, L.C., Miranda, J., Mergulhão, F.J.: Operation of biofilm reactors for the food industry. In: Computational Fluid Dynamics in Food Processing, pp. 561–590 (2019)
11. Kastillo, J.P., Martínez-Gómez, J., Villacis, S.P., Riofrio, A.J.: Thermal natural convection analysis of olive oil in different cookware materials for induction stoves. Int. J. Food Eng. **13** (3) (2017)
12. Martínez-Gómez, J., Guerrón, G., Riofrio, A.J.: Analysis of the "Plan Fronteras" for clean cooking in Ecuador. Int. J. Energy Econ. Policy **7**(1), 135–145 (2017)

13. Skinner, A.A., Lovers, H.O.: Ignition Coil. United States Patent, 7 (2013)
14. Espinoza, V.S., Guayanlema, V., Martínez-Gómez, J.: Energy efficiency plan benefits in Ecuador: long-range energy alternative planning model. Int. J. Energy Econ. Policy **8**(4), 52–54 (2018)
15. Rodríguez, D., Martínez-Gómez, J., Guerrón, G., Riofrio, A.: Impact of induction stoves penetration over power quality in Ecuadorian households. Revista ESPACIOS **40**(13) (2019)
16. Martínez, J., Martí-Herrero, J., Villacís, S., Riofrio, A.J., Vaca, D.: Analysis of energy, CO2 emissions and economy of the technological migration for clean cooking in Ecuador. Energy Policy **107**, 182–187 (2017)
17. Acurio, K., Chico-Proano, A., Martínez-Gómez, J., Ávila, C.F., Ávila, Á., Orozco, M.: Thermal performance enhancement of organic phase change materials using spent diatomite from the palm oil bleaching process as support. Constr. Build. Mater. **192**, 633–642 (2018)
18. Villacreses, G., Salinas, S.S., Ortiz, W.D., Villacís, S., Martínez-Gómez, J.: Environmental impact assessment of internal combustion and electric engines for maritime transport. Environ. Process. **4**(4), 907–922 (2017)
19. Martínez-Gómez, J., Ibarra, D., Villacis, S., Cuji, P., Cruz, P.R.: Analysis of LPG, electric and induction cookers during cooking typical Ecuadorian dishes into the national efficient cooking program. Food Policy **59**, 88–102 (2016)
20. Goodfellow. (2018)
21. Rosas, R.M.M., Genesca, M.M., Juan, X.R.: Aplicaciones dieléctricas del poliestireno cargado con neumáticos. Revista DYNA **88**(6), 652–662 (2013). (in Spanish)
22. Guerrero Salazar, C.A., González González, V.A.: Algunas aplicaciones de los plásticos en las industrias del empaque. Ingenierías **5**(17), 40–45 (2002). (in Spanish)
23. Beltrán, R.D., Martínez-Gómez, J.: Analysis of phase change materials (PCM) for building wallboards based on the effect of environment. J. Build. Eng. **24**, 100726 (2019)
24. Martínez-Gómez, J.: Material selection for multi-tubular fixed bed reactor Fischer-Tropsch reactor. Int. J. Math. Oper. Res. **13**(1), 1–29 (2018)
25. Villacreses, G., Martínez-Gómez, J., Quintana, P.: Geolocation of electric bikes recharging stations: city of Quito study case. Int. J. Math. Oper. Res. **14**(4), 495–516 (2019)
26. Chingo, C., Martínez-Gomez, J.: Material selection using multi-criteria decision making methods for geomembranes. Int. J. Math. Oper. Res. **16**(1), 24–52 (2020)

Performance Comparison of Two Electronic Controllers on an ARM Platform

William Montalvo(✉), Marcelo Ortega, and Eduardo Avilés

Universidad Politécnica Salesiana, UPS, 170146 Quito, Ecuador
wmontalvo@ups.edu.ec

Abstract. Nowadays, the advances in technology and the development of new processes have driven the engineering areas to implement new control strategies to optimize the use of resources, thus guaranteeing the quality of the final product. This document presents a comparative study of two electronic controllers; a modern incremental Reference Signal Tracking (RST) control and a conventional Discrete Proportional Integral and Derivative (PID) control, aimed at a temperature process belonging to a Control Plant Trainer (CPT) developed on an STM32F4 platform with Advanced RISC Machine (ARM) technology. In order to know the efficiency of the modern Incremental RST control, the operation comparison is made against a Discrete PID control under the same working conditions. The behavior of Incremental RST control against conventional PID control is statistically analyzed using data obtained both in simulation and in real time, through Matlab specialized Simulink software, together with the Waijung library for data delivery and acquisition.

Keywords: Diophantic, Integral Absolute Error (IAE) · Integral Time Absolute Error (ITAE) · Proportional, Integral and Derivative (PID) · Reference Signal Tracking (RST)

1 Introduction

Given the high levels of industrial competition, the increase in the price of resources and the internationalization of control operations, companies need to improve their processes through industrial automation and the use of new controllers.

Certain stages of an industrial process require a high level of security, accuracy and precision, which demands the implementation of advanced control techniques applied in controllers with better performance for data delivery and acquisition. For this purpose, the STM card of the 32F4 (STM32F4) family is proposed.

The evolution of embedded devices allows the inclusion of complex algorithms for data processing, storage capacity, speed of communication and response.

Within industrial automation there are several control systems to obtain the least error in the work processes. The PID control and the different hybrids that can be formed are the first option when automating a process. Although the PID controller is widely used, it presents a problem in the configuration and tuning of its parameters. An research is presented in [1], which concludes that the PID control, compared with the

M. Botto-Tobar et al. (Eds.): CI3 2020, AISC 1277, pp. 185–197, 2021.
https://doi.org/10.1007/978-3-030-60467-7_16

fuzzy control, is deficient to eliminate the error in the stable state of the system, which avoids to reach a high precision.

The control of temperature as a variable is relevant in the different fields of application: transportation of medical products, storage of living tissues and blood transfusion are examples of areas where PID-fuzzy control has been used [2].

Currently, some studies show the advantages of combining fuzzy control, neural networks and PID control to achieve an advanced control system, however, much of the data obtained is dependent on the controlled object and its results cannot be applied to real engineering.

A controller by pole assignment is the proposal of an optimized model for the solution of the above-mentioned problems, which reduces the complexity of implementation and does not need to acquire the precise model of the controlled object [3].

This paper proposes the comparative analysis of a modern controller by pole assignment, Incremental RST versus a conventional discrete PID controller, which will show that control stands out in efficiency on processes where the temperature variable is involved. The adequate control of the temperature or a working variable in an industrial process generates savings in resources, improving productivity without losing the quality of the final product. The Incremental RST controller stands out in its operation since, by means of an integrator in the control law, allows to generate an error in stationary state equal or approximated to zero generating a robust performance when facing perturbations [4]. Unlike the Discrete PID control, which has only feedback and a degree of freedom, the RST control stands out due to its feedback and prefeed, which filters the reference value and attenuates the over-pulses [5].

2 Methodology

2.1 Identification of the System

The black box method was used to recognize the mathematical model of plant operation with which data was collected over a sampling period of 0.01 s during 30 min of operation. The data obtained through the temperature sensor belonging to the EPC reflect the operation of the temperature led, which responds to a PWM with 50% duty cycle. The information is acquired through the STM card and, by using UART communication, the data is sent to the computer for storing, graphing and obtaining the numerical values that are processed with the Matlab System Identification Toolbox (IDENT), and thus establishing the mathematical model of the plant.

Matlab offers the possibility of obtaining the mathematical model in continuous time and its equivalent in discrete time. Figure 1 shows the response of the identified system together with the real response of the system, presenting a 90.62% of graphic similarity.

The transfer function obtained in continuous time has the form of Eq. (1):

$$G(s) = \frac{8.113s + 0.968}{2516s^2 + 325.4s + 1} \tag{1}$$

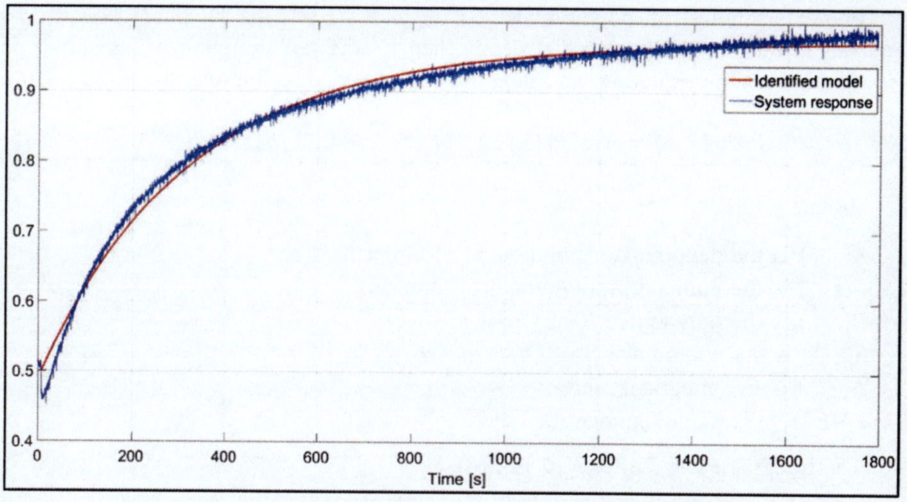

Fig. 1. System behavior characteristic curve.

The transfer function obtained in Discrete Time has the form of Eq. (2):

$$G(z^{-1}) = z^{-1}\frac{0.003209 - 0.002848z^{-1}}{1 - 1.878z^{-1} + 0.8787z^{-2}} \tag{2}$$

Calculation of Values for Incremental RST Control

Incremental RST control is an alternative option to conventional control, which bases the design of its control system on a polynomial equation approach.

The system to be controlled has as a transfer function:

$$G(z^{-1}) = \frac{B(z^{-1})}{A(z^{-1})} \tag{3}$$

The law of control is established by Eq. (4):

$$R(z^{-1})\Delta u(t) = T(z^{-1})y_i(t) - S(z^{-1})y(t) \tag{4}$$

The control law represents an alternative approach to design through the location of poles with a minimum order state observer, and Eq. (4) defines that control is made up of the blocks $\frac{S(z^{-1})}{R(z^{-1})}$ and $\frac{T(z^{-1})}{R(z^{-1})}$, which allows the controller to have two degrees of freedom [6].

As it is a system with two degrees of freedom, it can be evaluated using the "Monitoring and control with independent objectives" method. The $T(z^{-1})$ filter provides the monitoring behavior and the $R(z^{-1})$ and $S(z^{-1})$ filters provide the regulation behavior [7].

For determining the polynomials $R(z^{-1})$ and $S(z^{-1})$, the Diophant equation is equated to the desired polynomial determined by the pole assignment, as shown in Eq (5). In the same equation, an Integrator (Δ) is added to improve the accuracy [8].

$$PMF\left(z^{-1}\right) = A\left(z^{-1}\right)\left(1 - z^{-1}\right)R\left(z^{-1}\right) + \left(z^{-d}\right)B\left(z^{-1}\right)S\left(z^{-1}\right) \quad (5)$$

Where:

$A(z^{-1})$ is the denominator polynomial of Order n,
$B(z^{-1})$ is the numerator polynomial of Order n,
$R(z^{-1})$ is the polynomial R of Order n,
$S(z^{-1})$ is the polynomial S of Order n,
$\Delta(z^{-1})$ is the integrator, and
$PMF(z^{-1})$ is the assignment of Poles.

The Integrator has the form of Eq. (6)

$$\Delta\left(z^{-1}\right) = \left(1 - z^{-1}\right) \quad (6)$$

The polynomial corresponding to the pole assignment of Eq. (7) is composed of Eq. (8) and Eq. (9)

$$PMF\left(z^{-1}\right) = 1 + P_1 z^{-1} + P_2 z^{-2} \quad (7)$$

$$P_1 = -2e^{\varepsilon * wn * Tss} \cos\left(wn * Tss\sqrt{1 - \varepsilon^2}\right) \quad (8)$$

$$P_2 = -2e^{\varepsilon * wn * Tss} \quad (9)$$

The obtaining of the polynomial $T(z^{-1})$ is determined by the Eq. (10):

$$T\left(z^{-1}\right) = S_0 + S_1 + S_2 \quad (10)$$

In the design of the Incremental RST controller, it was defined a setting time $Tss = 3.25$ s and a damping factor $\varepsilon = 0.8$, so that the system output quickly reaches the set point, therefore the system poles are P1 and P2 detailed below:

$$wn = \frac{4}{\varepsilon * Tss} = \frac{4}{0.8 * 3.25} = 1.5385$$

$$P1 = -0.3524$$

$$P2 = -0.5841$$

This implies that the Diophantic equation would be established as follows:

$$\begin{aligned}1-0.3843z^{-1}-0.6105z^{-2} &= \left(1-1.878z^{-1}+0.8787z^{-2}\right)*\left(1-z^{-1}\right)*\left(R_0+R_1z^{-1}\right)\\ &\quad + z^{-1}\left(0.003209-0.002848z^{-1}\right)*\left(S_0+S_1z^{-1}+S_2z^{-2}\right)\end{aligned} \tag{11}$$

$$R\left(z^{-1}\right) = \left(0.001-1.1519z^{-1}\right)*10^3$$

$$S\left(z^{-1}\right) = \left(3.5978-7.1502z^{-1}+3.5542z^{-2}\right)*10^5$$

$$T\left(z^{-1}\right) = 175.7376$$

2.2 Discrete PID

The most widespread and well-known digital controller is the Discrete PID, which is the sum of three terms: Error proportional + Error integral + Error derivative [9].

Its discrete-time transfer function is given by the following equation:

$$K_p + k_i \times \frac{T_s}{z-1} + K_d \times \frac{1}{T_f + \frac{T_s}{z-1}} \tag{12}$$

Where:

K_p is the proportional gain of the PID controller in parallel represented by a dynamic system,
k_i is the integral gain of the PID controller in parallel represented by a dynamic system,
K_d is the gain derived from the PID controller in parallel represented by the dynamic system,
T_s is the filter time constant of the PID controller in parallel represented by the dynamic system, and
T_f is the sample time of the dynamic system [10].

Matlab Simulink Discrete PID block is used to calculate the coefficients. This block handles Euler's rectangular forward approximation method, which provides the following values:

$$K_p = 16.6,\ K_i = 0.235,\ K_d = 82.65,\ T_f = 12.69,\ T_s = 1$$

Figure 2 shows the block diagram in Matlab Simulink for Discrete PID Control.

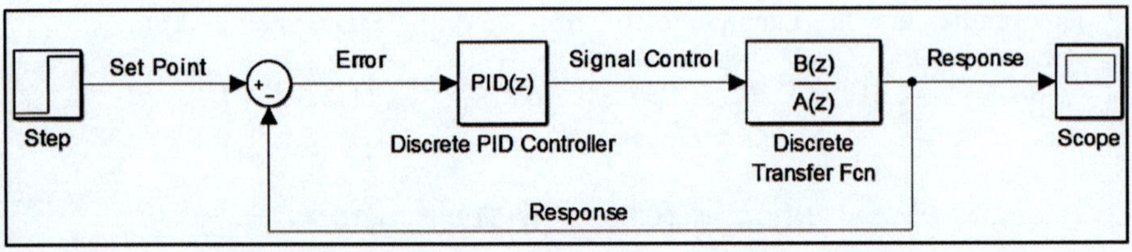

Fig. 2. Discrete PID control simulation.

2.3 Implementation of the Control in the Control Plant Trainer (CPT)

Figure 3 details the architecture applied for the implementation of the Incremental RST and Discrete PID controllers.

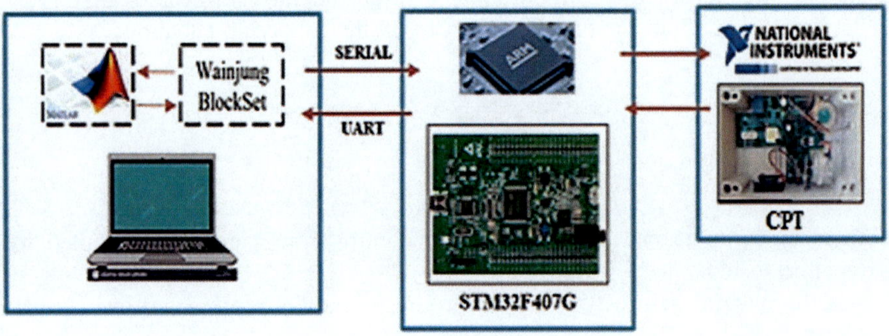

Fig. 3. Architecture for implementing Incremental RST and Discrete PID control

Figure 4 shows the blocks belonging to the Waijung library that allow the design of the experimental scheme for the control of the temperature variable by means of the Incremental RST controller implemented in the STM32F407G.

The platform where the Incremental RST control is implemented is a technology that requires the Waijung library to establish the communication between Matlab software and the STM32F407G card through serial communication, however, a UART module is used to receive the data in real time.

2.4 Performance Analysis of Electronic Controllers Using the Wilcoxon Statistical Model

The Wilcoxon statistical model allows comparing between two samples of results. Control performance indices such as: Integral Absolute Error (IAE) and Integral Time Absolute Error (ITAE), obtained from the Incremental RST and Discrete PID controllers, were considered to perform the comparative analysis in this study. The procedure to perform the Wilcoxon method is referenced in [11].

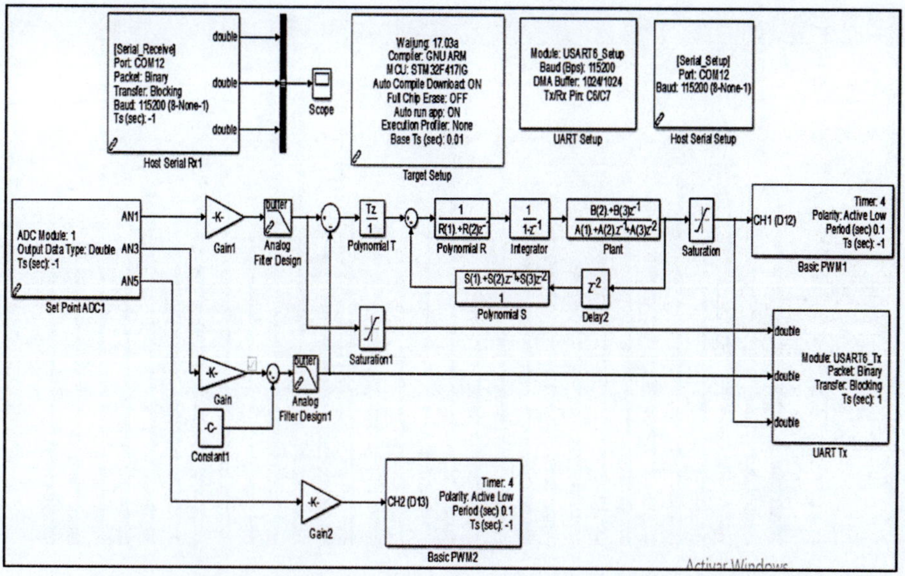

Fig. 4. Block program for implementing the Incremental RST control.

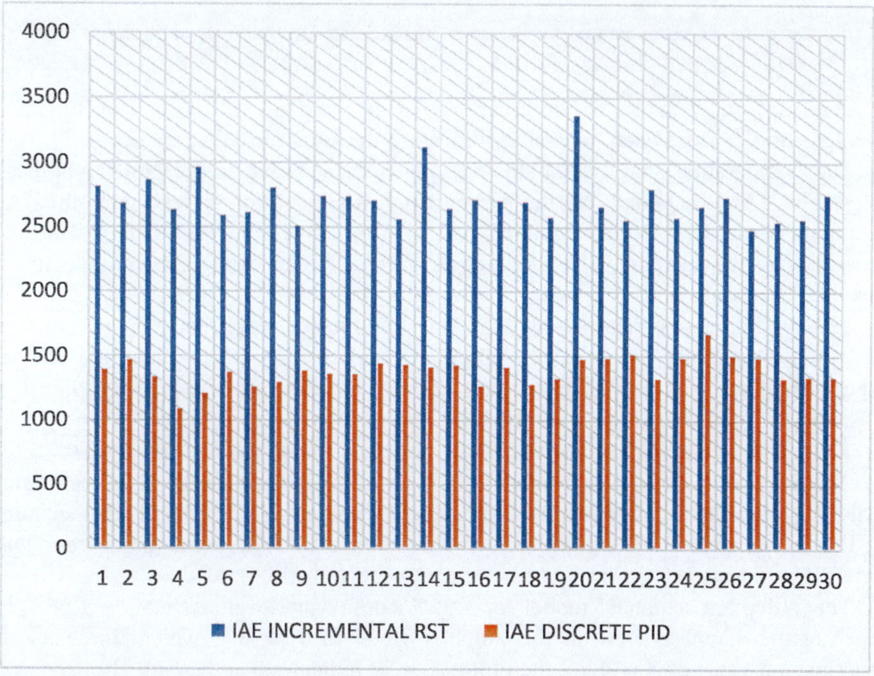

Fig. 5. IAE generated by RST Incremental and PID Discrete controllers.

Fig. 6. ITAE generated by Incremental RST and Discrete PID controllers.

Figures 4 and 5 show the work carried out by the plant with a total of 30 experimental samples from each controller indicating the IAE and ITAE respectively.

The comparative analysis for this research indicates that in the Null Hypothesis (H_o): The IAE generated by the Incremental RST control is less than the IAE belonging to the PID Discrete control, and as Alternative Hypothesis (H_a): The IAE generated by the Incremental RST control is greater than the IAE belonging to the Discrete PID controller.

The Wilcoxon statistical model for a 95% confidence level sets a $Z_a = 1.96$. The experimental samples show in the graph a value of $T = 464$, which indicates an experimental value of $Z = 4.76$; therefore, $Z > Z_a$ indicates that the Null Hypothesis is rejected (H_o) and the Alternative Hypothesis is accepted (H_a).

The comparative analysis for this research details that in the Null Hypothesis (H_o): The ITAE generated by the Incremental RST control is lower than the ITAE belonging to the Discrete PID control and as Alternative Hypothesis (H_a): The ITAE generated by the Incremental RST controller is higher than the ITAE belonging to the Discrete PID controller.

The Wilcoxon statistical model for a 95% confidence level sets a $Z_a = 1.96$. The experimental samples show in the graph a value of $T = 464$, which, indicates an experimental value of $Z = 4.76$; therefore, $Z > Z_a$ indicates that the Null Hypothesis is rejected (H_o) and the Alternative Hypothesis is accepted (H_a).

3 Results

3.1 Data Analysis Obtained from Simulation Belonging to the Incremental RST Control for Temperature Led Control by PWM

In order to know the appropriate response, the simulation is carried out together with the transfer function corresponding to the plant. The block diagram in Fig. 7 allows the system response to be adapted under the desired conditions. When guaranteeing the working values of the R, S and T variables, they are used in the real system.

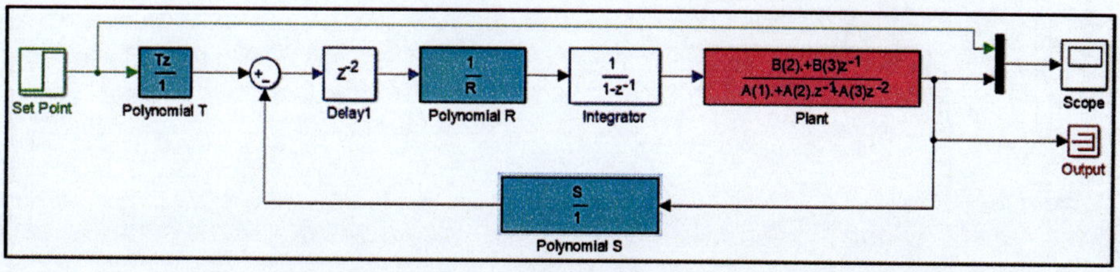

Fig. 7. Incremental RST control structure designed in Simulink.

Figure 8 represents the behavior of the Incremental RST control in relation to a Set Point of 50 °C, achieving the desired amplitude in a time of 150 s. Setting time was accepted, since by reducing it, the amplitude of the over-pulses is increased, which generates an excessively unstable response in the first seconds of work.

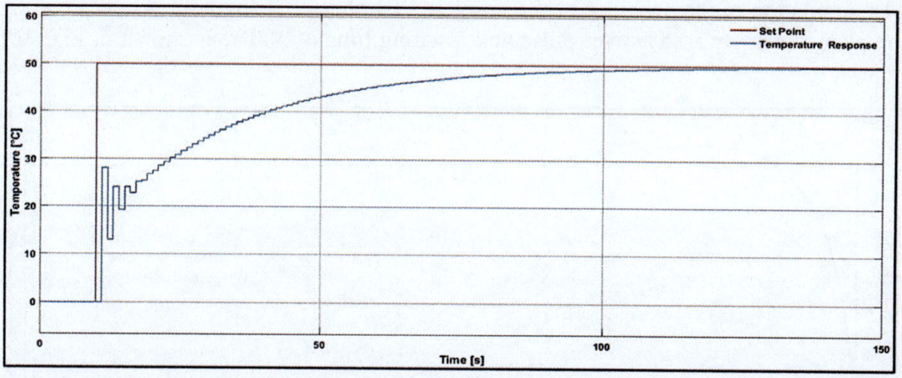

Fig. 8. Output signal of the Incremental RST control in response to the Set Point of value 50 °C

Analysis of data obtained from real time belonging to the Incremental RST control for temperature led control by PWM

Figure 9 details the real behavior of the plant in facing a Set Point of 50 °C, defining a setting time of 450 s. The first seconds of work do not reflect a response in the increase of temperature because the actuator is generating over-pulses of the controller in its design characteristics. Beyond this period, the temperature increase is progressive, justifying the temperature peaks with the increase or decrease of the work cycle in the PWM.

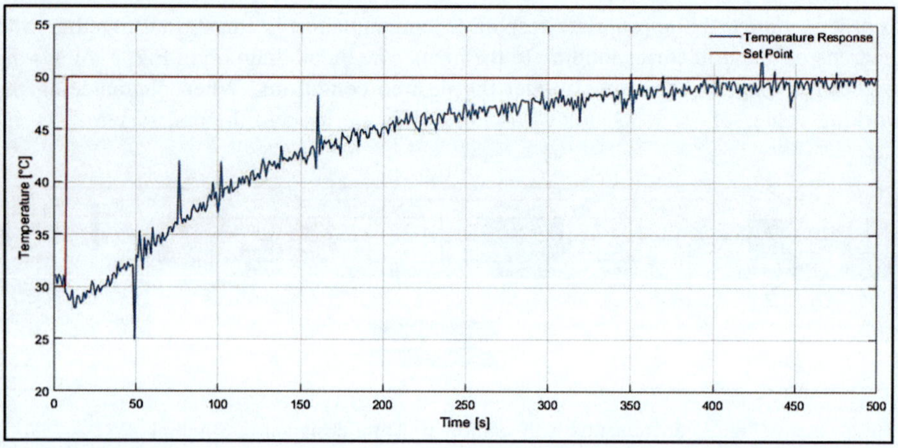

Fig. 9. Response of the Incremental RST controller: Plant temperature signal, temperature led control signal

3.2 Analysis of Data Obtained from Simulation Belonging to Discrete PID Control for Temperature Led Controller by PWM

The simulation of the behavior of the Discrete PID control facing a 50 °C Set Point of the plant shows an 8.85% over-pulse and a setting time of 300 s, as shown in Fig. 10.

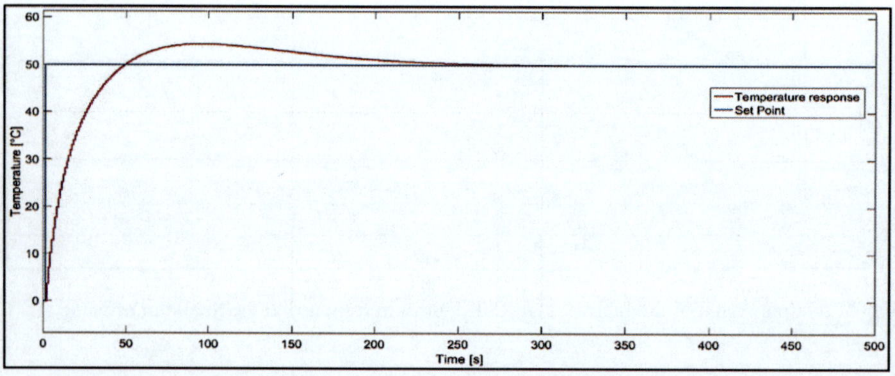

Fig. 10. Discrete PID control output signal in response to Set Point of value 50 °C

The aim was to reduce the setting time with the least amount of over-pulse, and to make it as robust as possible in order to obtain controlled growth in the temperature value in the transitory state.

Analysis of data obtained in real time belonging to Discrete PID control for temperature led control by PWM

Figure 11 details the real behavior of the plant facing a Set Point of 50 °C, defining a setting time of 350 s. The first seconds of work reflect a sudden response in the increase of temperature, due to the fact that the actuator is generating a PWM with a 100% work cycle, which is a characteristic of the controller design. After this period, the temperature increase is progressive, justifying the plant's behavior.

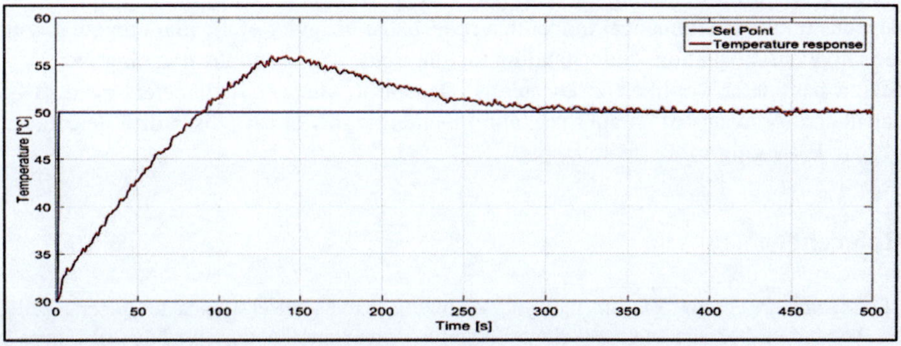

Fig. 11. Actual response of the Discrete PID Controller

4 Discussion on Controller Performance

According to [4], which establishes that the work of the Incremental RST controller has a good stability and accuracy as reflected in Fig. 9, its slow response leads to generate high values in the performance indexes compared to the Discrete PID controller as reflected in Figs. 6 and 7.

According to the study carried out in [11] which establishes that the values obtained from the simulation are similar in time and behavior with real data, this study was carried out based on a CDAQ-9172 card with the understanding that the control is done from the computer with the Matlab and LabView software. This paper analyzes an independent control applied on a STM32F4 card which generates the contrast between real and simulated data.

5 Conclusions

Through the analysis of the simulations by each controller, the efficiency of the Incremental RST control against the Discrete PID is observed both in setting time and in the IAE and ITAE values. In the design of the controllers, the aim was to avoid high amplitude over-pulses at the beginning of the work cycle, understanding that it generates a greater setting time for the two classes of control.

In comparison to the performance indexes obtained through simulation, it was determined that the Incremental RST control is more efficient, because the controller does not have over-pulses and mainly because the setting time is less than the Discrete PID as detailed in Figs. 8 and 10. However, a clear change of the IAE and ITAE performance indexes in favor of the Discrete PID control is observed as shown in Figs. 5 and 6. Due to the fact that they present a different setting time and rise time than the simulated one, these values are lower in the Discrete PID than in the Incremental RST, as shown in Figs. 9 and 11.

The Incremental RST control stands out in the design since it does not need a tuning method for obtaining the R, S and T vectors. This is due to the fact that it only depends on the mathematical model that reflects the behavior of the plant together with the Diophantic equation, understanding that the setting time and the damping factor are design parameters, unlike the Discrete PID control where its parameters need to be calculated by means of an approximation method which always discards a percentage of data belonging to the real sample.

References

1. Shaoming, L., Peng, Y., Fan, Y., Guoli, Z.: Investigación sobre el sistema de temperatura del horno de cal basado en un algoritmo PID difuso y un control óptimo, 14 July 2014. https://bibliotecas.ups.edu.ec:2095/document/6853097
2. Liu, Z., Chang, L., Luo, Z., Ning, F.: Design of vehicle-mounted medical temperature control system, 8 December 2016. https://bibliotecas.ups.edu.ec:2095/document/7774735
3. Jiahui, Z., Wenlan, W., Zhiming, L.: La optimización del sistema de control de temperatura de vapor basado en la asignación de polos, 18 July 2015. https://bibliotecas.ups.edu.ec:2095/document/7494447
4. García Jaimes, L.E., Giraldo Arroyave, M.: Controladores Avanzados en PLC. Reviste Politécnica **8**(14), 60 (2012)
5. Galdos, G., Karimi, A., Longchamp, R.: RST Controller Design by Convex Optimization Using Frequency-Domain Data, 2 September 2011. https://www.sciencedirect.com/science/article/pii/S1474667016454510
6. Viloria, C.A.R.: Desarrollo e implantación de técnicas control adaptativo en tiempo discreto para un péndulo simple, April 2009. http://bdigital.ula.ve/storage/pdftesis/pregrado/tde_arquivos/8/TDE-2012-08-28T22:10:06Z-1588/Publico/ratiacarlos.pdf
7. Bordry, H.T.F.: RST Digital Algorithm for Controlling the LHC Magnet Current, 28 October 1998. https://cds.cern.ch/record/377330/files/lhc-project-report-258.pdf
8. Ostertag, E., Godoy, E.: RST-controller design for sinewave references by means of an auxiliary diophantine equation. In: IEEE Conference on Decision and Control, no. 44, p. 6906, December 2005
9. Ramos Lara, R.: Sistemas Digitales en Tiempo Discreto, February 2007. https://upcommons.upc.edu/bitstream/handle/2117/6123/TEMA6.pdf2007
10. MathWorks: Access coefficients of parallel-form PID controller - MATLAB piddata - MathWorks América Latina. La.mathworks.com. https://la.mathworks.com/help/control/ref/piddata.html

11. Amat Rodrigo, J.: Prueba de los rangos con signo de Wilcoxon, January 2016. https://rpubs.com/Joaquin_AR/218464
12. Reyes Sierra, H.I., Montaña Ortega, M.F.: Modelamiento y Control Digital de Temperatura para un Horno Eléctrico, Bogotá (2010)

Bioelectricity Production with Organic Substrates, Nitrates and Lead Using High Andean Soils

Alex Guambo(✉), Cristina Calderón, Silvia Paña, and Magdy Echeverría

Grupo de Investigación y Desarrollo para el Ambiente y el Cambio Climático (GIDAC), Faculty of Science, Escuela Superior Politécnica del Chimborazo, Riobamba, Ecuador
aguambo@espoch.edu.ec

Abstract. The production of bioelectricity is linked to sintróficos processes, hydrolization and fermentation of complex organic compounds to be used like by-products for the generation of volts of direct current "Vdc". There is a great variety of substrates, ranging from pure compounds (organic and inorganic) to the complex mixture of organic matter present in wastewater, or inorganic matter of industrial waste.

Since 2016, the GIDAC Group has been studying the behavior of a bacterial strain from Páramo soils not intervened at altitudes of 4000 and 4200 masl as inoculum and producer of bioelectricity in different capacities of Microbial Fuel Cells "MFC" of simple configuration with the cathode partially exposed to air. The microbial consortium used in the organic mixture of organic residues of vegetables and fruits demonstrates a complete degradation in fruit waste, which suggests that the preference for waste is easily biodegradable. Of the MFCs positive correlations between the microbial abundance and the generation of electricity were not always observed, obtaining an average voltage of 330Vdc. However, when using inorganic compounds, lower voltage levels were obtained on average of 0.10 Vdc, but the voltage production had a significantly proportional relation to the concentration supplied, the same occurs at the highest concentration of heavy metals, voltage output shows a maximum of 228 Vdc, also achieving elimination efficiencies of 1.71% and 21.35% in 48 h, respectively, which indicates that certain microbial communities produce more bioelectricity if the appropriate substrate is supplied depending on the use and application of an MFC.

Keywords: Evaluation · Substrates · Bioelectricity

1 Introduction

Microbial Fuel Cells (MFC) have been of great scientific interest by their multiple applications, considering devises that transform the chemical energy that is present in substrates from potentially electrogenic microorganisms into electrical energy, these devises could reach be an alternative in the future to face problems of energy and treatment of wastewater with no conventional and even self-sustaining systems, due to

M. Botto-Tobar et al. (Eds.): CI3 2020, AISC 1277, pp. 198–208, 2021.
https://doi.org/10.1007/978-3-030-60467-7_17

its potential to degrade organic matter represented as substrate and at the time produce electric energy symbolized in volts of direct current (VDC). All this being possible due to MFC gives favorable conditions that makes the metabolic activity produced in one of its compartments liberate electrons to one electrode called anode instead to natural electron acceptor (oxygen), oxidizing microbial activity generating electrons, protons and CO2 [1]. The excess of protons in the cathodic section tend to react with oxygen present in the atmosphere, due to the latter chamber is aerobic, thus reaching to electrochemical equilibrium forming H_2O as final product.

1.1 Background

According to Kiely [2], the microbial community of bio electrochemical systems are very important in the bioelectricity production, due to certain microorganisms generate syntrophic processes, it means, they hydrolyze and ferment organic complex compounds and use byproducts for bioelectricity generation. So, during these years, scientists have dedicated themselves to the characterization of certain microbial communities, however, the microorganisms behave as different manner and depends on the used substrate. The used substrate for the bioelectricity generation is a key piece to obtain good results. There is a great variety of substrates, ranging from pure (organic and inorganic) to the complex mixture of organic matter present in wastewater or inorganic matter of industrial residues. Whatever of their nature, they serve as the main principal source of carbon to generate bioelectricity.

1.2 Substrates

As substrates or fuel source for bioelectricity production a wide variety of substrates are used from pure substances such as glucose or acetate to complex mixtures as leachate, whey, and organic and inorganic waste. In the majority of investigations, the used different nature substrate (organic, inorganic, synthetic or real), it has a potential influence in the structure and composition of the diverse microbial community that are formed to oxidize organic matter, and this has a direct relation with the efficiency and bioelectricity production [3]. Also, several exoelecrogenic bacteria cannot use any substrate [4]. Liu [5] used acetate that generated a potency of (800 mg/L) (506 mW/m^2 or 12.7 mW/L). Chaudhuri [6] used glucose, being one of the most common substrates, since bacteria had the capacity to oxidize glucose to CO2 and transfer the electrons quantitatively to a support such as graphite, resulting in long term electricity stable production. Huang [7] used synthetic wastewater in two different concentrations of xylose (0.5 mM to 1.5 mM), obtaining a maximum voltage of 55 $\pm$ 2.0 mV to 70 $\pm$ 3.0 mV respectively. Other investigations tend to use real substrates such as substrate of brewery where Feng [8] obtained a maximum potency of 205 mW/m^2, and if wastewater from paper industry were used, they obtain a maximum potency of 672 $\pm$ 27 mW/m^2 [9]. However, in both investigations, the common factor is they need to take into account other parameters as resistance of wastewater, conductivity of solution and storage capacity in the buffer that is a limiting factor at the moment to generate bioelectricity. In some cases, scientist use date syrup to obtain a maximum potency of 53.70 mW/m^2 at a concentration of 3 g/L, this substrate has high concentrations of glucose and is an acceptable substrate for the bioelectricity production [10]. Parot [11] used garden compost for the bioelectricity production, and also enriched it with acetate after the biofilm was formed, obtaining better results, indicating that certain microbial communities produce more bioelectricity if it is supplied with appropriate substrate.

1.3 Novelty

In 2016, Grupo de Investigación y Desarrollo para el Ambiente y Cambio Climático GIDAC of Escuela Superior Politécnica de Chimborazo studied the behavior of one bacterial culture of Pseudomonas spp., from high Andean soil of Parque Nacional Sangay, using as substrate different concentrations of nitrates that are considered as contaminants in wastewater, resulting in an increase of bioelectricity production gradually as the pollutant was added [12]. Actually, a great variety of microorganisms have been used; however, there is are not publications where they also use microbial consortia of the Andean soil. The next investigation mentions the interest of using the high Andean soil as source of electrogenic microorganisms and its contrast with organic and inorganic compounds as substrates in microbial fuel cells stablishing their difference in response in VDC.

2 Materials and Methods

2.1 Sampling

2.1.1 Physiochemical Analysis of Soil

The parameters determined in the soil samples were: Temperature (T), ph and electric conductivity (uS). These indicators were selected having account considered important factors for the investigation such as: the temperature that is a product of the internal energy of one body, the ph that controls many of chemical and biological activities, electric conductivity that is influenced by the concentration and the composition of dissolved salts, it means, the higher the value of electrical conductivity, the greater the salinity present, based on the methods described in UNE77303 rule (Table 1).

Table 1. Analyzed parameters and employed methods.

Parameters	Test methods
Temperature	Physic
ph	PEE/SFA/06 Internal method, EPA 9045D 2014 Reference method
Electric conductivity	2510B method

2.2 Construction of Microbial Fuel Cells

It was realized based on previous investigations, the material of which they were formed was acrylic and as electron exchange mediator (CEM) copper was used due to its conductivity electrical properties. Based on these aspects mentioned before, 8 cubic MFCs ($5 * 5 * 5\ cm^3$) of a single chamber were designed by printing them on 3 mm thick acrylic, which were correctly assembled and disinfected for later use.

2.2.1 Exchange Membranes

Each of MFCs are formed by two cameras; anodic and cathodic. In the first one called anodic the substrates referring to each experiment (organic substrates, nitrates and lead) were placed, and the second cathode chamber remained partially exposed to air. These cameras contain an exchange membrane; fulfilling the main function of energy transfer (anode and cathode) as in the formation of biofilms (anode) and as for the second membrane that was in the cathodic camera, it is responsible for exchange of protons to close the circuit and generate a continues current voltage.

2.2.2 Substrates

The High Andean soil from Casa Condor was introduced in the MFC, being its coordinates 1°31.929′S Latitude and 78°50.802′O Longitude. It was used three types of substrates for the bioelectricity production in MFCs. The first applied substrate was from a fruit and vegetable solution of 1000 ppm for the first MFCs. For the second case for the next two MFCs, it was prepared synthetic wastewater with a nitrate concentration of 50 ppm. Finally, the last substrate was lead in which was used two MFCs with a 20 ppm lead concentration. It is worth mentioning the total volume (substrate and inoculum) of each MFC was 125 ml, due to is the maximum capacity of volume that anodic camera allows per MFC, additional these MFC, there was used another two because was necessary to use the control group or reference target that were monitored by an electronic acquisition system.

Fig. 1. DAQ NI 6009 connection to MFCs

2.3 Data Acquisition

For data analysis and management, connections were made for each of the MFCs, which consisted of an output connection for the cathode chamber being the polarity positive and other for anodic camera as negative. These mentioned connections were coupled to a DAQ NI 6009 device and at the time to a computer (view Fig. 1), that use LabVIEW software for the data acquisition from bioelectricity generation from each MFC, with a reception time of 1 min for a duration of 15 days of monitoring.

3 Results and Discussion

3.1 Analysis of High Andean Soil

Table 2 shows the results of physio-chemical analysis of soils that are the object of study and is relevant and necessary to object as potentially sources of electrogenic microorganisms.

Table 2. High Andean soil analysis by Olsen method.

Soil 4000 m.a.s.l Soil type: Pajonal									
Repetitions #	Conductivity (uS)	Temperature (°C)	pH	Organic matter %	Calcium (cmol/kg)	Iron (mg/kg)	Magnesium (mg/kg)	Copper (mg/kg)	Zinc (mg/kg)
Average in 3 repetition	311.50	18.35	7.18	10.90	6.47	368.3	9.61	8.37	3.67

At this type of altitude, a marked difference in the analyzed parameters is observed, such as the presence of elevated cations, being an indication of the capacity of oxide reduction, in the same way a certain percentage of organic matter at sample depths of 40 cm is high for this case.

3.2 Microbial Fuel Cells Structure

Eight Microbial Fuel Cells of simple acrylic configuration of a thickness of 3 mm and a volumetric capacity of 125 ml were constructed, thus being a cube with measures of $5 * 5 * 5\ cm^3$ as shown in Fig. 2.

Fig. 2. Microbial fuel cell structure.

For electrode structure to be use in this investigation, metallic copper and graphite were used, which come in form of carbon fiber. For the electrodes selection it was made a preliminary essay in which was made 16 data captures in a lapse of 8 h, yielding the following average data that were: 0.23V for copper and 0.07V for carbon fiber, so the configuration that was propose for this investigation was the copper as connection for the electron transport from one camera to other (view Fig. 3).

Fig. 3. Microbial fuel cell with copper electrode

Table 3 shows the obtained results from the monitoring of different incorporated substrates in the anodic compartment as microbial degradation product where electrons was liberated and were trapped and registered in direct current voltage (VDC).

One microbial fuel cell resembles to a battery the emits electrical energy through a given voltage, its functionality refers to the transport of electrons since anodic to cathodic camera, by means a wide range of electrogenic microorganisms, which in this case were taken from Andean soil of 4000 m.a.s.l. As byproduct of this type of technology emanates water into the atmosphere as steam. With organic and inorganic substrates use according the results obtained in this investigation, we clearly highlight that one MFC, produce better results if it has substrates. To point out in Fig. 4, it can be seen that the substrate that most electrons meant by the microbial activity was organic substrates, being a mixture of fruits and vegetables in state of decomposition, considered in literature as "waste" and that in Riobamba-Ecuador very often they are discarded without a correct final disposition.

Determining the polynomic tendency lines for each substrate results, it is verified that exists a possible homogeneity between two substrates, which would be for the case of lead and for vegetable-fruits. According registered data on Table 3 exists a better production of bioelectricity with organic substrates, however, it was verified by means of a comparison test of means of averages for the homogeneity of variances. A p-value of 0.032 of less than 0.05 was obtaining using a Levene statistic indicating that the concentrations do not have homogeneous variances. Therefore, we chose to use a non-parametric means comparison test.

Table 3. Substrate valuation

Bioelectricity results (VDC)				
Day	Control	Vegetables	Nitrate	Lead
1	0.043	0.604	0.062	0.135
2	0.074	0.615	0.085	0.095
3	0.062	0.572	0.097	0.703
4	0.104	0.518	0.142	0.406
5	0.082	0.489	0.107	0.328
6	0.083	0.481	0.127	0.321
7	0.091	0.396	0.129	0.331
8	0.107	0.381	0.116	0.341
9	0.115	0.386	0.191	0.348
10	0.111	0.334	0.204	0.354
11	0.086	0.324	0.182	0.379
12	0.074	0.297	0.157	0.346
13	0.077	0.293	0.159	0.328
14	0.082	0.275	0.168	0.319
15	0.091	0.261	0.148	0.314
Sum	1.282	6.226	2.074	5.048
Average	0.09	0.42	0.14	0.34

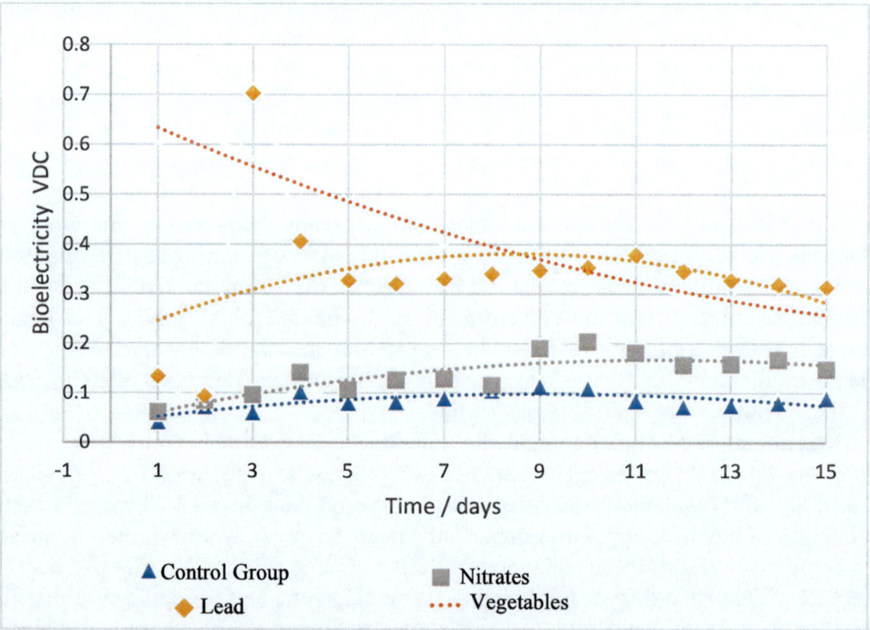

Fig. 4. Bioelectricity production by each cell.

3.3 Kruskall Wallis Test

This kind of test helped us to know is there was exist significant influence by the substrates in the bioelectricity production (Tables 4 and 5).

Approach:

$$H_o: \mu_1 = \mu_2 = \mu_3$$
$$H_1: \mu_i \neq \mu_j \; for \; some \; i \neq j \tag{1}$$

Resolution:

Table 4. Means comparison

	Ranks		
	Substrates	Data number	Average ranks
Bioelectricity	Vegetables	15	32.60
	Nitrite	15	9.40
	Lead	15	27.00
	Total	45	

Table 5. Statistical data for the mean comparison

Statisticians tests	
Title 1	Bioelectricity
Chi-square	25.490
GI	2
Asymptotic sig	0.000
a. Kruskal Wallis test	
b. Group variable: Substrates	

Rejection region:

If:

$$P\text{ - value } < 0.05, \text{ null hypothesis is rejected} \tag{2}$$

With the present analysis we can affirm that statistically significant differences have been found between the use of substrates, in the bioelectricity production that when not using it.

The Fig. 5 shows that there is a homogeneous group between the substrates of lead and vegetables, it indicates that their means are statistically equals, however, there exists a significant difference between organic substrates and metal with respect to the use of inorganic substrates in this case (nitrates). It should be mentioned that in addition to bioelectricity production, elimination efficiencies are also achieved for all cases of substrates tested in the first 48 h (View Table 6).

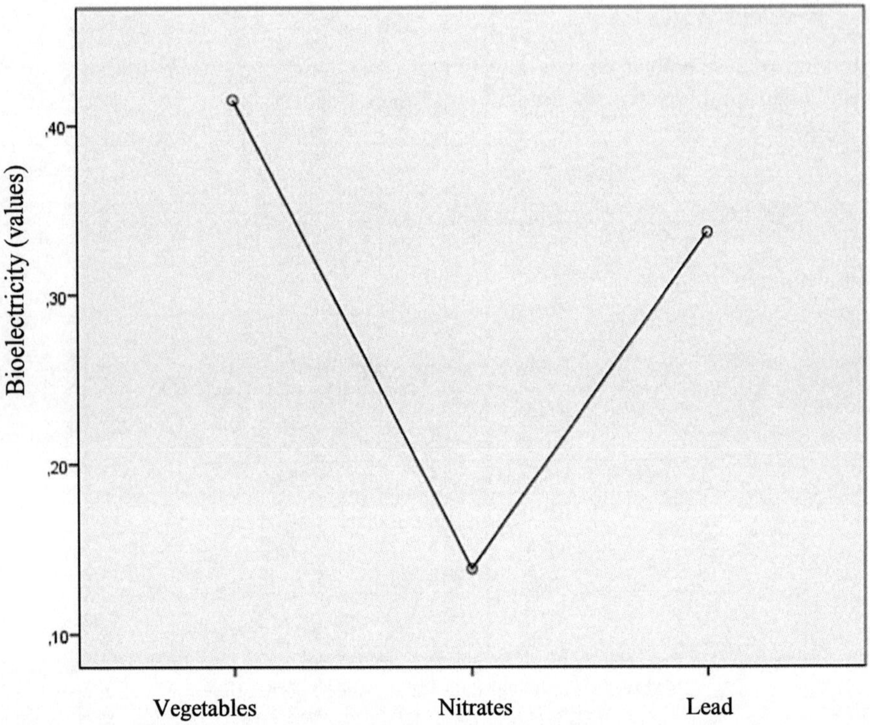

Fig. 5. Bioelectricity production with means statistically equals

Table 6 shows that the supplied concentrations are based on the maximum permissible limits according to the current Ecuadorian legislation, except in the case of lead due to its high level of tolerance by the potentially electrogenic microorganisms of the soil, also the monitoring time of voltage production and substrate degradation was 15 days. In the case of organic substrates, it does not apply because the degradation of organic matter can only be represented when it reaches the maximum voltage production and bioelectric stability [13].

Table 6. Elimination efficiencies

Type of substrate	Initial supplied concentration (mg/L)	Final concentration after experimentation (mg/L)
Nitrate	50	20.7
Lead	20	8.75
Organic substrate	1000	N.A

In the last decade, different alternatives for bioenergy production have been studied, so in this study soil samples originating from the high Andean moorland of Ecuador were used, being a little studied place, not intervened by humans, which gives it a great potential to find bioelectricity producing bacteria. By studying the behavior of the bacterial consortium with the inorganic substrate, it results in lower bioelectricity production compared to the organic substrate, the reasons for it may be suggested that organic substrates have biomolecules essential for bacterial development and bacteria can metabolize faster compared to an inorganic substrate (nitrates) only.

For higher bioelectricity production, it is important the electron conductive material that is used to avoid losses, in this study two materials were compared, the best conductor being metallic copper sheets compared to carbon fiber, although means a higher economic cost, production tripled so there is a greater cost-benefit ratio in this case. Finally, it should be noted that the production of bioelectricity is affected by the substrate used, and the conductive material used, so that an optimal model of patentable microbial fuel cell can be obtained with which the greatest amount of bioelectricity is generated at from wasteland bacteria.

4 Conclusions

The vegetable substrate, together with the bacteria from the moor soil, generated greater bioelectricity (0.42VDC) compared to other substrates used, possibly due to the availability of carbon and magnesium sources necessary for microbial metabolism.

The nitrate substrate produced the least amount of bioelectricity (0.14VDC), due to its chemical nature, which does not have the macro and micronutrients necessary for optimal bacterial development.

For the configuration of microbial fuel cells, the best material as an electron exchange mediator (CEMs) was metallic copper, having a better conductivity against carbon.

References

1. Revelo, D.M., Hurtado, N.H., Ruiz, J.O.: Celdas de Combustible Microbianas (CCMs): Un Reto Para La Remoción de Materia Orgánica y La Generación de Energía Eléctrica. Inf. tecnológica **24**(6), 7–8 (2013). https://doi.org/10.4067/S0718-07642013000600004
2. Kiely, P.D., Regan, J.M., Logan, B.E.: The electric picnic: synergistic requirements for exoelectrogenic microbial communities. Curr. Opin. Biotechnol. **22**(3), 378–385 (2011). https://doi.org/10.1016/j.copbio.2011.03.003
3. Lefebvre, O., Ha Nguyen, T.T., Al-Mamun, A., Chang, I.S., Ng, H.Y.: T-RFLP reveals high β-proteobacteria diversity in microbial fuel cells enriched with domestic wastewater. J. Appl. Microbiol. **109**(3), 839–850 (2010). https://doi.org/10.1111/j.1365-2672.2010.04735.x
4. Wrighton, K.C., Virdis, B., Clauwaert, P., Read, S.T., Daly, R.A., Boon, N., Piceno, Y., Andersen, G.L., Coates, J.D., Rabaey, K.: Bacterial community structure corresponds to performance during cathodic nitrate reduction. ISME J. **4**(11), 1443–1455 (2010). https://doi.org/10.1038/ismej.2010.66

5. Liu, H., Cheng, S., Logan, B.E.: Production of electricity from acetate or butyrate using a single-chamber microbial fuel cell. Environ. Sci. Technol. **39**(2), 658–662 (2005). https://doi.org/10.1021/es048927c
6. Chaudhuri, S.K., Lovley, D.R.: Electricity generation by direct oxidation of glucose in mediatorless microbial fuel cells. Nat. Biotechnol. **21**(10), 1229–1232 (2003). https://doi.org/10.1038/nbt867
7. Huang, L., Zeng, R.J., Angelidaki, I.: Electricity production from xylose using a mediatorless microbial fuel cell. Bioresour. Technol. **99**(10), 4178–4184 (2008). https://doi.org/10.1016/j.biortech.2007.08.067
8. Feng, Y., Wang, X., Logan, B.E., Lee, H.: Brewery wastewater treatment using air-cathode microbial fuel cells. Appl. Microbiol. Biotechnol. **78**(5), 873–880 (2008). https://doi.org/10.1007/s00253-008-1360-2
9. Huang, L., Logan, B.E.: Electricity generation and treatment of paper recycling wastewater using a microbial fuel cell. Appl. Microbiol. Biotechnol. **80**(2), 349–355 (2008). https://doi.org/10.1007/s00253-008-1546-7
10. Ghoreyshi, A.A., Jafary, T., Najafpour, G.D., Haghparast, F.: Effect of type and concentration of substrate on power generation in a dual chambered microbial fuel cell. In: Proceedings of the World Renewable Energy Congress – Sweden, 8–13 May 2011, vol. 57, pp. 1174–1181. Linköping University Electronic Press, Linköping (2011). https://doi.org/10.3384/ecp110571174.
11. Parot, S., Délia, M.L., Bergel, A.: Acetate to enhance electrochemical activity of biofilms from garden compost. Electrochim. Acta **53**(6), 2737–2742 (2008). https://doi.org/10.1016/j.electacta.2007.10.059
12. Guambo, A., Paña, S., Calderón, C., Echeverría, M., Recalde, C.: Environmental biosensor potential of microbial fuel cells for nitrate reduction. Sens. Transducers **217**, 23–27 (2017)
13. Logroño, W.N., Echeverría, M.M., Recalde, C.G., Graziani, P.: Bioconversión de Residuos Sólidos Orgánicos Con Suelos de La Región Amazónica y Alto Andina Del Ecuador En Celdas de Combustible Microbiano de Cámara Simple. Inf. Tecnológica **26**(2), 61–68 (2015). https://doi.org/10.4067/S0718-07642015000200008

Reduction of Ripple Current in DC-DC SiC Converter Using HIL System

Efrén Fernández(✉) and Diego Rojas

University of Azuay, Cuenca, Ecuador
efernandez@uazuay.edu.ec

Abstract. This research proposes an analysis for the reduction of ripple current in the output of DC-DC converter type V-I (voltage-current), with silicon carbide (SiC) devices. The frequency of operation of the devices is 100 kHz. For the implementation and control of the topology, a hardware in the loop application (HIL) is designed in LabVIEW + FPGA toolbox. The electronic board used for the generation of signals and control is the SBrio 9636 with FPGA Xilinx Spartan-6 LX45. This document aims to analyze the advantage of using SiC devices in this converter the increase of the frequency help to reduce the ripple in the output current. In addition, to showing the advantages of using HIL applications for the control of converters that help solve the limitations of high frequency operation in some conventional digital systems. Finally, an analysis of experimental results of topology to low and high frequency with SiC are presented.

Keywords: DC-DC converter · Silicon Carbide (SiC) · Hardware in the loop

1 Introduction

At present, there is a tendency for researchers to search to optimize topologies of converters with greater efficiency, higher switching frequency, cost reduction, weight and volume. With the advance of the technology, several devices have been appearing and contributing to solve these characteristics, one of these devices are the devices of silicon carbide (SiC).

The SiC technology has a major role in the design of power converters [1–4]; these devices have a better behavior at high operating frequency and increase the range of operation in temperature, compared to conventional silicon devices where the characteristics of operation are below these elements.

The increase in the switching frequency, on the one hand benefits at the operation and optimization level and on the other hand, it generates problems in the simulation and programming environment. The conventional programming tools used to generate on and off states in transistors are limited for frequency operations less than 50 kHz [5, 6]. Faced with this problem, it is necessary to use new programming and development platforms among which stand the FPGA and microcontroller systems.

The current-voltage converter V-I is presented in [7, 8], this type of topology allows to solve the current return problems in current source inverter (CSI). The investigations present in [8, 9] used this type of topology to regulate and control the input current to

M. Botto-Tobar et al. (Eds.): CI3 2020, AISC 1277, pp. 209–221, 2021.
https://doi.org/10.1007/978-3-030-60467-7_18

the inverter CSI. These works analyze this topology in control drives for applications in power electric traction systems. However, for their implementation they use components insulated gate bipolar transistors (IGBTs) with reverse-blocking (RB) capability to low frequency of switching, 15 kHz [8]. For the generation of the control and gate control signals they used a digital signal processor (DSP), this generates a limitation the CPU-based sample frequency is largely limited by the performance of its CPU [9]. This limitation contributes to the difficulty of simulations for high frequency applications using this type of system. To solve this problem, the use of an FPGA-based model for HIL simulations is proposed. In addition, the operation at low frequency in this type of converter generates a ripple in the output current that could then generate problems of THD harmonic distortion when coupled to a CSI.

This paper proposes the study for the reduction of the ripple in the output current in a topology of converter V-I with SiC devices at high switching frequency and show the advantages, ease of programming of a HIL application with FPGA for the control and activation of these devices.

This document contains the following sections: Sect. 2 presented the features, operation and control in simulation of SiC topology proposed. In addition, an analysis is carried out to reduce curling in the current. Section 3 present the implement of the HIL system the programming environment and the design strategy. The Sect. 4 presents the results obtained in an experimental validation. Finally, in Sect. 5, the conclusions of this paper are discussed.

2 V-I SiC Converter Topology

In several investigations proposed topologies of bidirectional DC-DC converters [10–13], efficiency and power loss analysis are presented. In this work are concentrated in the study of topologies based on silicon IGBTs and low frequency of operation. The V-I SiC converter for the analysis is show in the Fig. 1 and is a DC-DC topology used as a current regulator for CSI inverters. This topology helps solve the current return problem in the CSI inverter when applied to electric traction systems [7].

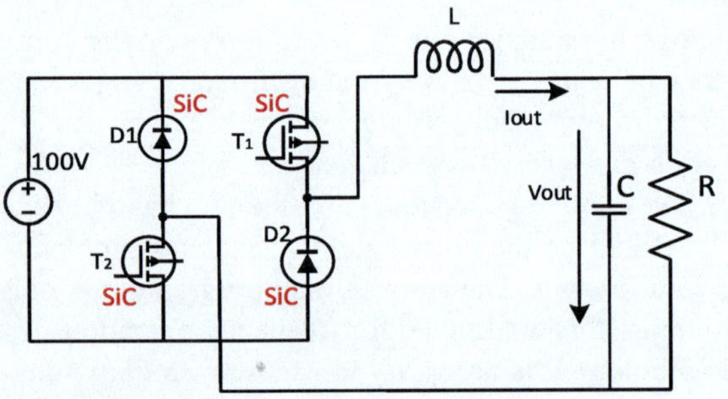

Fig. 1. Topology proposed for the analysis.

This topology has two situations. In the first situation (Fig. 2) the SiC mosfet's T_1 and T_2 are turned on and the battery voltage is supplied to charge the inductor [8–14]. The Mosfet T_2 close the circuit for the return of the current and the diodes D_1, D_2 are in non-active polarization and therefore do not work.

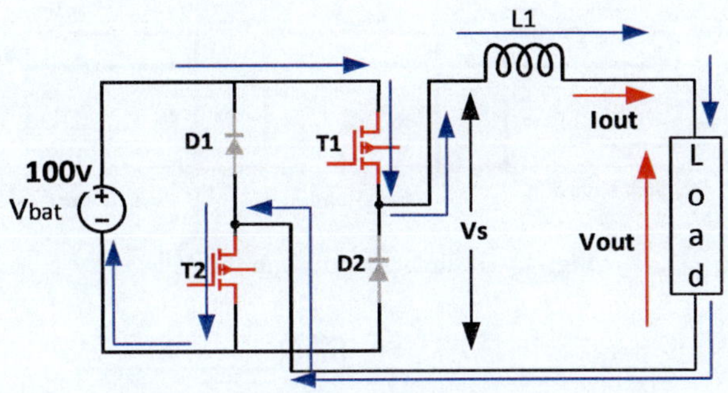

Fig. 2. First situation of operation in the V-I.

In the second situation Fig. 3 the SiC mosfet's are turned off the current returns the battery through of the diodes D_1 and D_2 [8–14].

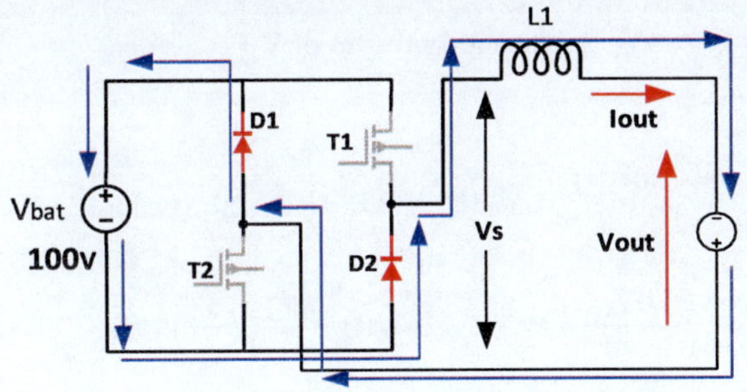

Fig. 3. Second situation of operation in the V-I.

The V-I converter works between first situation and second situation the turn On and turn Off of T_1 or T_2 to formed a dc link of current (Fig. 4), but for obtained a constant output current is necessary the design of a control in close loop of the current.

For the regulation of current a PI type control is implemented for the current loop and in this way regulate the amount of output current. The design of the control is presented as follow: The description of the circuit of the Fig. 5 shows the current and voltage on load RC. Also, for the analysis are considered internal resistance (R_{on}) of the SiC mosfet's, inductance (R_L) and capacitance (R_C).

The equations that are obtained of the circuit presented in Fig. 5 are expressed in (1) and (2) [8]:

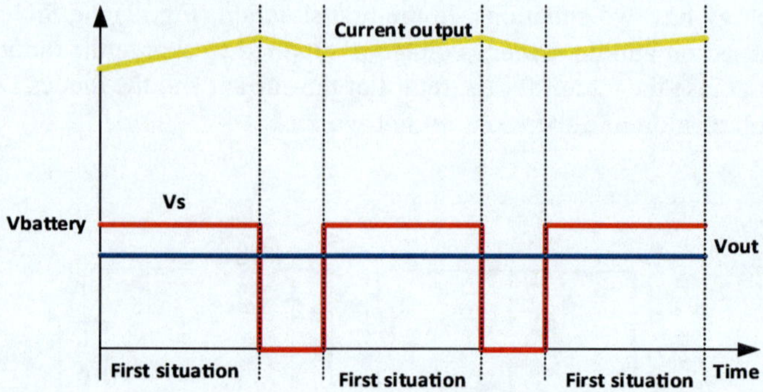

Fig. 4. Dc link of current in the V-I.

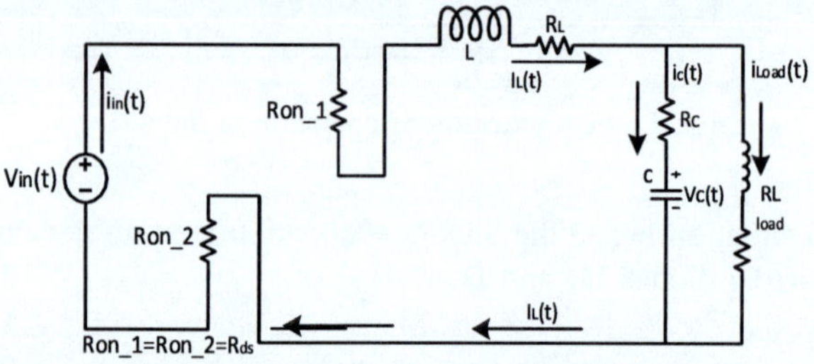

Fig. 5. Circuit equivalent of V-I converter.

$$L\frac{di_L(t)}{dt} = V_{in}(t)\ 2R_{ds}I_L(t)\ R_{Load}I_L(t)\ V_{out} \tag{1}$$

$$\frac{dV_C(t)}{dt} = \frac{\frac{R_{Load}}{R_{Load}+R_C}}{C} i_L(t) - \frac{\frac{1}{R_{Load}+R_C}}{C} V_C(t) \tag{2}$$

The Eqs. (1) and (2), can be represented in the equation of state space model expressed as (3), (4) and in its matrix function (5)–(6) [8]:

$$\dot{X}(t) = AX(t) + B \tag{3}$$

$$Y(t) = CX(t)\ +\ D \tag{4}$$

$$\begin{bmatrix} \frac{di_L(t)}{dt} \\ \frac{dV_c(t)}{dt} \end{bmatrix} = \begin{bmatrix} \frac{2R_{ds}+R_L+\frac{R_{Load}R_C}{R_{Load}+R_C}}{L} & -\frac{\frac{R_{Load}}{R_{Load}+R_C}}{L} \\ \frac{\frac{R_{Load}}{R_{Load}+R_C}}{C} & -\frac{\frac{1}{R_{Load}+R_C}}{C} \end{bmatrix} \begin{bmatrix} i_L(t) \\ V_C(t) \end{bmatrix} + \begin{bmatrix} \frac{1}{L} \\ 0 \end{bmatrix} V_{in}(t) \tag{5}$$

$$\begin{bmatrix} \frac{V_o(t)}{dt} \\ I_{in}(t) \end{bmatrix} = \begin{bmatrix} \frac{R_{Load} R_C}{R_{Load} + R_C} & \frac{R_{Load}}{R_{Load} + R_C} \\ 1 & 0 \end{bmatrix} \begin{bmatrix} i_L(t) \\ V_C(t) \end{bmatrix} + \begin{bmatrix} 0 \\ 0 \end{bmatrix} V_{in}(t) \tag{6}$$

With these expressions, we proceed to tune a Pi control in function of the transfer function considering the output with respect to the input of current. The result in simulations in Matlab-Simulink are shown in the Fig. 6.

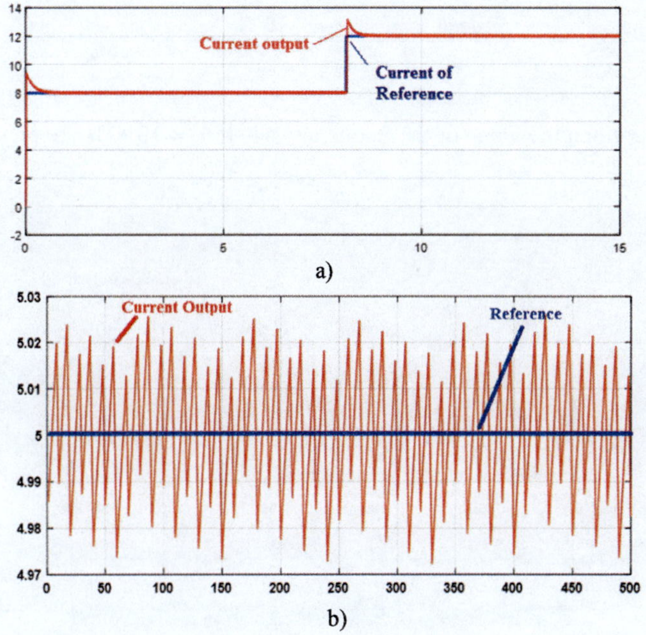

Fig. 6. Results in simulation of PI control. a) Tuning of control, b) Current of output to 5 A of reference and 100 kHz of switching.

The simulation analysis is development; the idea is to check the topology operation based on the frequency increase. The relation (7) is established. The values for the simulation are presented in the Table 1.

$$f_s = \begin{cases} f_{s1(vi)} = 10\,\text{kHz} \\ f_{s2(vi)} = 100\,\text{kHz} \end{cases} \tag{7}$$

Where, $f_{s1(v\text{-}i)}$ is the frequency of commutation of the converter using silicon devices and $f_{s2(v\text{-}i)}$ is the frequency for converter using SiC devices.

Table 1. Values of simulation

Parameter	Value
Battery voltage	100 V
Current input set	5 A
L_1	500 uH
Frequency switching	100 kHz
SiC mosfet	SCT2450KE
SiC Sckotty diode	C3D10065I
C load	1500 uF
R load	10 Ω

The results when the value of the frequency is half fs = 10 kHz are presented in the Fig. 7.

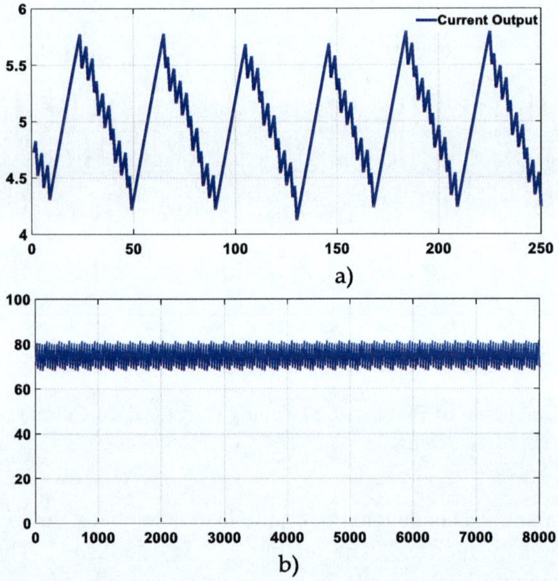

Fig. 7. Simulations to 10 kHz a) Simulation of current to 5 A, b) Simulation of voltage.

The results with the second option where the frequency of the V-I is fs = 100 kHz, the results shown in Fig. 8.

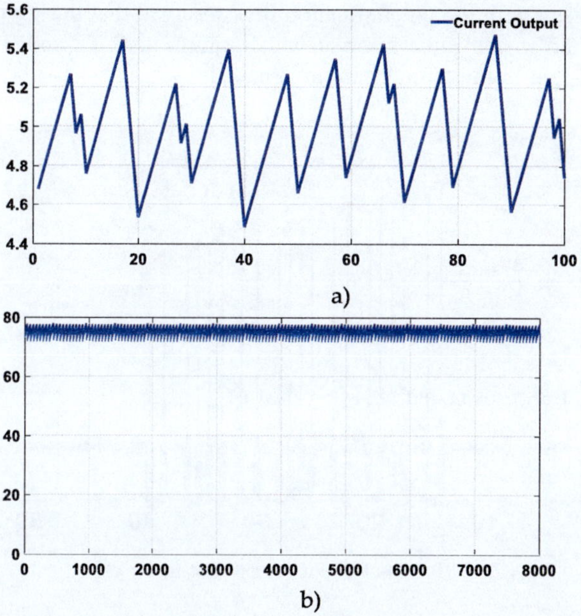

Fig. 8. Results in simulation to 100 kHz a) Simulation of current to 5 A, b) Simulation of voltage.

The results obtained indicate that when the switching frequency is higher, a lower ripple current is obtained in the converter. Under these conditions of switching frequency is essential the use SiC devices that are suitable for working at these ranges. When analyzing the results, is observed that the ripple is higher in the second condition (Fig. 9) because the On-Off times of the transistors T1 and T2 are better distributed.

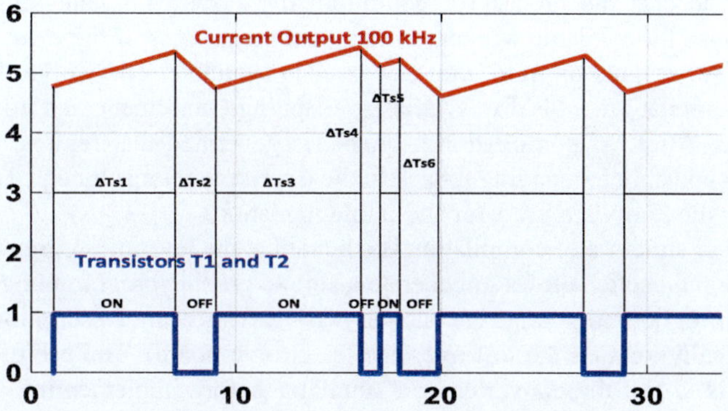

Fig. 9. Analysis of current out in 100 kHz.

The comparison with the other situation it is observed that during the time On-Off states are repeated in the same interval which causes that T1 and T2 are activated producing the ripple in the output current (Fig. 10).

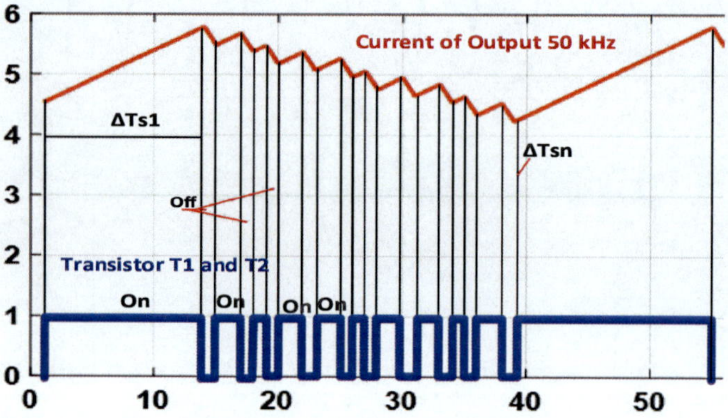

Fig. 10. Analysis of current out in 10 kHz.

3 HIL System

The programming environment used is LabVIEW FPGA module. Using this environment, you can implement an intuitive graphic code to define inputs and outputs to control a hardware without prior knowledge of complex programming tools [15].

The FPGA compilation in LabVIEW has three fundamental software components generating a modular solution for the compilation. This compilation system is shown in Fig. 11. The LabVIEW development environment is the site where the application is scheduled. Now of starting the simulation, this generates intermediate files that are sent through communication to web services to an FPGA compilation server. The compilation server accepts the process of compiling the system implementing LabVIEW FPGA and looks for available workers to carry out the process. If there are no available workers, the server puts on hold until a worker is available. The worker has installed the Xilinx compilation tools for synthesis, mapping, placement and routing of the FPGA design. After the compilation a bit file is generated that returns to the server where it is responsible for sending it back to the development machine and then loading it the card or the hardware used for the implementation.

This type of structure of compilation is a benefit at the level of programming is very easy to develop didactic, allows to generate a simple graphic programming language to an advanced programming language such as VHDL (Hardware Description Language) and automatically load on the test interface (electronic board). The use of this type of applications is to test the converter in a simulator before implementing it in the real process; this generates a great advantage because the tests can be performed without damaging the equipment or the components in this case when working with very sensitive variables, as are voltages and currents.

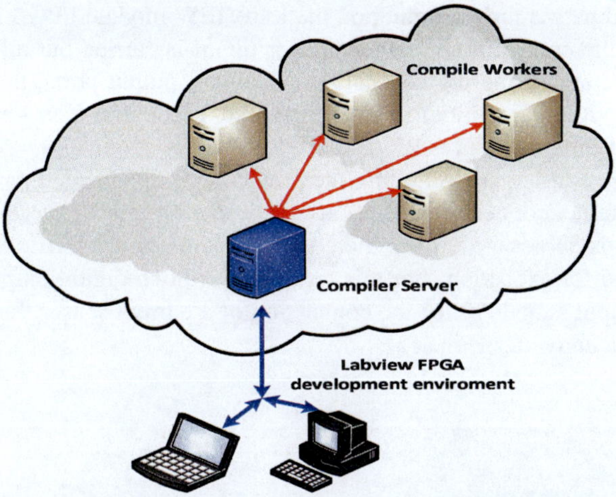

Fig. 11. FPGA compilation in LabVIEW.

The system to be implemented has three parts. The first part implements the design and tune of the PI controller of current of the converter in the LabVIEW FPGA module (Fig. 12). The second part involves validation in simulation of PI control. The third part is implemented and builds the programming environment for controlling and triggering SiC devices through online compilation. This implies the connection of the electronic card and configuration of the output inputs. In this part you must have the plant or test prototype that in our case is the V-I converter with SiC devices. The SBrio 9636 with FPGA Xilinx Spartan-6 LX45 electronic card is used to implement the system and LabVIEW 2016.

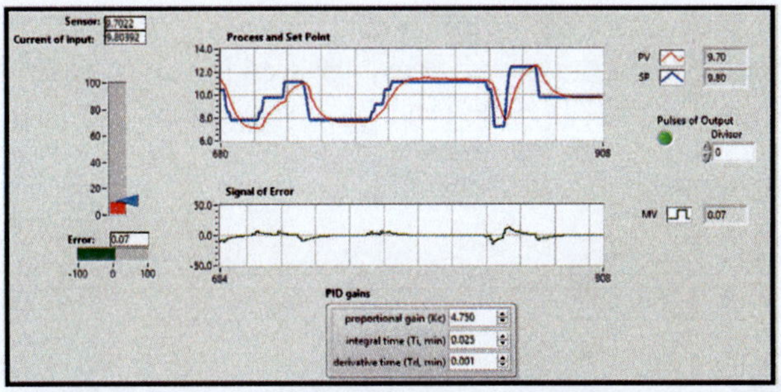

Fig. 12. First part validation in LabVIEW.

In the first part, the control is implemented and contains the data specific to the model and the plant to be controlled. The process consists in supplying the necessary data obtained in the PI tuning obtained in Sect. 2 and in this way validate it. The purpose is to simulate in this case the V-I converter in LabVIEW and to analyzed the behaviour of the output current as a function of a reference signal.

In the second part, a programming on the LabVIEW module FPGA environment is carried out and the control is implemented, for this it is carried out all the configurations as the selection of the digital analog and digital output ports, the clock configuration and the times are initialized of CPU and execution. The clock used is of 40 MHz and the time of CPU is 25 μs.

The last stage consists of compiling the previously implemented program. For this execution, an internet connection and an access account to the NI server are necessary. For this stage, it is necessary to have the converter connected to SBrio 9636 as well as the ports selected for execution. For this test, there are two digital outputs D0 and D1 and an analog input signal A0 for the connection of a current sensor that is installed at the output of the converter (Fig. 13).

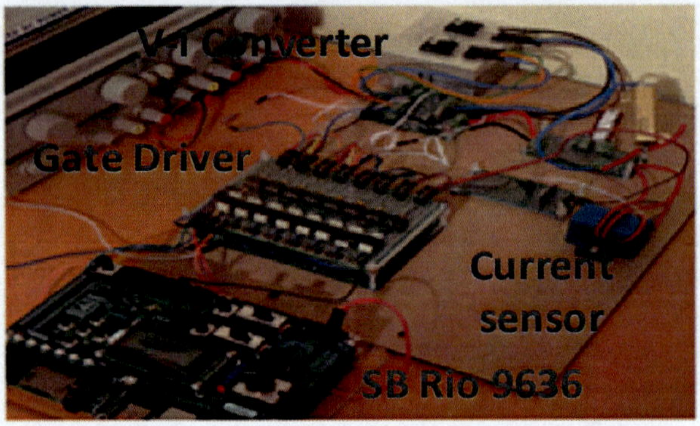

Fig. 13. V-I power converter implementation.

4 Experimental Validation

The results obtained when the power converter working at 100 kHz switching frequency are presented in Fig. 14. The results show that working at this frequency, a current output is obtained with less ripple and it allows to validate the control carried out with the HIL application.

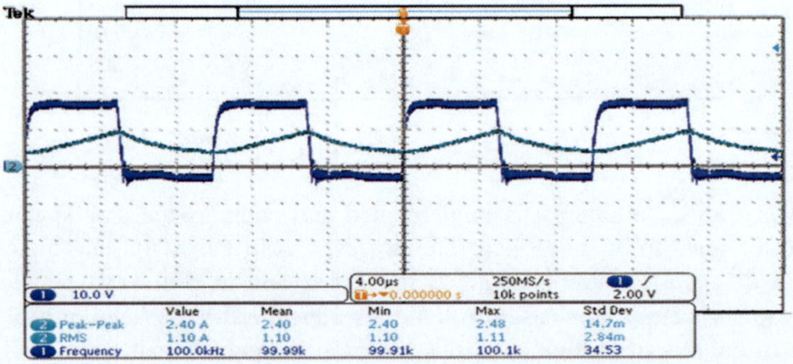

Fig. 14. Current output to 100 kHz of switching.

If the tests are carried out but with frequency of 10 kHz, the results shown in Fig. 15.

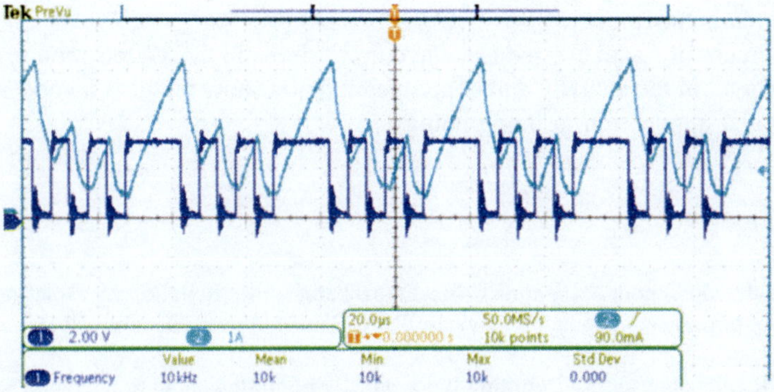

Fig. 15. Current output to 10 kHz of switching.

The results obtained demonstrated that reducing the frequency increases the ripple in the output current. This demostrate that the use of these SiC devices in this type of converter topology at a high switching frequency allows reducing the ripple in the output current. The HIL system allows the real-time control of the V-I converter. In addition, it allows verifying the behavior in real time of the converter (Fig. 16).

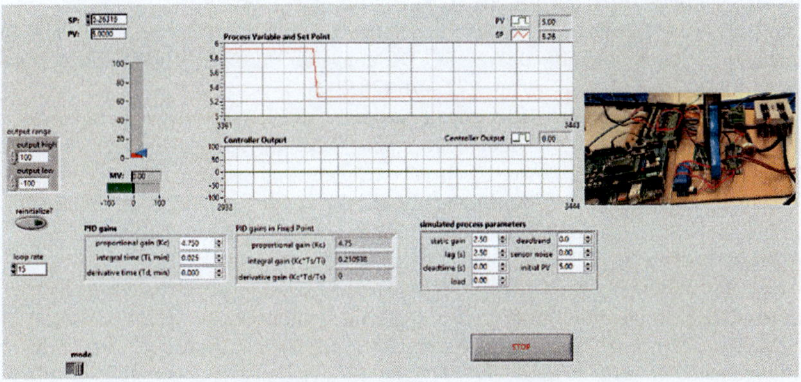

Fig. 16. HIL system in real time.

5 Conclusions

This paper presents an analysis for minimize the ripple in the current in the V-I topology implemented with SiC devices to 100 kHz or switching. The control and activation gate control is development with HIL system with FPGA programming. The result obtained demonstrate that the use of HIL in SiC power converter topology allow increase the frequency of operation, this parameter is limited in other applications of programming. Furthermore, with the help of a programming platform such as LabVIEW, it is possible to reduce the difficulty of programming through the FGPA module that implements this tool.

The increase of frequency reduces the current ripple in the V-I power converter, the results obtained in the analytic and experimental parts show that it is a good alternative for its use and application in this topology.

References

1. Millán, J., Godignon, P., Perpiñà, X., Pérez-Tomás, A., Rebollo, J.: A survey of wide bandgap power semiconductor devices. IEEE Trans. Power Electron. **29**(5), 2155–2163 (2014)
2. Singh, R., Sundaresan, S.: Fulfilling the promise of high-temperature operation with silicon carbide devices: eliminating bulky thermal-management systems with SJTs. IEEE Power Electron. Mag. **2**(1), 27–35 (2015)
3. Jahdi, S., Alatise, O., Fisher, C., Ran, L., Mawby, P.: An evaluation of silicon carbide unipolar technologies for electric vehicle drive-trains. IEEE J. Emerg. Sel. Topics Power Electron. **2**(3), 517–528 (2014)
4. Shang, F., Arribas, A.P., Krishnamurthy, M.: A comprehensive evaluation of SiC devices in traction applications. In: 2014 IEEE Transportation Electrification Conference and Expo (ITEC), pp. 1–5, 15–18 June 2014
5. Grégoire, L.A., Al-Haddad, K., Nanjundaiah, G.: Hardware-in-the-loop (HIL) to reduce the development cost of power electronic converters. In: India International Conference on Power Electronics 2010 (IICPE 2010), New Delhi, pp. 1–6 (2011)
6. Kang, R., Kim, S., Yang, I., Jeong, K., Kang, C., Kim, G.: The use of FPGA in HIL simulation of three phase interleaved DC-DC converter. In: 2012 IEEE Vehicle Power and Propulsion Conference, Seoul, pp. 772–776 (2012)
7. Tang, L., Su, G.J.: Boost mode test of a current-source-inverter-fed permanent magnet synchronous motor drive for automotive applications. In: 2010 IEEE 12th Workshop on Control and Modeling for Power Electronics (COMPEL), Boulder, CO, pp. 1–8 (2010)
8. Fernández Palomeque, E.E.: Optimization of a CSI inverter and DC/DC elevator with silicon carbide devices, for applications in electric traction systems. Tesi doctoral, UPC, Departament d'Enginyeria Electrònica (2019)
9. Ji, F., Fan, H., Sun, Y.: Modelling a FPGA-based LLC converter for real-time hardware-in-the-loop (HIL) simulation. In: 2016 IEEE 8th International Power Electronics and Motion Control Conference (IPEMC-ECCE Asia), Hefei, pp. 1016–1019 (2016). https://doi.org/10.1109/ipemc.2016.7512426
10. Su, G., Tang, L.: A current source inverter based motor drive for EV/HEV applications. SAE Technical Paper 2011-01-0346 (2011)
11. Lusignani, D., Barater, D., Franceschini, G., Buticchi, G., Galea, M., Gerada, C.: A high-speed electric drive for the more electric engine. In: 2015 IEEE Energy Conversion Congress and Exposition (ECCE), Montreal, QC, pp. 4004–4011 (2015)
12. Han, D., Noppakunkajorn, J., Sarlioglu, B.: Comprehensive efficiency, weight, and volume comparison of SiC- and si-based bidirectional DC–DC converters for hybrid electric vehicles. IEEE Trans. Veh. Technol. **63**(7), 3001–3010 (2014)

13. Attia, Y., Abdelrahman, A., Hamouda, M., Youssef, M.: SiC devices performance overview in EV DC/DC converter: a case study in a Nissan Leaf. In: 2016 IEEE Transportation Electrification Conference and Expo, Asia-Pacific (ITEC Asia-Pacific), Busan, pp. 214–219 (2016)
14. Su, G.-J., Tang, L.: Current source inverter based traction drive for EV battery charging applications. In: 2011 IEEE Vehicle Power and Propulsion Conference (VPPC), pp. 1–6, 6–9 September 2011
15. LabVIEW FPGA module Application Note (2017)

Power Flow Solution Combining Newton-Raphson and Fast Decoupled Methods

W. P. Guamán(✉), G. N. Pesántez, X. A. Proaño, E. M. Pérez, and W. V. Tigse

Universidad Técnica de Cotopaxi, Latacunga, Ecuador
wilian.guaman8956@utc.edu.ec

Abstract. In this article the Newton Raphson (NR) power flow solution method is developed from the initial conditions obtained starting with the Fast-Decoupled Load Flow (FDLF) method. It means, an algorithm is proposed that combines both procedures to resolve power flows more efficiently, thus decreasing the number of iterations for processing times approximately equal to those obtained in the NR method. The methodology developed is applicable only for systems where $R \ll X$, with a relationship between 1/5 y 1/10, common in systems higher than 200 kV. The proposed combined method (CM) has been validated through 4 case studies and for three different tolerance levels using MATLAB. Then, the compilation time between NR and CM for all evaluated cases between the two PCs shows variations of tenths of a second, practically negligible for the four case studies analyzed in this paper.

Keywords: Power flow · Newton-Raphson method · Fast decoupled load flow · Combined method · Power systems

1 Introduction

The analysis of power flows is used for the planning, operation, and economic programming of an Electric Power System (EPS) [1]. Solving an EPS in a stable state implies obtaining the voltages in all the busses of the system and with it, the power flows that circulate from the bus "i" to bus "j", it means, the power that flows through the elements of the EPS, such as transmission lines and transformers [2].

First of all, the complex voltages of the nodes are calculated using iterative techniques, since the problem of power flows requires the solution of a system of nonlinear equations. This article reviews the techniques of power flow solution, with special emphasis on the Newton-Raphson method (NR) and Fast Decoupled Load Flow (FDLF). Subsequently, both methods are combined to demonstrate that, the number of iterations is reduced with the combined method (CM) proposed for processing times equivalent to the NR method.

M. Botto-Tobar et al. (Eds.): CI3 2020, AISC 1277, pp. 222–233, 2021.
https://doi.org/10.1007/978-3-030-60467-7_19

2 Theoretical Background

The first algorithms used for the solution of power flows were based on the Gauss-Seidel (GS) method. This technique uses an initial assumption of voltage in module and argument to obtain the value of a particular variable $(Q_k; |V_k|; \delta_{V_k})$. The initial assumption value is replaced by the calculated value, and the process is repeated until the solution converges. However, this method requires a high number of iterations to find the solution, therefore, high computational processing times for a relatively poor convergence [3, 4].

Later, in 1967, Tinney y Clifford [1] develops the NR method for calculating power flows, this algorithm requires less computational memory, since convergence can be found in a few iterations, so it quickly replaces GS. Nowadays, NR is the most used approach since it provides reliable data, even in large systems, although it requires high computation times due to the calculation of its Jacobian matrix [2].

In the 70 s, the method FDLF was published [3], which increased the flow calculation speed. This technique is an approximation to the NR method [4], based on the decoupling that exists between the active power (P) and the voltage module < small voltage variations do not affect P > ; and reactive power (Q) with the voltage angle < small variations in the angle do not affect Q > [5, 6].

Recently, Lagace and collaborators [7] proposed to improve the convergence of NR by applying the Levenberg-Marquardt method. While members of the North Bangkok Institute of Technology [8] proposed a modified FDLD algorithm in 2005 to minimize calculation times and reduce the number of iterations. On the other hand, in [9] a comparison is made between the NR and FDLD methods to evaluate: tolerance, convergence, number of iterations, and computation time for the test cases of the IEEE with 9, 30 y 57 busses.

2.1 Newton-Raphson Method (NR)

It is an iterative method that has a quadratic convergence, and it is also a mathematically superior method to GS since it is less likely to diverge for problems with unconditioned cases [4]. The NR method is practical and efficient for large power systems, with several iterations independent of the system size to get the solution, but it is necessary to evaluate all the functions in each iteration [10]. Most power flow problems converge in less than ten iterations, although it requires more time and computer storage per iteration [6]. For any type of bus, the current equation in terms of its Admittance Matrix (Y_{Bus}) is expressed as:

$$I_i = \sum_{j=1}^{n} Y_{ij} V_j \quad (1)$$

Where:
I_i: Bus current i.
Y_{ij}: Admittance between busses i and j.

V_j: Bus voltage j.

By expressing Eq. (1) in polar form researchers have:

$$I_i = \sum_{j=1}^{n} |Y_{ij}||V_j| \angle \theta_{ij} + \delta_j \tag{2}$$

The complex power on the bus i is:

$$P_i - jQ_i = V_i^* I_i \tag{3}$$

Where:
P_i: Active Power on the bus i.
jQ_i: Reactive Power on the bus i.

Substituting I_i from (2) in the Eq. (3) and separating the real and imaginary part:

$$P_i = \sum_{j=1}^{n} |V_i||V_j||Y_{ij}| \cos(\theta_{ij} - \delta_i + \delta_j) \tag{4}$$

$$Q_i = -\sum_{j=1}^{n} |V_i||V_j||Y_{ij}| \sin(\theta_{ij} - \delta_i + \delta_j) \tag{5}$$

The Eqs. (4) and (5) constitute nonlinear algebraic equations in terms of independent variables that, for all cases, are found in per unit (p.u.) with angles given in radians. Therefore, for each loading bus (PQ) users have both equations and for each controlled voltage bus (PV) they have the equation given by (4). Then, by expanding the two equations in Taylor series around estimated initial conditions, assuming bus 1 as balance (Slack Bus) and omitting all higher-order terms, the result is:

$$\begin{bmatrix} \Delta P_2^k \\ \vdots \\ \Delta P_n^k \\ \hline \Delta Q_2^k \\ \vdots \\ \Delta Q_n^k \end{bmatrix} = \left[\begin{array}{ccc|ccc} \frac{\partial P_2^k}{\partial \delta_2} & \cdots & \frac{\partial P_2^k}{\partial \delta_n} & \frac{\partial P_2^k}{\partial |V_2|} & \cdots & \frac{\partial P_2^k}{\partial |V_n|} \\ \vdots & \ddots & \vdots & \vdots & \ddots & \vdots \\ \frac{\partial P_n^k}{\partial \delta_2} & \cdots & \frac{\partial P_n^k}{\partial \delta_n} & \frac{\partial P_n^k}{\partial |V_2|} & \cdots & \frac{\partial P_n^k}{\partial |V_n|} \\ \hline \frac{\partial Q_2^k}{\partial \delta_2} & \cdots & \frac{\partial Q_2^k}{\partial \delta_n} & \frac{\partial P_2^k}{\partial |V_2|} & \cdots & \frac{\partial P_2^k}{\partial |V_n|} \\ \vdots & \ddots & \vdots & \vdots & \ddots & \vdots \\ \frac{\partial Q_n^k}{\partial \delta_2} & \cdots & \frac{\partial Q_n^k}{\partial \delta_n} & \frac{\partial P_n^k}{\partial |V_2|} & \cdots & \frac{\partial P_n^k}{\partial |V_n|} \end{array}\right] \begin{bmatrix} \Delta \delta_2^k \\ \vdots \\ \Delta \delta_n^k \\ \hline |\Delta V_2|^k \\ \vdots \\ |\Delta V_n|^k \end{bmatrix} \tag{6}$$

The Jacobian Matrix (J) expresses the relationship between small changes in module and voltage argument versus small changes in active and reactive power. The elements of J are obtained from partial derivatives of (4) and (5) evaluated in $\Delta\delta_i^k$ and $\Delta|V_i^k|$, so:

$$\begin{bmatrix} \Delta P \\ \Delta Q \end{bmatrix} = \begin{bmatrix} H & M \\ N & L \end{bmatrix} \begin{bmatrix} \Delta \delta \\ \Delta |V| \end{bmatrix} \tag{7}$$

The terms ΔP^k y ΔQ^k are the difference (ε) between the specified values ($S_{gen} - S_{dem}$) for each iteration based on (4) and (5). Then, the new complex voltages in the knots are given by:

$$\delta_i^{(k+1)} = \delta_i^k + \Delta\delta_i^k \tag{8}$$

$$\left|V_i^{(k+1)}\right| = |V_i^k| + \Delta|V_i^k| \tag{9}$$

Where:
$\boldsymbol{\delta}_{\mathbf{i}}^{(\mathbf{k}+1)}$: Angle result i for the iteration k + 1.
$\left|\mathbf{V}_{\mathbf{i}}^{(\mathbf{k}+1)}\right|$: Voltage result i for the iteration k + 1.

2.2 Fast Decoupled Load Flow (FDLF) Method

The recognized coupling P-δ and Q-$|V|$, and the relative decoupling between both subproblems [11], it means that the numerical values of the blocks *N* y *M* from the Jacobian matrix expressed in (6), they are much smaller than the diagonal matrices *H* and *L*. Ignoring these blocks, and introducing additional simplifications, researchers get a model of two decoupled systems with matrices of constant coefficients:

$$\begin{bmatrix} \Delta P \\ \Delta Q \end{bmatrix} = \begin{bmatrix} H & 0 \\ 0 & L \end{bmatrix} \begin{bmatrix} \Delta \delta \\ \Delta |V| \end{bmatrix} \tag{10}$$

This technique usually has a convergence comparable to the NR method, at least for transport networks where the quotient *R/X* is quite small [6]. The principle on which the decoupling approach is based is supported on the fact that a change in the voltage angle **δ** in a bus, mainly affects the real power flow in the transmission lines and leaves the reactive power relatively unchanged. While, A change in the voltage magnitude |V| in a bus, mainly affects the reactive power flow in the transmission lines and relatively leaves the real power flow unchanged. So:

$$\frac{\Delta P}{V_i} = -B' \Delta\delta \tag{11}$$

$$\frac{\Delta Q}{V_i} = -B'' \Delta|V| \tag{12}$$

Where B' and B'' are the imaginary part of Y_{Bus} and in both cases, they are constant matrices, so they are built and factor themselves only once. This explains the remarkable reduction of the calculation effort, which has given this method a lot of popularity [11].

2.3 Combined Method (CM)

As evidenced in the two previously analyzed cases of resolution of power flows, being iterative methods, they require as a first step to establish initial conditions, it means, to assign values of magnitude and voltage angle of the nodes where the solution is intended. These values obviously influence the computation time, so that the further away from the solution are V_i^0 and δ_i^0, the algorithm will take longer to resolve the system.

Generally, $V_i^0 = 1$ y $\delta_i^0 = 0°$, since the voltages in the nodes approximate the nominal values in normal operating conditions. Similarly, the angles approach zero, the closer to the reference bus they are. Then, selecting remote values would significantly increase the number of iterations as well as the processing time, regardless of the method used.

The main advantage of FDLF lies in the speed to obtain complex voltages in the system busses. If both techniques are combined so that the first, second, or third FDLF solution represents the initial NR conditions, the iterations that NR initially would have had to develop to obtain the solution with the convergence level guaranteed by NR would be reduced.

3 Results and Discussion

Starting from the hypothesis that, the number of iterations of the combined method decreases concerning the exact NR method. For the development of this section, four cases of relevant bibliography were solved separately in the SEP study. The first three exercises are 5-bus ones obtained from books that deal with the analysis of Power flows: Glover and Sarma [10, p. 287], Grainger and Stevenson [6, p. 356] and Hadi Saadat [4, p. 334]. While the last exercise is based on the IEEE 9-bus test case [9, p. 516].

The validation of results was carried out by designing a power flow solution algorithm in Matlab for 5 and 9-bus. For all cases, the methods of NR, FDLF and, CM were applied, for tolerances (ε) of $1e^{-3}$; $1e^{-5}$ y $1e^{-7}$. Also, for the combined method, the initial conditions were obtained from the results of the first, second, and third iteration of FDLF.

It should be noted that, after the validation of results with the original exercises, the resistance values were modified in all cases except for Glover and Sarma, reducing the resistance value (R) and maintaining the inductive reactance (X), until the ratio is at least 1 to 5, to comply with the FDLF restriction.

3.1 Case 1. Example 6.11 – Glover and Sarma

In the system of Fig. 1, bus 1 constitutes the balance or Slack Bus node, while bus 3 is a PV Slack Bus, where the voltage module is known ($|V_3|^0 = 1{,}05$ p.u.). Finally, busses 2, 4, and 5 are load busses, whose status variables are unknown ($\bar{V}_2^k$, $\overline{V}_4^k$ y $\bar{V}_5^k$).

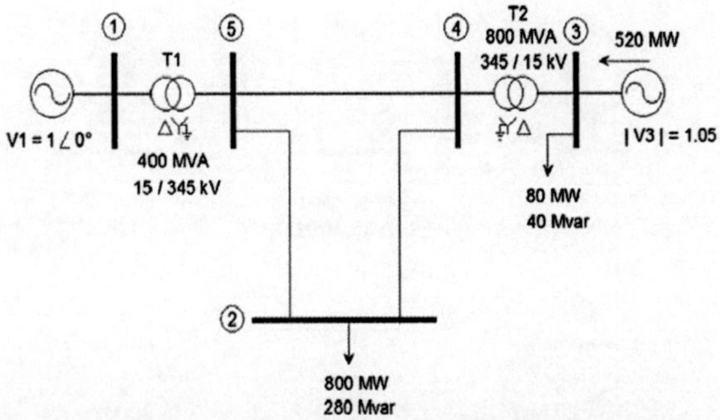

Fig. 1. Case 1–5-bus Glover and Sarma [10, p. 287]

3.2 Case 2. Problem 9.14 - Grainger and Stevenson

Figure 2 shows a 5-bus system, where the first is the Slack bus. Bus 5 is of the PV type with $|V_5|^0 = 1{,}00$ p.u. The remaining Slack busses (2, 3 y 4) constitute PQ type busses, and as in the previous case, they are unknown, therefore: $\overline{V_2}^k$, $\overline{V}_3^k$ y $\overline{V}_4^k$ y δ_5^k. Unlike case 1, this system has capacitor banks at nodes 3 and 4 of 18 and 15 MVar respectively. Note also, that the tap is not in the nominal position: t = 0,975

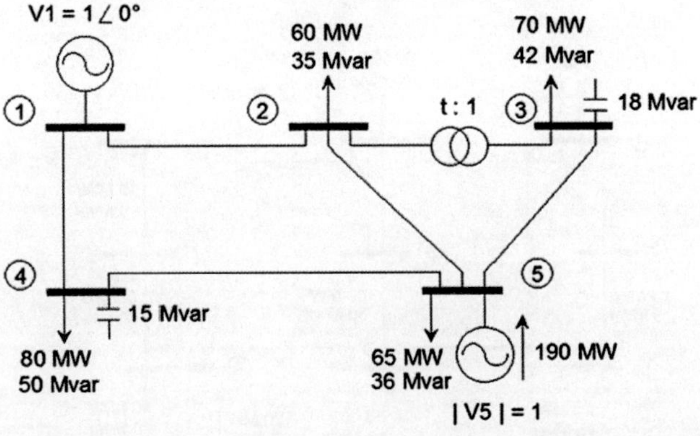

Fig. 2. Case 2–5 bus Grainger and Stevenson [6, p. 356]

3.3 Example 7.9 from the Saadat's Book

The exercise of this case raises a system with 3 generators on busses 1, 2, and 3 interconnected with each other, which are modeled as Slack, PV, and PV busses, respectively. Meanwhile, busses 4 and 5 are of the PQ type, with the following status variables as unknowns: $\overline{V}_4^k$, $\overline{V}_5^k$, δ_2^k and δ_3^k (Fig. 3).

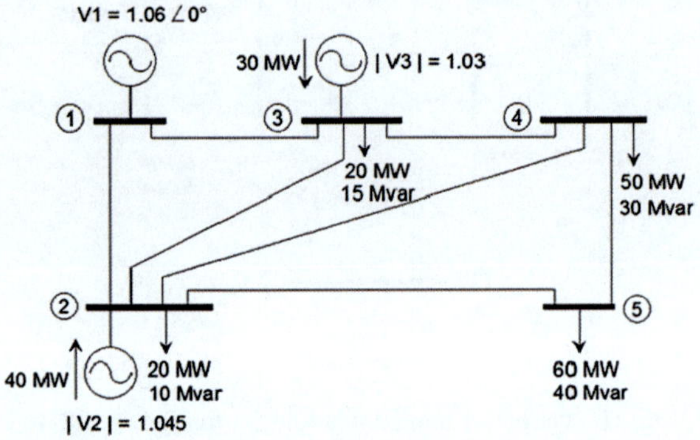

Fig. 3. Case 3–5 busses Saadat [4, p. 334]

3.4 Case 4 – Example 9 Bus IEEE

The sample shows a system of 9 busses, with three generators in busses 1, 5, and 8. The last two constitute controlled voltage nodes, while the first represents the Slack bus. With voltages of $\bar{V}_1 = 1{,}03\angle 0°$ p.u., $|V_5|^0 = 1{,}06$ p.u. y $|V_8|^0 = 1{,}01$ p.u. respectively. The remaining busses (2,3,4,6,7,9) are considered of the PQ type, which indicates that the iterative method should generate the solution of: $\overline{V}_4^k$, $\overline{V}_5^k$, δ_2^k, and δ_3^k (Fig. 4).

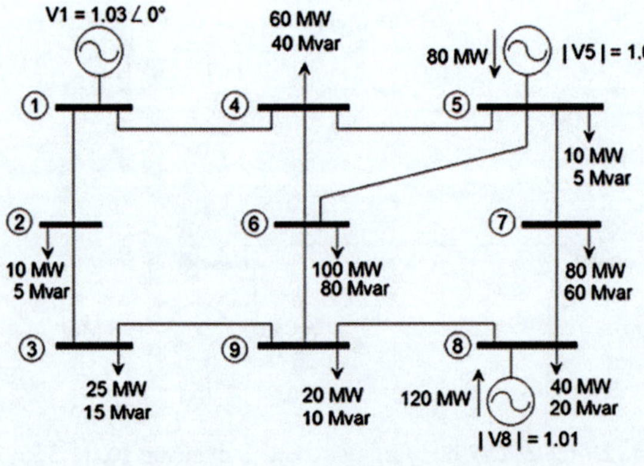

Fig. 4. Case 4–9-bus IEEE [9, p. 516]

3.5 Evaluation of the Number of Iterations

In Table 1, the comparison of the three methods for all cases with three tolerance levels is made, taking as initial conditions of NR, the results of the first iteration of FDLF. It is identified that for case 1, an iteration is reduced by using the combined method. A situation that is repeated with the proposed Grainger-Stevenson exercise.

Table 1. Comparison of the number of iterations for 1 FDLF

No	Case	Busses	1.00E-03			1.00E-05			1.00E-07		
			NR	FDLF	CM	NR	FDLF	CM	NR	FDLF	CM
1	Glover	5	4	13	3	5	22	4	5	30	4
2	Grainger	5	3	6	2	4	9	3	5	12	4
3	Saadat	5	2	5	2	3	7	2	3	10	3
4	IEEE	9	2	5	2	3	8	2	3	11	3

However, the situation changes in cases 3 and 4, where only 1 iteration is reduced for a tolerance of $1e^{-5}$, while in the other two tolerance values the number of iterations is identical to NR.

For the next stage, one iteration to the FDLF method was increased, so that the initial NR conditions will be the responses of the second iteration generate from FDR. In this case, the combined method increased its effectiveness, reducing the number of iterations by at least 1 for all cases. Even in Table 2 it can be evidenced that for case 1 with tolerances of $1e^{-3}$ and $1e^{-5}$ 2 iterations have been reduced.

Table 2. Comparison of the number of iterations for 2 FDLF

No	Case	Busses	1.00E-03			1.00E-05			1.00E-07		
			NR	FDLF	CM	NR	FDLF	CM	NR	FDLF	CM
1	Glover	5	4	13	2	5	22	3	5	30	4
2	Grainger	5	3	6	2	4	9	3	5	12	4
3	Saadat	5	2	5	1	3	7	2	3	10	2
4	IEEE	9	2	5	1	3	8	2	3	11	2

Finally, one iteration in FDLF is increased, thereby updating the initial NR conditions. In Table 3, it is evident that in case 1, 2 iterations are reduced for the three tolerance levels. This is not the situation in case 2, where just one iteration is reduced.

The situation in cases 3 and 4 is particular, although the reduction of one iteration for the tolerances of $1e^{-3}$ and $1e^{-7}$, for the level of $1e^{-5}$ there is a reduction of two iterations with respect to NR.

Table 3. Comparison of the number of iterations for 3 FDLF

No	Case	Busses	1.00E-03			1.00E-05			1.00E-07		
			NR	FDLF	CM	NR	FDLF	CM	NR	FDLF	CM
1	Glover	5	4	13	2	5	22	3	5	30	3
2	Grainger	5	3	6	2	4	9	3	5	12	4
3	Saadat	5	2	5	1	3	7	1	3	10	2
4	IEEE	9	2	5	1	3	8	1	3	11	2

3.6 Compilation Time Evaluation

Two PCs were used to evaluate the compilation times of the Matlab algorithms, the characteristics of each are presented below (Table 4):

Table 4. Specifications of the PCs used for compilation

PC	Manufacturer	Processor	RAM	Matlab version
1	Dell	Intel(R) Core (TM) i5-5200U CPU @ 2.20 GHz 2.20 GHz	4 GB	R2017a
2	Acer	Intel(R) Core (TM) i7-8750H CPU @ 2,20 GHz 2.21 GHz	16 GB	R2017b

Matlab provides a suite for performance analysis, in addition to timing functions. For this case, the commands "tic" and "toc" allow calculating the execution time of the algorithms NR and CM. Then, in the evaluation can only be compared both techniques, because FDLF compiles too quickly concerning the previous two, and the purpose of this paper is to analyze the proposal regarding the exact Newton Raphson method. The results obtained are shown in Figs. 5, 6 and 7 for tolerances of $1e^{-3}$; $1e^{-5}$ and $1e^{-7}$ respectively. In all cases, it is observed that the compilation times are practically identical for the same PC, the lines NR_1 and MC_1 correspond with the PC _1, analogously, subscripts 2 relate to the PC _2.

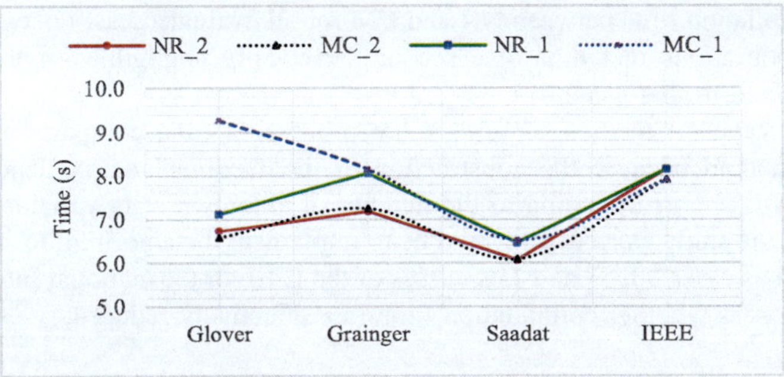

Fig. 5. Program compilation time for a tolerance of 1e-3 of PC1 and PC2

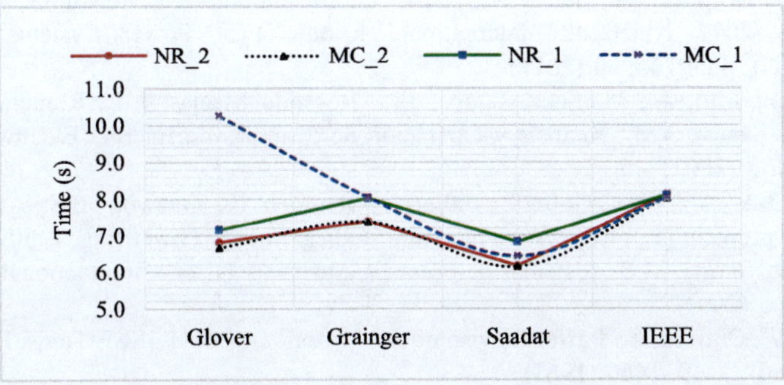

Fig. 6. Program compilation time for a tolerance of 1e-5 of PC1 and PC2

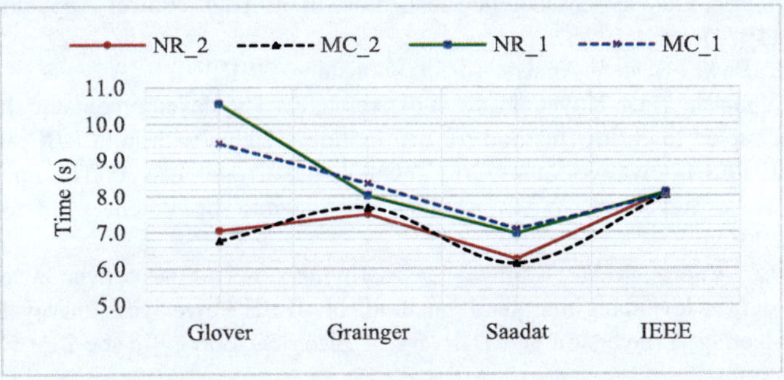

Fig. 7. Program compilation time for a tolerance of 1e-7 of PC1 and PC2

4 Conclusions

The Combined Method reduces the number of iterations concerning the exact Newton Raphson method and it is applicable to any system that meets the X/R restrictions for FDLF.

The compilation time between NR and CM for all evaluated cases between the two PCs shows variations of tenths of a second, practically negligible for the four case studies analyzed in this paper.

Newton Raphson method requires more processing time compared to the Fast-decoupled method because the construction of the Jacobian matrix depends on the partial derivatives corresponding to the number of unknown state variables.

Scope to the study carried out would be to implement the algorithm for much larger systems such as cases 30, 57 and 118 busses of the IEEE or the National Interconnected System, to assess whether computation times are effectively reduced.

References

1. Abokrisha, M., Diaa, A., Selim, A., Kamel, S.: Development of Newton-Raphson power-flow. In: 2017 Nineteenth International Middle East Power Systems Conference (MEPCON), pp. 976–980 (2017)
2. Gómez Exposito, A., Martínez Ramos, J.L., Rosendo Macías, J.A., Romero Ramos, E., Riquelme Santos, J.M.: Análisis y Operación de Sistemas de Energía Eléctrica. McGraw-Hill, Madrid (2002)
3. Afolabi, O.A., Ali, W.H., Cofie, P., Fuller, J., Obiomon, P., Kolawole, E.S.: Analysis of the load flow problem in power system planning studies. Energy Power Eng. **7**, 509–523 (2015)
4. Glover, D., Sarma, M.S.: Sistemas de Potencia Análisis y Diseño. International Thomson S. A., Mexico (2004)
5. Tinney, W., Clifford, E.: Power flow solution Newton' s method. IEEE Trans. Power Appar. Syst. **11**(86), 1449–1460 (1967)
6. Mueller, G., Komarnicki, P., Golub, I., Styczynski, Z., Dzienis, C., Blumschein, J.: PMU placement method based on decoupled. In: 9th International Conference on Electrical Power Quality and Utilisation, vol. 9, no. 11 (2007)
7. Stott, B., Alsac, O.: Fast desacoupled load flow. IEEE Trans. Power Apparatus Syst. 1(3), 859–869 (1974)
8. Saadat, H.: Power System Analysis. PSA Publishing (2010)
9. Ochi, T., Yamashita, D., Koyanagi, K., Yokoyama, R.: The development and the application of fast decoupled load flow method for distribution systems with high R/X ratios lines. In: 2013 IEEE PES Innovative Smart Grid Technologies Conference (ISGT), pp. 1–6 (2013)
10. Grainger, J.J., Stevenson, W.D.: Análisis de Sistemas de Potencia. McGRAW-HILL, México (2001)
11. Lagace, P.J., Vuong, M.H., Kamwa, I.: Improving power flow convergence by Newton Raphson with a levenberg-marquardt method. In: IEEE Power and Energy Society 2008 General Meeting: Conversion and Delivery of Electrical Energy in the 21st Century, PES, pp. 1–6 (2008)
12. Sriyawong, T., Sriyanyong, P., Koseeyaporn, P., Kongsakorn, P.: A modified fast decoupled power flow algorithm. Int. Energy J. **6**(1), 95–103 (2005)
13. Pan, Y., Yuan, Z., Chen, Y., Liu, B.: The real-time research of optimal power. Int. J. Adv. Comput. Sci. Appl. **6**(2), 78–82 (2015)
14. Hongfu, W., Shixia, M., Yi, W., Zhiqiang, Z.: An approximate power flow method. In: 2018 Internacional Conference on Power System Technology, vol. 6, no. 8, pp. 292–298 (2018)
15. MacIel, R.S., Padilha-Feltrin, A., Righeto, E.: 2006 IEEE PES Transmission and Distribution Conference and Exposition: Latin America, TDC 2006, pp. 1–6 (2006)

16. Wamser, R.J., Slutsker, I.W.: Power flow solution by the newton-raphson method in transient stability studies. IEEE Power Eng. Rev. **8**(103), 2299–2307 (1984)
17. Yubin, Y., Li, M.: Designs of fast decoupled load flow for study purpose. Energy Procedia **17**, 127–133 (2012)

Flexible Manufacturing System Oriented to Industry 4.0

David Trajano Basantes Montero[1](✉), Sylvia Nathaly Rea Minango[2], Daniel Isaías Barzallo Núñez[1], Carlos Gabriel Eibar Bejarano[3], and Paúl David Proaño López[3]

[1] Instituto Superior Tecnológico Central Técnico, Quito, Ecuador
dtbasantesm@gmail.com
[2] Universidad de las Fuerzas Armadas ESPE, Quito, Ecuador
[3] Universidad Técnica de Ambato, Ambato, Ecuador

Abstract. Currently, global competition, technological development and innvtion represent a challenge for companies, above all for manufacturing, as they are forced to reconfigure their processes for the growing market of personalized products. Industry 4.0 and manufacturing generate a transformation, and both manufacturing and information technologies have been integrated to create efficient systems of production, management and ways of doing business. The objective of the research is to analyze the influence of technological tools oriented to Industry 4.0 in the increase of the flexibility of a manufacturing system ap-plied, implementing vertical and horizontal integration systems, data analysis, cloud and simulation in production systems modular for obtaining data. The flexibility reached was determined according to the parameters of a flexible manufacturing system: variety of parts, programming change, error recovery and new parts, allowing to automate the manufacturing processes, as well as to attend in a timely manner the needs of the market, promoting the growth of the national industry in concepts of flexibility and Industry 4.0 applied.

Keywords: Manufacturing · Flexibility · Industry 4.0 · Industrial networks · Man machine interface · Modular production systems

1 Introduction

Recent advances in manufacturing allow information from all related perspectives to be closely monitored and synchronized between the physical level of the factory and computational cyberspace.

In addition, by using advanced information analysis, networked machines will be able to work more efficiently and collaboratively. This trend is transforming the manufacturing industry to the next generation, i.e. Industry 4.0 [1, 21].

Automation will have broad-spectrum effects across parts and sectors, although it is a global phenomenon, four economies (China, India, Japan, and the U.S.) account for just over half of total wages and nearly two-thirds of the number of associated employees with activities that are automated if proven technologies are adapted today [2].

M. Botto-Tobar et al. (Eds.): CI3 2020, AISC 1277, pp. 234–245, 2021.
https://doi.org/10.1007/978-3-030-60467-7_20

In Latin America 60% of jobs are capable of automating, in Ecuador 49% of manual processes are able to be replaced by high technology [3].

Industries such as manufacturing and agriculture include predictable physical activities that have high automation potential, but low wage rates in some developing countries can slow their adoption. Manufacturing companies must resist growing global competition in different strategic dimensions, such as production costs, product quality and product innovation [10].

Ecuador ranks 60th in the world based on industrial product, being the eighth industrial economy in Latin America. "The level of automation presented by industries is: in manual drive 48%, semi-automatic 27%, automatic 18%, and computerized 7%. The results presented make known the low level of technology that is present in SMEs" leaving aside the vertical integration of several components to implement a flexible and reconfigurable manufacturing system i.e. a smart factory. In industries of high levels of production due to the high demand, these methods become inefficient generating poor quality products, unreliable and increasing their cost [4] [5, p. 1].

Ecuador's industrial sector is characterized by a predominant presence of intensive branches in natural resources and labor, and less presence of intensive branches in engineering [6].

By 2013, engineering-intensive sectors explained just under 10% of employment and industrial value added. Labour-intensive branches, meanwhile, concentrated 41% of employment, and 17% of manufacturing value added. The central role is occupied by the natural resource-intensive branches that explained almost half of industrial employment in 2013, and 73% of sectorial value added [4].

The term Industry 4.0 refers to a new model of organization and control of the value chain throughout the product lifecycle and throughout manufacturing systems supported and made possible by information technologies [7].

In this transformation of companies, sensors, machines, workpieces and IT systems information technologies, can interact with each other to obtain more reliable forecasts, be able to configure themselves, and adapt to changes. Figure 1 shows a timeline of the evolution of the industry [8, 22, 23].

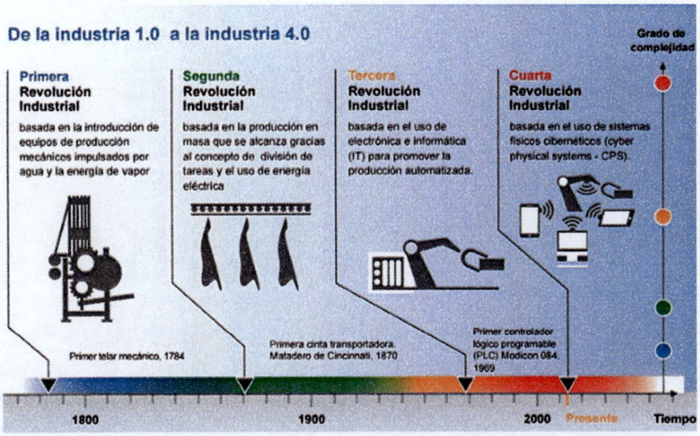

Fig. 1. Industry evolution. Source: [11]

2 Development

The project took a quantitative approach, because numbers of Industry 4.0 techniques were manipulated to exercise control over them and obtain quantifiable data to verify their influence on flexibility through modular production systems [17]. This data comes from the analysis of the parameters of a flexible system. Figure 2 graphically shows the methodology applied in the development of research.

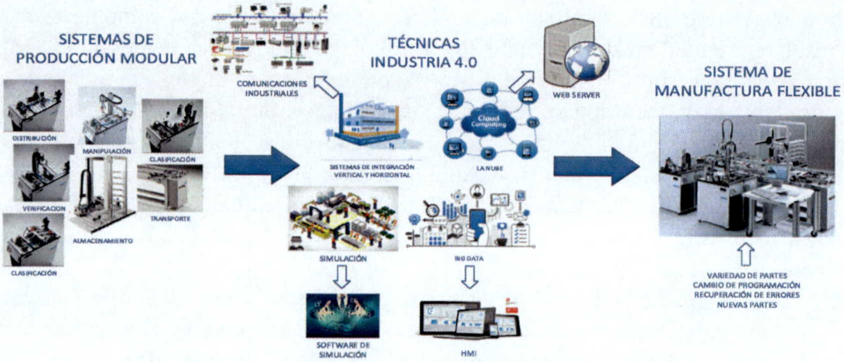

Fig. 2. Methodology to solve the problem. Source: The Author.

Considering the base operation of each modular production system, an observation sheet was generated as shown in Table 1.

Table 1. Observation sheet of the model manufacturing system.

Model manufacturing system state observation sheet					
Stations / Parameter	INCOMING MERCHANDISE	Process	Store	SALT MERCHANDISE	Transport
Current operating percentage	35%	5%	5%	27.5%	5%
Maintenance and Calibration	Magnetic, capacitive optical, similar resistive and pressure switches.	End-of-race, magnetic, capacitive optical and inductive sensors	End-of-stroke, magnetic sensors and encoders.	Magnetic, capacitive optical and inductive sensors	Barrier and inductive optical magnetic sensors.
Field level network	Doesn't apply	Doesn't apply	Doesn't apply	Doesn't apply	Red AS-interface in functional
Cell-level network	Non-functional Profibus DP network	Non-functional Profibus DP network	Non-functional Profibus DP network	Non-functional Profibus DP network	Non-functional Profibus DP network

(continued)

Table 1. (*continued*)

Model manufacturing system state observation sheet					
Stations / Parameter	INCOMING MERCHANDISE	Process	Store	SALT MERCHANDISE	Transport
Electropneumatic diagrams	There is no diagram	There is no diagram	There is no diagram	There is no diagram	There is no diagram
Operating manuals.	No manual available	No manual available	No manual available	No manual available	No manual available
Production time	It cannot be determined as it does not conclude the process	It cannot be determined as it does not conclude the process	It cannot be determined as it does not conclude the process	It cannot be determined as it does not conclude the process	It cannot be determined as it does not conclude the process
Variety of production	It cannot be determined as it does not generate any type of product	It cannot be determined as it does not generate any type of product	It cannot be determined as it does not generate any type of product	It cannot be determined as it does not generate any type of product	It cannot be determined as it does not generate any type of product

Source: The Author

Based on the flexibility parameters set, the next current system situation is determined in Fig. 3.

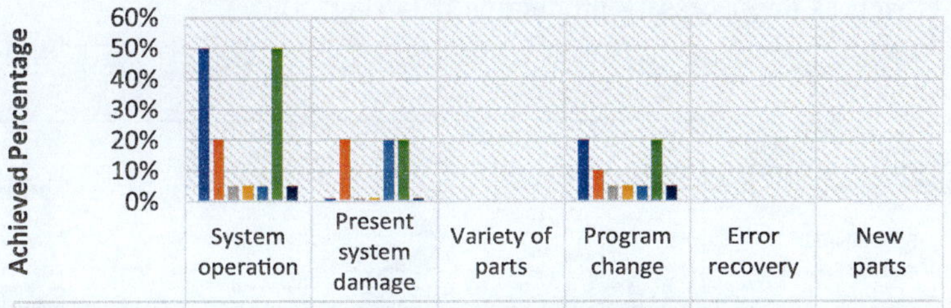

	System operation	Present system damage	Variety of parts	Program change	Error recovery	New parts
Distribution	50%	1%	0%	20%	0%	0%
Check	20%	20%	0%	10%	0%	0%
Handling	5%	1%	0%	5%	0%	0%
Machined	5%	1%	0%	5%	0%	0%
Storage	5%	20%	0%	5%	0%	0%
Classification	50%	20%	0%	20%	0%	0%
Transport	5%	1%	0%	5%	0%	0%

Fig. 3. Initial situation analysis. Source: The Author.

2.1 Horizontal and Vertical System Integration

The horizontal and vertical integration of the model manufacturing system as part of the applicable Industry 4.0 technologies was achieved through the implementation of industrial communications such as Bus AS-I, Profibus DP, Industrial Ethernet, Profinet such as shown in Fig. 4 [12–14].

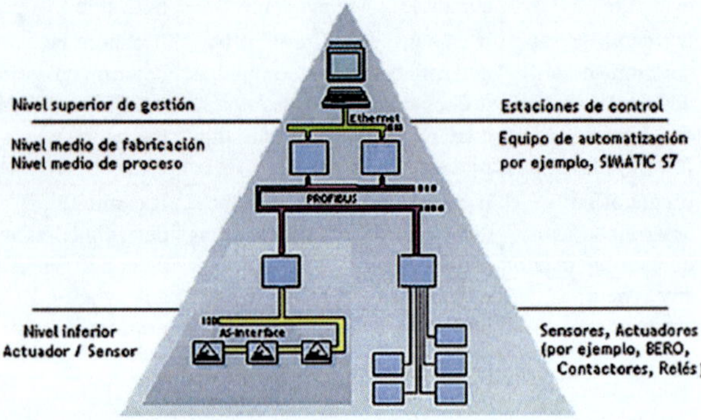

Fig. 4. Different levels of industrial communication. Source: [9]

2.2 Data Analysis

To perform a data analysis oriented to Big Data, a human machine interface was developed from which you get control and visualization of the main tasks of each MPS station, as well as the necessary production data (Fig. 5).

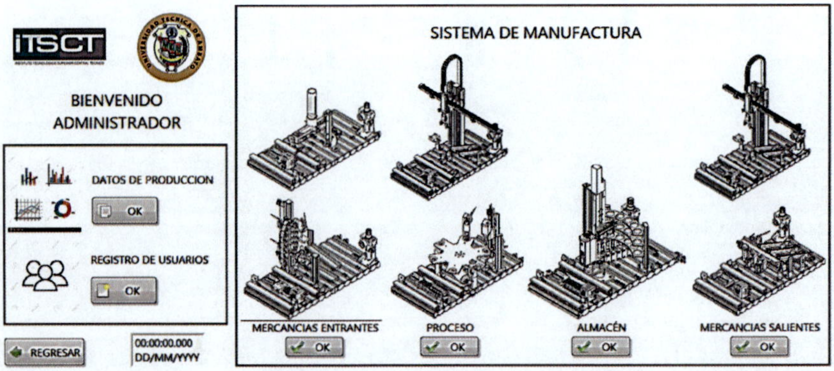

Fig. 5. HMI Manager window.

2.3 The Cloud

As part of the research it was proposed to integrate Industry 4.0 technologies, with the cloud being one of the functions that best fits the development of the manufacturing system model [24], it was chosen to develop a web application on the GoDaddy cloud platform dedicated to independent small businesses that counts with more than 18 million customers around the world and manage more than 77 million domain names.

Within GoDaddy's website builder, the graphical interface was designed, and data that will be available in the cloud through the Google Drive platform was established (Fig. 6).

Fig. 6. Web Server Configuration.

2.4 Simulation

One of the techniques oriented to Industry 4.0 is the use of these simulation systems, in this case the staff directly related to the operation of the manufacturing model system can obtain a generalized view of the operation of each of the stations considered.

Festo's Ciros software is a powerful industrial development tool of 3D simulation for process and factory automation, provides among its practical examples the simulated environment of a manufacturing system MPS 507-FMS valid as simulation of our implemented manufacturing model system (Fig. 7).

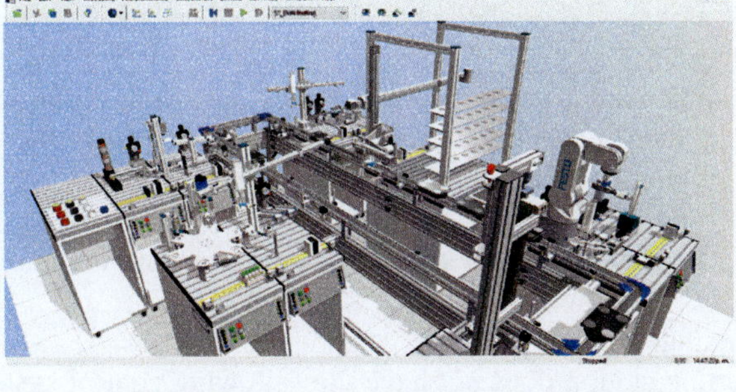

Fig. 7. Simulation manufacturing system.

3 Results

The tests were developed taking into consideration two important phases of the project, in the first instance the previously established base functioning of each of the MPS was verified and their partial integration by analyzing the flexibility of the system, and then depending on the techniques of Industry 4.0: Horizontal and Vertical Integration, Big Data, The Cloud and Simulation applied, determine the increase in flexibility achieved according to the established parameters of the manufacturing model system [20].

3.1 Step 1

After the initial situation analysis, maintenance performed, drawing and diagram surveying, the configuration and base programming implemented in MPS systems, the operation of the manufacturing system based on the analysis of the flexibility parameters (Fig. 8) [15].

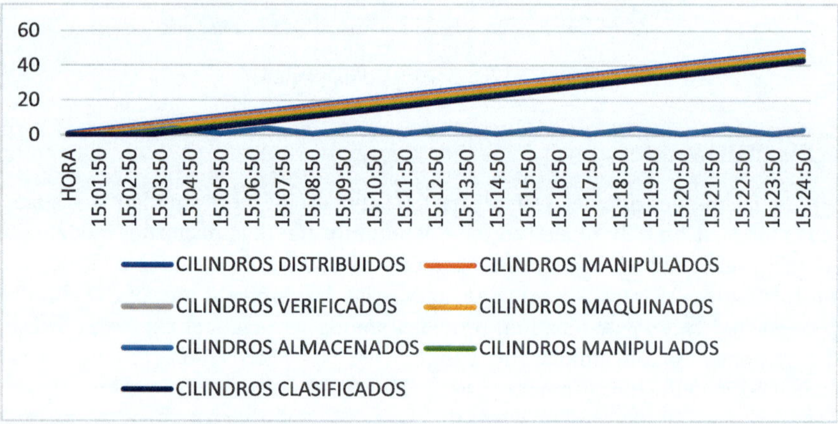

Fig. 8. Production of the manufacturing system.

Analysis of Variety of Parts
Due to the base implementation of the system it is only possible to have one type of variety of cylinders due to the initial setup implemented in such a way that:

Material Type: Unique

Size: Unique

The machining station obeys operator commands requiring the implementation of a control loop described below:

Type of machining: Unique

Finally, the sorting station does not have the following control system necessary to classify the colors of the cylinders:

Color: Unique

Programming Change
There is no cell-level network type so the lack of horizontal communication between production stations prevents the manufacturing system from supporting programming changes, the system cannot be reconfigured automatically.

Error Recovery
Since the base manufacturing system implemented does not have a vertical communication between production levels, also a human interface, control and supervision machine, it cannot set error alarms so that the operator can act according to the case.

New Parts
Due to the lack of information technologies implemented and the zero-production analysis for the potential market, it empowers the impossibility and unnecessary action of introducing new designs to existing ones.

3.2 Step 2

Following phase one and analysis of the results obtained, Industry 4.0-oriented technologies are implemented in the manufacturing model system and its flexibility is determined against the established parameters obtaining the following results:

Analysis of Variety of Parts
After the implementation of the necessary control systems and M2M Machine-to-machine communication as part of the technologies oriented to Industry 4.0 within the horizontal and vertical integration implemented in the manufacturing model system specifically providing the necessary communication to production stations through the Profibus network, it is possible that depending on the operator's requirements the system can produce nine varieties of cylinders (Table 2) (Figs. 9, 10 and 11).

Table 2. Variety of parts.

Size	Material	Machining	Color
Great	Plastic	Hammered	Red
		Drilling	Red
		Hammered/Taladrated	Red
	Metal	Hammered	Silver
		Drilling	Silver
		Hammered/Taladrated	Silver
Small	Plastic	Hammered	Black
		Drilling	Black
		Hammered/Taladrated	Black

Fig. 9. Machined cylinders.

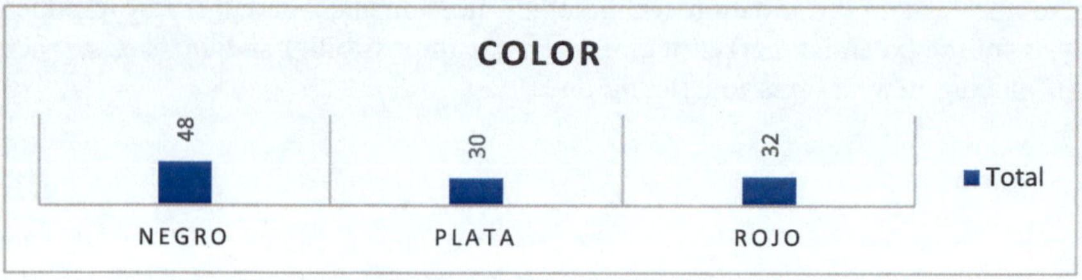

Fig. 10. Cylinders classified by color.

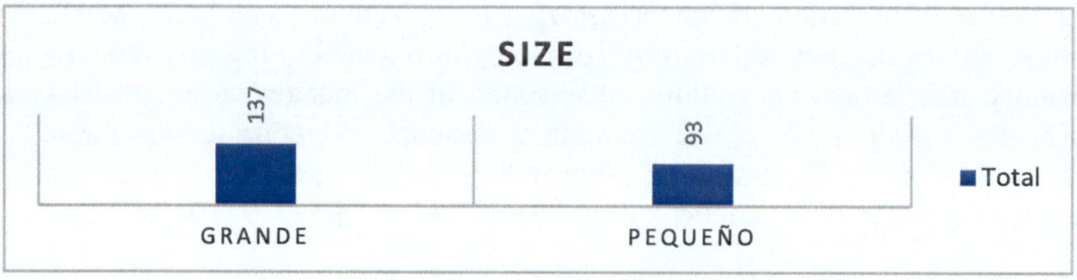

Fig. 11. Cylinders classified by size.

Schedule Change
Based on M2M and HMI implemented in the manufacturing model system through AS-i, Profibus, Profinet, and Industrial Ethernet networks it is possible to change the programming according to the operator's requirements through the control interface [16, 18, 19].

From the operator's perspective, the facility required for systems to be put up or stopped and to provide all relevant production information, it refers to determine the change in production schedule based on the variety of parts analyzed.

Error Recovery With the implementation of the HMI the operator has at his disposal alarms in the face of possible failures or errors that may arise in the system to act accordingly with the possibility of stopping and isolating the affected system from the process.

The error-recovery capability provided by the susceptible programming change is one of the most important parameters in terms of flexibility level.

New Parts

Finally, the need to introduce new parts is closely linked to the analysis of production data generated through the integration of the manufacturing model system, as well as real-time access of such data through the cloud. The system currently provides that possibility efficiently. Not forgetting the importance of a previous feasibility analysis in terms of new parts manufactured using simulation systems specifically specific to the one presented above by the manufacturing model system.

It is important to note that the model manufacturing system was subjected to several tests prior to obtaining the results presented, the flexibility levels are established compared to a base programming implemented following the manufacturing line of a single product, however, the stated objective focuses directly on the impact that the development and application of Industry 4.0-oriented technologies generated on the system (Fig. 12).

Fig. 12. Manufacturing system.

In summary, the flexibility analysis of the manufacturing system is presented below (Table 3):

Table 3. Test summary

	Variety of parts	Result	Industry technologies 4.0
Step 1	Variety of parts	11%	N/A
	Changing programming	No	N/A
	Error recovery	No	N/A
	New parts	No	N/A
Step 2	Variety of parts	100%	Vertical and horizontal integration of systems
	Changing programming	Yes	Human machine interface
	Error recovery	Yes	Vertical and horizontal integration of systems, Human machine interface
	New parts	Yes	Data analysis, cloud, simulation

4 Conclusions

- Having reached a range of capacity in terms of 100% of parts compared to the initial 11% with the horizontal and vertical integration of the system determines that this technology is essential to meet one of the established flexibility parameters, the other parameters being purely qualitative their results are expressed in levels of compliance and ease of application, through the human machine interface, which provides the ability to control the entire system from a device in this case a PC, plus all the acquired data generated by each station, each area and each sensor from the office, cell plant and field levels respectively fully integrates the system focused on the feasibility of programming changes and error recovery.
- In determining the elements involved in the system, a total of 50 actuators are discussed between indicators, valves, parameterized outputs and motors. On the other hand, a total of 150 input signals of pressure switches, magnetic, optical, inductive, capacitive, encoders, pushbuttons, selectors and parameterized inputs that without proper handling through the acquisition of data implanted their control, visualization analysis and logging would not be possible, this underpins the relative ease with which a change of programming is feasible, even immediate detection of failures.
- The analysis of new parts to be introduced will be determined by the use of proposed simulation tools, since substantial savings in implementation issues to determine the feasibility of new parts produced is the main factor analysis.
- The influence of Industry 4.0-oriented technology tools of the model manufacturing system was determined, through quantitative and qualitative analysis with respect to established parameters resulting in a significant increase in the flexibility of the system.
- The application of new technological tools in terms of the development of manufacturing systems directly incentivizes the growth of the domestic industry in concepts of flexibility in the manufacture of products, as well as the capacity to analysis of these possibilities within laboratories equipped with scale systems and simulation techniques for the development of higher technical education.

References

1. Lee, J., Bagheri, B., Kao, H.-A.: A cyber-physical systems architecture for industry. Manufact. Lett. **3**, 18–23 (2015)
2. Manyika, J., Chui, M., Miremadi, M., Bughin, J., George, K., Willmott, P., Dewhurst, M.: Un Futuro que Funciona: Automatización, Empleo Y Productividad, p. 7. Mckinsey Global Institute (2017)
3. Mundial, B.: Dividendos digitales. Informe sobre el desarrollo mundial (2016)
4. Garzón, N., Kulfas, M., Palacios, J.C., Tamayo, D.: Evolución del sector manufacturero ecuatoriano 2010–2013 Tipologías estadísticas y dinámicas de las manufacturas. Cuaderno de Trabajo N.1 (2016)

5. Sánchez, V., Pizarro, D.: Diagnóstico del nivel de automatización en las pequeñas y medianas industrias de la ciudad de Cuenca. INGENIUS, Revista de Ciencia y Tecnología (2010)
6. Katz, J., Stumpo, G.: Regímenes sectoriales, productividad y competitividad internacional. Revista de la CEPAL Nro. 75, Santiago de Chile (2001)
7. Del Val Román, J.L.: Industria 4.0: la transformación digital de la industria. Conferencia de Directores y Decanos de Ingeniería Informática (2016)
8. Gerbert, P., Lorenz, M., Rüßmann, M., Waldner, M., Justus, J., Engel, P., Harnisch, M.: Industria 4.0: El futuro de la productividad y el crecimiento en las industrias manufactureras. Boston Consulting Group (2015)
9. Guerrero, V., Yuste, R., Martinez, L.: Comunicaciones industriales (2010)
10. Wang, S., Wan, J., Zhang, D., Li, D., Zhang, C.: Towards smart factory for Industry 4.0 a self-organized multi-agent system with big data based feedback and coordination. Comput. Netw. **101**, 158–168 (2016)
11. Miranda, A.: Industria 4.0. NC Tech (2016)
12. García Moreno, E.: Automatización de Procesos Industriales, Universitat Politècnica de València (1999)
13. Smith, C.A., Corripio, A.B.: Control Automático de Procesos Teorí y Practica. Limusa, Mexico DF (1991)
14. Daneri, P.A.: PLC Automatización y Control Industrial. Hispano América S.A, Buenos Aires (2008)
15. Ebel, F., Idler, S., Prede, G., Scholz, D.: Neumática Electroneumática Fundamentos. Festo Didactic GmbH & Co, Denkendorf (2010)
16. Siemens, A.G.: Profinet Descripción del sistema. Siemens AG, Nurnberg (2012)
17. Kalpakjian, S., Schmid, S.R.: Manufactura, ingeniería y tecnología. Pearson Educación (2008)
18. Siemens A&D: SIMATIC PCS 7 – SIMATIC IT – Integration. Applicantion & Tools (2006)
19. Guerrero, V.: BUS AS-I Configuración y programación de una red (2005)
20. VDMA German Engineering Federation: Guiding principles for the implementation of Industrie 4.0 in small and medium sized businesses. Guideline Industrie 4.0 (2016)
21. Brettel, M., Klein, M., Friederichsen, N.: The relevance of manufacturing flexibility in the context of Industrie 4.0. Procedia CIRP **41**, 105–110 (2016)
22. Soler, S., Wollschlaeger, M.: Control as an Industrie 4.0 component network-adaptive applications for control. In: 22nd IEEE International Conference on Emerging Technologies and Factory Automation (ETFA), pp. 1–4 (2017)
23. Mätzler, S., Wollschlaeger, M.: Interchange format for the generation of functional elements for Industrie 4.0 components. In: IECON 2017 - 43rd Annual Conference of the IEEE Industrial Electronics Society, pp. 5453–5459 (2017)
24. Ali Khan, W., Wisniewski, L., Lang, D., Jasperneite, J.: Analysis of the requirements for offering Industrie 4.0 applications as a cloud service. In: 2017 IEEE 26th International Symposium on Industrial Electronics (ISIE), pp. 1181–1188 (2017)

Levelized Cost of Storage (LCOS) Considering the Reliability of Battery Life

Daniel Andagoya Alba(✉), Ximena Guamán Gavilanes, and Daniel Isaías Barzallo Núñez

Instituto Superior Tecnológico Central Técnico, Quito, Ecuador
andagoyaalba@gmail.com

Abstract. Over the last few years, electricity generation sources have diversified and diverse primary energy sources are increasingly used, including sun and wind, however, these types of energy need a storage system that allows manage them in a better way. The following work presents a study to determine the value of the Levelized Cost of Storage considering the reliability of the battery life used in the storage system. For this study, the mathematical developments carried out in previous research were considered, both in the calculation of the Levelized Cost of Storage as for calculating battery life with a certain reliability percentage. This study used Weibull's probability function to determine the reliability of battery life as well as all the parameters necessary for the calculation of the Levelized Cost of Storage. The calculation processes were carried out in computer programs, especially with regard to iterative processes. The results show that the value of the Levelized Cost of Storage increases as you have a higher percentage of reliability, which means that a storage system might have the same value of the Levelized Cost of Storage with different percentage of reliability, which shows an additional parameter to compare various storage systems.

Keywords: Levelized cost of storage · Battery life · Weibull probability function · Storage systems

1 Introduction

Over the last few years the electrical system has had some changes in relation to the power flow, it has gone from having a known topology with no interaction with consumers and with vegetative growth to having electric systems with application of distributed generation technologies in which users take an important role within the generation of electricity not only as consumers but also as power generators.

Most of the Distributed Generation is based on renewable generation technologies, this allows to have a certain local energy autonomy in addition to the environmental benefits that come with using renewable primary sources such as sun or wind, however it presents some disadvantages, especially in the unmanaging of these kind of sources energy, that is, they depend on climatic conditions that cannot be accurately predictable, this provokes to an intermittecy in the generation of energy. In addition, the demand variability not necessarily coincide with the variation of generation, causing

M. Botto-Tobar et al. (Eds.): CI3 2020, AISC 1277, pp. 246–256, 2021.
https://doi.org/10.1007/978-3-030-60467-7_21

the generation and demand system to not have the balance necessary to maintain themselves [1].

At this point, the alternatives to be able to decouple the generation of the demand are very important. The principal strategy to use this kind of energy are the storage systems, that allow to manage the balance between the demanded energy and the produced energy. Currently there are many storage technologies with various characteristics that can be used for electrical power systems.

Recently the concept of LCOS (Levelized Cost Of Storage) has been developed as a tool that allows to carry out a comparative analysis between different types of storage systems, the final result of this calculation is the dollar value for each kWh or MWh stored, which characterizes the behavior of the storage system over the lifetime of the installation. This data is relevant as it allows to determine within a wide range of options, the one that presents the best technical and economic performance in a facility intended for energy storage [2].

For the calculation of LCOS, various technical and economic parameters are taken into account, among which is the battery life that will be used within the storage system. However, this data is not an exact parameter, but rather a probabilistic one, which makes the final data of the LCOS also is a probabilistic too, because the most of the parameters used in the calculation depend on the life of the battery used [3].

This project aims to determine the influence of the battery life used in the storage system in the final calculation of the LCOS, this will allow to determine the final value of this parameter with a percentage of reliability based on life useful battery used, it will also allow to compare different storage systems with different reliability percentages or even the same storage system with different probabilities of reliability for the value obtained from the LCOS.

2 Materials and Methods

In this work for the calculation of LCOS is used the mathematical formulations developed in previous researches, in most cases the parameters used are those related to the initial investment, operating costs and residual value, recharge cost, storage system replacement costs and installation life time. Emphasis is placed on calculating battery life since the others parameters within the storage system dependent on this value.

2.1 LCOS (Levelized Cost of Storage)

The LCOS tool is defined as a comparative calculation between different storage system technologies in terms of average cost per store kWh or MWh, depending on both technical and economic parameters. The mathematical expression developed for the calculation of LCOS is defined according to Eq. (1) [3–5].

$$LCOS = \frac{I_0 + \sum_{n=1}^{N} \frac{Cv_n}{(1+d)^n}}{\sum_{n=1}^{N} \left(E_{DayOp}.days_{op}\right)} \quad (1)$$

The parameters of Eq. (1) are:

- $LCOS$ = Levelized Cost Of Storage [$/kWh].
- I_0 = Initial investment [$].
- Cv_n = Types of costs [$].
- d = Discount rate or update rate [%].
- N = Installation life [years].
- E_{DayOp} = Energy stored per day [kWh]
- $days_{op}$ = Operation days per year.

2.1.1 Initial Investment

The investment refers to the money that would result as the cost of putting the facility into operation in year zero. That is, it is the money spent at the time of installation of the storage system [3–5].

2.1.2 Cost Types

The costs taken into consideration by the LCOS are:

- Operation and maintenance costs.
- Residual value.
- Cost of recharging.
- Costs of storage system replacement.
- Cost of other elements replacements [3–5].

2.1.3 Discount Rate

The discount rate or discount type or capital cost is a financial measure that is used to know the current value of a future payment [3–5].

2.1.4 Installation Life

The installation lifetime is a variable parameter that depends on the analysis to be carried out. Normally the default values are 10, 15, 20 or 25 years. However, these values can be determined according the way to use that will be given to the system [3–5].

2.1.5 Stored Energy

For this analysis the energy stored by the system is related to the life of the battery used. Battery life will also depend on environmental factors, the way that it is used and the discharge depth to which the battery is subjected during its life [3–5].

2.2 Calculating the Life of a Battery

This work analyses the calculation of battery life based on the Number of Cycles-Residual Capacity curve. This curve allows to graphically observe the battery's behavior over its lifetime and the variation in its storage capacity based on the number of cycles it has performed. The residual capacity of a battery depends on the Depth of Discharge (DoD) with which it will be used over its lifetime [6] (Fig. 1).

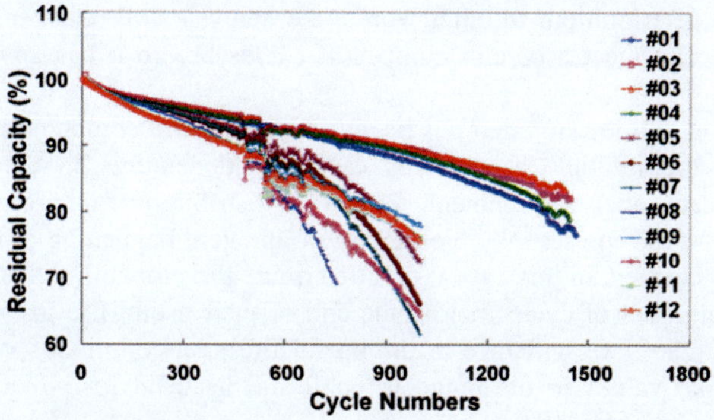

Fig. 1. Cycles - Residual Capacity characteristic of a Lithium Ion battery. Source: [6]

The area under the curve of these graphs in technical terms represents the energy that the battery will be able to store over its all useful life, for this reason, to calculate the battery life should start with the calculation of the surface that is covering this curve, and then using the Eq. (2), calculate the battery life [6–8].

$$Life_{bat} = \frac{C_{bat}.N_{cycles}}{\frac{E_{bat}.days_{op}}{\eta_{bat}}} = \frac{Area\ bajo\ la\ curva}{\frac{E_{bat}.days_{op}}{\eta_{bat}}} \quad (2)$$

The parameters of Eq. (2) are:

- C_{bat} = Battery's capacity [kWh o MWh].
- N_{cycles} = Number of cycles.
- E_{bat} = Energy stored by the battery per day [kWh o MWh].
- $days_{op}$ = Operation days per year.
- η_{bat} = Battery performance.

2.2.1 Battery Life

In engineering, the lifetime of an element refers to the time that the element can be used before it has anomalies or meets a specified parameter indicating compliance with a defined duty cycle.

Mathematical theory has been developed mainly under industrial requirements, especially applied to the life prediction of electrical equipment. These techniques are mainly inferred according to the determination of probabilistic distributions whose parameters are determined according to the specific characteristics of each element analyzed [9].

Weibull's probability distribution was proposed as an alternative for estimating the devices life time. This distribution is currently used in many applications due to its great versatility and simplicity of its formulation [10, 11].

2.2.2 Battery Life for a Given Reliability Percentage

The estimate of battery life is determined by the number of cycles that the battery could meet before its deterioration, in other words, the analysis will result in the reliability that the battery can meet a certain number of cycles before it has any kind of malfunction [12].

Weibull's distribution function has been used to perform component life reliability studies due to the multiple shapes this can take, depending on the characteristic parameters of each analyzed element. Weibull's distribution function is a statistical model that allows to represent the probability of an event happening or not happening after a time has elapsed, in this case it will determine the probability that a battery will meet a certain number of cycles before the end of their useful life [6].

For these studies, we will take to the useful life as a continuous variable T, this variable will take values in the range [0; ∞], and its behavior will depend on the probability function studied.

The Weibull distribution is determined by two main parameters, scale parameter (α) and shape parameter (β), these parameters define the type of probability distribution that represents the analyzed element.

The battery life is estimated according to the analysis of a Weibull probability function, the final result will be the number of cycles the battery will perform for a given reliability percentage. Equation (3) calculates the value in per unit of reliability that a battery would achieve a certain number of cycles based on the Weibull function's parameters [6].

$$F\left(Cicl_{[p.u]}\right) = e^{-\left(\frac{Cicl_{[p.u]}}{\alpha}\right)^{\beta}} \tag{3}$$

The parameters of Eq. (3) are:

- $F(Cicl_{[p.u]})$ = Probability based on the number of cycles in per unit..
- $Cicl_{[p.u]}$ = Number of cycles per unit.
- α = scale parameter.
- β = shape parameter.

The number of cycles per unit for a given reliability will be calculated using Eq. (4) resulting from the Eq. (3).

$$Cicl_{[p.u]} = \alpha\left[-ln\left(\Pr\left(Cicl_{[p.u]}\right)\right)\right]^{\frac{1}{\beta}} \tag{4}$$

With this value in per unit we can determine the final cycles that the battery will need to achieve for a required percentage of reliability.

As the maximum cycles determined in the Cycles - Residual Capacity curve was the value taken as the basis, then the final cycles ($Cicl_{[end]}$) for a given reliability percentage will be calculated according to Eq. (5). This equation takes the values of the maximum cycles that the battery would have in its entire lifespan ($Cicl_{[fmax]}$) and the maximum cycles that the battery would have for a certain reliability in per unit ($Cicl_{[fpu]}$) resulting from the Eq. (4).

$$Cicl_{[end]} = Cicl_{[p.u]} . Cicl_{[max]} \quad (5)$$

With the determination of the final cycles that the battery will have for a certain percentage of reliability, area under the curve of the Cycles – Residual Capacity graphic is calculated, for which in this work the Riemann's integral is used through iterative calculations.

With the above parameters calculated you can use the Eqs. (2) to calculate the battery life for a given reliability percentage.

2.3 Study Scenarios

The LCOS calculation requires various input data depending on each system to be analyzed, to obtain this data has been used in the studies carried out by LAZARD in its annual study of the Levelized Cost of Energy and Levelized Cost of Storage published in November 2018. [13]

The scenario chosen to perform the analysis will be "Commercial & Industrial (PV +Storage)" due to the data presented by the study in this operating scenario. The data in this scenario are detailed in Table 1.

Table 1. LAZARD "Commercial & Industrial (PV+Storage)" scenario data.

Life of the project	Power (MW)	Useful capacity per day (MWh)	Cycles per day	Days operating per year
20	0.5	2	1	350

Source: [13]

Table 2. General data for LCOS analysis.

Parameter	Value
Installation life	20 years
Annual inflation	0%
Power in/out for one day	2000 KWh
Number of days per year in operating state	350
Battery recharge price	0.1$/kWh
Annual electricity price inflation	1,5%

Source: [13]

It is also estimated that the cost of electricity will increase by 1.5% per year (Table 2).

2.4 Data from Storage Systems

Three storage technologies were discussed in this work because they present the necessary information to perform the LCOS calculation with a percentage of reliability (estimated number of cycles that the battery will perform before present any damage).

These three technologies are:

- Lithium
- Lead
- Advanced Lead

The cost used is the average of those indicated in the LAZARD study [13]. Table 3 shows the cost of both the storage system and the additional items.

Table 3. Cost of each type of LAZARD storage.

Type of storage	Storage system cost $/kWh			Cost of added items $/kW		
	Minimum	Maximum	Used	Minimum	Maximum	Used
Lithium	409	572	491	191	292	242
Lead	384	417	401	191	255	223
Advanced Lead	463	537	500	191	292	242

Source: [13]

The total investment cost will result to multiply the cost per kWh by the total capacity of the storage system (Table 4).

Table 4. Total investment of each type of storage.

	Total investment	
Type of storage	Storage system	Added elements
Lithium	$981 000	$120 750
Lead	$801 000	$111 500
Advanced Lead	$1 000 000	$120 750

Source: [13]

In this case, inflation of the price of storage systems will be given the value of −1%. To facilitate calculations, general inflation will be taken as 0%. That way, it can be said that storage systems reduce 1% price due to technological improvement. For the parameters of the Weibull distribution, the following values $\beta = 2.45$ and $\eta = 0.71$ were used.

For the calculation of the LCOS it is necessary to determine the number of cycles that the battery will perform during its useful life. This parameter can be defined by the Cycles-Residual capacity curve or by a linear relationship in which only the total cycle value is required during battery life. In this analysis, a linear relationship will be defined based on the total number of cycles defined in previous studies. Table 5 shows the number of cycles that the batteries analyzed will perform (Tables 6 and 7).

Table 5. Total estimated cycles for each storage type.

Type of storage	Cycles
Lithium	3000 (80% DoD)
Lead	1200 (50% DoD)
Advanced Lead	2000 (70% DoD)

Source: [14, 15]

Table 6. Efficiency and operating and maintenance costs of each type of storage technology.

Battery type	Round-trip efficiency (%)	OyM variable ($/kWh)	OyM Fixed ($/kW)
Li-ion	91	0.004	10
Lead-Acid (valve-regulated)	80	0.005	5
Advanced Lead-Acid	92	0.005	5

Source: [16]

Table 7. LCOS for different reliability percentages for LEAD, LEAD ADVANCE and LITHIUM storage technology.

Reliability	LCOS [$/kWh]		
	LEAD	LEAD ADVANCE	LITHIUM
95%	398.0	357.6	249.6
90%	346.2	303.9	223.3
85%	346.2	285.8	210.2
80%	312.0	271.0	206.6
75%	312.0	259.8	206.6
70%	287.2	259.8	206.6
65%	287.2	250.3	206.6
60%	269.0	250.3	206.6
55%	269.0	242.1	206.6
50%	269.0	242.1	206.6
45%	269.0	235.1	206.6
40%	254.5	235.1	206.6
35%	254.5	228.9	206.6
30%	254.5	228.9	206.6
25%	254.5	228.9	206.6

Source: Own.

3 Results

The results obtained for these three storage technologies can be compared according to the diagram in Fig. 2.

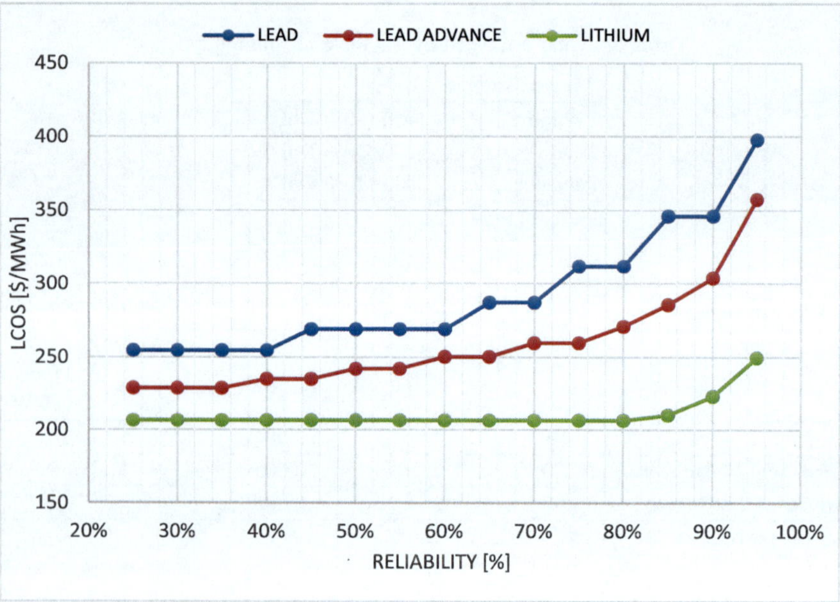

Fig. 2. LCOS comparison for three different storage technologies with different reliability percentages. Source: Own.

4 Analysis of Results

The LCOS calculation depends on some technical and economic parameters that determine the cost for each unit of stored energy. The LCOS results shown are based on the calculation of the battery useful life for a certain percentage of reliability, this percentage of reliability has been calculated through a Weibull probability function with parameters of shape and scale previously determined in previous works. The final results in the scenario studied, show that the LCOS value is proportional to the percentage of reliability required, this is because as the percentage of reliability increases the battery useful life decreases, which makes the costs associated with the battery useful life increased too, in this way the value of the LCOS is also increased. The variation of the LCOS value is mainly noted in the high reliability values, however, it can be noted that the LITHIUM battery not only presents the lowest LCOS values but also, less variation in different reliability percentage, which shows that this technology is the best option within the three kinds of storage technologies studied.

5 Conclusions

The LCOS calculation allows to compare the value of $/MWh stored for various storage technologies, this value depends directly on the battery life used in the system so, by estimating the battery life for a given reliability, the LCOS value with that percentage of reliability will be calculated implicitly.

The LCOS calculation depends on various technical and financial parameters, however, the main parameter of each system is the battery life, in this work the incidence of the reliability of the battery life in the LCOS calculation was determined. This research has demonstrated that the higher percent of reliability, the value of the LCOS increases.

The Curves of the Number of Cycles-Residual capacity allows to estimate the number of cycles that the battery will perform in its useful life. Additionally, this curve can be used to calculate the number of cycles that the battery could complete for a given reliability probability.

Previous studies propose the Weibull′s Probability function as the best option for estimating the battery′s useful life for a certain percentage of reliability, this probability function depends on two factors α (scale factor) and β (form factor). These factors can be estimated based on measurements of the number of cycles and the Residual capacity that the battery performs over its useful life. Additionally, this process by requiring a repetitive calculation process is ideal for performing it in a computational tool.

Developing the LCOS calculation for a given reliability percentage presents one more comparison parameter that would help to select a storage system more suitable for a given facility not only for storage technologies but also for the same technology with different characteristics in its operation.

Bibliography

1. Beltran, H., Tomás García, I., Alfonso-Gil, J.C., Pérez, E.: Levelized cost of storage for li-ion batteries used in PV power plants for ramp-rate control. IEEE Trans. Energy Convers. **34** (1), 554–561 (2019). https://doi.org/10.1109/tec.2019.2891851
2. Schmidt, O., Melchior, S., Hawkes, A., Staffell, I.: Projecting the future levelized cost of electricity storage technologies. Joule **3**(1), 81–100 (2019). https://doi.org/10.1016/j.joule.2018.12.008
3. Lai, C.S., McCulloch, M.D.: Levelized cost of electricity for solar photovoltaic and electrical energy storage. Appl. Energy **190**, 191–203 (2017). https://doi.org/10.1016/j.apenergy.2016.12.153
4. Belderbos, A., Delarue, E., Kessels, K., D'haeseleer, W.: Levelized cost of storage — introducing novel metrics. Energy Econ. **67**, 287–299 (2017). https://doi.org/10.1016/j.eneco.2017.08.022
5. Jülch, V., et al.: A holistic comparative analysis of different storage systems using levelized cost of storage and life cycle indicators. Energy Procedia **73**, 18–28 (2015). https://doi.org/10.1016/j.egypro.2015.07.553
6. Eom, S.-W., Kim, M.-K., Kim, I.-J., Moon, S.-I., Sun, Y.-K., Kim, H.-S.: Life prediction and reliability assessment of lithium secondary batteries. J. Power Sources **174**(2), 954–958 (2007). https://doi.org/10.1016/j.jpowsour.2007.06.208

7. Dufo-Lopez, R., Bernal-Agustín, J.L.: Techno-economic analysis of grid-connected battery storage. Energy Convers. Manage. **91**, 394–404 (2015). https://doi.org/10.1016/j.enconman.2014.12.038
8. Jiménez Cañizares, M.: Estimation of battery life in small electrical systems (2018)
9. Salazar Moreno, R., Aguilar Rojano, A., Figueroa Hernández, E., Pérez Soto, F.: Applications of weibull distribution in reliability engineering. https://scholar.googleusercontent.com/scholar?q=cache:RNuY_7RhYKkJ:scholar.google.com/+Aplicaciones+de+la+distribuci%C3%B3n+weibull+en+ingenier%C3%ADa&hl=es&as_sdt=0,5&as_vis=1. Accessed 28 Feb 2020
10. Chiodo, E., Lauria, D., Fabrizi, V., Ortenzi, F., Sglavo, V.: Battery design based upon life cycle statistics, pp. 7.1.1–7.1.1 (2014). https://doi.org/10.1049/cp.2014.0876
11. Chiodo, E., Lauria, D., Pagano, M., Pede, G., Vellucci, F.: Experimental performances and life cycle estimation of hybrid electric storage systems. In: 2013 International Conference on Clean Electrical Power (ICCEP), pp. 614–619 (2013). https://doi.org/10.1109/iccep.2013.6586943
12. Andrenacci, N., Pede, G., Chiodo, E., Lauria, D., Mottola, F.: Tools for life cycle estimation of energy storage system for primary frequency reserve. In: 2018 International Symposium on Power Electronics, Electrical Drives, Automation and Motion (SPEEDAM), pp. 1008–1013 (2018). https://doi.org/10.1109/speedam.2018.8445314
13. Ray, D.: Lazard's Levelized Cost of Energy Analysis—Version 4.0, p. 28 (2018)
14. Sabihuddin, S., Kiprakis, A.E., Mueller, M.: A numerical and graphical review of energy storage technologies Energies **8**(1), 172–216 (2015). https://doi.org/10.3390/en8010172
15. San Martín, J.I., Zamora, I., San Martín, J.J., Aperribay, V., Eguía, P.: Energy storage technologies for electric applications. Renew. Energy Power Qual. J. 593–598 (2011). https://doi.org/10.24084/repqj09.398
16. Brinsmead, T., Graham, P., Hayward, J., Ratnam, E., Reedman, L.: Future energy storage trends: an assessment of the economic viability, potential uptake and impacts of electrical energy storage on the NEM 2015–2035 (2015)

Management and Control Strategy of Battery-Supercapacitor Vehicular Powertrain System

Livio Miniguano[1], Henry Miniguano[2(✉)], Santiago Illescas[1], Andrés Cuasapaz[1], and Ricardo Rosero[1]

[1] Instituto Superior Tecnológico Sucre, Av. 10 de Agosto N26-27 y Luis Mosquera Narváez, Quito, Ecuador
[2] Universidad Carlos III de Madrid, Calle Madrid, 126, 28903 Getafe, Madrid, Spain
dminiguano@tecnologicosucre.edu.ec

Abstract. This Research Work presents the use of a hybrid power source system with the battery and supercapacitor in a powertrain electric vehicle. The efficiency and performance of these systems are important, in terms of saving energy and increasing autonomy. Therefore, the control of DC-DC converters is designed through a feedforward technique and power management, through the splitting method of power demand. This allows controlling the power sources with the battery at the lowest frequency, and the supercapacitor at the highest frequency to provide the fast dynamic of power demand.

To confirm the performance of powertrain, a simulation is carried with Simulink of MATLAB software that demonstrates the robustness and flexibility under ECE-15 drive-cycle test.

Keywords: Power management · Splitting frequency · Battery · Supercapacitor · Powertrain electric vehicle

1 Introduction

Sustainability throughout the development of human activities is vital to ensure survival, where the basis of life must be preserved for future generations [1]. Therefore, it is necessary for a progressive energy transition, from fossil to renewable energies.

Car manufacturers are gradually increasing the electrification of traction systems until they reach their maximum power, for the development of emission-free vehicles.

In the literature, power supply settings with unidirectional or bidirectional converters, to better manage the energy are shown in each source [2].

In [3] and [4] different control strategies, the advantages and disadvantages of each can be identified, and it is possible to integrate and design a control strategy that brings together some of these advantages present in the control strategies analyzed.

The main contribution of this Research paper is the proposal of a general, practical and effective power management application with splitting frequency technique through Simulink.

M. Botto-Tobar et al. (Eds.): CI3 2020, AISC 1277, pp. 257–266, 2021.
https://doi.org/10.1007/978-3-030-60467-7_22

The next sections are organized as follows: Sect. 1 describes the powertrain structure of an electric vehicle. Section 2 shows a description of control and management. Section 3 shows the results of the simulation using the ECE15 dynamic driving cycle.

2 Structure of a Hybrid Power Source

2.1 Architecture Description

Hybrid energy storage systems using both a battery and supercapacitor have many advantages like the high energy density of batteries, as well as the high-power density of supercapacitors. In addition, the battery does not allow more than ten of thousands of charge-discharge cycles, while supercapacitors can work with over half-a-million cycles [5]. In [6], for powertrain in an electric vehicle, the connection of the power sources can be seen as a micro DC network, which contains power generation, storage, and consumption units, each of which are connected by a power management circuit to a common DC bus. There are many settings for the powertrain, active or passive connection of energy storage systems, with or without dc-dc converters to connect the batteries and the ultracapacitors. The best topology between performance and cost, is using a dc-dc converter connected to the supercapacitors [7], shown in Fig. 1, which solves the problem of voltage variations of the ultracapacitors by placing a dc-dc converter, allowing an almost soft current to flow from the battery. The main advantage of the topology is the use of one dc-dc converter, which allows flexibility in the independent control of the supercapacitor.

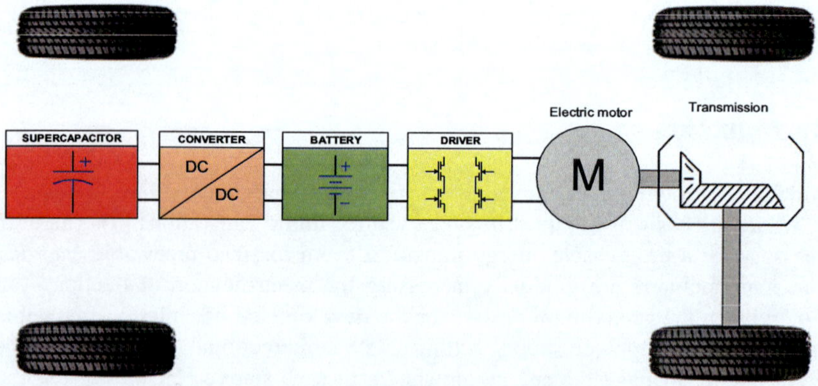

Fig. 1. The powertrain of a parallel semi active hybrid power source.

2.2 Battery and Ultracapacitor Modeling

The lithium-ion battery model proposed in [8] is used because it requires only a few data obtained from the manufacturer's data sheet (battery discharge curve). This prototype is based on the modified Shepherd model, and can reflect with sufficient accuracy the current and voltage characteristics at a macro level, which are important for system-level simulations. This model is shown in Fig. 2; its implementation has been represented on a Simulink library.

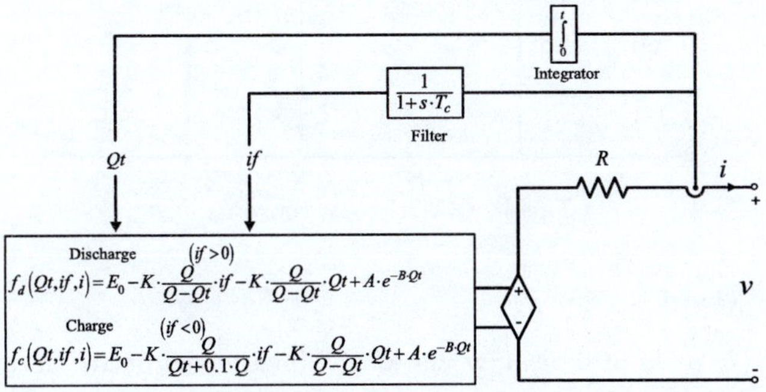

Fig. 2. Battery model.

The electrochemical supercapacitor model proposed in [9] uses the Stern-Tafel model to describe the behavior of the equivalent non-linear capacitor. This model reproduces the capacitance of the double-layer capacitor, related to non-linear diffusion dynamics. For this purpose, the supercapacitor model combines both Helmholtz capacitance and Gouy-Chapman capacitance (GCC), as described and implemented on Simulink.

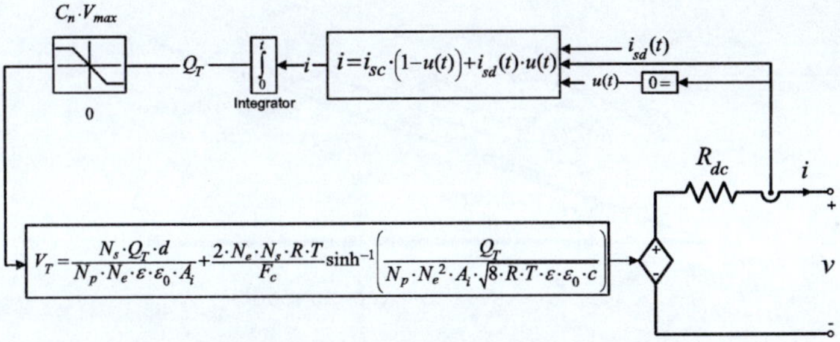

Fig. 3. Supercapacitor model.

2.3 Power Converter Modeling

The supercapacitor is connected to the DC/AC converter through a DC/DC converter, in order for bidirectional energy to flow to the DC bus voltage. The DC/DC boost converter topology shown in Fig. 4 is used in the application described here for elevating the voltage in one direction, as well as reducing the regenerative deceleration of the electric vehicle. The converter is regulated with the feed forward method as described in [10].

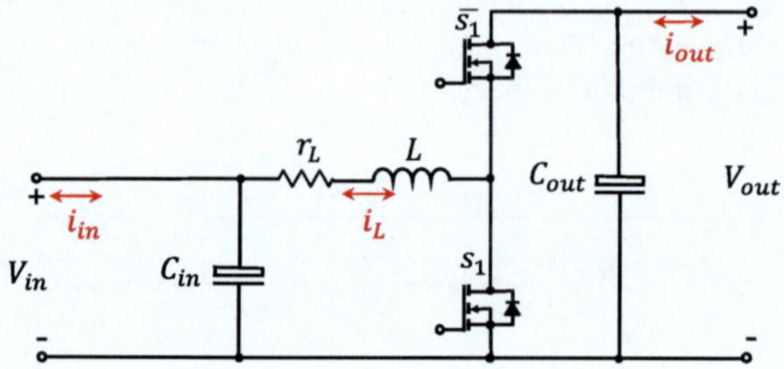

Fig. 4. DC/DC boost converter.

2.4 Vehicle Model

The total tractive force of the equivalent longitudinal model of the electric vehicle in motion is described by

$$F_{tr} = F_{aero} + F_{grade} + F_{rr} + F_i \tag{1}$$

Where F_{tr} is the total tractive force, F_{rr} is the rolling resistance force, F_{grade} is the grade resistance force, F_{aero} is the wind resistance force, F_i is the acceleration resistance force [11], as illustrated in Fig. 5.

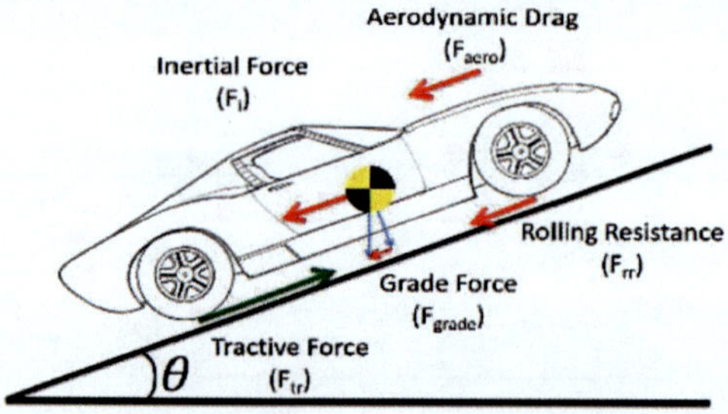

Fig. 5. Forces in a vehicle model.

The equivalent electric power of the electric vehicle is expressed by

$$P_{Load} = \left(C_{rr} \cdot m \cdot g \cdot \cos\alpha + m \cdot g \cdot \sin\alpha + 0.5 \cdot \rho \cdot C_d \cdot A_f \cdot v^2 + m \cdot \frac{dv}{dt} \right) \cdot v \tag{2}$$

Where m is the vehicle mass, g is the acceleration due to gravity, C_{rr} is the rolling resistance coefficient, α the road angle, v the vehicle speed, A_f the frontal area, C_d the drag coefficient and ρ the air density.

3 Control and Management of Hybrid Power Source

The energy management mechanism is based on the technique described in [12], which uses the function of the frequency breakdown in order to find out the value of each power source, as shown in Fig. 2. This control technique allows the decomposition of power, where input power is separated into two components, one of low frequency and the other of higher frequency. In this application, the fast transients of high power are absorbed by the supercapacitor, and the rest of power is provided by the battery (Fig. 6).

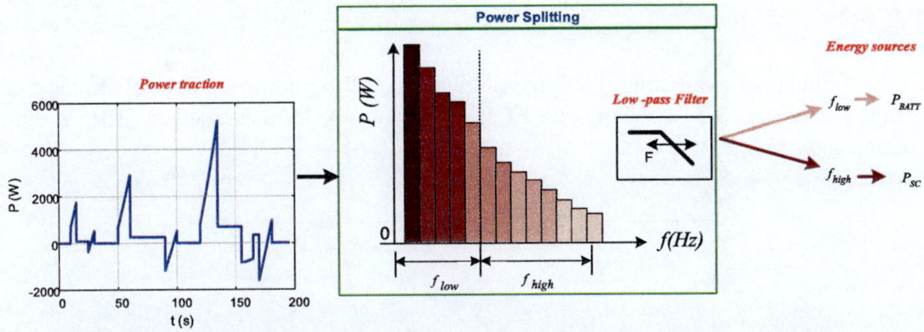

Fig. 6. Block diagram of control and management of hybrid power source.

In the Research paper in hand, the proposed power management system is based on a power splitting frequency, with the purpose of assigning transient power to the supercapacitor and the rest of power to the battery, as depicted in Fig. 3 (Fig. 7).

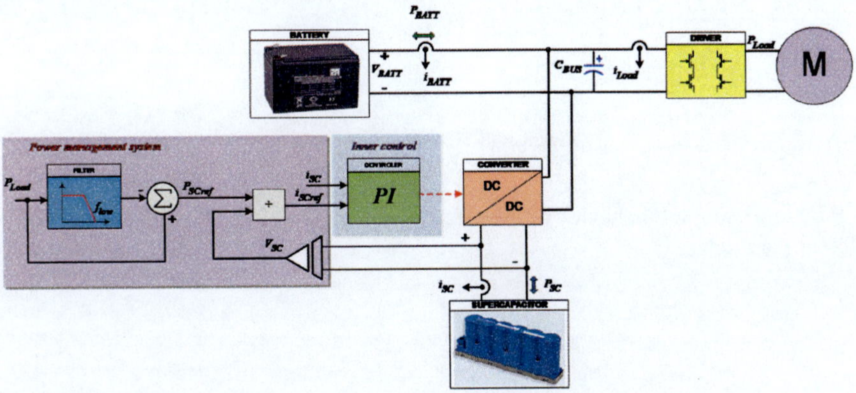

Fig. 7. Block diagram of control and management of hybrid power source.

The following Table 1 gives a summary of all parameters of energy sources used in the Simulink simulation model.

Table 1. Electrical parameters of the battery and supercapacitor

Parameters	Value	Units
Rated power	270	kW
Battery voltage	345	V
Battery capacity	246	Ah
Supercapacitor voltage	240	V
Supercapacitor capacity	15	F

As a means of evaluating the dynamic response of the proposed control strategy, a standard drive cycle is applied. The ECE-15 driving cycle, described in [13], is diagrammed on Fig. 8.

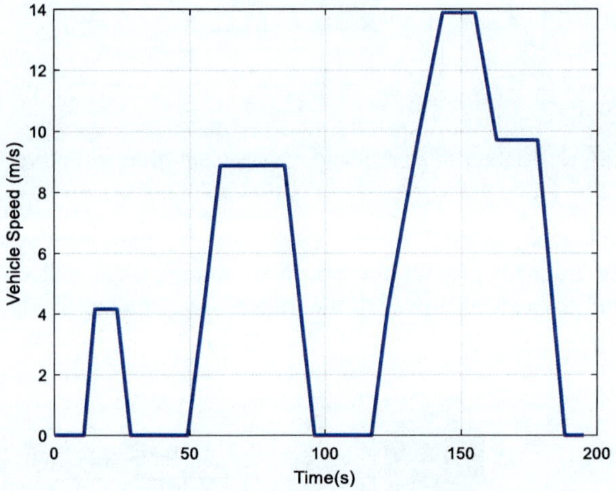

Fig. 8. ECE 15 driving cycle.

For simulating the behavior of the electric vehicle, Table 2 provides the data of the different parameters.

Table 2. Parameters of the electric vehicle.

Parameter	Value	Units
Air density	1.223	Kg/m^3
Drag coefficient	0.24	
Vehicle frontal area	2.34	m^2
Vehicle mass	2155	Kg
Rolling resistance coefficient	0.0084	

The overall system is depicted on Simulink, as illustrated in Fig. 9, where the model of the electric vehicle produces the required power. The supercapacitor provides fast power through the DC/DC converter, while the battery supplies smoothed power.

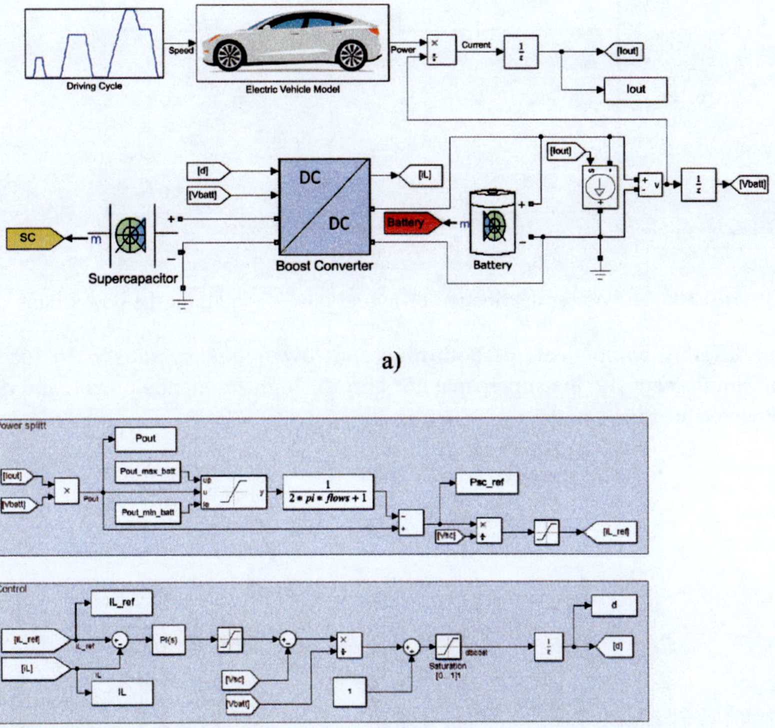

b)

Fig. 9. a) Block diagram of control and management of the hybrid power source in Simulink and b) frequency split and control.

4 Simulation Results

The proposed control has been portrayed on Simulink, where the selection of cut-off frequency is the ratio between power density and energy density [14]. Its value is that of 300 MHz with a low-pass filter. The corresponding results are shown in Fig. 10, where the battery voltage that corresponds to the dc bus voltage is stable, and the supercapacitor voltage is in turn more potent because of the fast dynamic.

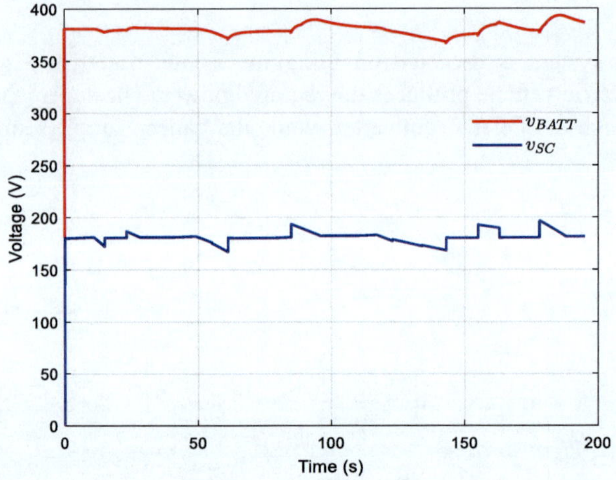

Fig. 10. The power distribution of hybrid power sources in the powertrain

The different components of both high and low frequency satisfy the load powertrain. Simultaneously, the supercapacitor absorbs high-frequency power, and delivers it as depicted in Fig. 11.

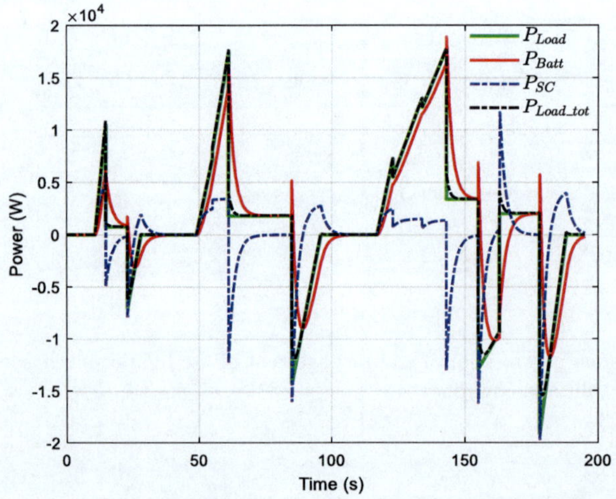

Fig. 11. The power distribution of hybrid power sources in the powertrain

5 Conclusions

In this contribution, the results obtained through the simulation on Simulink demonstrate the effectiveness of the technique for control of power management, through the frequency domain of battery and supercapacitor. A simple description and simulation were used to verify power distribution of the powertrain of the electric vehicle and share the appropriate power.

In addition, the simulation of a complete power distribution system of a hybrid electric vehicle, including two power sources, was portrayed. Nonetheless, the control techniques used may be applicable to a larger number of sources, while achieving a stable system, and adequate distribution of the energy, and complying with the specifications of the system.

References

1. Töpler, J., Lehmann, J.: Hydrogen and Fuel Cell. Springer (2015)
2. Schaltz, E., Rasmussen, P.O.: Design and comparison of power systems for a fuel cell hybrid electric vehicle. In: Conference Record - IAS Annual Meeting IEEE Industry Applications Society, pp. 1–8 (2008)
3. Thounthong, P., Raël, S., Davat, B.: Energy management of fuel cell/battery/supercapacitor hybrid power source for vehicle applications. J. Power Sources **193**(1), 376–385 (2009)
4. Azib, T., Larouci, C., Chaibet, A., Boukhnifer, M.: Online energy management strategy of a hybrid fuel cell/battery/ultracapacitor vehicular power system. IEEJ Trans. Electr. Electron. Eng. **9**(5), 548–554 (2014)
5. Castaings, A., Lhomme, W.: Comparison of energy management strategies of a battery/supercapacitors system for electric vehicle under real-time constrains. Appl. Energy **163**, 190–200 (2015)
6. Aharon, I., Kuperman, A.: Topological overview of powertrains for battery-powered vehicles with range extenders. IEEE Trans. Power Electron. **26**(3), 868–876 (2011)
7. Cao, J., Emadi, A.: A new battery/ultracapacitor hybrid energy storage system for electric, hybrid, and plug-in hybrid electric vehicles. IEEE Trans. Power Electron. **27**, 122–132 (2012)
8. Tremblay, O., Dessaint, L.A.: Experimental validation of a battery dynamic model for EV applications. World Electr. Veh. J. **3**(2), 289–298 (2009)
9. Oldham, K.B.: A Gouy-Chapman-Stern model of the double layer at a (metal)/(ionic liquid) interface. J. Electroanal. Chem. **613**(2), 131–138 (2008)
10. Dominguez, X., Camacho, O.: A fixed-frequency sliding-mode control in a cascade scheme for the half-bridge bidirectional DC-DC converter. IEEE Ind. Electron. **16**(4), 1603–1609 (2016)
11. Liu, X., Diallo, D.: Cycle-based design methodology of hybrid electric vehicle powertrain: application to fuel cell vehicles. IEEE Veh. **12**(7), 1853–1857 (2009)

12. Georgious, R., Garcia, J., Garcia, P., Sumner, M.: Analysis of hybrid energy storage systems with DC link fault ride-through capability. In: ECCE 2016 - IEEE Energy Conversion Congress and Exposition Proceedings (2016)
13. Raga, C., Barrado, A., Lazaro, A., Quesada, I., Sanz, M., Zumel, P.: Driving profile and fuel cell minimum power analysis impact over the size and cost of fuel cell based propulsion systems. In: Proceedings of the 2015 9th International Conference on Compatibility and Power Electronics CPE, pp. 390–395 (2015)
14. Snoussi, J., Ben, S., Faouzi, M.: Sizing and control of onboard multisource power system for electric vehicle. In: IEEE STA, no. 2, pp. 347–352 (2019)

Security

Proposal for a Secure Architecture for the Internet of Things on a Smart Campus

William Villegas-Ch[1(✉)] and Xavier Palacios-Pacheco[2]

[1] Universidad de Las Américas, Av. de los Granados E12-41 y Colimes esq., Quito, Ecuador
william.villegas@udla.edu.ec

[2] Universidad Internacional del Ecuador, Av. Simón Bolívar y Av. Jorge Fernández, Quito, Ecuador

Abstract. Currently, emerging technologies such as the internet of things, big data, cloud computing, allow society to create intelligent environments. Intelligent environments are based on the interaction of people with technology directly without the need for intermediary devices such as servers or computers. A sample of this is the smart campuses where there is a large concentration of people who develop different activities consume many resources. The objective of emerging technologies is to facilitate the development of these activities in sustainable environments where good management of resources is prioritized and lived in harmony with nature. However, companies that offer these technologies with the aim of taking advantage of the sale of devices, forget something as important as information security. Smart campuses use this technology as the main component in their architectures for data acquisition. This becomes a problem when the security of the devices is poor because they can become a gateway to cybercriminals. This work proposes an architecture that improves the security of the network of an intelligent campus that relies on the internet of things for its administrative and academic management.

Keywords: Internet of Things · Security of the information · Smart campus

1 Introduction

Currently, the use of technology is common in society, to the point that it depends on it for the development of most of its activities. These activities are not limited to work environments, technology is integrated into both companies and homes [1]. This has allowed the quality of life of people to improve considerably. This evolution has increased with the approach and use of new emerging technologies [2]. Its integration and convergence converts a traditional environment into an intelligent one as is the case of several universities worldwide [3]. Universities integrate technologies such as the internet of things (IoT), data analysis through big data and the potential of cloud computing to improve all their management. With this integration, universities become smart campuses whose focus is based on the proper management of resources, the improvement of the learning model and better administrative management [4].

M. Botto-Tobar et al. (Eds.): CI3 2020, AISC 1277, pp. 269–280, 2021.
https://doi.org/10.1007/978-3-030-60467-7_23

Each of these characteristics depends specifically on the ability that technologies have to interact with each other. Within an intelligent campus, these technologies depend on each other, in the specific case of the IoT, this is responsible for monitoring all areas of the campus [5]. The information generated from these events is stored in both private and public clouds depending on the circumstances or policies of each campus. The information stored goes into a phase of data analysis through the development of big data frameworks that take care of everything that involves the analysis of the information. In general, these components are the fundamental pillars for a university campus to be considered intelligent. But, it is necessary to clarify that basically the infrastructure that supports an intelligent campus is the same that manages a traditional university campus [6].

In other words, most universities that have seen the need to become an intelligent campus update their infrastructure so that the requirements of new technologies are supported. Previous work defines the requirements that each of these has, the IoT being the most demanding technology for the number of devices it handles [7]. The IoT by concept can include a large number of devices and adapts to any environment. Its requirements are essentially established in network resources; it is here that a network is in danger when there is no design focused on data security in an infrastructure that supports IoT [8]. It is necessary to remember that the network infrastructures before the heyday of IoT met other conditions and shielded the information through usage policies, as well as, a security model applied to the network. In this model, the use of a firewall, antivirus, information management rules, etc., guarantees to some extent the security and availability of information.

The IoT evades this relative network security by managing an architecture where its main requirement is free to access to the Internet. Similarly, IoT devices mostly do not have the ability to integrate antivirus or respond directly to a firewall [1]. These factors have compromised the security of smart campuses that have been violated by cybercriminals. They use IoT weaknesses to access services on campus and compromise information. Even several IoT devices store user data, which makes them the main targets of computer attacks.

With this background, it is possible to ask the following questions: Is an intelligent campus that uses IoT as one of its pillars able to guarantee information security? And is it possible to create an architecture that improved information security in the use of IoT?

The information has become digital gold and its importance for every environment is fundamental. By this reference, the integration of IoT in intelligent environments is a reason for research by the academy to guarantee information security to residents. The problem in information security begins with companies that offer IoT products. They are more interested in selling their devices and do little or nothing to improve computer security on them. At the moment there is no clear picture regarding security on IoT platforms. Countries such as the United States and several of the European Union have recently worked on the creation of standards and regulations that IoT products must meet in order to be able to go to market [9].

This work proposes a secure architecture within a university campus that has an IoT platform for the management and control of different events. This architecture guarantees the security of the information generated in the IoT, as well as the internal security of the campus. For this, an analysis of the infrastructure of the campus is carried out, in addition, the critical points in the management of IoT are identified. To fulfill these objectives, the technology implemented within the campus is used as the framework for big data [10]. This framework is used in the analysis of the logs of several network security teams. This allows detecting the patterns present in a computer attack and being able to act quickly and effectively in the event.

This work is divided in the following way, in Sect. 2 the method that led to the construction of the proposal is presented, in Sect. 3 it is carried out the discussion where the benefits of the proposed architecture are established, finally, in Sect. 4 the conclusions and recommendations are developed based on the approach to the research problem.

2 Method

To design an architecture that guarantees information security, it is necessary to clearly understand where it is applied. This work is developed in a smart campus that requires that all systems that allow its residents to interact with it be considered. Next, each of the phases and components that have been used for the development of this work is detailed.

2.1 Architecture of a Smart Campus

As a starting point, the path to be followed during the security implementation is identified [10]. This article describes, in general, the architecture that gives greater depth to the relationship between IoT and security.

On the university campus, its administrative structure, the diversity in its inhabitants, the increasing consumption of resources, etc., make it comparable in scale with a small city [11]. With this, what we want to make clear is that both the campuses and the smart cities are currently experiencing similar problems in the management of IoT. This is why this proposal seeks to solve security in a smart campus in the first instance and then it is replicated to a larger environment.

A smart campus to be considered as such must include the ability to meet the needs of its residents automatically. In addition, this capacity must be done in a sustainable way, in total harmony with nature. This requires that there is a good use of resources for which it is necessary that the information and communication technologies (ICT) come into play. These technologies allow the interaction of people with the devices in a transparent way without there being a team involved. Figure 1 shows the architecture to which an intelligent campus responds. Each of the stages of this architecture composes a specific task and on which the following stages depend.

Data Acquisition Layer. At this stage, the IoT is responsible for the acquisition of data through several devices. These devices have the ability to detect countless variables and act autonomously in the event of any event [12]. For example, the detection of temperature within a classroom is carried out by means of sensors, where the system based on the readings made can open and close the blinds or activate an air conditioner.

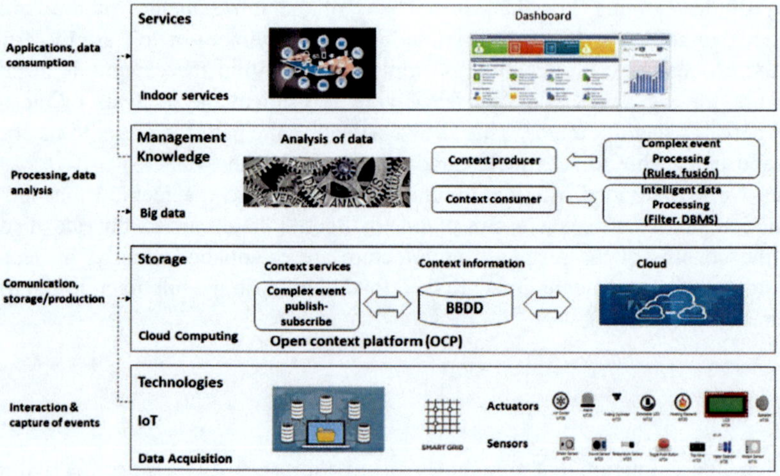

Fig. 1. Layered architecture of a smart campus [10].

These are common activities in automated systems, the difference is that in an intelligent campus, in addition to creating the action, these devices send information to a cloud to be analyzed [13]. The versatility of IoT has made it possible to establish robust physical security systems on campus through the use of facial recognition cameras, systems that handle resources [14].

Devices that are part of the IoT system by concept, need access to the Internet to communicate with other devices or to store data. This feature within a campus compromises data security, as well as the institution's services. For example, in a traditional service the user makes the request to go out to the network, the request goes through different controls that go from the first filtering in the core layer through access lists (ACL), [15]. Subsequently, it goes through the filtering of a proxy or a firewall, this independently and according to the user's permissions, may or may not access the service [16]. Instead, in an IoT service, such as an inventory control system for cleaning supplies. The system is responsible for making requests to suppliers each time a product is terminated, for which it needs a free Internet access connection [17]. Even this system has the ability to make payments directly. This means that the system or device responsible for placing the order has the financial data of a user or the smart campus. It is here that there are doors open to possible attacks.

Figure 2 shows the architecture of a three-layer IoT environment [18]. The first is the perception of where all the devices that are responsible for monitoring the different environments or executing actions are located. The network layer supports connectivity between devices and there are multiple communication protocols that are responsible for linking the devices to the cloud [25]. The most commonly used protocols are HTTP, MQTT 3.1/3.1.1 and the restricted application protocol (CoAP).

As the figure indicates, the IoT architecture is basic and its network components are well defined. However, for the effective functioning of the IoT on a university campus,

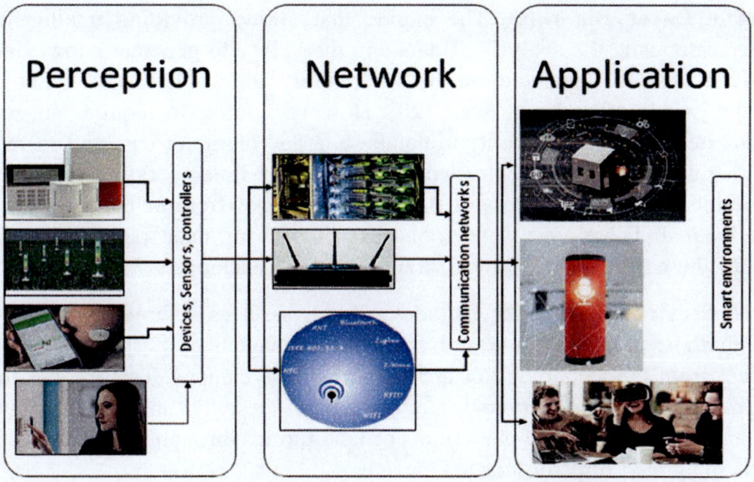

Fig. 2. Basic architecture of an IoT environment

it is necessary to transform the topology. This network must be flexible, scalable and secure by the number of devices that are integrated into the network.

Storage and Computer Layer in The Cloud. This layer gives storage to all the data that is generated in the data acquisition layer. The smart campus uses three methods to store data, this depends on the functionality of the systems [19]. The methods are local storage, private clouds and public clouds. To determine the functionality of each system, a brief analysis of the data generated by each of them has been carried out. Table 1 shows the criteria under which the functionality of the systems and the model under which they are stored have been classified.

Table 1. Types of systems and storage on a smart campus

System	Type of data	Storage
LMS	Learning management	Local/ public cloud
Inventories	Administrative management	Local/ private cloud - public
Financial	Administrative management	Local
Academics	Learning management	Local/private cloud
Air conditioning	Domotics	Private cloud/public cloud
Virtual assistants	Administrative management	Public cloud

The first column catalogs the systems according to the type of service. The data type column does not refer to a database, what you are looking for is to define what the system provides. The third column details the type of storage; several systems are considered hybrids because it uses more than one of the available models.

Knowledge Layer Big Data. The engine that allows providing intelligence to a university campus is the analysis of data and the ability to generate knowledge about any activity that is generated on it. This ability allows decisions based on the knowledge gained after the analysis [20]. However, doing so requires an extremely robust architecture with the ability to handle a large volume of data and in very short periods of time. This task is entrusted to the big data frameworks that can deal with these volumes of data regardless of the source it comes from or the format in which they are located. There are several tools that allow a big data framework to be integrated into the campus, for example, frames such as Hadoop or Spark [21].

Layer of Services. This layer includes all the services offered by an intelligent campus, both in administrative and academic management [22]. The services are transparent from the technology or architecture of the campus. This implies that they must guarantee the use, quality and safety in each of these. As main services in a smart campus are the reservation of classrooms or laboratories through virtual assistants, this service is integrated directly into IoT devices.

2.2 Security Analysis in IoT

In order to reach this stage, it has been necessary to know the environment where IoT is applied since this depends on the measures that can be taken to improve information security [23]. The weaknesses and strengths of a domestic environment with an intelligent campus are incomparable both in infrastructure and in the volume of risks. For this reason, campus technology administrators are working on standards and policies that allow them to reduce the number of incidents that are generated. These incidents are computer security failures that have leaked into the campus network through IoT devices [24].

In general, the security of the objects connected to the IoT is targeted by all attackers. Kaspersky Lab's report revealed that IoT devices were attacked with more than 120,000 malware variants during the first half of 2018. There is no data yet to compare them with the first quarter of 2019, however, when compared with 2017 data this figure is triple the amount of IoT malware [25]. This makes malware the most dangerous trend for IoT devices. One of the most popular vectors of attacks and device infection seeks to get Telnet password [26]. If the percentage of attacks based on the service is presented in 2018, Telnet with 75.40%, SSH with 11.59% and other attacks with 13.01%.

The attacks are designed to test the defenses of the devices, for this multiple attack vectors are used. An attack vector is a route that the attacker uses to take advantage of the device. This means that the attacker can use the device for something other than its purpose. The most common attack vectors in IoT devices that are exploited to exploit weaknesses are weak passwords, lack of encryption, backdoors.

2.3 Secure Architecture

Once the different layers and devices that are part of the smart campus have been described, an architecture is established that guarantees the security of IoT. Figure 3 shows an architecture that provides security to the IoT layer to protect campus data. The architecture starts from the network infrastructure that the campus already has, the architecture updates the network and reuses certain services to improve attack detection.

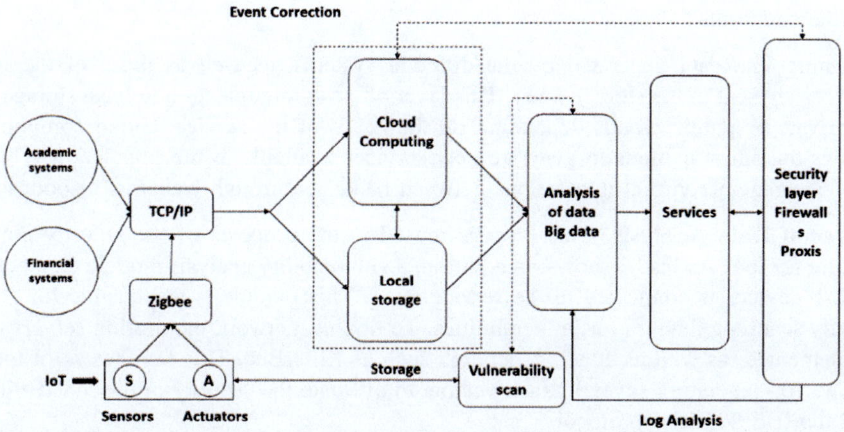

Fig. 3. Secure IoT architecture with Zigbee management and event correction

Acquisition of Data by IoT. One of the layers that need to be modified to improve security is responsible for the acquisition of data through IoT devices. To achieve this, a specification of a mesh network for local wireless networks (WLAN) is incorporated into the architecture. This specification is Zigbee, which is based on the 802.15 specifications. It operates on the physical specification of IEEE 802.15.4 radio and in unlicensed radio frequency bands, including 2.4 GHz, 900 MHz and 868 MHz. Zigbee provides high data throughput in applications where the duty cycle is low consumption [27]. The ZigBee standard provides security to the communications network with the implementation of embedded algorithms.

The security of transmissions and data are key points in ZigBee technology. ZigBee uses the security model of the MAC sublayer, which specifies four security services. The first is access control, the device maintains a list of the devices tested in the network. Then this data is encrypted, for which they use encryption with a 128-bit code [28]. The third service is the integration of frames to protect data from being modified by others. Finally, there are the update sequences to verify that the frames have not been replaced by others. The network controller checks these update frames and their value to see if they are expected. It all depends on the final device, where the decision is of the administrator to provide it with more or less security.

Zigbee Integration with TCP/IP. Once Zigbee has been included, it is integrated into the transmission control protocol/ internetworking protocol (TCP/IP) network. The integration between Zigbee and TCP/IP requires a monitoring algorithm, this usually depends on a server located at one end of the wireless nodes [56]. To establish communication with the data network, algorithms are generated using the TCP/IP libraries and the IP addressing of the network is configured. Socket communication is configured to open connections and transfer information through the designated port when creating sockets. The algorithms allow you to open connections through sockets where TCP acts as a client. The initial connection is established through the IP address of the TCP server [29]. This server listens to the activity through a common logical port for the endpoints.

Storage. The data generated by the different systems, as well as those of the IoT devices are sent to the storage layer. This layer offers communication to local storage or to private or public clouds depending on the needs of the service. Cloud computing allows the smart campus to generate new services available to the population. These services generate virtual interactions between residents through webcasts or podcasts.

Layer of Data Analysis. This layer is reused in the proposal of the security architecture for IoT. As Fig. 3 shows, in addition, a vulnerability analysis module connected to IoT devices is connected to the service layer. This module is responsible for constantly scanning devices for vulnerabilities. To do this, network monitoring servers are implemented, as well as attack generators such as KillerBee. This framework of tools allows for generating several attack vectors to evaluate the security in the IoT devices and the IEE 802.15.4 protocol.

Security Layer. The security layer is made up of several devices that guarantee network and data security. Security is based on the smart campus network that focuses on three important aspects, computer security, Intranet security and network security to which it is interconnected [30]. The campus security plan includes certain elements necessary to ensure that the network functions properly. These elements include an architecture based on data protection through five parameters.

The first parameter is authentication; authentication is the process where users are verified before letting them into the system. This is done by comparing the information received with that stored in a database. The second parameter is access control, which has the function of interrupting unauthorized access to any resource. This means that you do not have permission to use, modify or delete any resource. To achieve this control, each unit must first authenticate itself to determine whether or not it has the privileges for the activity it wishes to perform. The third parameter is confidentiality and Integrity that refers to ensuring that the information is not disclosed to unauthorized persons. This implies protection against passive attacks. Integrity refers to the data being transmitted without suffering any modification, alteration, deletion, duplication, etc. The fourth parameter is the non-repudiation that is responsible for verifying that both parties are who they say they are so that neither can deny the communication. The fifth parameter is encryption, basically, it refers to the mechanisms that allow maintaining the confidentiality of the data. In addition to cryptography, digital signatures are also used. These two options allow you to encrypt the information so that it is only understandable for the elements with the appropriate access [31]. To execute the aforementioned requires an encryption algorithm and a key management scheme.

As for the devices that allow network security, the campus manages firewalls with high performance. These devices allow filtering packets when they pass between their interfaces [32]. Firewalls filter IP packets based on criteria such as the source IP address, the destination IP address, the source TCP/UDP port and the destination TCP/UDP port.

3 Discussion

The approaches that address security through ad-hoc solutions are not applicable in an environment where the number of devices is very large. Therefore, the proposal of a secure architecture in this work is totally valid because it treats security in a reliable, scalable and proactive way. The base is in the integration of all layers of the intelligent campus, generating cycles of design, production and deployment. Safety from design implies verification of hardware and software since its manufacture. The objective is that the products used in a smart campus have the least possible number of vulnerabilities. This work presents a dynamic architecture for IoT management where the architecture processes several inputs and delivers a single output. The output is the one used to infer data on the platform and that is used to act on it. The architecture integrates the management of data analysis that was already part of the smart campus to take security measures "mitigation" and for the analysis of security "detection module".

After achieving the correct execution of the vulnerability analysis tests, the integration tests are carried out, in order to verify the security in the network and the components collaborate with each other satisfactorily. On the one hand, in the IoT, there are those devices that do not allow to be configured safely or that have received support from the manufacturers. In these cases, the architecture allows us to monitor its behavior, not analyzing the device itself but the connections it makes.

Within the analyzed devices it has been detected that most of these present vulnerabilities by administrators who do not manage security-focused configurations. Ports and services enabled by default were detected within the vulnerability audit module. Even some obsolete or not being used. In addition, many users and passwords were found by default that is not changed, for which it was used with Shodan. Among the common attacks on IoT devices is DDoS attacks and in addition, there are network attacks to the Zigbee protocol. This low power network is specifically designed to send small and critical messages from any IoT device. Due to the need to save energy, the IoT device is activated and sends the data asynchronously to the network, then returns to sleep mode. This creates an extremely small window for hackers to enter the network and take control of the device. As a result, devices that work with Zigbee on the smart campus have a low probability of being victims of a network attack.

4 Conclusion

The IoT is revolutionizing all the environments of society allowing each device, object and person to connect to the Internet. With that massive presence of things interconnected by all sites, IoT offers interesting and complicated security challenges that must be investigated before being applied to intelligent environments. In this work, we have proposed an architecture that covers the entire network environment from the IoT endpoint to the multiple clouds, to integrate each security element. This approach allows threat analysis to be shared in real-time and scaled to the entire distributed network. This process reduces the necessary detection windows and provides the automatic repair required for current threats.

Smart environments can defend against DDoS attacks in various ways. These include improving the infrastructure of your networks and ensuring full visibility of the traffic entering or leaving your networks. This can help detect DDoS attacks, as well as ensure mitigation in the environment. Finally, it is necessary to have a defense plan against DDoS attacks, keep it updated and test periodically.

References

1. Santoso, F.K., Vun, N.C.H.: Securing IoT for smart home system. Proceedings of International Symposium of Consumer Electronics ISCE, 5–6 August 2015 (2015). https://doi.org/10.1109/ISCE.2015.7177843
2. Butz, P., Tauscher, B.: Emerging technologies: chemical aspects. Food Res. Int. **35**, 279–284 (2002). https://doi.org/10.1016/S0963-9969(01)00197-1
3. Luo, L.: Data acquisition and analysis of smart campus based on wireless sensor. Wireless Pers. Commun. **102**(4), 2897–2911 (2018). https://doi.org/10.1007/s11277-018-5314-4
4. Liu, M., Li, L.: The construction of smart campus in universities and the practical innovation of student work. In: International Conference on Information Management & Management Science, pp. 154–157 (2018). https://doi.org/10.1145/3277139.3278307
5. Alvarez-Campana, M., López, G., Vázquez, E., Villagrá, V.A., Berrocal, J.: Smart CEI moncloa: an IoT-based platform for people flow and environmental monitoring on a smart university campus. Sensors (Switzerland). **17**, 2856 (2017). https://doi.org/10.3390/s17122856
6. Li, H., Shou, G., Hu, Y., Guo, Z.: WiCloud: innovative uses of network data on smart campus. In: ICCSE 2016 - 11th International Conference on Computer Science and Education, pp. 461–466 (2016). https://doi.org/10.1109/ICCSE.2016.7581624
7. Zanella, A., Bui, N., Castellani, A., Vangelista, L., Zorzi, M.: Internet of things for smart cities. IEEE Internet Things J. **1**, 22–32 (2014). https://doi.org/10.1109/JIOT.2014.2306328
8. Tan, L., Wang, N.: Future internet: the Internet of Things. In: ICACTE 2010 - 2010 3rd International Conference on Advanced Computer Theory and Engineering, Proceedings, pp. 376–380 (2010). https://doi.org/10.1109/ICACTE.2010.5579543
9. Zucchetto, D., Zanella, A.: Uncoordinated access schemes for the IoT: approaches, regulations, and performance. IEEE Commun. Mag. **55**, 48–54 (2017). https://doi.org/10.1109/MCOM.2017.1600617
10. Villegas-Ch, W., Palacios-Pacheco, X., Luján-Mora, S.: Application of a smart city model to a traditional university campus with a big data architecture: a sustainable smart campus. Sustainability **11**, 2857 (2019). https://doi.org/10.3390/su11102857
11. Villegas-Ch, W., Molina-Enriquez, J., Chicaiza-Tamayo, C., Ortiz-Garcés, I., Luján-Mora, S.: Application of a big data framework for data monitoring on a smart campus. Sustainability **11**, 5552 (2019). https://doi.org/10.3390/su11205552
12. Li, S., Xu, L.D., Wang, X.: Compressed sensing signal and data acquisition in wireless sensor networks and internet of things. IEEE Trans. Ind. Inform. **9**, 2177–2186 (2013). https://doi.org/10.1109/TII.2012.2189222

13. Villegas-Ch, W., Palacios-Pacheco, X., Ortiz-Garcés, I., Luján-Mora, S.: Management of educative data in university students with the use of big data techniques. Rev. Ibérica Sist. e Tecnol. Informação **E19**, 227–238 (2019)
14. Lee, J., Kao, H.A., Yang, S.: Service innovation and smart analytics for Industry 4.0 and big data environment. Procedia CIRP. **16**, 3–8 (2014). https://doi.org/10.1016/j.procir.2014.02.001
15. Zhu, Q., Wang, R., Chen, Q., Liu, Y., Qin, W.: IOT gateway: bridging wireless sensor networks into Internet of Things. In: Proceedings - IEEE/IFIP International Conference on Embedded and Ubiquitous Computing, pp. 347–352 (2010). https://doi.org/10.1109/EUC.2010.58
16. Christin, D., Reinhardt, A., Mogre, P.S., Steinmetz, R.: Wireless sensor networks and the Internet of Things: selected challenges. In: Proceedings of the 8th GI/ITG KuVS Fachgespräch Drahtlose sensornetze, pp. 151–157 (2009)
17. Lee, C.K.M., Na, C.M., Kit, N.C.: IoT-based asset management system for healthcare-related industries. Int. J. Eng. Bus. Manage. **7**, 7–19 (2015). https://doi.org/10.5772/61821
18. Yashiro, T., Kobayashi, S., Koshizuka, N., Sakamura, K.: An Internet of Things (IoT) architecture for embedded appliances. In: IEEE Region 10 Humanitarian Technology Conference, pp. 314–319 (2013). https://doi.org/10.1109/R10-HTC.2013.6669062
19. Aazam, M., Khan, I., Alsaffar, A.A., Huh, E.N.: Cloud of Things: integrating Internet of Things and cloud computing and the issues involved. In: Proceedings of 2014 11th International Bhurban Conference on Applied Sciences and Technology, IBCAST 2014, pp. 414–419 (2014). https://doi.org/10.1109/IBCAST.2014.6778179
20. Oktug, S.F., Yaslan, Y.: Big data and smart city platforms. In: 2017 25th Signal Processing and Communications Applications Conference (SIU), pp. 1–4 (2017). https://doi.org/10.1109/SIU.2017.7960555
21. Saraladevi, B., Pazhaniraja, N., Paul, P.V., Basha, M.S.S., Dhavachelvan, P.: Big data and Hadoop-A study in security perspective. Procedia Comput. Sci. **50**, 596–601 (2015). https://doi.org/10.1016/j.procs.2015.04.091
22. Atif, Y., Mathew, S.S., Lakas, A.: Building a smart campus to support ubiquitous learning. J. Ambient Intell. Humanized Comput. **6**(2), 223–238 (2014). https://doi.org/10.1007/s12652-014-0226-y
23. Oh, S.R., Kim, Y.G.: Security requirements analysis for the IoT. In: International Conference on Platform Technology and Service, pp. 1–6 (2017). https://doi.org/10.1109/PlatCon.2017.7883727
24. Lin, J., Yu, W., Zhang, N., Yang, X., Zhang, H., Zhao, W.: A survey on Internet of Things: architecture, enabling technologies, security and privacy, and applications. IEEE Internet Things J. **4**, 1125–1142 (2017). https://doi.org/10.1109/JIOT.2017.2683200
25. Dietz, C., Castro, R.L., Steinberger, J., Wilczak, C., Antzek, M., Sperotto, A., Pras, A.: IoT-BotNet detection and isolation by access routers. In: International Conference on the Network of the Future, NOF 2018, pp. 88–95. IEEE (2018). https://doi.org/10.1109/NOF.2018.8598138
26. Zhang, T.: The Internet of Things promoting higher education revolution. In: International Conference on Multimedia and Security, pp. 790–793 (2012). https://doi.org/10.1109/MINES.2012.231
27. Al-Fuqaha, A., Guizani, M., Mohammadi, M., Aledhari, M., Ayyash, M.: Internet of Things: a survey on enabling technologies, protocols, and applications. IEEE Commun. Surv. Tutorials. **17**, 2347–2376 (2015). https://doi.org/10.1109/COMST.2015.2444095
28. Swan, M.: Sensor mania! the Internet of Things, wearable computing, objective metrics, and the quantified self 2.0. J. Sens. Actuator Networks. **1**, 217–253 (2012). https://doi.org/10.3390/jsan1030217

29. Li, J.Q., He, S.Q., Ming, Z., Cai, S.: An intelligent wireless sensor networks system with multiple servers communication. Int. J. Distrib. Sens. Netw. **2015**, 960173–960181 (2015). https://doi.org/10.1155/2015/960173
30. Babar, S., Stango, A., Prasad, N., Sen, J., Prasad, R.: Proposed embedded security framework for Internet of Things (IoT). In: 2011 2nd International Conference Wireless Communication Vehicular Technology, Information Theory and Aerospace & Electronic Systems Technology Wireless VITAE 2011, pp. 1–5 (2011). https://doi.org/10.1109/WIRELESSVITAE.2011.5940923
31. Palmer, D., Fazzari, S., Wartenberg, S.: Defense systems and IoT: security issues in an era of distributed command and control. In: Proceedings of the ACM Great Lakes Symposium on VLSI, GLSVLSI, pp. 175–179 (2016). https://doi.org/10.1145/2902961.2903038
32. Kirichek, R., Koucheryavy, A.: Internet of Things laboratory test bed. Lect. Notes Electr. Eng. **348**, 127–140 (2016). https://doi.org/10.1007/978-81-322-2580-5

A Comprehensive Study About Cybersecurity Incident Response Capabilities in Ecuador

Roberto O. Andrade[1], Daniela Cordova[1], Iván Ortiz-Garcés[2], Walter Fuertes[3], and María Cazares[4(✉)]

[1] Faculty of Informatics, Escuela Politécnica Nacional, Quito, Ecuador
roberto.andrade@epn.edu.ec
[2] Universidad de las Américas, Quito, Ecuador
[3] Computer Science Department, Universidad de las Fuerzas Armadas Espe, Sangolquí, Ecuador
[4] IDEIAGEOCA Researchgroup, Universidad Politécnica Salesiana, Quito, Ecuador
mcazares@ups.edu.ec

Abstract. The increase of security threats and attacks in Ecuador, motivates the implementation of security incident response teams (CSIRT) in the different Ecuadorian organizations in different domains: academic, military, financial, public sector and critical infrastructures. NIST has developed a set of steps that are necessary to establish the CSIRT. The purpose of this study is to develop an analysis of each of the steps proposed by NIST in the Ecuadorian context, to identify the current status of response capabilities to security incidents in Ecuador and analyze possible actions to improve these capabilities.

Keywords: CSIRT · Incident response · Cybersecurity skills · Cybersecurity metrics

1 Introduction

The study developed in the year 2016 by the Organization of American States (OAS) and Inter-American Development Bank (IDB), based on the indicators of the Cyber Security Capability Maturity Model (CMM), about the state of cybersecurity in Latin America, presented two relevant conclusions:

- Four of five countries lack cybersecurity strategies to protect their data and critical infrastructure.
- Latin American region presents "potentially devastating" vulnerabilities for people, economies, and critical infrastructure.

This study mentions that countries such as Uruguay, Brazil, Mexico, Argentina, Chile, Colombia and Trinidad and Tobago, have an intermediate level of cybersecurity maturity, but are quite far from the levels of maturity presented in countries such as the United States, Israel, Estonia and Republic of Korea (OAS&IDB 2016). After the year 2016, the level of cybersecurity maturity in Latin America has not presented a substantial improvement, and the number of cyber-attacks has increased from 33 attacks

M. Botto-Tobar et al. (Eds.): CI3 2020, AISC 1277, pp. 281–292, 2021.
https://doi.org/10.1007/978-3-030-60467-7_24

per second in 2017 to 9 attacks per second in 2018 (Kaspersky 2019) and different types of cybersecurity attacks have been presented in the Latin American region since the year 2016 (Kaspersky 2019; Symantec 2018), among which we can cite:

- Interception attacks,
- Phishing,
- System abuse devices,
- Cyber forgery,
- Computer fraud,
- Child pornography,
- Offenses against intellectual property.

LACNIC (2019), mentions that phishing leads the cyberthreat ranking in Latin America and Caribbean representing 60% of recorded attacks and the report in the statistical portal statista (Pasquali 2019) indicates that the five Latin American countries that are targeted by phishing attacks are: Brazil 28.28%, Guatemala 20.34%, Chile 20.1%, Venezuela 19.89%, and Ecuador 19.55%; a second study presented by OAS related to security in the banking sector in Latin America, mentions that 9 out of 10 banks are targets of cyber-attacks, and mentions that the most of attacks on banking systems are malware in 24% and phishing in 22% (OAS 2016). So, Phishing is one of the most relevant attacks in Latin American.

The analysis presented in IEEE Innovation "Three Reasons Why Latin America is Under Cyber Attack" (IEEE 2019) and the study of the OAS-IDB "Are We Ready in Latin America and the Caribbean" (OAS&IDB 2016), agree in that the most relevant reasons why Latin America Region is growing in cybersecurity attacks are:

- Lack of coordinated capacity to respond to cybersecurity incidents; although the countries of the region are developing Cyber Emergency Response Teams (CERTs) and Computer Security Incident Response Teams (CSIRTs) to handle cyber-attacks are still at an intermediate level of preparedness.
- Lack of Public awareness to publicized the dangers of the Internet, and of structured cybersecurity programs.

These two reasons led us to consider in this study the following research questions related to the ability to respond to incidents of cybersecurity and the training of security specialists in Ecuador:

- RQ1. What is the situation of Ecuador in the context of cybersecurity?
- RQ2. Ecuador has the capabilities to respond to cyber-attacks?
- RQ3. What profiles are necessary to effectively respond to cyber-attacks?

The contribution of this work is establish a baseline for define strategies that allow enhanced the state of cybersecurity in Ecuador, for which this study is structured as follows: Sect. 2 presents the research methodology used for establish a baseline about cybersecurity incident response capabilities in Ecuador, Sect. 3, shows a state of cybersecurity in Ecuador, according metrics define for international organizations. Section 4, shows the capabilities of incident response in Ecuador analyzing each step of guidelines NIST 800-61 against the initiatives or strategies made in Ecuador. Finally, Sect. 5 presents conclusions and future works in this context.

2 Research Methodology

The research methodology used in this work follows the following steps:

1. Define the research questions.
2. Define the sources of primary and secondary studies.
3. Present the results of the research carried out.

To proceed with the analysis of these research questions, the following relationship with the sources of studies has been defined: For RQ1, the state of cybersecurity in Ecuador is analyzed according to reports issued by international organizations. For the RQ2, the structure of response to cybersecurity incidents proposed by international organizations. For the RQ3, the profiles of the specialists that are required to handle an effective response to cybersecurity incidents are analyzed according to standards or best practices issued by international organizations.

We will establish a baseline of cybersecurity incident response in Ecuador, for this we analyze each step of the NIST 800-61 against initiatives or strategies establish in Ecuador in this context in basis of information collected for answers the three research questions.

3 State of Cybersecurity in Ecuador

Reports issued by international organizations such as the National Cyber Security Index (NCSI 2019) ranks to Ecuador in 82nd place of 130 countries, and the ITU Global Security Index (ITU 2019) ranks to Ecuador in position 98 of 175. Each of these indexes measures globally the preparation of countries to prevent cybersecurity threats and manage incidents (NCSI), as well as the commitment to cybersecurity Index (GCI 2019) to the five pillars supported by the Cybersecurity Agenda (GCA ITU-GCA: Global Cybersecurity Agenda (GCA 2019): legal, technical, organizational, capacity development and cooperation. In the case of the GCI, this index has measured to countries in aspects such as legislation, agencies in charge of cybersecurity, public policy among others, where the results of Ecuador have been condensed in Table 1.

Table 1. Index GCI of Ecuador. Version 2018 are still in a draft

Índex	Year	Score
GCI	2014	17
GCI	2015	17
GCI	2017	64
GCI	2018*	98

This result allows us observe that during the years 2014 and 2015 Ecuador maintained a score of 17 in terms of the degree of commitment to cybersecurity, and between the years 2017 and 2018, Ecuador dropped its score from 64 to 98, this in part to the economic changes of the country, the change of government, and the elimination of intelligence and security entities. The study presented in 2015 by the Organization of American States, "Report on Cybersecurity and critical infrastructure in the Americas" (OAS 2016), on the perception of preparedness for response to cybersecurity incidents, defines Ecuador as a State "NOT PREPARE", so that detection, protection and response capabilities should be immediately considered.

4 Capabilities of Response to Cybersecurity Incidents in Ecuador

The National Centre for Science Information-NCSI uses five aspects to assess countries' cybersecurity:

- Identification of cybersecurity threats at the national level.
- Identification of cybersecurity measures and capacities.
- Selection of important and measurable aspects.
- Development of cybersecurity indicators.
- Grouping of cybersecurity indicators.

Each aspect includes other criteria that are analyzed and entered by a member of the NCSI project who also includes the evidence for the assessment. One of the indicators evaluated is the "INCIDENT AND CRISIS MANAGEMENT" in which four aspects are evaluated: Cyber incidents response, Cyber crisis management, Fight against cybercrime, and Military cyber operations. Related to the Cyber Incident Response, NCSI evaluates that the country has a unit specialized in national-level for cyber incident detection and response; this unit is designated as single point of contact for international cyber security coordination. Regarding this aspect, Ecuador complies through the establishment of ECUCERT in the year 2014.

Have a structure to respond to cybersecurity incidents allows organizations to quickly detect possible incidents, identify weaknesses and quickly restore operations that may be affected. To define the security incident response structure, NIST (2019), in the recommendation 800-61 "Computer Security Incident Handling Guide", presents a guide for establish response teams to cybersecurity incidents:

1. Create an Incident Response Policy.
2. Develop Incident Response and Reporting Procedures.
3. Establish Guidelines for Communicating with External Parties.
4. Define Incident Response Team Services.
5. Select a Team Structure and Staffing Model.
6. Staff and Train the Incident Response Team.

4.1 Incident Response Policy and Reporting Procedures

Regard to points 1 and 2, the definition of policies and procedures are required for the handling of response to security incidents, and these can be aligned to different phases defined by organisms such as:

- NIST, establishes a set of four phases: preparation, detection and analysis, containment, eradication and recovery, and post-incident activity.
- ENISA, determines a set of six phases: incident report, report registration, triage, incident resolution, incident closure, and post-analysis.
- SANS, defines a set of six phases: preparation, identification, containment, eradication, recovery, and lessons learned.
- ITIL, presents a set of five phases: incident detection and recording, classification and initial input, investigation and diagnosis, resolution and recovery, and incident disclosure.

Regarding this point in Ecuador, the institutions of the public sector adopted the scheme similar to the ISO 27000 proposed by the Governmental Scheme of Information Security –EGSI (MINTEL 2019), while the educational institutions adopt COBIT or ITIL (UETIC 2018).

4.2 Guidelines for Communicating with External Parts and Team Services

About points 3 and 4, are directly related to the mission of the organization in which the process of managing security incidents is established, as well as the correct identification of the stakeholders. The structure of a CSIRT is intimately related to the sector to which it will provide incident response services, for which the National Institute of Standards and Technology Review No. 2, NIST 800-61, allows us to identify some branches, within which find:

- Communications Sector.
- Electric sector.
- Financial Services Sector.
- Information Technology Sector.
- Research and Education Sector.

The structures of each CSIRT can vary according to the sector it serves, but in general the standard shows a structure like the following:

- Centralized structure.
- Distributed structure.
- Coordinated structure.

Each structure can be formed with own personnel of the organization, subcontracted personnel or subcontracting of the incident response service. One of the factors important to consider the structure CSIRT is the staff, which should be included in the organizations considering the response to Incidents. The creation of cybersecurity incident response teams that cover different domains of Internet Service Provider, Academy Institutions, Army, and Private Organizations, in Figs. 1, 2 and 3, and Table 2, presented the established CSIRTs in Ecuador until the middle of 2019, of which few are currently recognized as members of First.org.

Relevant data that can be obtained, is that only half of the CSIRT established in Ecuador are members of FIRST, also that most of them were established in the year

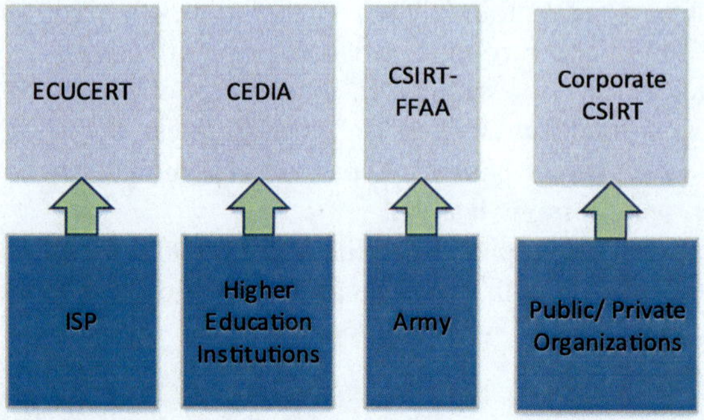

Fig. 1. Response teams in Ecuador by sector.

Team	Official Team Name	Country
BLUE HAT CERT	Blue Hat Cyber Security Incident Response Team	EC
CERT Radical	CERT RADICAL ALTERNATIVAS DE AVANZADA ALTRADICALAVAN CIA. LTDA.	EC
CSIRT Telconet	Centro de Respuesta a Incidentes de Ciberseguridad de Telconet	EC
CSIRT-CEDIA	CSIRT-CEDIA	EC
CSIRT-EPN	CSIRT-EPN	EC
EcuCERT	Centro de Respuesta a Incidentes Informáticos de la Agencia de Regulación y Control de las Telecomunicaciones	EC
MAINTLATAM CSIRT	MAINTLATAM CSIRT	EC

FIRST follows the Internationa Olympic Committee (IOC) country name listings.

Fig. 2. Response teams in Ecuador recognized as members of First (2019).

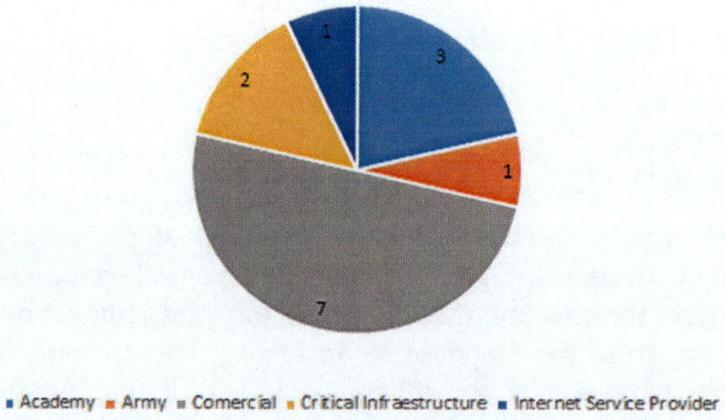

Fig. 3. CSIRTs in Ecuador by sector.

2019. The 50% of the CSIRTs implemented are associated with commercial purposes, focused on offering incident response services to companies. At the academic level

Table 2. CSIRTs of Ecuador.

Name	Type	Community	FIRST Member
CSIRT-CEDIA	Academy	CEDIA members	Yes
CSIRT-EPN	Academy and Research	Students, Teachers, Research, Authorities from EPN	Yes
Army	Army	CSIRT FFAA	No
National EcuCert	Government	Internet Service Provider	Yes
CSIRT UTPL	Academy	Students, Teachers, Research, Authorities from UTPL	No
CERT Radical	Commercial	Institutions that are clients of the Radical Group.	Yes
SSCorp-CSIRT	Commercial	Government, Private and Public sectors	No
CSIRT-GMS	Commercial	Institutions that are clients of the GMS.	No
CSIRT Telconet	Commercial	Telconet Group and Clients	Yes
CSIRT Telefónica Ecuador	Commercial	Clients	No
MAINTLATAM CSIRT	Commercial	Clients	Yes
Blue Hat CSIRT	Commercial	Clients	Yes
CSIRT Energía	Critical Infrastructure	Institutions associated with resources and energy domain	No
CSIRT CELEC EP	Critical Infrastructure	Institutions associated with Generation and Transmission of Electric Power	No

there are three, equivalent to 21.7% of the total number of CSIRTs created in Ecuador, if we consider the number of higher education institutions affiliated with CEDIA (56) the value of academic CSIRTs is only 5.4%. No financial CSIRT is specifically registered or identified in Ecuador, this service is provided through the Commercial CSIRTs, for example CSIRT-Telconet LATAM. There are only five CSIRTs linked to the public sector, among which 1 is included in the academic domain, 1 in the army and 3 in critical infrastructures or public agencies.

In Ecuador, there is currently no official communication process between Incident Response Teams, state organizations, public and private entities. To establish an adequate communication process, the first step is to establish the related actors; in Table 3, we present some of the actors that we consider relevant for each type of CSIRT in Ecuador based on interviews conducted with organizations in the different academic, army and national domains. In academic organizations, incident response management involves collaborative work among the objective community that includes students, researchers, professors and authorities. In the case of internal escalation, it requires interaction with technical areas such as IT Services and Security Management; there are security incidents that require legal actions that require the support of the legal department. The dissemination of a cybersecurity incident requires appropriate dissemination designs to transmit the real state without alarming the community, so the contribution of the Public Relations department is required. For certain incidents, the support of external actors is required, such as the Internet service provider. In the case of Higher Education Institutions in Ecuador, interaction is required as actors such as CEDIA and ECUCERT. Is worth mentioning that in the academic domain, basic education institutions are not yet integrated into the context of security incident response. In the case of CSIRT National, the objective community is society and organizations under public administration, for this there is a direct relationship with Internet service providers, for the articulation of regulations and regulations ECUCERT in coordination with MINTEL and ARCOTEL; the CSIRT has external relations such as the OAS and the FIRST. As of 2018, the ECUCERT and CEDIA established cooperation agreements, which allow improving collaboration processes to respond to cybersecurity incidents. The most of CSIRT are recently established in the year 2019, so a process of communication and collaboration between cybersecurity incident response teams has not been formally established.

Table 3. Stakeholders CSIRT Ecuador.

CSIRT	Internal Parts	External Parts
Academy	IT Services, Security Management, Legal Department, Human Resources, Public Relations	CSIRT-CEDIA, Ecucert, Judicial Police, Dinardap
Army	IT Services, Security Management, Legal Department, Human Resources, Public Relations	CSIRT FFAA, Interior Minister
National	IT Services, Security Management, Legal Department, Human Resources, Public Relations	Judicial Police, Government, Interior Minister, Dinardap
Comercial	IT Services, Security Management, Legal Department, Human Resources, Public Relations	Enterprises
Critical Infraestructure	IT Services, Security Management, Legal Department, Human Resources, Public Relations	Ecucert, Minister of Energy and Non-renewable Resources, Interior Minister

4.3 Team Structure and Staffing Model

Regard to point 5, NIST recommends the roles determined for a CSIRT are:

- Incident Response Coordinator
- Incident Response Handler
- Insider Threats
- Law Enforcement
- Office of General Counsel
- Officer

Members of CSIRTs must present technical and social skills (soft-skills) to provide an effective response to security incidents such as: system administration, network administration, programming, technical support, intrusion detection, problem solving skills and critical thinking. According to Wall Street Journal the main differentiator between more and less effective cybersecurity professionals is the soft skills and not the technical skills; therefore, the appropriate selection of CSIRT staff requires a methodological and formal selection process. The staff selection model based on competency is appropriate in environments that require organizational flexibility and cover positions with low offer of professionals, this model is useful for skill gap analysis because allows compare between available and needed competencies. U.S Department of Labor define competency as: "A cluster of related knowledge, skills, and abilities that affects one´s job, correlate with job performance can be measured against well-accepted standards and can be improved through training, development, and experience". So, a competency model is a list of competencies needed to develop a specific job. U. S department (DOL 2015) proposes a competency model framework to identify the industry-specific skills and competences that staff require. The model proposed by U.S department consists of a pyramidal arrangement; each layer represents the level of specificity and specialization of staff. The framework consists of nine layers grouped into three clusters: foundational competencies, industry-related competencies, and occupation-related competencies.

- Tier 1. Personal Effectiveness Competencies: integrity, reliability, and adaptability.
- Tier 2. Academic Competencies: reading, writing, critical thinking, and formal education.
- Tier 3. Workplace Competencies: teamwork, planning, and organizing.
- Tier 4. Industry Technical Competencies: awareness or understanding for specific job tasks.
- Tier 5. Industry-Sector Functional Areas: crossroad security-industry skills and industry-sector skills.
- Tier 6. Occupation-Specific Knowledge Competencies.
- Tier 7. Occupation-Specific Technical Competencies.
- Tier 8. Occupation-Specific Requirements.
- Tier 9. Management Competencies.

In the case of Ecuador, job profiles in cybersecurity proposed by NIST have not yet been formalized in relation to cybersecurity incidents response, especially in the public sector, we have proposed in this study the identification of job descriptors related to the profiles of CSIRT specialists that could be adopted in Ecuador. To establish the job descriptors, the process shown in the Fig. 4 has been followed, the process considers the objective of the unit in which the activities will be carried out, as well as the objective of job-role, then proceed with the identification of the responsibilities and products associated with the job-role and finally define the competence profile.

4.4 Staff and Train the Incident Response Team

Regarding to step number six, the training for CSIRT members is important focus not only on technical skills, so also on soft-skills such as: teamwork, integrity and

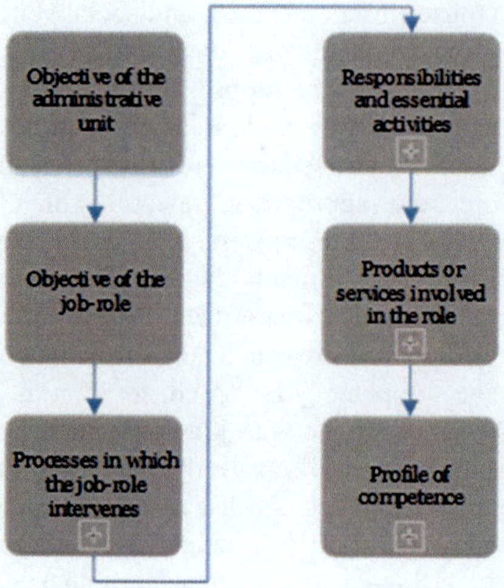

Fig. 4. CSIRTs in Ecuador by sector.

communication; in the Table 4 we show a set of competences that are important to be considered in the training. In the case of Ecuador, some formal cybersecurity courses exist on the master's degree level, as well as certifications in cybersecurity, but it is important to strengthen the soft-skills; for this aspect is important to increase the number of simulation training such as: flag capture, tabletop or cyber-drills, these types of training are limited in the country.

Table 4. Competencies dimensions in cybersecurity.

Competencies dimensions	Related concepts
Workplace competencies	Teamwork, Planning and organization, Innovation and Strategic thinking, Problem Solving and Decision making, and Working with tools
Academy competencies	Security fundamentals, Business foundation, Critical and analytical thinking, STEM literacy, Reading and writing, and Communication
Personal effectiveness competencies	Interpersonal skills and teamwork, Integrity, Initiative, Adaptability and Flexibility

5 Conclusions and Future Works

The commitment and preparation of Ecuador in the cybersecurity context has improved significantly in recent years, but the degree of maturity of Ecuador about the state of cybersecurity in the different sectors and domains provided by a model such as CMM is not known jet; so, is important strengthen the creation of national standards and the creation of CSIRTs on different sectors of society.

Is important enhanced the specific knowledge of the members of the national CSIRTs teams and include within the formal undergraduate academic education subjects related to incident response, malware analysis, reverse engineering, computational thinking and problem solving, can significantly improve the response to cybersecurity incidents.

Are important implement strategies for proactive responses to security incidents at the national level and implement cybersecurity laboratories that allow the generation of own methodologies in response to incidents through experimental practice. Promote the joint and coordinated of the work of all the CSIRTs formally implemented at the national level is necessary for adequate incident response actions.

Understand their organization's cybersecurity workforce is still low in Ecuador, so is necessary enhanced the training and define profiles according to needs of organization in related to protect their information, customers, and networks.

References

Cichonski, P., Millar, T., Grance, T., Scarfone, K.: Computer Security Incident Handling Guide NIST (2019). https://www.nist.gov/publications/computer-security-incident-handling-guide. Accessed 29 June 2019

IEEE Innovation at Work: Three Reasons Why Latin America is Under Cyber Attack - IEEE Innovation at Work (2019). https://innovationatwork.ieee.org/latin-america-is-under-cyber-attack/. Accessed 29 June 2019

ITU-GCA: Global Cybersecurity Agenda (GCA) (2019). https://www.itu.int/en/action/cybersecurity/Pages/gca.aspx. Accessed 29 June 2019

ITU-GCI: Global Cybersecurity Index (2019). https://www.itu.int/en/ITU-D/Cybersecurity/Pages/global-cybersecurity-index.aspx. Accessed 29 June 2019

Kaspersky: 33 ataques por segundo: Kaspersky Lab registra un aumento del 59% en ataques de malware en América Latina (2019). https://latam.kaspersky.com/about/press-releases/2017_33-attacks-per-second-increase-in-malware-attacks-in-latin-america. Accessed 29 June 2019

LACNIC: LACNIC NEWS: Cybersecurity Trends in Latin America and the Caribbean (2019). https://prensa.lacnic.net/news/en/cybersecurity/cybersecurity-trends-in-latin-america-and-the-caribbean. Accessed 29 June 2019

NCSI: National Cyber Security Index (2019). https://ncsi.ega.ee/. Accessed 29 June 2019

NICSS: Cybersecurity Workforce Development Toolkit| National Initiative for Cybersecurity Careers and Studies (2019). https://niccs.us-cert.gov/workforce-development/cybersecurity-resources/cybersecurity-workforce-development-toolkit. Accessed 29 June 2019

Organization of American States –OAS: The State of Cybersecurity and Critical Infrastructure in the Americas - Trend Micro PL (2015). https://www.trendmicro.com/vinfo/pl/security/news/cyber-attacks/the-state-of-cybersecurity-and-critical-infrastructure-in-the-americas. Accessed 29 June 2019

Organization of American States (OAS) and Inter-American Development Bank (IDB): Cybersecurity-Are-We-Prepared-in-Latin-America-and-the-Caribbean (2016). https://www.cybersecobservatory.com/2016/11/12/cybersecurity-ready-latin-america-caribbean. Accessed 29 June 2019

Organization of American States –OAS: OAS Report Reveals that 9 out of 10 Banks in Latin America and the Caribbean Suffered Cyber Incidents during the last year (2019). http://www.oas.org/en/media_center/press_release.asp?sCodigo=E-056/18. Accessed 29 Jun 2019
Pasquali, M.: Infografía: ¿Qué países latinoamericanos sufren más intentos de phishing?. Statista Infografías (2019). https://es.statista.com/grafico/18427/intentos-de-phishing-en-america-latina/. Accessed 29 June 2019

Software

Development of an App for Monitoring Heart Rate in People Who Practice Regular Physical Activity

Julio A. Mocha-Bonilla[1(✉)], Brayan Fabricio Punina Chimborazo[1], Kevin Israel Mocha Altamirano[2], and Dennis José Hidalgo Alava[1]

[1] Facultad de Ciencias Humanas y de La Educación, Unidad de Investigación y Desarrollo, Universidad Técnica de Ambato, Ambato, Ecuador
KIMA6499@estudiantes.edu.ec, {bpunina9028,dj.hidalgo}@uta.edu.ec

[2] Unidad Educativa Guayaquil, 18H00087, Ambato, Ecuador
ja.mocha@uta.edu.ec

Abstract. During the review of the supporting literature, in the different databases such as Scielo and Scopus, we considered the updated information of the contributions that the authors make about the different mobile applications designed for the practice of physical activities. Therefore, a low-cost app was designed that allows the calculation of heart rate, using a sensor SEN-11574 and an Arduino plate that sends data in real time to the application through a Bluetooth module HC-05, the app has a database where you can record the user's results and export them to an Excel file. For the development, the ADDIE methodology was used, which allowed the structured development of the application. Finally, the implementation of the resource is evaluated using the TAM technology evaluation model. The results found show a high degree of interest in the users towards the developed mobile application, obtaining significant results against reliable and high quality instruments.

Keywords: Mobile application · Heart rate · Arduino plate · Pulse monitor

1 Introduction

In recent years apps have had a constant evolution, that is why they help to strengthen physical practice, according [1] in his study, "Impact of mobile apps on physical activity", presents research related to the effect of mobile applications on physical activity, based on various reviews of previous research, the study counted with the implementation of a meta-analysis to verify the impact of mobile applications are immersed in everyday life. Mobile applications are of great importance in all areas of physical activity, according to [2] the importance of a technological tool for information management in sport is presented, where the perception of the technical staff of a high performance volleyball team is analyzed, a technological tool for data collection is presented which consists of a web page and its application works from mobile devices.

M. Botto-Tobar et al. (Eds.): CI3 2020, AISC 1277, pp. 295–306, 2021.
https://doi.org/10.1007/978-3-030-60467-7_25

This way, the importance of controlling the heart frequency during the practice of physical exercise, as well as during the moments of rest, appears. The data provided by this tool is very useful to know the number of times the heart contracts in a minute. The World Health Organization mentions the use of the so-called 'mHealth' or 'mobile health' and defines it as "the use of mobile devices and wireless technologies to support both health and medical goals".

The lack of training towards Bluetooth technology leads to the continued use of wired and expensive devices. For this reason, a project has been carried out to establish certain requirements; both hardware to be able to perform a heart pulse monitoring, and software to perform a wireless communication through Bluetooth, to be able to implement it in a low consumption Android platform, which includes a database for its respective operation, registration and control.

The storage and transfer of information is of great importance when performing physical activity as [3] mentioned the implementation of a mobile application in an ad hoc environment for information exchange, presented the development of an Android mobile application capable of creating an ad hoc environment for data exchange, the product of research was the implementation of a mobile application to establish connections between devices and broadcast technique to transmit information.

Information and communication technologies (ICT) can complement, enrich and transform physical activity habits [4]. Given the evident scientific literature related to the use of apps, the need to implement a low-cost device, designed by researchers to monitor heart rate in university students who practice regular physical activity, arises.

In this way, the design of a device and an app for the collection of heart rate data is proposed. The application was designed for the collection of data related to heart rate, the automation of the system allowed to carry out the process of development of the System, providing data of each one of the people who make regular physical activity, in short, the FC data can be visualized to establish the values with the table established by the World Health Organization.

2 State of the Art

The following is a review of the literature related to apps applied to sport and physical activity. Technological and scientific progress worldwide is promoting the use of sports apps, currently has been introduced in the field of sport, both elite and grassroots, according [5] the use of smart phone applications to start and maintain physical activity, realizes that fitness applications based on smart phones help users to start a physical exercise regime. The use of technology in sports is proliferating thanks to advances in facilitating its operation; in this sense [6] they conducted a study on the influence of an app on the adherence to sports practice, the publication presents a protocol whose main objective was to publicize the adherence to physical exercises of certain users of sports centers through the support of a mobile application. Likewise, curricular sports activities are immersed, which at present are implemented by mobile applications, so we agree with [7] those who have studied the practice of sports in high school students through m-health. The research presents multiple applications that help to practice sports, as well as how to maintain a healthy physically active life.

There are numerous ways of exercising to get away from a sedentary life, one of them is outdoor sports, which is better accompanied by a mobile application that allows the monitoring of physical activity. In this sense [8], they present the implementation of a software based on mobile computing that allows the capture, control and monitoring of physical variables related to the practice of cycling; the software developed is a portable and adaptable tool, allowing the configuration of different routes in a personalized way for the sportsman or woman.

Every day, mobile applications are modified to guide people correctly, so that they can become an excellent tool for health care [9]. The study conducted with 158 university students mentioned that a third of the surveys were unaware of the existence of mobile health applications, and that 4 out of every 10 people use an application to monitor their physical condition and stressed that they had not paid for it.

The mobile applications take care of health, ranging from nutrition to the control of physical activities, according to this research has been conducted on Mobile application for improving physical performance in the gym using augmented reality, which presents the development of a mobile application for a gym, the methodology used was agile coming to design and create new tools that facilitate physical activity [10]. At the same time, mobile applications "eHealth-UTA" have been designed for low cost health care in order to establish the body condition of people, through the automation of the system in the taking of real data, it is concluded that differentiated programs of integral health supported by technology are necessary at present [11].

Finally, the contents of physical activities are of great importance and what better way to present them in a mobile application, since today everyone has devices or tablets, in this sense [12] proposes the use of apps for the Physical Education class by content areas, concluding that the use of classifying the apps by content areas provides the possibility of adapting them to other content areas depending on the contents of the students and their contexts.

3 Methodology

A quantitative approach was used to develop this study, which has allowed us to evaluate the heart rate behavior shown by university students who perform regular physical activity when using the designed mobile application.

3.1 Instruments and Methodological Development

The research project used the ADDIE methodology [13], which allows for the analysis, design, development, implementation and evaluation of the mobile application, with the objective of facilitating the collection of information related to heart rate. The ADDIE methodology is an interactive instructional design process, which allows the designer to lead back to any previous phase, that is, the final product of one phase is the starting product of the next phase.

The heart rate measurement system was designed using a SEN-11574 sensor [14], which provides real-time information on the heart rate of people who are physically active. This required an Arduino ONE board, which is connected to a wireless module.

The data transmission is sent via a Bluetooth HC-05 module and the user's data is displayed numerically in the application designed on the open source MIT App Inventor platform for the Android operating system [15]. The procedure used is shown below.

3.2 Development Phase

General Scheme of the System

For the calculation of the heart rate, a non-invasive measurement is used, that is, one that does not penetrate the user's skin. This is done by means of a Bluetooth module where the sensor is read on the Arduino Uno card, with this method the data of each person is displayed [15]. The system operation scheme can be seen in Fig. 1.

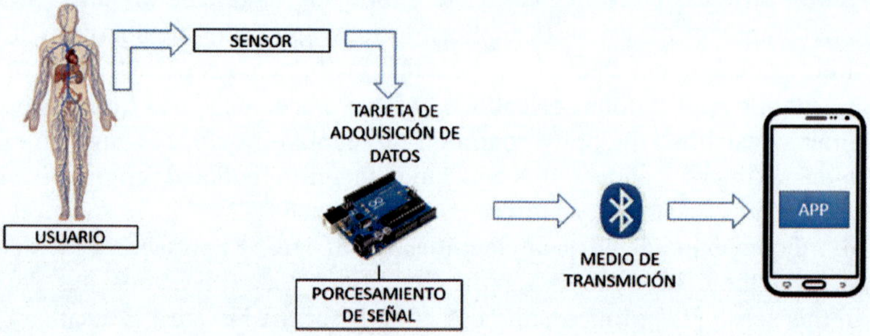

Fig. 1. General scheme of the system

3.3 Programming the Android Application

A design and block programming environment was used that allows dragging and dropping elements (data or functions), thus focusing on a single programming logic, leaving aside the syntax of the programming languages (points, commas, parentheses, etc.). All the labels, buttons and other elements that make up the graphic interface of all the screens of the PPP were placed in the design environment, as shown in Fig. 2.

The mobile application has a screen that allows the calculation of heart rate, using the PHOTOPLETISM [16] technique, which consists of applying a light source to one side of the finger. For the calculation of the heart rate, it was applied by means of a voltage peak count that is represented in 60 s, equivalent to the beats per minute (bmp).

3.4 Heart Rate Meter Main Components

The development of new programmable logic cards, allows the design of reliable, compact and important features at low cost [17], then it is released each of the electronic components that were used to develop the application and heart rate meter, which will send data in real time to the designed mobile application.

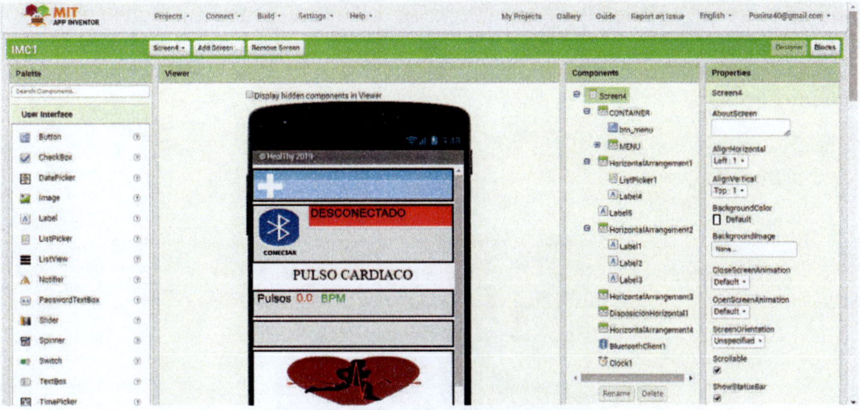

Fig. 2. Elements that make up the Heart Rate screen

Sensor SEN-11574

It is a non-invasive device, is composed of an ambient light sensor and a bright green led, it is used to capture live heart rate data, contains an active filter to make the output signal more defined and only needs 4 mA of current consumption at 5 V, so it is ideal for working with the mobile application designed.

Features (Fig. 3).

- Diameter: 16 mm
- Total thickness: 3 mm
- Supply voltage: 3–5 V
- Current consumption: 4 mA at 5 V

Fig. 3. Sensor SEN-11574

Bluetooth Module HC-06

It is a wireless protocol that allows establishing a connection between two devices, the HC-06 modules are very popular for apps with Arduino, since it can be inserted into a

Protoboard and wired directly to any microcontroller, one of the main advantages is its small size and its considerable range of transmission and reception of data.

Features (Fig. 4).

- Supply voltage 3.6 V–6 V
- Led status indicator

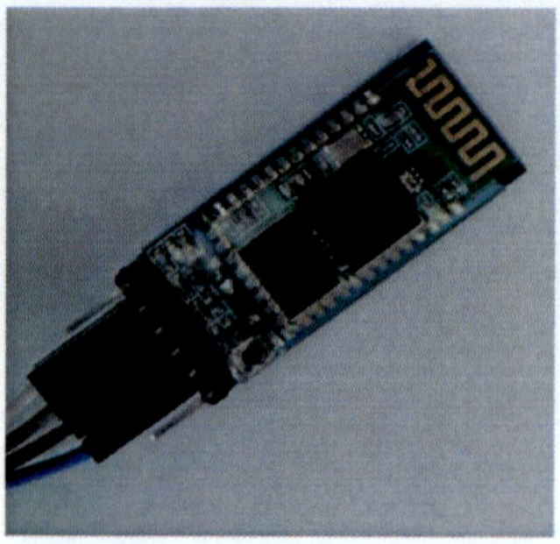

Fig. 4. Bluetooth module HC-06

Arduino Plate One

It is an electronic board based on free hardware and software, easy to use, you can build circuits and program them using this board, it has 14 digital inputs/outputs, a USB connector for programming, a power connector and a reset button.

Features (Fig. 5).

- Microcontroller: ATmega328
- Voltage: 5 V
- Analog inputs: 6

Fig. 5. Arduino Plate One

Shield Plate

Circuit board that is placed on the Arduino board and is attached to it by means of the coupling of its pins, without the need for another external connection. The function of this board is to expand the capacity of the main board and is compatible with Arduino Uno and Arduino Mega.

Features (Fig. 6).

- Ports: 6 (prepared for direct sensor connection)
- Dimensions: 57.2 * 53.5 mm

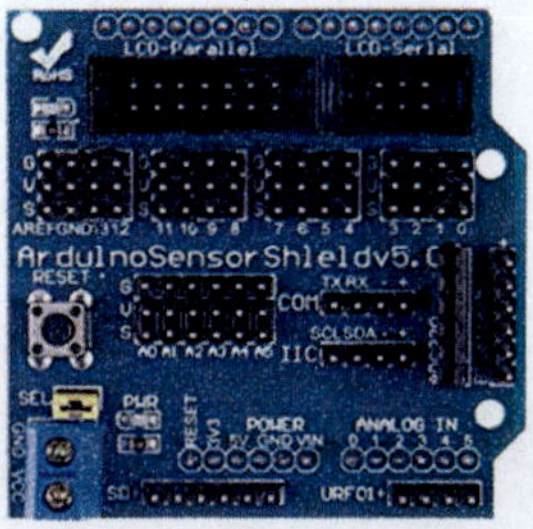

Fig. 6. Shield plate

Heart Rate Meter Assembly

Once the programming was saved on the Arduino UNO board, a Shield board was used which is located on top of the aforementioned board, in order to avoid external connections and damage to the Bluetooth, pulse sensor (Fig. 7).

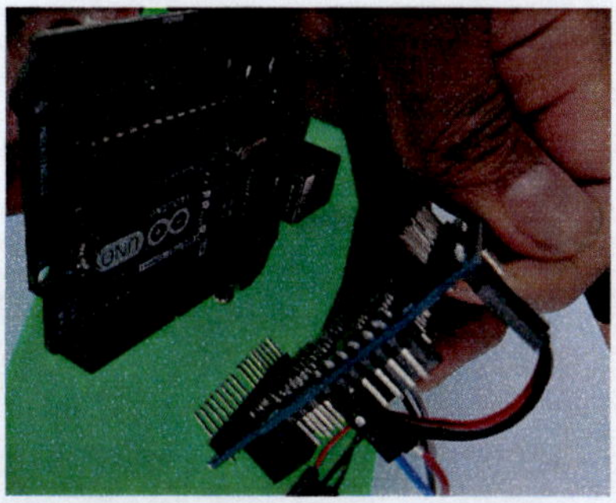

Fig. 7. Connection between the Arduino and Shield Plate

The sensor is connected to the Shiel board on pins URF01 (VCC, A0, GND), the Bluetooth module is connected to the pins of the Bluetooth interface (VCC, GND) for data transfer to pins 11 and 12 respectively (Fig. 8).

Fig. 8. Pin connection

To avoid using PC power to turn the component on, a 9-V battery with its respective on/off switch was used (Fig. 9).

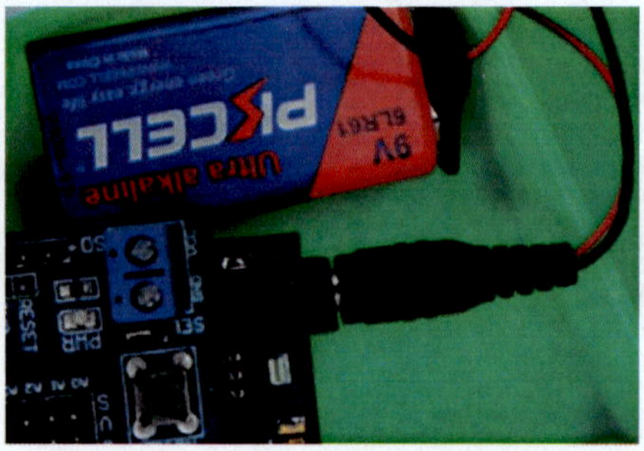

Fig. 9. Connection between battery and Arduino Board

Once all the connections between the components have been established, they are placed in a housing for adequate protection of the instrument (Fig. 10).

Fig. 10. Finished instrument with all its components attached

3.5 Participants to Test the App

For our study, we considered a sample of 20 eighth semester students of the Physical Culture Department of the Technical University of Ambato.

4 Results

The implementation was carried out with university students from the eighth semester of the Physical Culture career at the Technical University of Ambato, where the use of the application and the heart rate meter was socialized. The students had to download and install the APP on their mobile devices. To do so, they had to enter the address https://punina40.wixsite.com/misitio, where the installer is hosted.

Once installed, the participants proceeded to test the heart rate function; to take the data, they approached the meter and put their index finger on the sensor for 20 s, to verify the data displayed on the APP.

To validate the designed resource, the TAM technology acceptance model was used, which consists of 13 questions and 5 categories, the data were tabulated using the SPSS statistical package with the Cronbach's Alpha method. According to Campo & Oviedo, (2005) the Cronbach's alpha coefficient was described in 1951 and is used to measure the reliability in which the items of the instrument are correlated. Alpha values above 0.7 or 0.8 are sufficient to guarantee reliability.

The TAM model (Technology Acceptance Model), helps us to know the degree of acceptance that students had towards the designed application, the categories that are part of this model are: ease of use, perceived utility, attitude of use, intention of use and accessibility. The survey was structured according to a Likert type scale with 7 items; where "Very satisfactory" is equivalent to 7 and "Very unsatisfactory" corresponds to 1.

The TAM model includes 3 questions that refer to the ease of use of the mobile application, which shows a high number of highly satisfactory responses from a total of 20 respondents, which highlights the application and the appropriate interface used.

For the category Perceived Utility, 2 questions were used. The results determine a high number of highly satisfactory responses, since the contents shown are useful for physical practice and also that they had a pleasant experience when using the application.

In the category Attitude of Use, which consists of 2 questions and refers to the attitude of students after using the application, respondents say that it is highly satisfactory, because it motivates the presentation of data through a mobile device, saying that they would again use the PPP designed for the area of physical activity.

The category Intention of Use of the TAM evaluation model was used in 3 questions. Out of a total of 20 respondents, it was determined that the application has a highly satisfactory impact on the practice of sports, and they also consider that it would be beneficial in other physical sports activities.

Finally we have the category Accessibility, 3 questions were taken into account that helped to obtain the data illustrated in figure N° 11, after seeing the results of the 20 respondents with a highly satisfactory level of response, it can be stated that they do not present problems at the time of obtaining and accessing the application.

Figure 11 shows an example of the data provided by the mobile application, compared to a BEURER branded heart rate monitoring device.

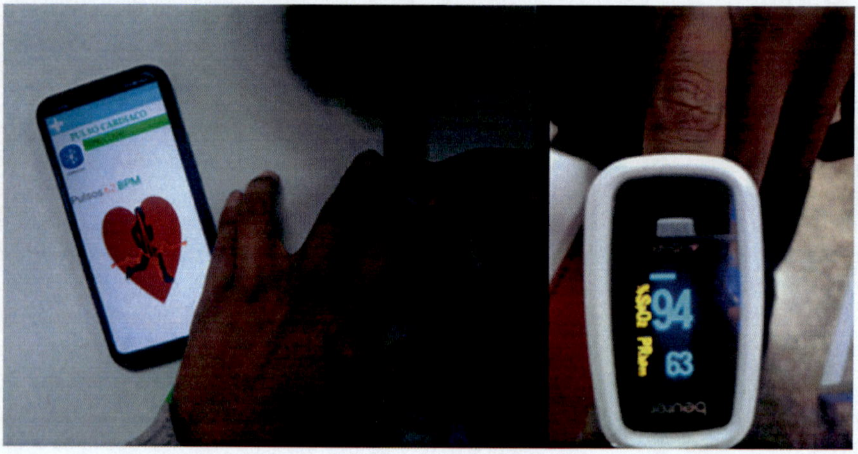

Fig. 11. Application data collection versus a high-quality device

5 Conclusions

After analyzing the literature supporting various tools that allow the design and development of mobile applications, he was inclined to use MIT APP Inventor, because it is a free software, accessible and provides little complexity in the programming code.

With the development of the app it was possible to capture in real time the heart rate of the users, the information is recorded in an Arduino plate, where the data is processed and prepared to be sent to the designed app, later it is saved in the application's own database.

No prototype has been able to surpass the existing devices on the market for the control of heart rate, therefore, it is necessary to continue improving the application to obtain more precise data, because an erroneous data can lead to an incorrect diagnosis. Since, in the integral health aspect related to the support of mobile applications, the complement of an intervention must be advised by a professional in the field of medicine, in this sense we agree with [18] who conclude that the vast majority of future applications, as they are standardized and improved, could be a very useful tool for society and the health care system.

In future research, it will be possible to visualize the creation of a mobile app for the Android system where the data of the instrument can be evidenced versus a reliable instrument existing in the market whose brand is recognized.

Acknowledgement. The authors would like to thank the Technical University of Ambato (UTA) and the Directorate of Research and Development (DIDE) for their support for the successful execution of this work through the research project entitled "MOBILE TECHNOLOGY, VIRTUAL REALITY AND VIDEO ANALYSIS OF MOVEMENT IN THE FIELD OF PHYSICAL ACTIVITY AND SPORT", code PFCHE13.

References

1. Aznar Díaz, I., Reche, C., Del Pilar, M., Trujillo Torres, J.M., Romero Rodríguez, J.M.: Impacto de las apps móviles en la actividad física: UN meta-análisis. Retos **36**, 52–57 (2019)
2. Fernández-Echeverría, C., Mesquita, I., Conejero, M., Moreno, M.P.: Importancia de una herramienta tecnológica en la gestión de información en el deporte. Percepción del staff técnico de un equipo de voleibol de alto nivel. Revista Iberoamericana de Ciencias de la Actividad Física y el Deporte **7**(3), 57–70 (2019)
3. Guingla, D.A.P., Marin, M.J.F., Castillo, L.E.B., Andrade, C.J.B.: Implementación de una aplicación móvil en un entorno ad hoc para el intercambio de información. Revista Científica ECOCIENCIA **5**(5), 1–25 (2018)
4. Guillen Pereira, L., Herrera Camacho, A.P., Ale de la Rosa, Y.: ICT technological tools as an alternative element for the development of the physical component. Retos-nuevas tendencias en educacion fisica deporte y recreacion (34), 222–229 (2018)
5. Aroni, A.L., Castillo, E., Sousa, C., Machado, A.A., Tenenbaum, G.: Aplicaciones de teléfonos inteligentes utilizadas para iniciar y mantener la actividad física: Un análisis exploratorio/Smartphone applications used for initiating and maintaining physical activity: an exploratory analysis. Revista de Psicología del Deporte **27**(4), 89 (2018)
6. Valcarce, M., Díez Valgo, C.: Influencia de una app en la adherencia a la práctica deportiva: protocolo de estudio. Revista de Educación Motricidad e Investigación (11) 16–34 (2018)
7. García, A.M.R., Montoro, M.A., Lucena, M.A.H.: Estudio de la práctica deportiva en el alumnado de bachillerato a través del M-Health. TRANCES. Transmisión del Conocimiento Educativo y de la Salud (1), 461–472 (2016)
8. Alarcón-Aldana, A.C., Urrutia-Pinilla, J., Callejas-Cuervo, M.: Aplicación Móvil para la Administración de Variables Físicas en Ciclismo al Aire Libre. Información tecnológica **27** (4), 175–182 (2016)
9. Carmona, E.M., Salas, P.E.R., Lau, M.G.C., Ramírez, T.D.: Uso y consumo de aplicaciones móviles en salud por jóvenes mexicanos. In: Conference Proceedings EDUNOVATIC 2017: 2nd Virtual International Conference on Education, Innovation and ICT, p. 1-1215. Adaya Press (2018)
10. Rueda Díaz, C.F., Niño Pinto, Á., Arboleda Mazo, W.H., Escobar, L.E.: Aplicación móvil para la mejora del rendimiento físico en gimnasio utilizando realidad aumentada. Doctoral dissertation, pp. 1–82 (2017)
11. Guerrero, J.S., Jimenez, L.A., Poveda, M.P., Barona-Oñate, R.V., Guerrero, A.G.S.: Analysis of the body composition index and basal metabolic rate through the mobile application eHealth-UTA. In: 2018 International Conference on eDemocracy & eGovernment (ICEDEG IEEE), pp. 386–391 (2018)
12. Bermejo, J.P.: Propuestas de uso de Apps para la clase de Educación Física por áreas de contenido. Revista Pedagógica ADAL (33), 6–11 (2016)
13. Arifin, Z., Gunawan, S.: Design and testing impact attenuator of formula SAE FG17 Garuda UNY car. J. Phys.: Conf. Ser. **1387**(1), 012091 (2019)

14. Santos, N.S.N., Motoyam, S.M.S.: A large scale platform using WBAN technology for patient monitoring. IEEE Lat. Am. Trans. **16**(3), 705–711 (2018)
15. Villamil, X., Guarda, T.: App Móvil Desarrollada con Metodología Ágil para IoT Controlada desde una Red LAN/WAN con Placa de Desarrollo de Hardware Libre (Arduino). Revista Ibérica de Sistemas e Tecnologias de Informação (17), 379–392 (2019)
16. Gil Borrallo, F.: Sistema de medida de tensión arterial en dedos con fotopletismografía: sistema PPG, pp. 1–112 (2018)
17. Haz, L., Calle, W., Moran, M.E.F., Carcelen, J., Cortez, A., Nunez-Unda, A.: Design of smart stretchers and vital signs monitoring system for reduced-mobility patients. In: 2018 13th Iberian Conference on Information Systems and Technologies (CISTI-IEEE), pp. 1–5 (2018)
18. San Mauro Martín, I., González Fernández, M., Collado Yurrita, L.: Aplicaciones móviles en nutrición, dietética y hábitos saludables: análisis y consecuencia de una tendencia a la alza. Nutrición hospitalaria **30**(1), 15–24 (2014)

Mobile Applications as Digital Support Material for the Inclusion of Students with Special Educational Needs

Marco Antonio Checa Cabrera[1(✉)]
and María Amparo Freire Cadena[2]

[1] Universidad UNIANDES, Ibarra, Ecuador
ui.marcocheca@uniandes.edu.ec
[2] Unidad Educativa Presidente Velasco Ibarra, Ibarra, Ecuador

Abstract. The right to access a regular education by students with special educational needs has been met in a limited way, because the Ecuadorian educational institutions have not the necessary support elements that strengthen the learning of these students. The present study focuses on the use of mobile applications that allow an educational inclusion of this vulnerable group of society. The mixed quantitative-qualitative method was applied, analyzing the data with the descriptive statistics and the hierarchical cluster method and the surveys applied to the teachers of the eighth grade of basic education of the Educational Unit "Presidente Velasco Ibarra" of Ibarra City, on the use of the two applications developed to be evaluated (WOOW ABC and READSIGNS). The results show that mobile applications allow greater participation and educational inclusion on the part of students with special educational needs, both in the academic activities of the classroom, as in the subjects taught by teachers of the regular education system.

Keywords: APP · Mobile applications · Educational inclusion · Digital support material

1 Introduction

According to UNESCO [14] Ensuring that everyone has an equal opportunity for educational progress remains a challenge worldwide. Sustainable Development Goal 4 on Education and the Education 2030 Framework for Action emphasize inclusion and equity as laying the foundations for quality education.

Educational Inclusion is related to the guarantee of the right of all children, adolescents, youth and adults to a quality education, through access to the regular education system, in all its levels and modalities; propitiating good living with the recognition of the diversity of people [12]. It is hoped that with the educational inclusion of all the inhabitants of a country, they will enjoy a quality life with equal access to the space of opportunities [11].

M. Botto-Tobar et al. (Eds.): CI3 2020, AISC 1277, pp. 307–318, 2021.
https://doi.org/10.1007/978-3-030-60467-7_26

Hence, in Ecuador and other countries of the world, educational systems have opted for educational inclusion in primordial State policies so that all people can access their right to quality education in formal education systems.

In the document "Ecuador en cifras", the National Council of Disability of Ecuador (CONADIS) has calculated 12.14% of the national population has a type of disability, that is 1,608,334 people. Being this vulnerable group of society that have fewer opportunities to access a quality education that promotes Ecuadorian formal education.

In this context there are students who have special educational needs (SEN) because they have hearing impairment or mild intellectual disability, which limits them to be able to enter an formal Educational Institution because of the way in which the Ecuadorian educational system is designed without considering the diversity of characteristics and capacities that has each student, although something has been advanced about it [6].

The educator has an important participation in this model of educational inclusion, since he has the responsibility that his teaching be meaningful and reaches all his students with quality, independently of his diverse capacities. But the educational support materials most used to achieve participation in the classroom whether they are graphic, auditory or visual, they do not allow an active intervention of all the students due to their different capacities to catch the information. It is evident that the elements before are not suitable to achieve an educational inclusion that is so much desired in the Ecuadorian formal education.

An action that the teacher must do is the use of technology as ICT facilitate access to digital materials adapted to the needs of the student to access information as quickly and effectively as possible [8].

It is unquestionable that technology has helped in various aspects of people's lives, be it socially, economically and educationally, it should be stressed at this point that the use of smartphones and their applications related to education has increased worldwide; it is then evident that mobile technology can become an essential component so that educational inclusion is truly for all and thus facilitate the teacher digital support material that allows to involve all students with or without special educational needs.

It is for this reason that two educational mobile applications projects aimed at students with disabilities were developed:

WOW ABC: Mobile Application (APP) aimed at helping in the field of educational reinforcement for children with Down Syndrome with the use of touch screens thinking about the special educational needs they have [6] (Fig. 1).

Fig. 1. Main interface of the WOOW ABC Application.

READSIGNS: It is an APP that translates digital books into sign language in real time, helping in this way that children with hearing and language disabilities can access the information that is available in the primary source of books, in digital format [2, 10] (Fig. 2).

Fig. 2. Main interface of the READSIGNS Application.

This document will show how these applications helped educators to involve all their students with or without special educational needs in activities that take place in the classroom, thus offering the right conditions for the teaching process to be truly for all with the help of this digital support.

2 Methods

The study was carried out at the "Presidente Velasco Ibarra" Educational Unit in the city of Ibarra, it is the institution that receives the most children with special educational needs, there are students with mild Down Syndrome as well as with hearing disabilities (with knowledge sign language).

To meet the objective, the strategy was to use the WOW-ABC application [6] (APP that is aimed at this group of students, but with Down Syndrome) in children of the first year of basic education, distributed in four parallels. In the 2019–2020 school period, a total of two hundred students have enrolled, of whom thirty has mild Down Syndrome; hen, the learning level of children with Special Educational Needs (SEN) in the topic of color and image recognition with respect to their classmates was evaluated, and in this way, individual differences in the cognitive aspect could be found [1] that both groups could have; to finally reach the conclusion of the degree of inclusion that this APP allowed on the topic previously raised.

The next thing was to work with the APP READSIGNS with the eighth-grade students of basic education [10] with hearing impairment (with knowledge of sign language) in the subject of Language and Literature, in the theme of comprehensive reading. This level has a total of 148 enrolled students, of which twenty have hearing disabilities with knowledge of sign language, on which the evaluation was carried out that determined the level of reading comprehension in the two groups of students and reached infer from the point of view of educational inclusion.

Finally, the satisfaction of the use of the proposed applications and the perception of their validity as didactic material by the Teachers was taken into consideration, applying to the students with and without SEN the topics mentioned above with the proposed digital tools.

The groups identified to perform the present study is the following (Table 1):

Table 1. Population involved in the study.

Group	Number
Teachers first and eighth year of Basic Education	11
Number of First Year Students Basic Education with Mild Down Syndrome	30
Number of Eighth Year Students Basic Education with Hearing Impairment	20

With the mixed quantitative qualitative method, an analysis was obtained of the data obtained from surveys and interviews applied to the teachers of the institution who work with students who have special educational needs, to arrive at a measurement that values the use of these mobile applications in activities in the classroom by all students.

With the descriptive-multivariate statistic, a global description of the problem was made and of the way in which mobile applications created learning conditions so that students could have a better educational inclusion in the Institution.

3 Results

Gikas and Grant concluded that the information was accessed quickly through mobile devices and allowed closer collaboration between students, thus increasing the knowledge of the subject, the motivation and involvement of students in this process [5]. This is the purpose of this study, around the level of inclusion that students with SEN have during the exposition by the Teachers in the subjects with which they work, taking as test entities the first and eighths of basic education and mobile applications developed for this purpose.

The objectives to fulfill are the following:

- Know the level of inclusion and learning improvement offered by the WOW-ABC and READSIGNS APPs in students with SEN in the development of the topics proposed for this purpose; infer in the end, between learning with traditional educational resources and with the use of these applications.

- To determine the degree of satisfaction with the use of the proposed applications and the perception of their validity as didactic material by the tutor teacher of the selected subjects.
- To know the perception of the students in their learning with the use of the APPs in class [9].

3.1 Degree of Inclusion of Students with NEE

WOW-ABC

In order to obtain accurate results that help to make an efficient descriptive analysis, the WOW-ABC application was installed on the smartphones of all students with and without SEN of the first year, in the matter of Discovery and Understanding of the Natural and Cultural Environment, on the subject of color and image recognition; The teacher Alicia Loza designed and applied the rubric to evaluate the proposed topic with and without the use of the application.

The rubric is made up of the following criteria:

- Identification of elements
- Selection of elements by color.
- Interest in the activity
- Characteristics of the el

Will be evaluated according to the following scores:

- Excellent (4 puntos)
- Good (3 puntos)
- Suitable (2 puntos)
- Moderate (1 punto)

Table 2. Results of the rubric.

Criterion	Without the WOW-ABC APP		With the WOW-ABC APP	
	Evaluation with SEN (Average)	Evaluation without SEN (Average)	Evaluation with SEN (Average)	Evaluation without SEN (Average)
Identification of elements	1.3	3.1	3.4	3.7
Selection of elements by color	2.4	3.5	4	4
Interest in the activity	0.7	2.4	3.5	4
Characteristics of the elements	1.3	3.5	3	3.8
Total	5.7 points out of 16 possible	12.5 points out of 16 possible	13.9 points out of 16 possible	15.5 points out of 16 possible

Observations found:

- WOW-ABC can also be used by students without SEN, so the two groups were evaluated together.
- From the above, the students without SEN considerably improved the subject of image and color recognition in a more attractive and fun way.

READSIGNS

Work was carried out in the same way with READSIGNS in the eighth year of the Educational Unit "President Velasco Ibarra", in the subject of Language and Literature with the theme of comprehensive reading (using two books: " Las garras de la Luna Llena " [3] and the digital "La culpa es de la vaca" [7]), the rubric that Professor Pastora Carlosama provided for this purpose was then evaluated, which is proposed by Andújar.es [13].

The rubric is made up of the following criteria:

- Identification of the class and purpose of the Text.
- Determination of the subject of the text.
- Identification of the text structure.
- Inference between data and ideas in the text.
- Reflection on the content and forms of the text.

Will be evaluated according to the following scores:

- Fully Achieved (4 puntos)
- Not totally (3 puntos)
- With difficulty (2 puntos)
- Not gotten (1 punto)

Once the two literary works have been read by the twenty students with and without SEN of the eighth grade students of basic education, the rubric is applied and then the average of the obtained marks is calculated to know the improvement or not of the group of job:

Table 3. Results of the rubric.

Criterion	Without the READSIGNS APP		With the READSIGNS APP	
	Evaluation with SEN (Average)	Evaluation without SEN (Average)	Evaluation with SEN (Average)	Evaluation without SEN (Average)
Identification of the class and purpose of the Text	2.3	3.1	3.4	3.7
Determination of the subject of the text	3.3	3.8	4	4
Identification of the text structure	1.7	2.7	3.1	3.4
Inference between data and ideas in the text	2.2	3.1	2.9	3.3
Reflection on the content and forms of the text	2.4	2.4	3.2	2.8
Total	11.9 points out of 20 possible	15.1 points out of 20 possible	16.6 points out of 20 possible.	17.2 points out of 20 possible

Observations found:

- It was not necessary to evaluate the students without SEN with the use of the application, because they do not need for the educational activity.
- Students with SEN without the application need a translator, who are generally the parents who help the student to read and sometimes to understand.
- Instead, with the use of the application they omitted the translator and the APP allowed a direct reading with the text, improving their reading comprehension.

With these results it can be inferred that there is a clear cognitive difference: 6.8 points for the first years (subtracting the totals in column 2 and 3 from Table 2) and **3.2** for the eighths (subtracting the totals in column 2 and 3 from Table 3), determined by the scores obtained when applying the rubrics with traditional educational re- sources. In contrast, the cognitive difference found: 1.6 points for the first years (subtracting columns 4 and 5 from the totals in Table 2) and 0.6 points in the eighth years (subtracting columns 4 and 5 from the totals in Table 3) is minimal, it shows that learning has been improved using the proposed mobile applications.

It should be noted, in addition, that not only students with SEN from the first years improve their scores but also the other group of classmates, because they also make use of the application, allowing them to make school activities more entertaining.

3.2 Satisfaction of the Use of the Proposed Applications and the Perception of Its Validity as a Teaching Material

Satisfaction of the Use of the Proposed Applications
For this purpose, an AD HOC questionnaire was developed with dichotomous and open-ended questions, consisting of twenty items, of which not all were considered in the present study.

The most important questionnaire questions for the present analysis were:

a. Do all students participate using mobile applications during classroom activities?
Here 63.63% of teachers perceive that all students participated in the proposed activities carried out with the applications.
b. How do students with SEN participate in school activities?
This participation has been active by students with or without SEN, which is 54.54% of the surveyed teachers.
c. Do mobile applications as support material allow more participation by students with or without SEN, compared to traditional materials?
81.81% of the teacher's state that these applications considerably improve the participation of students with SEN in classroom activities, thus complementing, from the point of view of teachers, with the results obtained in Sect. 3.1.
d. According to your academic perception, does the use of mobile applications in classroom activities allow or not a greater educational inclusion?
Most teachers (72.72%) agree that these applications do significantly help to improve the inclusion of students with SEN in the regular educational system.

The Perception of Its Validity as a Teaching Material

In order to visualize the problem in all its complexity and from a complete approach, in this work the multivariate clustering statistic method have been applied. An analysis was also carried out with which the results obtained can be observed objectively; for which the method based on the cluster analysis or hierarchical groups of elements to be studied was applied [4].

Cluster's multivariate technique

For this, a rating system was generated for the most important previously proposed questions, which help the descriptive-multivariate statistical analysis according to their type of answer:

Table 4. Rating based on the type of response.

Answer	Quantitative value
If you answer negatively	1
If you have doubts	2
If you answer positively	3

The cluster's multivariate technique allows you to classify objects into groups (clusters), which are homogeneous and heterogeneous with each other [15]. Taking this concept into consideration, the questions of the survey applied to each teacher will be evaluated based on a quantitative qualification based on their answers according to Table 4, the purpose of this analysis will be to group of teachers with homogeneous criteria (despite being different in their academic perception) about the variables identified and defined by each question to determine conclusions about how mobile applications help as digital support material and that allow the educational inclusion of students with SEN. (See Table 5).

Table 5. Relationship of the question of the survey with the variable.

Question	Name
1	Participation
2	Form of participation
3	Mobile applications with respect to other materials
4	Teaching perception

Each of the eleven teachers evaluated the variables in Table 5 among the values established in Table 4. The results are shown in Fig. 3.

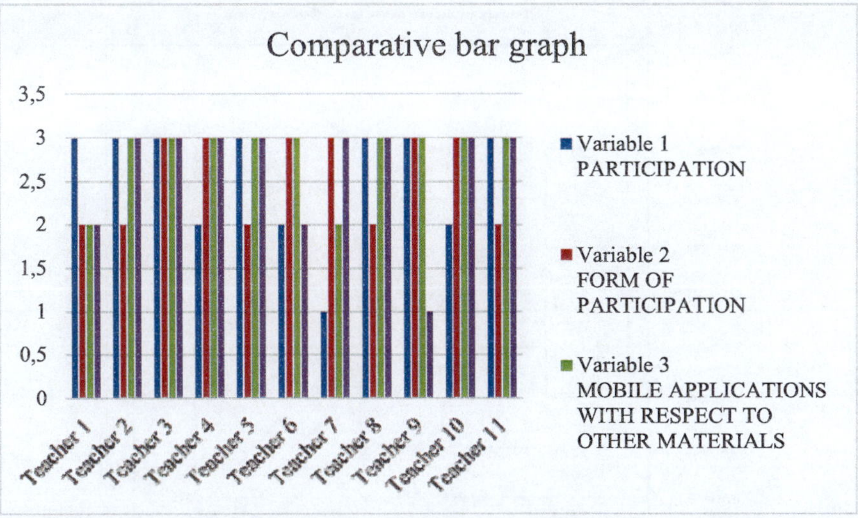

Fig. 3. Comparative bar graph showing the frequency of responses of all teachers with the quantitative rating.

Most of the teachers propose positive answers to the questions asked about the participation of students with SEN and how mobile applications contribute to the fulfillment of educational inclusion by this vulnerable group of society.

Analysis of hierarchical cluster

The data in Fig. 3 was used in the SPSS software ver. 25 for the generation of the dendrogram for its respective analysis.

Figure 4 shows how the stages 1 to 4 there is no distance, indicating that most groups of teachers (cluster) have positive opinions on the variables analyzed. The following stages from 5 to 8 are not so distant, indicates that the teachers of these clusters have similar opinions with few differences in their criteria on the mobile applications and their use in students with SEN. Very few teachers in stages 9 and 10 have very different points of view on the variables analyzed.

It is concluded then that the perception on the part of the teachers is HOMOGENOUS and POSITIVE regarding the validity of the studied mobile applications, and recommend its use as didactic material in the exposition of contents to students with or without SEN.

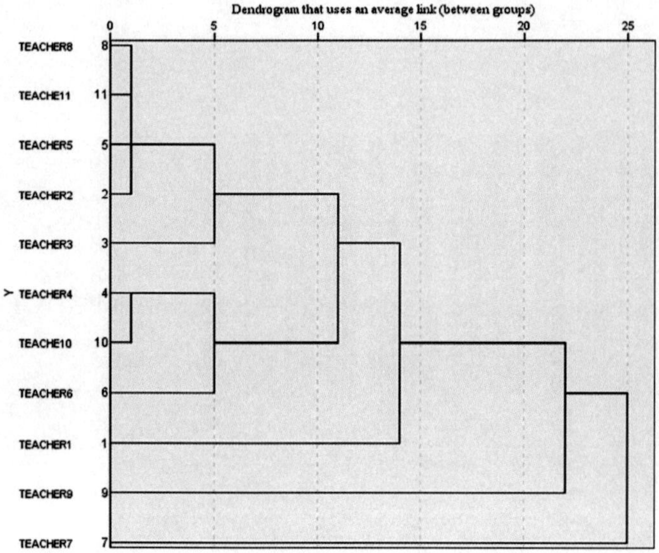

Fig. 4. Dendogram showing the clusters obtained from the hierarchical analysis of Fig. 3.

4 Discussion

It will be analyzed from the point of view of the three objectives proposed in the results section of this article.

The results obtained in the first objective on the degree of inclusion, it was determined that the WOW-ABC application can be used by children with and without SEN; READSIGNS does the same with the eighth grade students with hearing impairment, not needing a translator to read digital books and having a direct understanding of the author's ideas; thus achieving homogeneous educational resources and equalizing the school conditions of the two groups, thus reflecting a more effective educational inclusion with minimal cognitive difference.

In the second objective, with the application of the questionnaire and the statistical analysis of clusters, it is determined that the studied mobile applications become a good alternative that helps the participation of students with SEN, since the perception of teachers is unanimous and positive.

Finally, in the third objective, an analysis is made of the perception of learning obtained by the students themselves, with the use of the applications studied, inferring that the reception of knowledge is more effective compared to traditional educational resources such as the use of printed images, videos, texts, blackboards, projectors that, not being interactive, considerably limit the skills of children with and without SEN; the same happens with eighth-year students with hearing disabilities, since they are limited in reading comprehension due to the translation errors of the person in charge of it, because sign language requires a previous interpretation of the text, and then it is transmitted to the person with this disability, which means that the ideas of the author of said book can be misinterpreted.

5 Conclusions

The descriptive statistic and the hierarchical cluster analysis helped to evidence based on the analysis of Group Three, that each time the teachers have positive criteria on the use of technological tools and in this case of mobile technology, smartphones and their applications; much more, if these are dedicated to the work of students with disabilities and allow greater inclusion in regular education.

It was possible to demonstrate that with mobile applications if educational inclusion can be made to be present in Ecuadorian regular education by students with SEN; However, these applications must be adapted and updated both in their content and in the processes in a permanent manner, so that it is perfected as its use becomes more massive, making this right more and more fulfilled in Ecuadorian educational institutions.

It could also be assured of the results obtained, that mobile applications students with SEN or do not, work in a better way, considering them as educational support material, unlike the traditional elements: graphic, audio and video.

It should be noted and differentiate that when talking about mobile applications aimed at students with SEN, does not refer to ICTs widely used in education, this because these APPs are developed for a specific disability that students may have, as was the case of the present study: WOOW ABC and READSIGNS, the first one dedicated to children with Down Syndrome and the other to people with hearing disability.

Mobile applications such as the projects mentioned above are developed to reduce increasingly the gap between those who easily access information or those who are disadvantaged for various reasons, including disability.

Mobile applications not only allow educational inclusion, but also propose the proper use of smart devices, be they smartphones, tablets or even Smart TV, which with these APPs can be used to the maximum in obtaining information.

It is also concluded that, to achieve full inclusion, both students with SEN or not and their teachers, are willing to use these technological tools that are developed more and more in the educational field, as well as, which companies or institutions dedicated to the Research will refine the existing ones and develop applications with new approaches, contents and activities that adapt the different types of existing disability.

References

1. Down21.org Homepage. La importancia de comprender las diferencias individuales en el síndrome de Down. https://www.down21.org/revista-virtual/1738-revista-virtual-2017/revista-virtual-sindrome-de-down-diciembre-2017-n-199/3139-articulo-profesional-comprender-las-diferencias-individuales-en-el-sindrome-de-down.html. Accessed 19 Mar 2020
2. El Norte Homepage. APP premiada por History Channel. https://www.elnorte.ec/ibarra/app-premiada-por-history-channel-FE262117. Accessed 09 Mar 2020
3. Gauthier, B.: Las Garras de la Luna Llena. Susaeta, Paris (1995)
4. Gemperline, P.: Practical Guide To Chemometrics. EUA: Taylor and Francis (2006)

5. Gikas, J., Grant, M.: Mobile computing devices in higher education: student perspectives on learning with cellphones, smartphones & social media. Internet High. Educ. **19**, 18–26 (2013)
6. Guamán, A.: Sistema informático basado en pantallas táctiles para la enseñanza de los niños con síndrome de Down del Instituto de Educación Especial Ibarra (IEEI), trabajo de fin de grado, UNIANDES, Ibarra, Ecuador (2016)
7. Gutierrez, L., Trujillo, J.: La culpa es de la Vaca. Intermedia, Bogotá (2000)
8. Ineverycrea Homepage. Acciones que todo docente debe realizar para fomentar la educación inclusive. https://ineverycrea.mx/comunidad/ineverycreamexico/recurso/acciones-que-todo-docente-debe-realizar-para/43553a2d-8ee7-4855-b879-4d1f567a5c1a. Accessed 09 Mar 2020
9. Kortabitarte, A., Gillate, I., Luna, U., Ibáñez-Etxeberria, A.: Las aplicaciones móviles como recursos de apoyo en el aula de Ciencias Sociales: estudio exploratorio con la app Architecture gothique/romane en Educación Secundaria. http://www.ub.edu/histodidactica/images/documentos/pdf/alex.pdf. Accessed 19 Mar 2020
10. Márquez, A.: Aplicación móvil en Android para lectura de libros digitales en lenguaje de señas de personas sordomudas iletradas. trabajo de fin de grado, UNIANDES, Ibarra, Ecuador (2017)
11. Méndez, L., Moreno, R., Ripa, C.: Adaptaciones curriculares en educación infantil. Narcea, Madrid (2006)
12. Ministerio de Educación del Ecuador Homepage. Escuelas inclusivas. https://educacion.gob.ec/escuelas-inclusivas/. Accessed 09 Mar 2020
13. Orientacionandujar.es Homepage. Rúbrica para evaluar la comprensión lectora. https://www.orientacionandujar.es/2019/05/19/rubrica-para-evaluar-la-comprension-lectora/. Accessed 19 Mar 2020
14. United Nations Educational Homepage. Scientific and Cultural Organization: Inclusion in education. https://en.unesco.org/themes/inclusion-in-education. Accessed 09 Mar 2020
15. Universidad de Granada Homepage. Métodos de análisis multivariante: análisis cluster. http://wpd.ugr.es/~bioestad/guia-spss/practica-8/. Accessed 09 Mar 2020

Web Application for the Management of Reagents, Based on MEAN Stack Tools

Marco V. Guachimboza-Villalva[1(✉)], Víctor H. Guachimbosa-Villalba[2], Héctor Alberto Luzuriaga Jaramillo[1], and Javier Sánchez-Guerrero[3]

[1] Facultad de Contabilidad u Auditoría, Universidad Técnica de Ambato, Av. Los Chasquis entre Payamino y Guayllabamba, Ambato, Ecuador
{marcovguachimboza, ha.luzuriaga}@uta.edu.ec

[2] Facultad de Ingeniería en Sistemas, Electrónica e Industrial, Universidad Técnica de Ambato, Av. Los Chasquis entre Payamino y Guayllabamba, Ambato, Ecuador
victorguachimbosa@uta.edu.ec

[3] Facultad de Ciencias Humanas y de la Educación, Universidad Técnica de Ambato, Av. Los Chasquis entre Payamino y Guayllabamba, Ambato, Ecuador
jsanchez@uta.edu.ec

Abstract. This paper presents the results of the implementation of a web application for the management of reagents of the Faculty of Accounting and Auditing at the Technical University of Ambato, the objective of this application is to provide teachers and staff in charge of academic management a tool for the creation, control, validation and monitoring of reagents, as well as the use of these reagents in the creation of a complex examination instrument as a modality of degree. The web development methodology is based on JavaScript that involves tools that are linked in the MEAN stack (Mongo-Express-Angular-Node), an analysis and use of these 4 technologies will be made together for the creation of a distributed application using the same JavaScript language in all its phases and layers. The results will reflect the implementation of a functional web application with a simple inter-face, which highlights the reliability, security and availability of information. Finally, the conclusions, limitations and future works of the present project are presented.

Keywords: Web application · Academic management · Reagent management · Stack MEAN · JavaScript

1 Introduction

In the demanding and changing 21st century, scientific, economic, social and technological changes have developed in an environment of globalization with the need to make reforms and updates to the management and administration systems of organizations. In the case of higher education institutions, they cannot be oblivious to the different changes that have taken place in science, communication and technologies that provide a significant tool for the management of their processes.

M. Botto-Tobar et al. (Eds.): CI3 2020, AISC 1277, pp. 319–333, 2021.
https://doi.org/10.1007/978-3-030-60467-7_27

This article details fundamentally the design and application of a system for the management of the information of reagents that are developed for the process of complex examination as a viable alternative within the processes of degree in the Faculty of Accounting and Audit of the Technical University of Ambato in Ecuador. The system is based on open and free code technologies for its later adaptation in other faculties of the university that requires a control and management of the reagents, giving importance to the reliability of the data, as well as real time responses and the use of asynchronous transactions.

The problem addresses the need to have an application that helps teachers to create, control and monitor the reagents produced in each academic period. Subsequently, these questions must go through a validation or peer review phase, as a precedent to the creation of an evaluation instrument that serves for the complex examination process, which is one of the current degree modalities presented by the Academic Regime Regulations.

The objective of the project is the implementation of an application in a web environment, with an architecture that allows communication and presentation of data to the user automatically and also the system facilitates the maintenance, scalability and flexibility of the growth of information, which in turn will allow any academic unit to develop and customize modules that suit their needs.

This paper also explores the use of the MEAN web development toolkit (named for its constituent parts: MongoDb, Express.js, Angu-lar.js, and Node.js) to develop web-based back-end services and front-end user interfaces.

2 State of the Art

2.1 Reagents Management

According to article 21 of the Academic Regime Regulation, a modality of qualification is the "Complex Examination", the same that must be articulated to the profile of exit of the career, and to the demonstration of the learning results or competences of the students of the careers of degree or postgraduate programs [1].

Also within [2], Art. 22 establishes that the degree exam consists of a complex theoretical or theoretical-practical examination through the application of one hundred (100) randomly selected multiple-choice reagents, which will last between one and three hours.

Reagent
In implication of the previous regulation, [3] mentions us that: "A reagent is the approach of a situation that requires solution, that proposes actions or provokes reactions translated into answers, whose degree of success can be an indicator of the learning result obtained or of the knowledge retained by the evaluated".

Management
Management is the process by which a variety of essential resources are managed in order to achieve the objectives of the organization.

In the case of the Faculty of Accounting and Auditing, the management of reagents has to do with the process where the Titling Unit of each career, is responsible for controlling, administering and coordinating the application of the reagents. This is what is established in several literals of Art. 6 of the Regulation for Obtaining the Third Level Degree at the Technical University of Ambato [2].

2.2 MEAN Stack

In web development, traditionally has been used technologies that are very popular, with extensive features and years of development such as JAVA servlets, PHP (Hypertext Preprocessor), ASP.NET (Active Server Pages). Technologies that have their own weaknesses in relation to application performance [4]. However, in recent years a new set of tools has appeared that form a Stack for the development of web applications, known as MEAN Stack or simply MEAN.

MEAN stack is a powerful full-stack JavaScript solution that includes four main components: MongoDB as the database, Express as the web server framework, AngularJS as the web client framework and Node.js as the server form-silver [5]. MEAN provides client and server components for interactive web applications.

MongoDB

MongoDB, is an open-source NoSQL database whose nomenclature comes from the word “humongous” (gigantic), this database stores documents in collections, in format derived from JavaScript Object Notation (JSON). NoSQL data bases are less rigid and structured than relational databases, this lack of structure often simplifies prototyping and facilitates development, making these databases tend to be faster [6].

MongoDB allows to work with many technologies and to be coupled to the development of modern applications due to its ease of working with data and provides:

- Easy to work with data in a natural and intuitive way, while providing ACID (Atomicity, Consistency, Insulation and Durability) guarantees to ensure data integrity.
- Fast that achieves great performance without much work.
- Flexibility to adapt and make changes quickly.
- Versatile as it supports a wide variety of data and queries.

Express.js

Express is a flexible and lightweight platform for creating and organizing server- side web applications using NodeJS. Express allows things to run faster as it immerses itself in the code of your application and hides much of Node’s internal workings [7]. It then provides a thin layer of basic web application features, which do not hide the features of Node.js that we know and that make it so easy to develop our applications. Express allows to instantiate web servers and receive HTTP requests in a simple way, it is also possible to have a set of directories with a standard architecture, and organize the file paths for the application views.

Angular.js

Angular.js (or Angular), is an open source MVC (Model-View-Controller) framework, maintained by Google and used for client-side development in single-page SPA web applications [8].

For [6], a SPA (Single-Page Applications), represent the latest evolution in web design, as it recovers all the necessary code in a single page load or dynamically loads as needed. When users interact with the application, data is sent and received from the server via Ajax requests. The user experience is then smoother than reloading pages, which will make it look like a native application.

Node.js

Node.js is an open source environment based on the JavaScript programming language to be applied in the back-end and based on the V8 JavaScript engine of Google Chrome browser and event oriented, non-blocking, which makes it very fast to create web servers and use in real time [9].

If you need live interaction and real-time results, Node.js is an excellent tool that has a very good compatibility with data delivery to and from the web server [10]. The clients communicate with the server, and it returns information to them, which can be from a web page, data stored in a database, message transmitted by another client or anything else.

3 Methodology

3.1 Requirements Analysis

As a first phase, a requirements analysis was carried out with the objective of creating a friendly, agile, safe application that allows for the elaboration, control, validation and application of the reagents for the complex examination process as a modality of degree in the faculty.

Then, based on the General Guidelines for the application of the Instructions for the Degree Modalities of the Faculties of the Technical University of Ambato [11] issued by the University Council, the format and characteristic of each of the types of reagents were investigated, especially point 18, which indicates how the complex exam should be elaborated, clarifying point 18.2 as follows:

The reagents will be multiple choice with a single correct answer, each question will consist of a statement and four options (A, B, C, D). The types of reagents may be:

a) Direct Questionnaire Reagents
b) Ordering Reagents
c) Element Choice Reagents
d) Column Ratio Reagents
e) Multi-reactive

Subsequently, an analysis is made of the set of tools that will allow the implementation of the technological architecture based on the MEAN stack of JavaScript.

3.2 Technology Used

With the rise of JavaScript as a programming language, a new trend is emerging in application development from start to finish. The so-called MEAN stack with end-to-end development tends to use JS in all parts of a current web application: Frontend, Backend and Database.

Bretz et al. [6], attributed that using these Javascript-based technologies, it is possible to create a production-level application using only HTML and CSS. The combination of MongoDB, Express, AngularJS and Node.js (MEAN) has become so popular and were created with the goal of providing scalable and high performance applications.

Essentially, the MEAN architecture works as follows:

1. We developed the front-end (client side) with AngularJS, the user accesses an application through the browser,
2. Angular.js sends the request through API REST (Post, Put, Get and Delete) for Express.js in JSON format,
3. Express converts this data in BSON format and sends it to MongoDB, this API, built with NodeJS and the express framework makes a CRUD (Create, Read, Update and Delete) to the MongoDB database.
4. MongoDB returns BSON data to Express.js,
5. The Express.js converts the data to JSON and returns to the Angular.js, and
6. Angular.js returns the data to the user.

The whole procedure happens very fast and dynamically. Figure 1 presents a MEAN architecture approach.

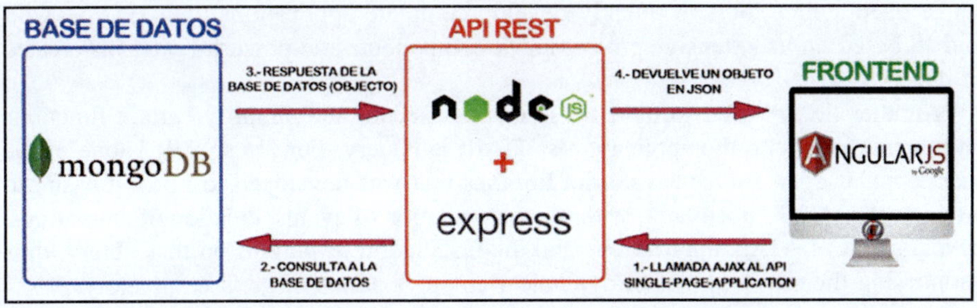

Fig. 1. MEAN JavaScript architecture [12]

The MEAN architecture works with the interaction of each of its components. Angular.js is not a library but a JavaScript framework that transforms the HTML extension into a more expressive and readable format [13]. Moreover, it has allowed us to produce a highly semantic HTML, that is, when it reads it clearly understands what it does or what each thing is used for.

The implementation of AngularJS in the front-end is made up of controllers, and each of them covers part of the application. Angular allows you to manage what is known as SPA (Single Page Applications), and one of its main components is routing, which defines if a route is valid, plus the view and controller to use for that route. The element that provides the connection between the controller and the view is the "$scope". In Angular there are also directives that provide us with an encapsulation of reusable code blocks that are used within HTML. There are many useful directives already included in this framework, however it was possible to create our own directives. With a NodeJS server, we got a scalable application that was able to re-mediate the performance problems due to the high demand of users, since it is an implementation in which the number of users can grow with time. Besides that it can evolve, offering more and more services to its users, this tool has an execution oriented to events instead of multithreading, it supports websockets to report events in real time using SocketIO. Express, framework for NodeJS that allowed to develop a web application with high performance speed, includes what is generally expected from frameworks: working with routes, models, views. But, in addition, Express also covers one of the needs of web development, which is to implement the mechanisms of HTTP. Express greatly facilitates the creation of REST web services.

Finally, the incorporation of MongoDB as a NoSQL database works with high volume data with high performance. The loading of data and interaction with our MongoDB database makes our web application very agile thanks to the use of objects in JSON for storage.

3.3 Other Bookstores

Interface interactions were developed with AngularJs, and Bootstrap, one of the world's most popular front-end component libraries for the creation of responsive websites, was used for the visualization and design of the application. Bootstrap, is a set of open source tools to quickly develop complete applications with HTML, CSS and JS based on its extensive pre-compiled components and powerful plug-ins created in jQuery [14].

Another library used to develop scripts to execute and simplify certain functions and interactions with the application's HTML is jQuery. For [15, p. 19] is one of the most popular open source JavaScript libraries that was developed to make it easier to navigate the HTML document, with the manipulation of events, creation of animations, development of AJAX applications and the installation of plugins on that library, thus simplifying the development of dynamic web applications of great complexity.

3.4 Front-End Design

Once the necessary requirements of the project were obtained and the tools used were analyzed, its design was proceeded with the help of some modeling tools. As a first instance, the composite diagram of the front-end was designed.

The design of the front-end allows concentrating all the development work in an architecture based on the AngularJS framework, so that each entity has a set of components and services. The components are blocks of the user interface with a specific functionality that is to say sections of the front-end. In them the business logic is written, they compose each other to form the view of the app. On the other hand, the services contain all the code that connects the components for data exchange, which exe- cute through HTTP calls to the API developed on the server side.

Figure 2 illustrates the component diagram where each component represents the resources of the front-end related to the application modules, which are linked to the main requests made from each of its services.

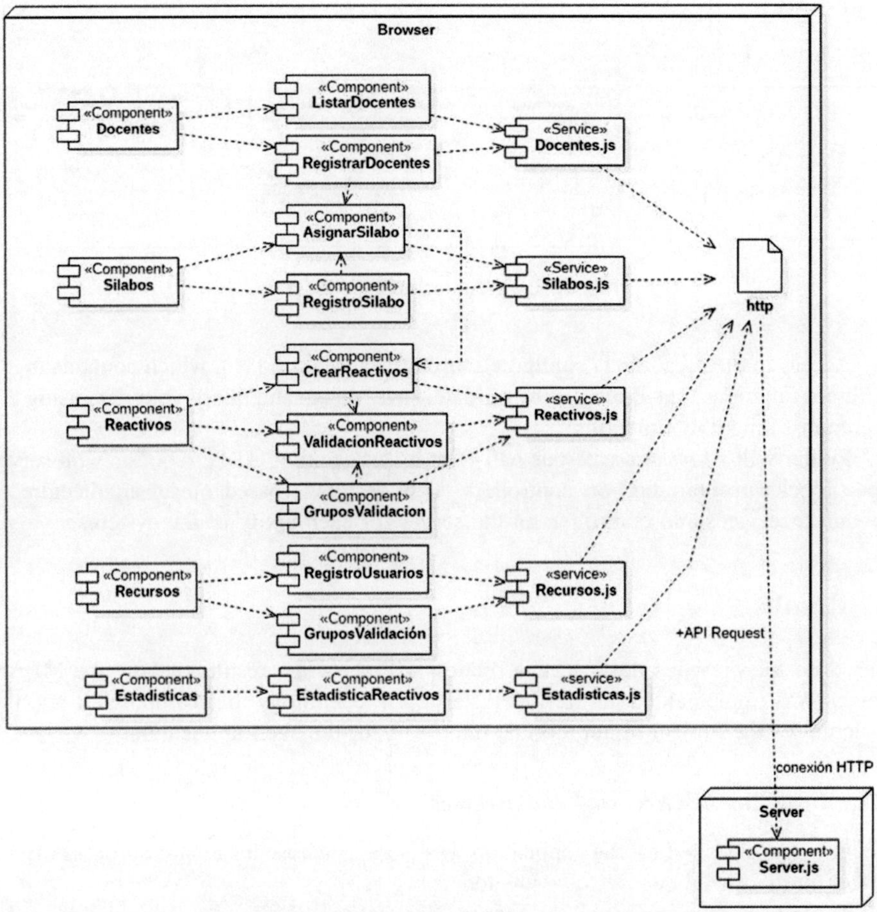

Fig. 2. Front-end component diagram.

As a result of the above diagram, the back-end structure was designed. The server developed on a NodeJS platform with ExpressJS, configures the access to the database, resources and administration of the only execution thread, to meet the HTTP requests of a large number of users at the same time. This approach provides a server with the ability to handle a large number of requests and connections with maximum efficiency. The Router allows to attend the requests declared through functions according to the HTTP request (GET, POST, PUT, etc.) and thus respond to the user with the resource that comes from the server. Figure 3 shows the division of components and how the server handles HTTP calls.

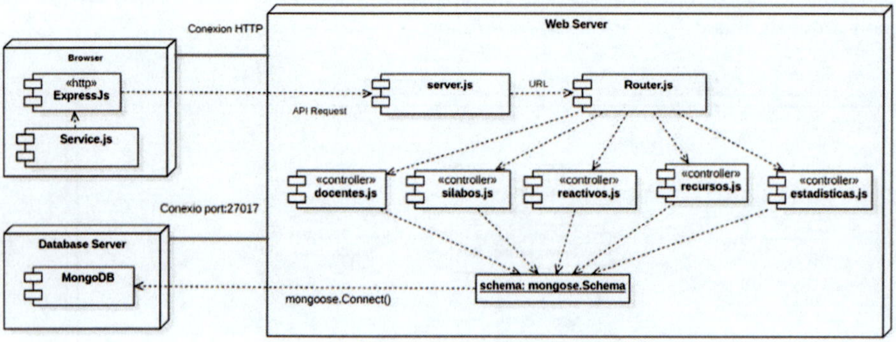

Fig. 3. Back-end component diagram

Access to the database is configured through Mongoose [16], which contains many different functions that allow you to validate, save, delete and query your data using the common MongoDB functions.

Router will allow to create our API-Rest by associating HTTP requests with server code blocks programmed on controllers. This design is based on an architecture of modules, services and controllers in the server for each entity of the system.

4 Results

Based on the previous designs, this section will show the results of how the MEAN Satack was implemented, as well as install each technology that composes it and the structure of the project in appropriate interfaces for the Web application.

4.1 Installing MEAN Application Tools

The first tool installed for the application was Node.js whose installation process can be found on its official site: http://www.nodejs.org [17].

The next step was to install the MongoDB database. The MongoDB installer can be found on its official website [18]: https://www.mongodb.com.

In the next step, all the necessary packages for the structure of a MEAN application were installed and a NPM (Node Package Manager) was used. NPM is a package manager for Javascript, which is used to install and manage versions of packages and open source libraries in Node.js.

To install all packages or modules, you must first select a directory of your preference and then accessed via terminal. All installed technologies will be contained in a folder called node_modules in the chosen directory. Table 1 describes the installation sequence of the different packages for the MEAN structure.

Table 1. Commands for package installation

Command Line	Task to be executed
1. npm install express	Express.js package installation
2. npm install mongodb	Installing the MongoDB package
3. npm install mongoose	Installation of the mongoose package, which will be used to access MongoDB.
4. npm install angular	Installing the Angular.js package

Since the application needs several packages, a better way to manage the dependencies of these packages was necessary. For this purpose, NPM allows you to create a configuration file called package.json in the root folder of your application. In the package.json file, you can define various meta-data properties of your application, including properties such as name, version, author of your application, application dependencies and many others [5]. The command line for creating the package.json file is as follows: npm init.

After entering npm init command line, the console interacts with the user to fulfill some application requirements, the creation process is shown in the Fig. 4.

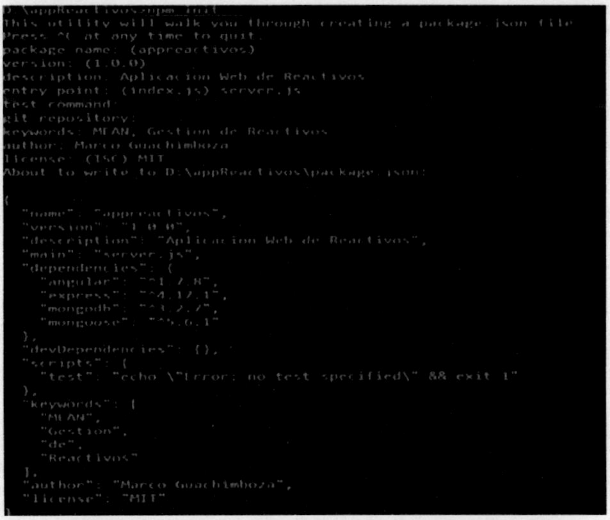

Fig. 4. Creating the package.json dependencies file

After creating the package.json file, the application dependencies were installed using the following command: npm install.

4.2 WebServer Configuration with Node.js and Express.json

The creation of a web server with Node.js is the essential part to structure and configure the application. In this way, Fig. 5 presents the necessary code to configure the webserver.

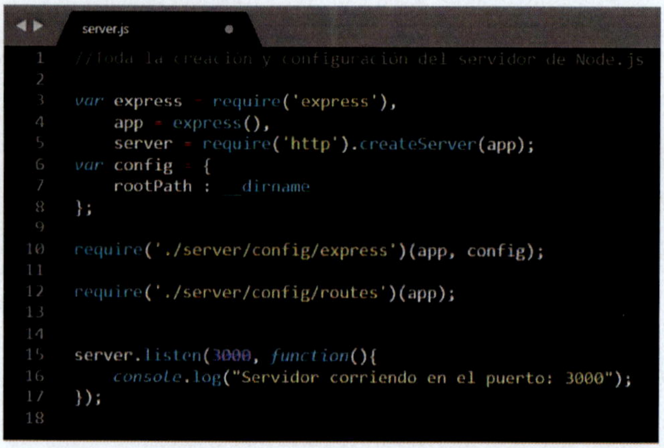

Fig. 5. WebServer creation with Node.js and Express.js

The first two lines require the Express module and create a new Express application object. Then, the function createServer instance the variable http and creates the web server, also configuring the paths. The listen () method is used to tell the Express application to listen to port 3000. The webserver can be executed through the npm start command line. Figure 6 shows the return that the terminal receives when the application is executed.

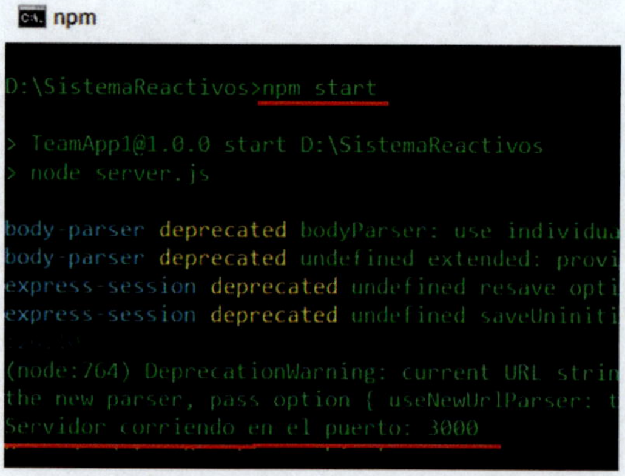

Fig. 6. Running the application with Node.js

The application can now be accessed through the address of the machine where it was installed and the configured port.

4.3 Application Architecture

The definition of the application architecture is very important. In our application the structure was created based on the MVC pattern in which specific folders can be created to place JavaScript files in a certain logical order. Files that can be accessed by users must be in a public folder. The files that belong to the control of the application must be in a directory that cannot be accessed directly by the browser (Fig. 7).

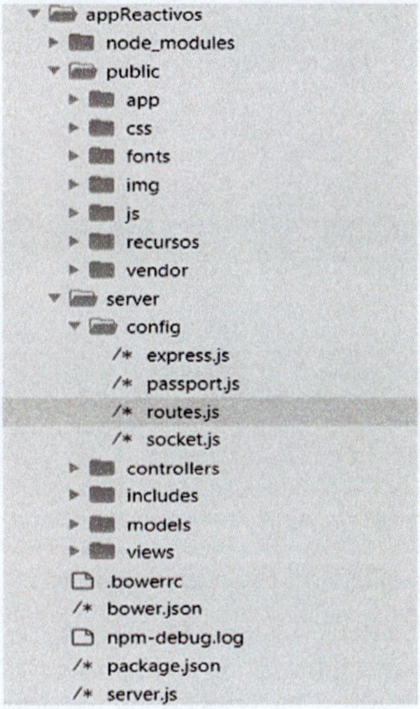

Fig. 7. Application architecture

4.4 Application Interfaces

As first interface we have the login for users. There are several levels of user, a user administrator, user teacher, user validator and user coordinator, the first one has access to all the modules and the rest of users will have access to parts of the module according to their assigned roles (Fig. 8).

As another example we will consider the presentation of the main interface. Figure 9 shows the navigation menu of the different modules and components of the application.

The Teaching Module allows you to create the teachers who will teach the different chairs, where it is mandatory to enter the identity card, names and surnames. This module also presents a detailed list of teachers.

Fig. 8. User login interface

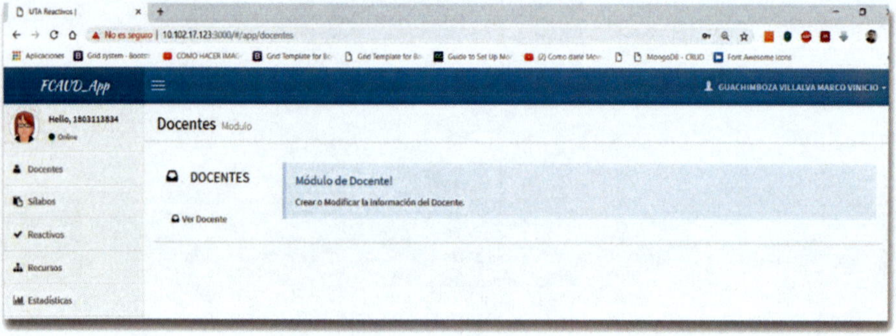

Fig. 9. Main navigation panel

Module Sílabos, will enter information of each one of the subjects (Sílabo) by career, to later make an assignment by teacher and semester or academic period.

Module Reagents, allows teachers to create one or several reagents (questions) per syllable, according to the guidelines defined by law and the higher education institution (direct questionnaire, list of columns). In this section, the process of validation of reagents is also carried out, as well as the printed instrument for the complex exam as a degree modality.

Module Resources, allows the registration of the different users and their respective roles, also assigning the validation groups for the respective process.

Module Statistics, generates statistics and reports of the creation, validation and control of activities related to the reagents.

Figure 10 specifies another important interface of the web application, the creation of reagents by teachers, who must first select the Syllable to which the reagent belongs.

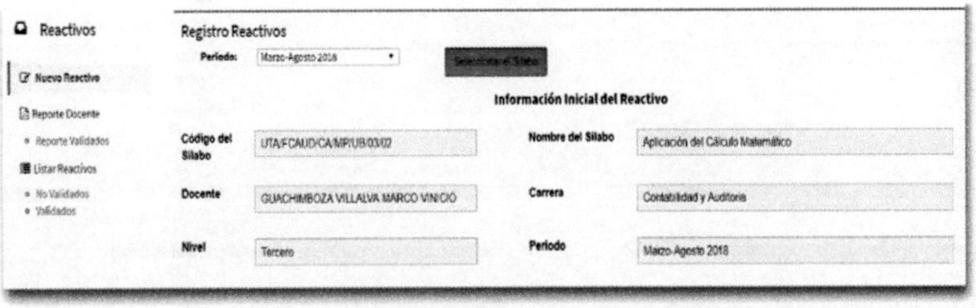

Fig. 10. Reagent creation, Syllable selection

With the information from the Syllable, the new reagent is generated and configured (Fig. 11).

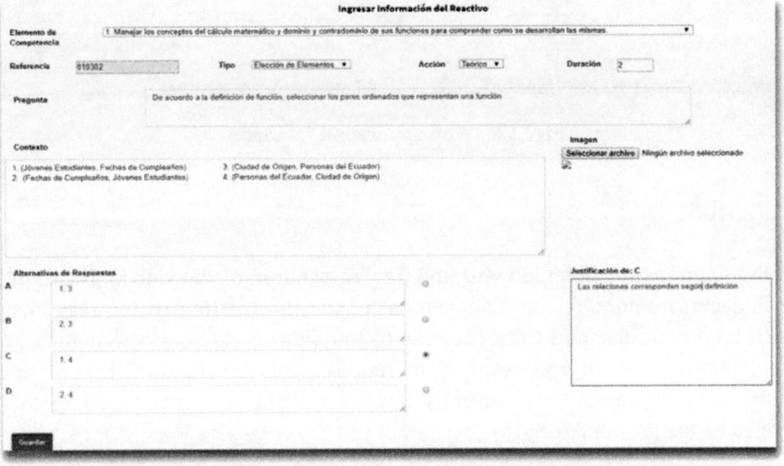

Fig. 11. Creation of reagent, entry of new reagent information

Among others of the most outstanding interfaces of the Project is highlighted in Fig. 12 and that represent some modules of the developed application.

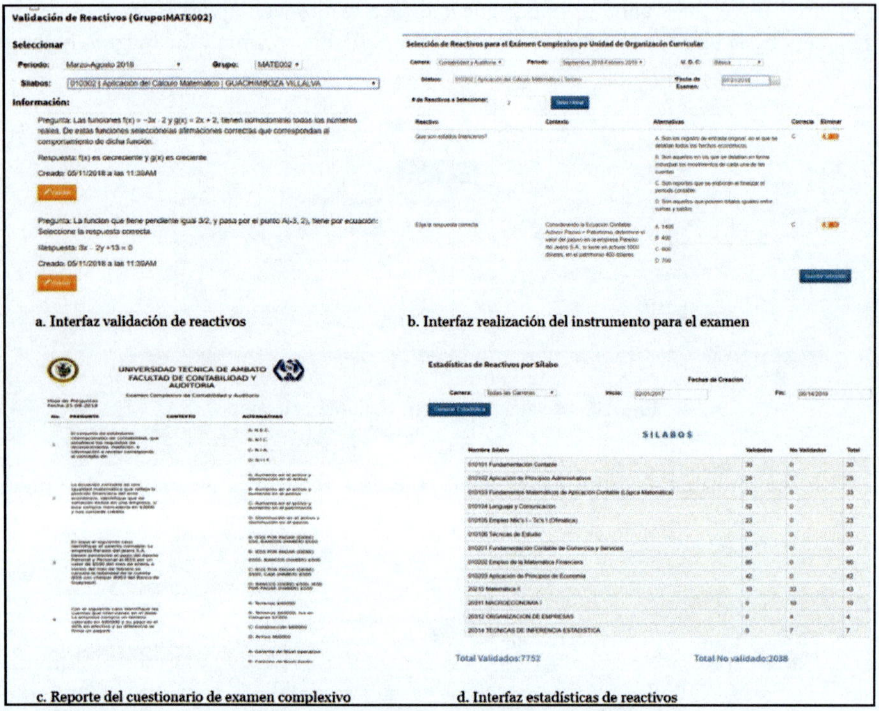

Fig. 12. Web application interfaces

5 Conclusions

One of the main results obtained with the implementation of the web application, is the remarkable performance in the management of the information of the reagents, since there is reliable, precise and timely access to the data.

The ability to record and respond to requests for information through the web application has improved exponentially, optimizing the resources used in the management of reagents.

The system is based on open and free code technologies that allow the production of high quality software with a minimum investment of economic resources for the payment of patents for proprietary software.

An application developed using the technologies that compose the MEAN Stack generates better productivity and performance results, since it is notorious that these technologies can provide robust and large-scale applications, being a viable solution to support the growing number of users and the scalability of data.

A future research derived from the present work is the application at the level of the 10 academic units of the university and that is an open service for all types of examinations.

References

1. Reglamento de Régimen Académico, Consejo de Educación Superior. Gaceta Oficial del Consejo de Educación Superior, Quito (2013)
2. Reglamento para la Obtención del Título de Tercer Nivel, de Grado en la Universidad Técnica de Ambato, Consejo Universitario, Ambato: RESOLUCION: 1967-CU-P-2018 (2018)
3. Altamirano, J.: El Examen Complexivo: una modalidad de titulación, 1ra. Edición ed. Comunicación Solutions, Quito, Ecuador (2015)
4. Nirgudkar, N., Singh, P.: The MEAN stack. Int. Res. J. Eng. Technol. **04**(05), 3237–3239 (2017)
5. Haviv, A.: MEAN Web Development: Master Real-Time Web Application Development Using a Mean Combination of MongoDB, Express, AngularJS and NodeJS. Packt Publishing, Birmingham (2014)
6. Bretz, A., Ihrig, C.: Full Stack JavaScript development with MEAN. SitePoint, Australia (2014)
7. Sevilleje, C., Holly, L.: MEAN Machine a Beginner's Practical Guide to the JavaScript Stack. S.L: Leanpub (2015)
8. Poulter, A., Johnston, S., Cox, S.: Using the MEAN stack to implement a RESTful service for an internet of things application. In: Proceedings IEEE World Forum on Internet of Things Conference (2015)
9. Dura, M.: Aplicación web para el seguimiento online de lecturas basadas en el stack MEAN, Valencia (2015)
10. Bangare, S.L., Gupta, S., Dalal, M., Inamdar, A.: Using Node.Js to build high speed and scalable backend database server. Int. J. Res. Advent Technol., 61–64 (2016)
11. Lineamientos Generales para la aplicación de los Instructivos de las Modalidades de Titulación de las Facultades de la Universidad Técnica de Amba, Consejo Universitario, Ambato: Resolución: 0888-CU-P-2015 (2015)
12. Moya, R.: MEAN (Mongo-Express-Angular-Node) Ejemplo de Aplicación Web (Parte II), 24 July 2014. https://jarroba.com/mean-mongo-express-angular-node-ejemplo-de-aplicacion-web-parte-ii/
13. Jain, N., Mangal, P., Mehta, D.: AngularJS: a modern MVC framework in JavaScript. J. Glob. Res. Comput. Sci. **5**(12), 17–23 (2014)
14. Bootstrap Team. Bootstrap - The most popular HTML, CSS, and JS library in the world (2019). https://getbootstrap.com/
15. Castillo, A.A.: Curso de Programación Web: JavaScript, Ajax y jQuery, 2 edn. IT Campus (2017)
16. Learnboost. Mongoose: Schemas (2019). https://mongoosejs.com/docs/guides.html
17. Node.js Foundation. NodeJs (2019). https://nodejs.org
18. MongoDB, Inc.: The database for modern applications (2019). https://www.mongodb.com

FSplines: A Software for Linear Stability Analysis of Thin-Walled Structures, Version 2.0

Ángel Chicaiza[1(✉)], Luis Prola[2], Pedro Gala[3], Cristhian Chicaiza[4], and Marcia Ortiz[5]

[1] Facultad de Ciencias Socio Ambientales, Universidad Regional Amazónica Ikiam, Tena EC150101, Ecuador
angelchicaizal@gmail.com
[2] School of Technology and Management, Polytechnic Institute of Leiria, 2411-901 Leiria, Portugal
[3] School of Technology and Management, INESC Coimbra, Polytechnic Institute of Leiria, 2411-901 Leiria, Portugal
[4] Sede Académica El Pangui, Universidad Estatal Amazónica, UEA, Puyo EC160101, Ecuador
[5] Ministry of Education of Ecuador, Archidona, Ecuador

Abstract. *FSplines* is a (geometrically) linear stability analysis tool of thin-walled structures with open section (useful for cold-formed steel profiles), that enables obtaining the bifurcation stresses (critical stress, load and moment) and the respective buckling modes by the Finite Strip Method (*FSM*).

The Finite Strip Method: (i) allows analyzing prismatic steel members (commercial structural profiles), (ii) is an alternative to the Finite Element Method (*FEM*), and (iii) has some important advantages over *FEM*. In the present article, two variants of the *FSM* are presented: (i) the Semi-Analytical Finite Strip Method (*SAFSM*), where use is made of trigonometric functions and (ii) the Splines Finite Strip Method (*SFSM*), employing spline functions. The *SAFSM* has the advantage of being less time consuming. Its main restriction is the fact that it only allows modelling simple supported members (pinned restrained). The *SFSM* most important advantage is the ability to model members with all kinds of boundary conditions. This method is, however, more time consuming.

It is worth noting that the bifurcation analysis, performed by the computer application FSplines, is required for the design of cold formed members according to the specifications of international standards.

FSplines 2.0 is the second version of the computer application here presented. In this second improved version more cross-sections are available, and more section properties are presented.

Keywords: Cold formed structural profiles · Open cross-section thin walled structures · Buckling modes · FSplines 2.0 · Bifurcation stresses · Linear stability analysis · Finite Strip Method

M. Botto-Tobar et al. (Eds.): CI3 2020, AISC 1277, pp. 334–349, 2021.
https://doi.org/10.1007/978-3-030-60467-7_28

1 Introduction

FSplines version 2.0 is a software that mainly performs the computation of bifurcation stresses and buckling modes of prismatic thin-walled members by the *SAFSM* and *SFSM* and that incorporates new implementations.

FSplines 2.0 eases the amount of work that designers and civil engineering researchers must do in getting bifurcation tensions and the respective buckling modes of open cross-section thin-walled structural members with different boundary and loading conditions. This version of the software improves the previous one [1] because more types of cross-sections and more section properties are implemented (taking as reference the work of [2]), such as: St. Venant torsion constant, the principal moments of inertia, the distance from centroid to shear center and warping constant.

FSM formulation used by FSplines version 2.0, follows Cheung [3], being a sequel of the work developed by Prola [4] and the version 1 that was developed by Chicaiza [5].

In the second section of this paper, the problems to be solved by *FSplines* 2.0 in the context of linear stability analysis are presented. The third section is devoted to presenting the *FSM* concepts. The fourth section is used to present the *FSplines* version 2.0 basic structure and in the fifth section the software's interface is also presented. In the sixth section, the *FSplines* 2.0 performance is compared with two other informatic applications - *CUFSM* [6] and *NASTRAN In-CAD* [7]. Several numerical examples are used with that purpose. The seventh section is devoted to presenting general conclusions.

2 Elastic Bifurcation Stresses of Thin-Walled Members

The present paper is focused in the computation of bifurcation stresses (critical stresses, loads or moments) and the respective buckling modes (local, distortional and global) [4] for thin-walled members with open cross-sections (generally cold formed profiles), using the Semi-Analytical Finite Strip Method (*SAFSM*) and the Splines Finite Strip Method (*SFSM*).

The critical stress is the lowest bifurcation stress value of all possible instability buckling modes of on structural element (column, bema or beam-column) [8]. Any of the buckling modes generate excessive deformations and consequently leads to the collapse of the structure [9]. In addition, the calculation of elastic bifurcation stresses is a necessary requirement for the modern design of cold-formed structures by the Direct Strength Method (*DSM*) [10]. The different buckling modes (Fig. 1), that can take place in thin-walled and open-section structural members are the following:

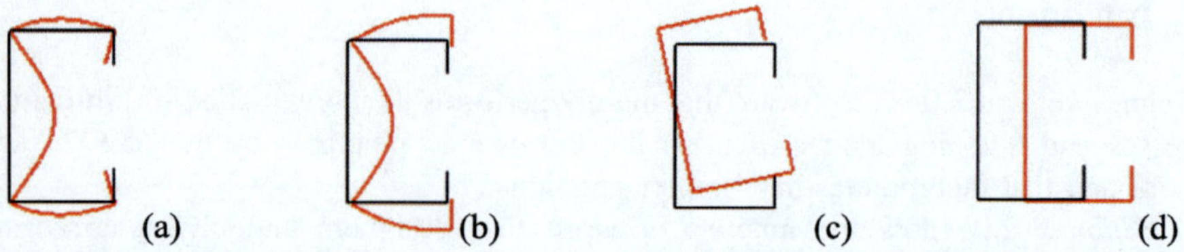

Fig. 1. Buckling modes in a column C subjected to compression: (a) Local, (b) Distortional, (c) Flexo-Torsion (d) Flexural [5].

(i) Local mode (Fig. 1a): the plates suffer deformations, while the lines of bends of the element cross-section remain in the same position. In other words, the cross-sections do not suffer any rigid-body motion [4, 9, 11].

(ii) Distortional mode (Fig. 1b): the cross section experiments a distortional deformation. In this buckling mode some of the cross-section fold lines may undergo a displacement from its original position, unlike what happens in the local mode [4, 12].

(iii) Global mode: it includes the bending (classical Euler's stability problem), flexural-torsional and the lateral-torsional mode. The cross-section moves transversely to the longitudinal direction of the member and/or rotates retaining its original shape [4, 12]. The cross-section suffers a rigid-body translation as shown in Fig. 1d. In the flexotorsional mode (Fig. 1c), the structural members simultaneously experiment a rigidbody translation and a rigid-body rotation. The original cross-sectional shape is however kept [11]. Figure 2 depicts 3D configurations of the different buckling modes for a column under compression.

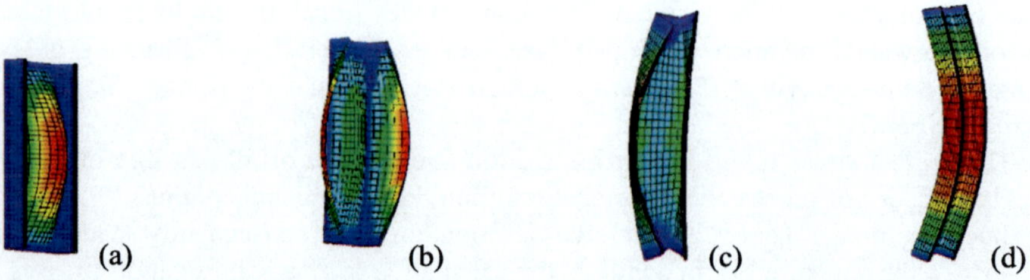

Fig. 2. Buckling modes in a C-type profile subjected to compression: a) Local, b) distortional, c) flexo-torsional, d) flexural.

According to Ádány and Schafer [9], when the *FSM* is used, the buckling modes can be identified more easily in comparison to other methods like *FEM*. This is because in the *FSM* the critical bifurcation stress is a function of the buckling length, i.e. the "signature curve" (i.e., the critical load vs length) depicted in Fig. 3a. Note that, in Fig. 3a, different colors are used to depict curves computed with distinct half-wave lengths (which can be obtained through the *SAFSM*). The black line represents the

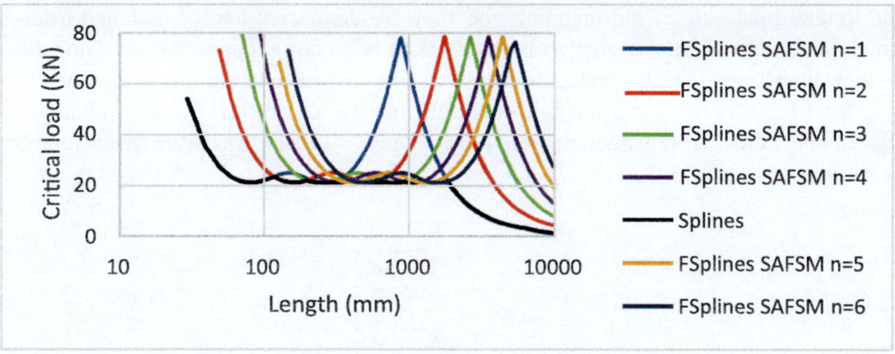

(a)

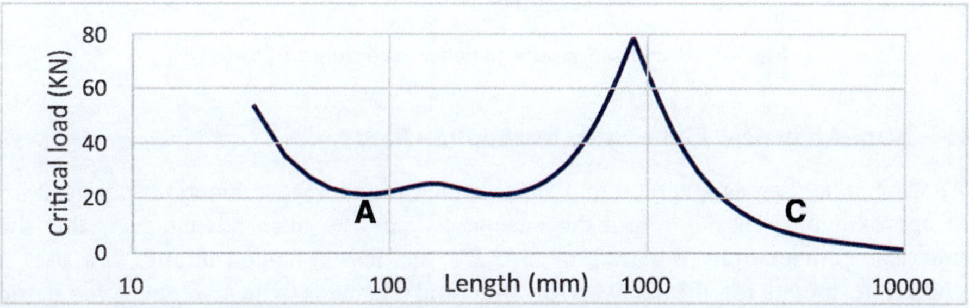

(b)

Fig. 3. Bifurcation loads of: a) the different buckling modes and b) a single half-wavelength VS Length of a C-type profile subjected to uniform compression.

minimum value for all half-wave lengths (which can be obtained through the *SAFSM*, where all the half-wave lengths are grouped; or directly by the *SFSM*).

In this paper, only curves corresponding to single half-wavelength, see Fig. 3b, are presented when *SAFSM* is being used because the others are curves of the same type obtained by its horizontal translations.

One can observe, in Fig. 3b, that the first minimum (point A) of the "signature curve" (for a single half-wavelength) is typically related to the local modes. The second minimum (point B) is typically associated with the distortional type of modes and, at last, the descending branch of the curve, for higher lengths, is associated to global modes.

3 Finite Strip Method

FSM is a useful tool for the analysis of structural problems, being a valid alternative to the *FEM* [13] and presenting important advantages relatively to the *FEM*, particularly for the analysis of prismatic elements.

FSM requires a discretization procedure in which a generic member is divided into several strips, that are parallel to each other. The strips widths are kept constant along the longitudinal axis of the member, and they are connected and constrained to each other. This means that the displacement fields of two contiguous strips are compatible in their interfaces, see [3, 4]. Regarding the kinematic of each strip of the member, it is worth noting that the superposition of the membrane and flexural (out of plane) displacement fields is considered, see [14]. Figure 4 depicts a discretized member according to the *FSM*.

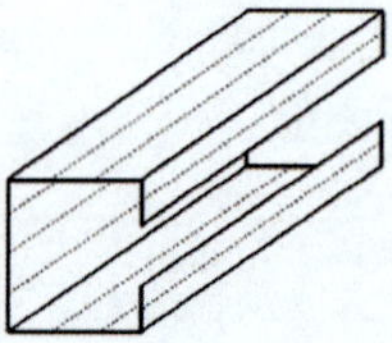

Fig. 4. Discretization of a member according to FSM [4]

3.1 Semi-Analytical Finite Strip Method (*SAFSM*)

SAFSM can be seen as a variant of *FSM* in which trigonometric functions are used for the approximation of the strip's displacement fields. Its main advantage is that the numerical computations required by *SAFSM* are less time-consuming. It's use is however restricted for the analysis of structural elements with a regular geometric configuration (such as thin-walled prismatic profiles) with simple boundary and load conditions [3, 4].

3.2 Splines Finite Strip Method (*SFSM*)

SFSM can be seen as a variant of *FSM* in which splines functions are used for the approximation of the strip's displacement fields, see [15]. It is important to clarify that there are several types of splines functions, which are used according to the calculation requirement. This paper addresses the basic cubic 3 -Spline function, see [16], proposed by Cheung [17] and Fan [18].

The *SFSM* with $_3$-Splines functions maintain the transverse interpolation polynomials and replaces the longitudinal trigonometric functions [16] by a linear combination (summation) of basic cubic functions [10, 16] of equal length of section [15]. An advantage of *SFSM*, in comparison with *SAFSM*, is its ability to give better approximations to the displacement fields in the longitudinal direction [18]. Additionally, the *SFSM* allows considering more complex boundary conditions (for example, fixedfixed, fixed-free, etc.) and a wider range of loading conditions [6].

In *SFSM*, the finite strips have four degrees of freedom along each station (longitudinal division of the finite strip) denoted: "u", "v", "w" y "θ" [9]. For the in-plane behavior (displacements "u" and "v"), a 2D plane stress condition is assumed, while for the out-of-plane behavior (displacement "w" and "θ") [3, 4]. As usual, the displacement functions are obtained from the product of nodal displacements by the shape functions (both in the longitudinal and transversal direction) [10].

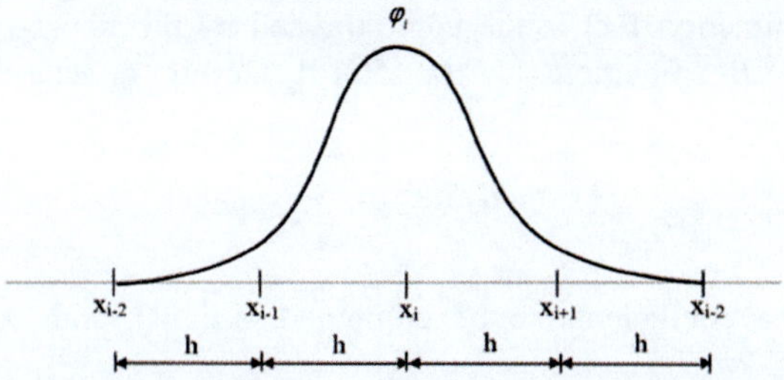

Fig. 5. Development of the basic cubic function B_3S in a generic domain [4]

In the longitudinal direction, the local cubic spline function (B_3S) is divided into four segments (represented by the letter "h") of equal length, but with different combinations of polynomials in each segment. The function mentioned is depicted in Fig. 5 [4, 10]. Equation (1) defines the spline function depicted in Fig. 5.

$$\varphi_i(x) = \begin{cases} 0 & x \leq x_i - 2h \\ (x - x_i + 2h)^3 & x_i - 2h \leq x \leq x_i - h \\ h^3 + 3\left\{h^2(x - x_i + h) + h(x - x_i + h)^2 - (x - x_i + h)^3\right\} & x_i - h \leq x \leq x_i \\ h^3 + 3\left\{h^2(x_i + h - x) + h(x_i + h - x)^2 - (x_i + h - x)^3\right\} & x_i \leq x \leq x_i + h \\ (x_i + 2h - x)^3 & x_i + h \leq x \leq x_i + 2h \\ 0 & x \geq x_i + 2h \end{cases} \tag{1}$$

An arbitrary function f(x), with domain $a \leq x \leq b$, can be approximated by using 3 functions as the one presented in Eq. (1) and depicted in Fig. 5. The after mentioned domain is subdivided into m parts with lengths h = (a − b)/m. Note that $(m + 3)$ "stations" are defined. The interpolation schemes with 3 functions introduce additional intervals at each end, that are added to the m subintervals, see Fig. 6 and [4].

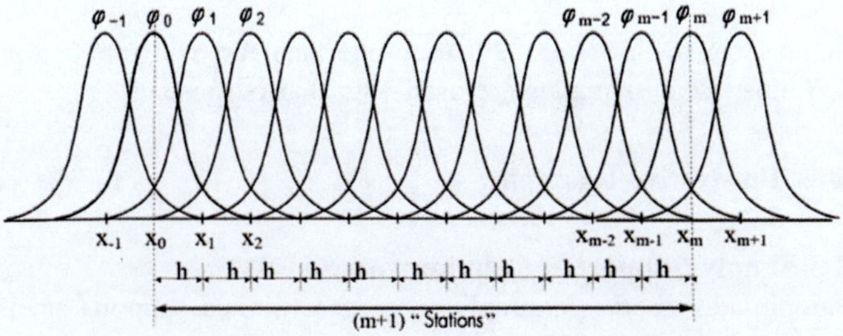

Fig. 6. Series of functions B_3 Splines combined linearly [4].

The arbitrary function f(x) is the approximated by the linear combination (summation) of (m + 3) B_3S functions [4, 10, 15, 19], see Fig. 6, according to Eq. (2),

$$f(x) \approx B_3S(x) = \sum_{i=-1}^{m+1} \alpha_i \varphi_i(x) \tag{2}$$

where α_i are some coefficients to be computed, see [4], and $\varphi_i(x)$ are the local B_3S functions.

4 Structure of FSplines Software

FSplines was programmed with Visual Basic and using the Fortran routines *ALESA* and *ALESPL* provided by Prola [4]. For the numerical computation of eigenvalues and eigenvectors (i) the sub-space method is applied when positive matrices are defined; and (ii) the "Lapack" subroutines are used when tensile and compression stresses take place simultaneously. OpenGL and OpenTK libraries were used for generating *FSplines* graphical outputs. Figure 7 shows the general flow-chart of *ALESA*, *ALESPL* [4] and *FSplines* [5].

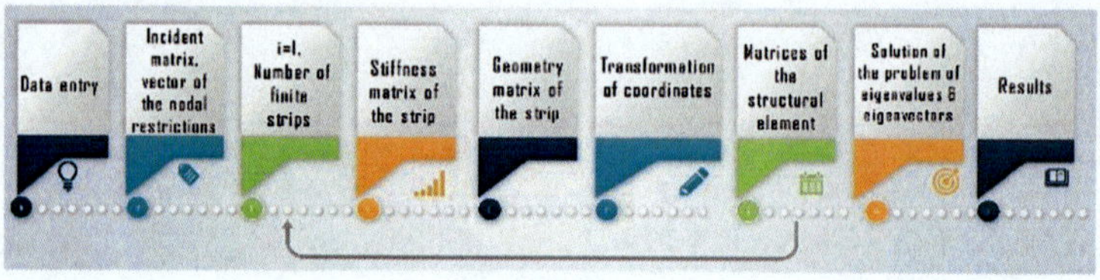

Fig. 7. General flow diagram of the *ALESA*, *ALESPL* [4] and *FSplines* [5].

5 Use of *FSplines* 2.0 Interface

This section presents the *FSplines* 2.0 user interface for data input. Some analysis results of two illustrative examples are used with that purpose.

5.1 FSplines Illustrative Examples

Analysis of a Simply Supported Column with SAFSM

The first example addressed is a simply supported (pinned support) steel C-column, under uniform compression. *SAFSM* was used more examples are detailed in the references [1, 5] and other examples will be incorporate in the *FSplines* 2.0 user's Manual. The following steps illustrate the use of the software's interface:

I. For the idealization of the problem, one has the following steps:
   1. Selection of the type of element to analyze (See Fig. 8a). The first option was selected since, in this case-study, the member is under uniform compression.

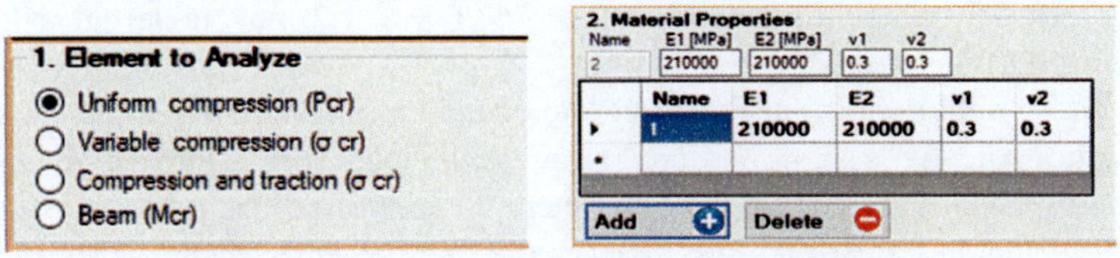

(a) (b)

Fig. 8. (a) Selection of the type of the element to analyze (in this case is a column used for the validation of results); (b) Entering the material properties

   2. Input of the properties of the material, see Fig. 8b. The software requires the input of the following parameters: elastic modulus in the two directions (E1 and E2) and the corresponding Poisson coefficients v1 and v2. For isotropic structural materials like steel one has E1 = E2 = 210000 (MPa); v1 = v2 = 0.3.

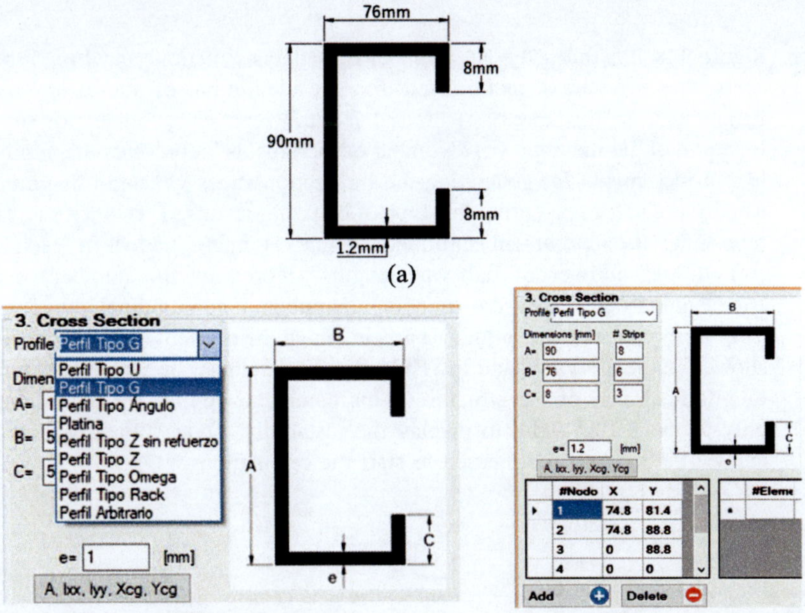

(a)

(b) (c)

Fig. 9. (a) Selection of profile type (cross-section) (b) Cross-section with dimensions, (c) Input of the dimensions in (*mm*) *FSplines 2.0*.

3. Selection of the type of profile (shape of the cross section - See Fig. 9b); input of the dimensions and the definition of the number of strips for each plate, see Fig. 9c. After entering this information, the "Add" button must be activated and the coordinates (with respect to the wall' mid surface) of the profile are then automatically computed. In this example, see Fig. 10, a C-profile ("Perfil Tipo G") with dimensions 90 × 76 × 8 × 1.2 mm (external edges, see Fig. 9a) is selected.
4. Stress field definition. For each one of the cross-section's plates, the software automatically assigns a unit compressive load (per width of the plate) as showed in Fig. 10a. On the other hand, in case of beams, it becomes necessary to generate a flexural stress field as the one presented in Fig. 10b. In the new version is possible to see the applied stresses (red line).

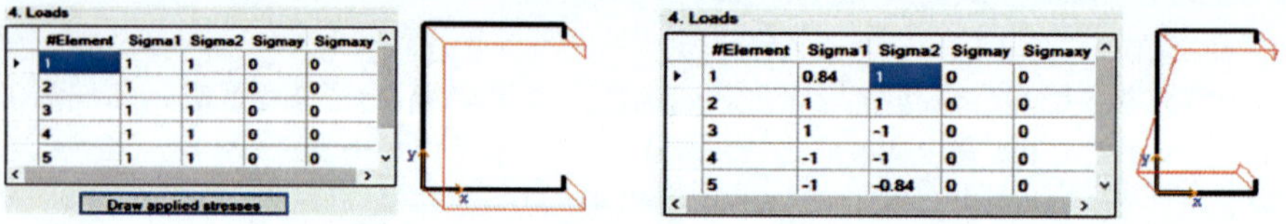

(a) (b)

Fig. 10. (a) Section destined to the entrance of stresses distribution in (a) columns and, (b) beams.

5. Figure 11a illustrates the selection of the analysis method and Fig. 11b presents the software's menu used for the definition of the discretization parameters. In this menu (i) "Initial length" represents the length (in millimeters) of the member, (ii) "Number of increments" represents the number of length increments for generating the bifurcation stress vs length diagram that usually is use for presenting this type of buckling results., (iii) "Increase value" represents the length (in millimeters) that is being added in each step, (iv) "Initial number of half-wavelengths" represents the number of half-wavelengths defined by the user, (v) "Number of longitudinal divisions" in *SAFSM* it represents the number parts in which the element will be divided and showed in the 3D-view; and in *SFSM*, it's the parameter of discretization in the longitudinal axis of the structure (This parameter applies for the analysis – only to the *SFSM* - and to display the instability modes). Finally, the "Calculate" button must be pressed to start the calculations.

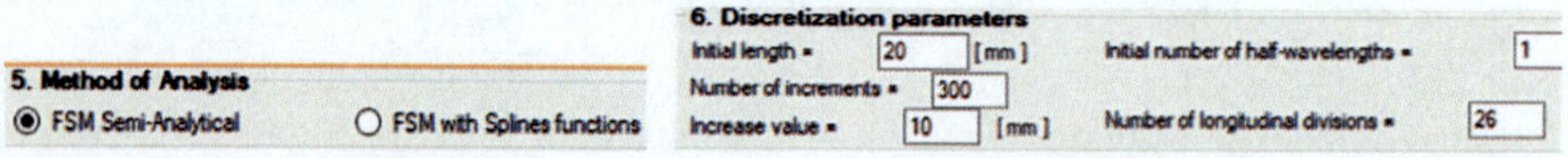

(a) (b)

Fig. 11. (a) Section destined to the selection of the method of analysis; (b) Space destined to the input of calculation and discretization parameters of the member.

II. For numerical computations, one has.

1. Section properties (See Fig. 12); in the second version more section properties of the cross section are presented.

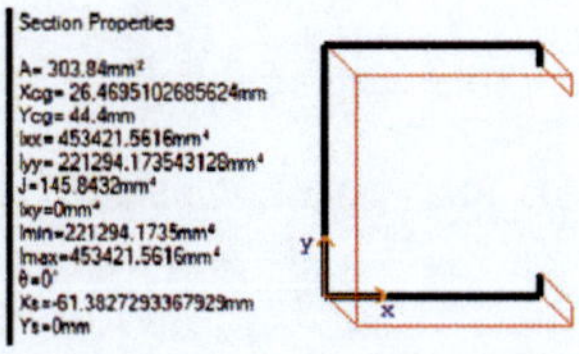

Fig. 12. Form and properties of the cross section.

2. Bifurcation stress vs length diagram using *SAFSM* with one half-wavelength (see Fig. 13).

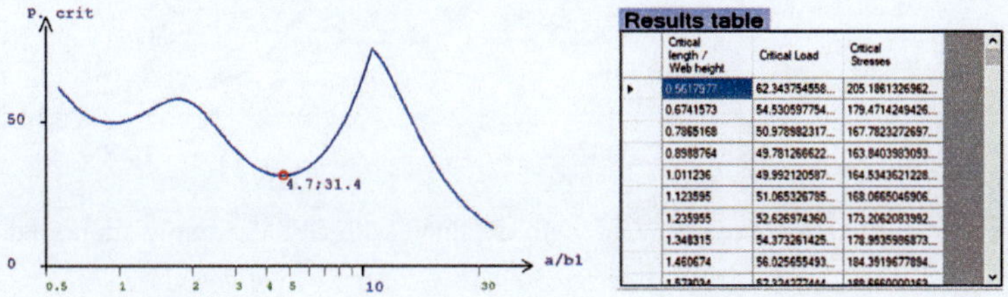

Fig. 13. Bifurcation stress vs length diagram computed with *FSplines 2.0.*

3. Information processed in Excel exported of the *FSplines* 2.0, See Fig. 14. The member length, the critical stress and (i) load (for columns) or (ii) moment (for beams).

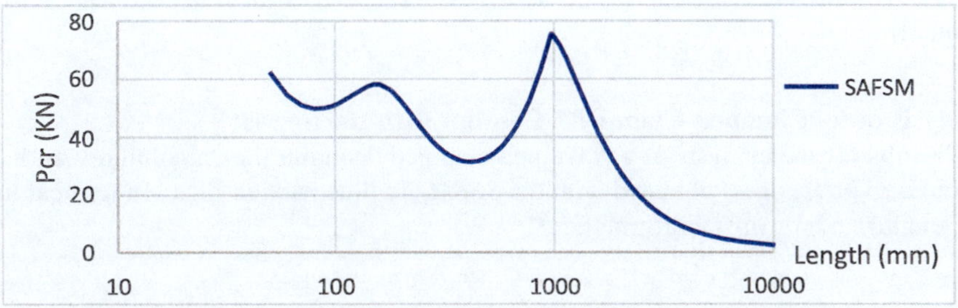

Fig. 14. Bifurcation stress VS length diagram exported to Excel (office-windows)

4. Instability bifurcation mode (See Fig. 15).

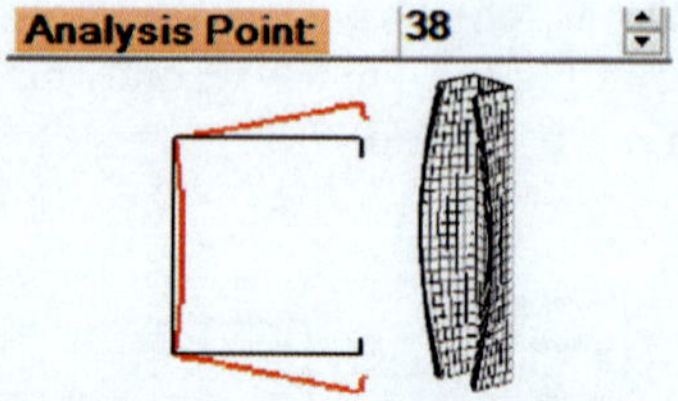

Fig. 15. Deformed in 2D and 3D obtained of *FSplines* 2.0 (analysis for 420 mm column length).

The results obtained by the *SFSM* are shown in Fig. 16 and Fig. 17.

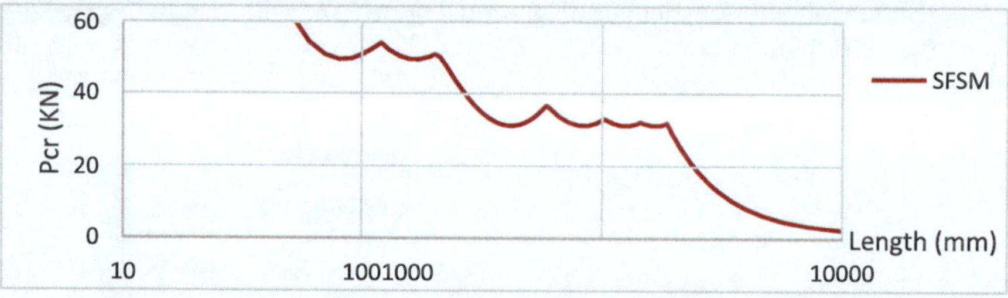

Fig. 16. Curve Bifurcation stresses VS length obtained with SFSM, simply supported column.

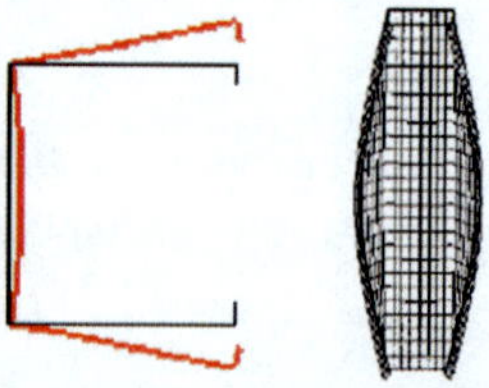

Fig. 17. Deformed in 2D and 3D (analysis for 420 mm column length, where distorcional modo is critical)

Analysis of a "Clamped-Clamped" Column with the *SFSM*

The results for the analysis of a "clamped-clamped" column (i.e., a column with fixed rotations in both edges) obtained with the *SFSM* are illustrated in Figs. 18 (critical load VS length) and 19 (instability mode).

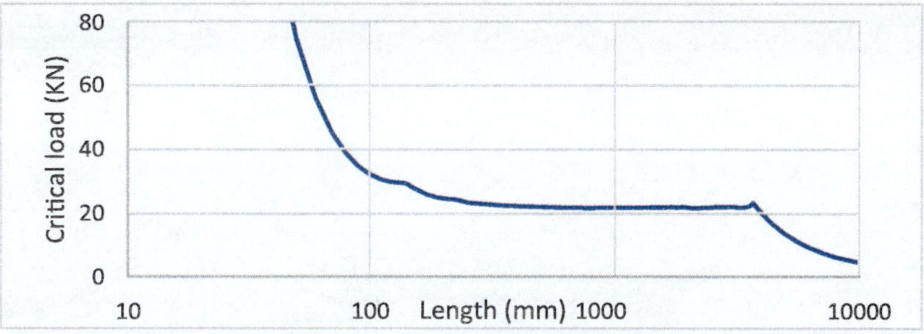

Fig. 18. Curve Critical load VS Length obtained with SFSM; fixed-fixed column.

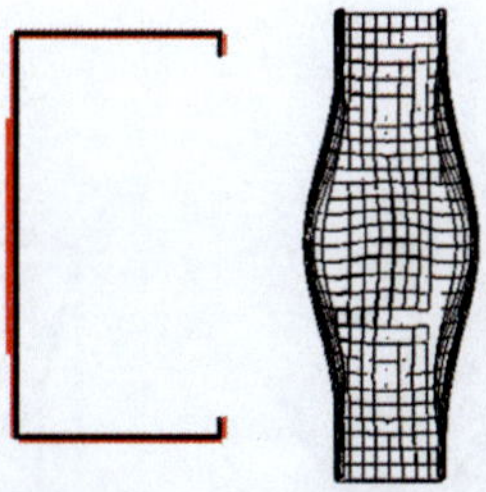

Fig. 19. Deformed in 2D and 3D (length 300 mm).

6 Comparison of the Results

Convergence analysis studies performed by Prola [4], concerning the cross-sectional discretization, proved that the use of models a employing the division of webs into 8 strips; the division of flanges into 6 strips and the division of stiffeners into 3 strips conduct to accurate results.

6.1 *CUFSM* Results

The stability results obtained with *SAFSM* for simply supported C-column with dimensions 101 × 51 × 5.5 × 1 mm (external dimensions) are compared, in Figs. 20 (critical loads) and 21 (critical modes) with simulations obtained with software *CUFSM* [6].

Note that the results obtained with *FSplines* [5] and *CUFSM* [6] match. For instance, in the case of a column with a length of 80 mm, *CUFSM* [6] gives a critical load factor of 101.62, and the stress's value obtained in this point by *FSplines* is the same.

The results of stability analysis obtained by *FSplines* 2.0 with *SFSM* for fixed (clamped) C-column (128 × 78 × 5.5 × 3 mm) are compared, in Figs. 22 and 23, (critical loads and modes) with simulation made by means *CUFSM* [6].

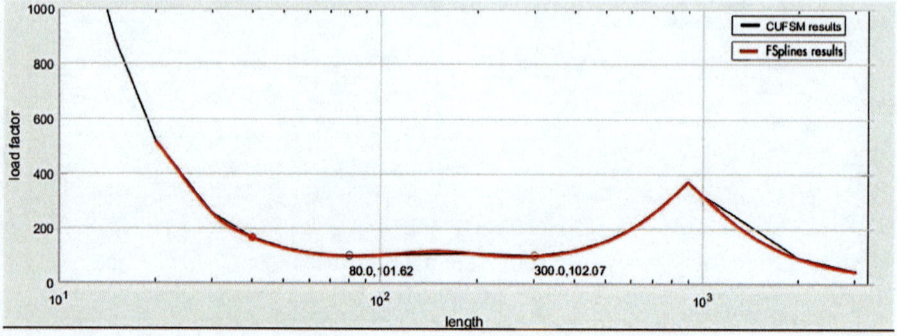

Fig. 20. Comparison between critical local obtained by *CUFSM* [6] e *FSplines* [5] programs.

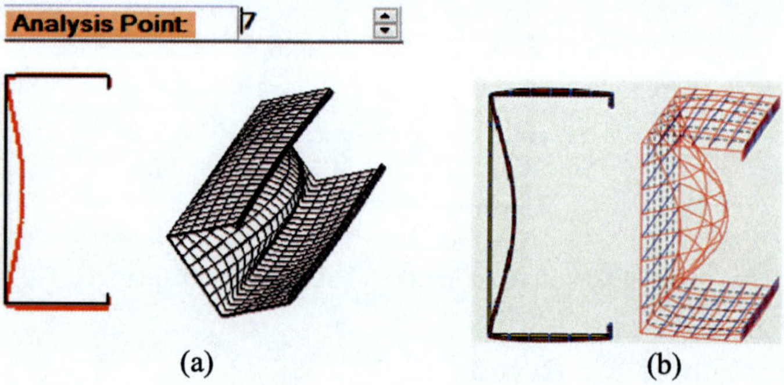

Fig. 21. 2D and 3D instability local mode (length 80 mm) by (a) *FSplines* [5], (b) *CUFSM* [6] programs

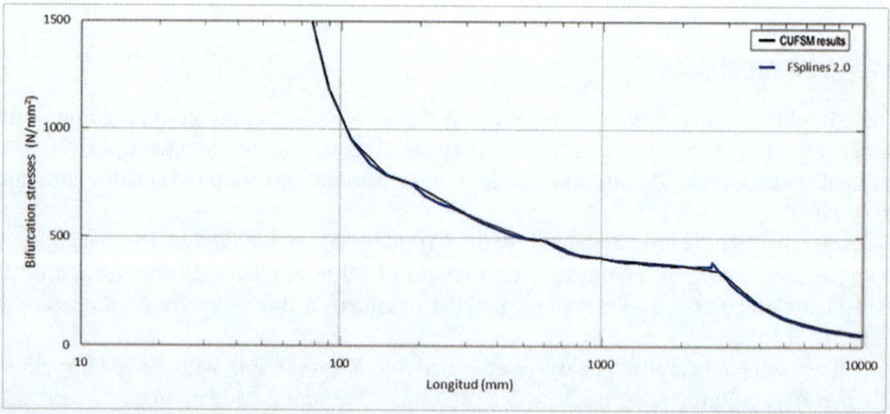

Fig. 22. Comparison between critical loads obtained by *FSplines 2.0* [5] and *CUFSM* [6] (for fixed column)

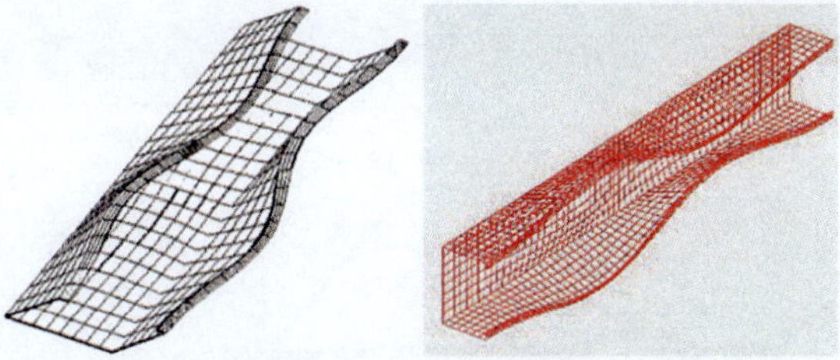

Fig. 23. 2D and 3D instability mode by *FSplines 2.0* [5] and *CUFSM* [6] (length = 790 mm).

6.2 *NASTRAN in-CAD* Results

The stability results given by *SFSM* for a simply supported C-column with dimensions 101 × 51 × 5.5 × 1 mm are compared, in Fig. 24 and 25 (critical loads and modes), with the results given by a Finite Element Analysis performed with *NASTRAN In-CAD* [7].

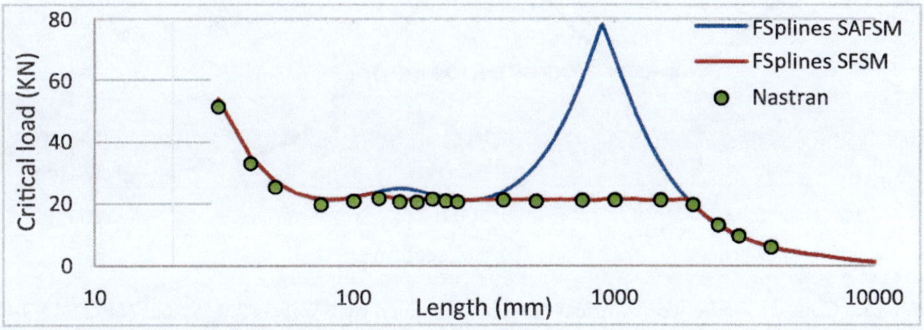

Fig. 24. Comparison between critical loads obtained by *FSplines 2.0* [5] and *NASTRAN In-CAD* [7] softwares.

The stability results given by *SFSM* for simply supported C and Z beams (both with dimensions 128 × 78 × 11.5 × 3 mm) are compared, in Fig. 25, with the results given by a Finite Element Analysis performed with *NASTRAN In-CAD* [7].

Note that, as depicted by Fig. 24 and 25, *FSplines*' results match the ones obtained with *Nastran In-CAD* [7].

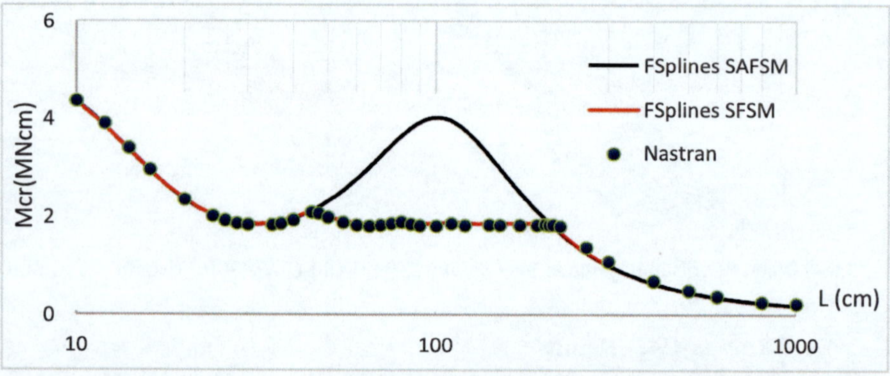

(a) C section

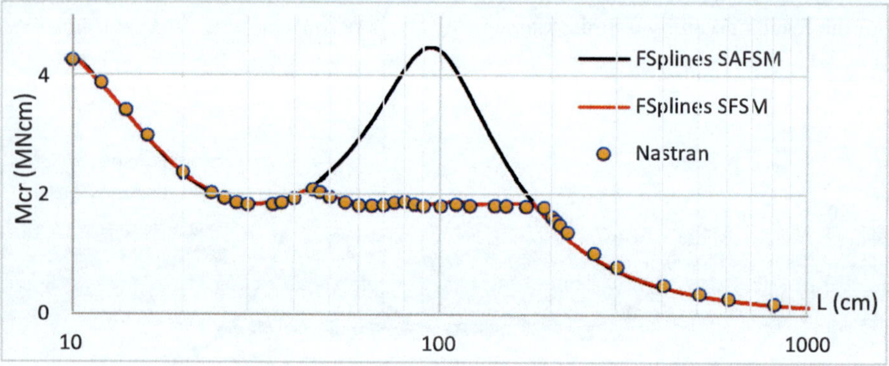

(b) Z section

Fig. 25. Comparison between Critical moments obtained by *FSplines* 2.0 [5] and *NASTRAN In-CAD* [7] programs.

7 Conclusions

FSplines 2.0 is a software that allows performing linear stability analysis (critical loads/moments and instability modes) of thin-walled structures (for example cold formed steel beams and columns) subjected to different structural stress fields and with different support conditions by the *FSM*.

FSplines 2.0 is a suitable tool for use of designers and researchers, whose results match the ones given by other numerical tools. The results obtained with the software will allow to apply the *DSM* to design of cold-formed steel structural members (beams and columns).

References

1. Chicaiza, Á., Prola, L., Martínez, L., Graça, P.: FSplines: um aplicativo para análise linear de estabilidade. In: XII Congresso de Construção Metálica e Mista, pp. 879–888 (2019)
2. Chunting, X., Seaburg, P., Crain, R., Lou, A.: On the computation of the cross-section properties of arbitrary thin-walled structures. In: 16th International Specialty Conference on Cold-Formed Steel Structures, pp. 602–615. University of Missouri, Rolla, Orlando (2002)
3. Cheung, Y.K.: Finite Strip Method in Structural Analysis. Pergamon Press, Oxford (1976)
4. Prola, L.: Estabilidade Local e Global de Elementos Estruturais de Aço Enformados a Frio. Ph.D. thesis, Instituto Superior Técnico, Universidade Técnica de Lisboa, Lisboa (2001)
5. Chicaiza, Á.: FSplines: una aplicación informática para análisis lineal de estabilidad de perfiles abiertos de pared fina. Master's thesis, Instituto Politécnico de Leiria, Leiria (2018)
6. Schafer, B., Ádány, S.: Buckling analysis of cold-formed steel members using CUFSM: conventional and con-strained finite strip methods. In: 18th International Specialty Conference on Cold-Formed Steel Structures, pp. 39–54. University of Missouri, Rolla, Orlando (2006)
7. Autodesk. Nastran In-CAD. https://www.autodesk.com/education/free-software/nastran-in-cad. Accessed 10 Jan 2020
8. Silva, S.: Cálculo de cargas críticas globais e deslocamentos de segunda ordem em estruturas reticuladas. Revista Portuguesa de Engenharia de Estruturas **1**, 21–35 (2007)
9. Ádány, S., Schafer, B.: Buckling mode classification of members with open thin-walled cross-sections. In: 4th International Conference on Coupled Instabilities in Metal Structures, Rome, pp. 27–29 (2004)
10. Ajeesh, S., Jayachandran, S.: Identification of buckling modes in generalized spline finite strip analysis of cold-formed steel members. Thin-Walled Struct. **119**, 593–602 (2017)
11. CIRSOC: Reglamento argentino de elementos estructurales de acero de sección abierta conformados en frío. Reglamento CIRSOC 303, Buenos Aires (2009)
12. Dubina, D., Ungureanu, V., Landolfo, R.: Design of Cold - Formed Steel Structures, 1st edn. ECCS - European Convention for Constructional Steelwork (2012)
13. Zienkiewicz, O., Cheung, Y.: The Finite Element Method in Structural and Continuum Mechanics (1967)
14. Gonzalez, G.: El método de la banda finita para el análisis de vigas cajón. Universidad Politécnica Salesiana, Cuenca (2010)
15. Van, G., Menken, C.: The spline finite-strip method in the buckling analyses of thin-walled structures. Int. J. Numer. Methods Biomed. Eng. **6**(6), 477–484 (1990)
16. Lau, S., Hancock, G.: Buckling of thin flat-walled structures by a spline finite strip method. Thin-Walled Struct. **4**(4), 269–294 (1986)
17. Fan, S., Cheung, Y.: Analysis of shallow shells by spline finite strip method. Eng. Struct. **5** (4), 255–263 (1983)
18. Fan, S.: Spline finite strip in structural analysis. Ph.D. thesis (1982)
19. Azhari, M., Hoshdar, S., Bradford, M.: On the use of bubble functions in the local buckling analysis of plate structures by the spline finite strip method. Int. J. Numer. Methods Eng. **48** (4), 583–593 (2000)

Scrum with eXtreme Programming: An Agile Alternative in Software Development

S. Barahona Rojas(✉), L. Pucha Guzmán, P. Villamarín Coronel, and A. Yunga Benítez

Instituto Superior Tecnológico Sudamericano,
Miguel Riofrío, 156-26 Loja, Ecuador
gmsandrab@gmail.com

Abstract. The growing demand for software products worldwide, in various areas of knowledge, means that software development teams are forced to use agile development methodologies that optimize time and resources. However, using SCRUM in software projects contributes to management, but it does not describe the development process; Due to the agile eXtreme Programming (XP) methodology, this process is provided in detail. This article focuses on the analysis of a large amount of research related to these approaches, to propose to the software development teams a framework validated by experts that combines these two powerful strategies, showing in detail the minimum technical deliverables in each phase of software development; complementing with other suggestions obtained from the expertise of the evaluators of the proposal. The first section details the fundamental concepts of SCRUM; Next, the XP methodology is described, then, it focused on analyzing previous studies of the combination of these strategies and then theoretically generating a detailed three-stage framework that combines SCRUM strategies with XP fortitude, which is employable in any type of software project; along with the minimum documentation required to expedite development work. Then the results obtained from the evaluation rubric are concluded and, finally, the open research directions on this topic are defined.

Keywords: SCRUM · XP · Extreme programming · Agile methodology · Software development

1 Introduction

Software development is a complex process that has evolved over time. In the 1960s, after the "software crisis" [1], software engineering was born. It is a science that establishes principles and foundations to develop and maintain quality computer products, which entails not only the needs and satisfaction of the customer, but also the operational characteristics of the software, its ability to bring about change, and its adaptability to new environments [2].

To this end, over the years, computer science has established different processes that guide software development, starting with traditional or conventional methodologies, which are rigid, predictive, and process-oriented. These have little

M. Botto-Tobar et al. (Eds.): CI3 2020, AISC 1277, pp. 350–361, 2021.
https://doi.org/10.1007/978-3-030-60467-7_29

communication with the customer, even though extensive documentation is generated [3]. Compare that model to agile methodologies, which are much more flexible and able to integrate the customer as part of the team. In addition, its flexibility allows for deliveries to be made in reasonable time, while only essential or minimum documentation is required [4].

Through a study conducted in 2015 [5], in which three agile methodologies were evaluated, according to the guidelines in accordance with the ISO/IEC 29110 standard, which contains two processes: Project Management and Software Implementation; it states that SCRUM presents a high level of process compliance and that XP has at least a moderate level. SCRUM is a framework for developing complex products. It integrates several processes and management-oriented techniques, generating quality results in short iterations, during which the development team uses events, artifacts and associated rules [6]. Extreme Programming is a software development methodology that focuses the entire team on common and achievable objectives, applying its own principles and values, which are adapted according to the context of the project. The team focuses on productivity and working together to develop quality software at a sustainable pace [7].

In the first section of this study, we detail the fundamental concepts of SCRUM; in the second section we describe the XP methodology; in the third section we focus on analyzing previous studies that combine these strategies; in the fourth section we theoretically propose a detailed framework in three stages, which blends the SCRUM strategies with the robustness of XP, which can be used in any type of software project; and finally, the minimum documentation required to speed up the development work. The paper concludes with the results and then provides a direction for future research on this topic.

2 SCRUM

SCRUM, a project management strategy, is not an acronym. It's based on the game of rugby. SCRUM's origin dates back to 1986, when Ikujiro Nonaka and Hirotaka Takeuchi, made reference to it as a new development process that had been employed at successful manufacturing companies in Japan and the United States [4]. But in the early 1990s, Jeff Sutherland and Ken Schwaber, established a working framework with formal rules, to be used for the development of complex product software [8].

The primary objective of applying this agile and flexible framework is the rapid return on an investment of time, through the application of an iterative process based on roles, events, artifacts and rules. And the principles behind SCRUM are built on the pillars of continuous inspection, transparency of process, and adaptation, should any deviations from acceptable limits occur [9].

A SCRUM team should be composed of a self-organized, multifunctional group. It includes the following roles: the Product Owner, who represents the stakeholders, and is responsible for detailing the list of requirements or needs of the software product; and the Scrum Master, who in turn is responsible for communicating to the Development Team project plans and deliverables of the product, also interacting with other Scrum Masters to enhance the efficiency of the applicability of this methodology within the organization. The responsibilities of the Development Team are to create and incrementally deliver the software components that are to be implemented in a production environment at the end of each "Sprint" [10].

Events at SCRUM include: the Sprint, which is a reiterating process, focused on the strengthening of the project output, and usually lasts 30 days or less. For the execution of the Sprint, it is first required to hold a Sprint Planning Meeting, in which the scope of the work to be done during the iteration is detailed; the Daily Scrum, daily meetings between the Scrum Master and Development Teams, which do not last more than 15 min and in which topics such as completed tasks, tasks in process and pending tasks are addressed; The Sprint Review, an event held at the end of each Sprint to evaluate the fulfillment of the Sprint Backlog; and finally, the Sprint Retrospective, consisting first of a report on the completed Sprint, and subsequently, proposed improvements to be implemented in the next iteration [11].

The required artifacts in SCRUM are: the Product Backlog, which lists the functional and nonfunctional requirements of the project; the Sprint Backlog, which reprints the prioritized elements of the requirements list that will be developed in each Sprint; and the Product Increment, which is a usable software version [9].

3 Extreme Programming (XP)

Extreme programming is constituted as a very powerful agile software development methodology [12] used by a vast majority of computer product developers. XP is a work strategy created by Kent Beck, Ward Cunningham and others [7]. This technique adequately combines peer programming, unit testing and integration, reconstruction and process modernization tests, improving teamwork and integration with the end user.

3.1 Phases

The XP methodology is one of the most widely used agile methodologies [16]. The bases of its functionality rests on four clearly-identified phases [7]:

Planning. - In this phase, information is collected through two methods, a tool called user stories and face-to-face communication with the client. New requirements can be added to the user stories at any point throughout the software development process, for a better estimation of the resources required for completion [13].

Design. - The design uses principles such as keeping the design simple, generating operational prototypes, and using a systematic approach to prioritization through CRC cards [14].

Coding. - The first step of the coding phase is for the developers to write the unit tests. These tests allow the programmers to verify the requirements of each user history. The coding phase is usually done in pairs, two coders working together to create the source code for each user story.

Testing. - During this phase, four types of testing are run: unit tests, integration smoke tests, regression tests, performed each time a new module is added, and acceptance tests, performed with the client's participation.

4 Unified Modelling Language (UML)

UML (Unified Modeling Language), created by Grady Booch, James Rumbaugh, and Ivar Jacobson, standardizes models with the ability to communicate object-oriented processes in an understandable way [15].

The ISO/IEC 19501:2004, states that UML is a graphical language for visualizing, specifying, building and documenting the artifacts of a software-based system [16].

There are a variety of artifacts that are used throughout the software development process: the domain model, use case model, vision, supporting specifications, glossary, design model, software architecture document, data model, implementation model, software development plan, testing model and development framework [17].

5 Related Work

The projects developed under an agile context, by combining SCRUM and XP, have a software estimation precision of 85%, according to the results obtained in the research developed by [13], it's evident that when one applies diverse techniques such as Poker and Delphi to estimate the work and then predict the success of a software project, encouraging results can be obtained.

According to the authors of the research cited in [18], who performed a study of the combination of XP, with its characteristics and phases pointed out, and SCRUM, with its events and artifacts, a mixture of the two methods contributes substantially to an improvement in development. In addition, their use in tandem avoids generating too much information, while still directly involving the client in the team's work during the development cycle. The authors state that the blending of these two methods also allows teams to make a range of new changes during a Sprint, without compromising the functionalities already implemented.

It is also important to consider an investigation of both similarities and differences between SCRUM and XP, as cited in reference [19], which confirmed that it is possible to combine both methodologies in a manner that's effective; as long as the context and the nature of the software product has first been carefully analyzed and that the project development is organized so that the methodology applied best fits the conditions of the environment of the software.

One of the fundamental factors to obtain efficient results when using a combination of SCRUM and XP, lies in the experience of the development team. So, those team members who lack sufficient experience would best serve the client by studying good practices of applicability of these methodologies, based on patterns collected from known problems/solutions [20].

6 Theoretical Proposal

This proposal, which seeks to defend a means by which software can best be developed has been abstracted from a set of accredited and published research papers and, in addition, complemented by recorded experiences in software development, being derived from the projects of the entities who were researched.

Given this background, an eclectic software development process is indeed proposed, based on a combined usage of the SCRUM methodology, XP, and UML artifacts. Having briefly considered the basic notion of SCRUM's naming origin (taken from the game of rugby), its three-phase process will now be analyzed: PreGame, Game, and PostGame.

6.1 PreGame

This phase begins with the determination of the stakeholders' ideas or goals, regarding the software, which are relayed by the owner of the product (Product Owner). Next, this Product Owner 'collects' needs or requirements, taken from user stories, which are descriptions of the user's perspective on the functionality of the product. From this information, a product stack (Product Backlog) is generated, where the, and as a UML artifact the general diagram of use cases and ends with the formation and training (Training) of the development team (Development Team).

6.2 Game

This is the core phase of development, in which the XP methodology now becomes involved. It starts with the planning of the iterations (Sprint Planning); from the functionalities of the product stack; where the tasks, responsibilities of the development team and work schedule are specified, generating the iteration stack (Sprint Backlog), as well as determining the Domain Model and the physical and logical architecture of the software.

Each iteration stack determines a usable functionality for the user. This generates a development cycle or iteration (Sprint); it should not take more than four weeks and a team of developers of between two and six people. In addition, there should be daily work meetings (Daily Sprint) no longer than 15 min in which "progress made, problems encountered, and work to be developed" are all addressed and recorded on a progress control panel. The execution of a development cycle involves the XP methodology with its phases: Planning, Design, Coding and Testing; the first phase is contemplated in the planning of the iterations, in the second phase the User Story, prototypes of interfaces and process diagram are documented; while in the third phase it is coded in pairs, documenting the API and source code; and, in the fourth phase the advances are validated by means of the execution of unitary tests and of integration of the software, registering them in a log of errors/defects; and, finally to obtain an increase of deliverable software so that it is tested in a production environment.

6.3 PostGame

Once the Sprint is completed, the owner of the project, the project manager (Scrum Master) and the development team intervene, validating (Sprint Review) that the software results satisfy the requirements defined in the product stack, through acceptance tests; which will allow updating the error/defect log. In case of not meeting the user's specifications, it returns to the game phase for the respective corrections. After that, after the software increment is accepted, an analysis (Sprint retrospective) is performed between the Project Manager and the development team, to consider them for improvement in the next Sprint. Finally, the release of the finished product is carried out, in which the formal documentation is generated with all the generated software artifacts and the formal delivery of the finished product is made.

The Game and PostGame process will be executed repeatedly until the execution of all the requirements, having been defined in the product's stack, is completed. Figure 1 graphically details the proposed agile software development methodology, based on the combination of Scrum with Xtreme Programing, using UML modeling.

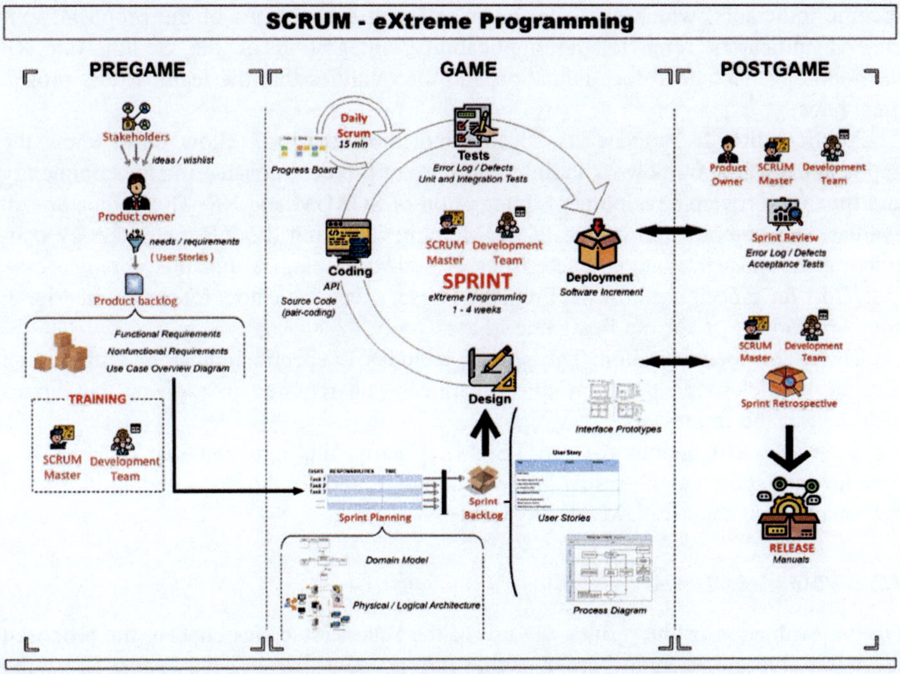

Fig. 1. Software development proposal: SCRUM combined with XP

7 Proposal Evaluation

For the validation of the proposed framework: SCRUM and XP; The use of an evaluation rubric with the Likert scale has been chosen, whose general (4) and specific (39) indicators have been proposed based on the research objective, to verify if the proposal is valid in environments real software development and whether it is necessary to deepen the arguments or redesign the proposal according to the criteria and expertise of the evaluators.

The proposal has been evaluated by twenty experts, including representatives of software development companies at a local and national level (Ecuador), and independent developers of software products; who after having commented on the proposal were interested in learning about this combination, since many of them have already been applying them in their work as a development methodology.

7.1 Evaluation Criteria

The criteria which are part of the evaluation rubric have been formulated as general and specific indicators, which evaluate the phases and components of the proposal; four general indicators refer to the applicability, integration of the Scrum and XP methodology, technical documentation and the qualities that the team of this project must-have.

Applicability: It includes six specific indicators that will allow us to know the applicability of the framework in different types of projects, considering the complexity and time of software development. Integration of SCRUM and XP: This indicator will evaluate the combination of the SCRUM framework and the XP methodology considering the most relevant characteristics of both technologies. For this, three specific indicators have been used for the Pregame Phase; seven indicators for the Game Phase; and, two indicators for the Postgame Phase.

Technical Documentation: This section includes 17 specific indicators, with these it pretends to determine the minimum documentation required in each of the Phases proposed in the framework.

Scrum Team Qualities (Scrum TEAM): Finally, this indicator that includes five specific validation parameters will reveal the importance of the skills and attitudes that the members of the SCRUM TEAM must-have.

7.2 Valuations Levels

For the evaluation of the results, the use of the Likert scale was chosen; the proposed levels are five, in ascending order, taking into account that 1 is the lowest and 5 the highest, as detailed: 1) Totally disagree; 2) Disagree, 3) Indifferent; 4) Agree and 5) Totally agree; criteria that have been raised since "people tend to react in a certain way to the item, not only for the content and format but also for their style of facing this type of task" [21].

7.3 Instruments Used for the Rubric Evaluation

Table 1 details the evaluation carried out on the proposed framework, taking into account the four general evaluation criteria and their respective specific criteria, as well as the aforementioned Likert scale assessment levels.

Table 1. Evaluations rubric

Applicability					
Valuation level	1	2	3	4	5
Short term projects (1–6 months)					
Medium term projects (6–12 months)					
Long-term projects (12 months)					
Low complexity (CRUD Apps with basic administration modules)					
Medium complexity (Apps that deploy APIs or Services)					
High complexity (Integrated systems, expert systems, AI implementation, BigData, ERP systems, etc.)					
SCRUM and XP integration					
Valuation level	1	2	3	4	5
Phase 1 - Pregame:					
Importance of work team training					
Importance of the initial participation of the Stake Holders for the identification of requirements					
Relevance of the Product Owner as spokesperson for the Stake Holders					
Phase 2 - Game:					
Allocation of 1 to 4 weeks for each Sprint					
15 min time for each Daily Scrum					
Implementation of the XP methodology for each Sprint					
Use of dashboard tools to control activities of each Sprint					
Relevance of Sprint Planning for each Sprint.					
Relevance of User Stories for the control of each Sprint					
Importance of pairwise encoding in each SPRINT					
Phase 3 - Postgame:					
Relevance of the Product Owner in the Sprint Reviews					
Importance of the Sprint Retrospective for the timely identification of functional coding errors					
Technical documentation					
Valuation level	1	2	3	4	5
Phase 1 - Pregame:					
Product Backlog: Functional requirements					
Non-functional requirements					

(*continued*)

Table 1. (*continued*)

Applicability					
General diagram of use cases					
Phase 2 - Game:					
Sprint planing: Domain Model					
Physical/logical architecture					
Sprint backlog: User Stories					
Process diagram					
Interface prototype					
Sprint: Progress dashboard					
API					
Source code					
Unit tests					
Integration testing					
Errors/Defects Log					
Phase 3 - Postgame:					
Sprint review: Updating of errors/defects Logbook					
Acceptance Tests					
Release: Manuals					
SCRUM team qualities					
Valuation level	1	2	3	4	5
Importance of effective communication between Scrum Team members					
Relevancia de la experiencia del SCRUM Master					
Empowerment of the SCRUM Team to the assignment of tasks					
Opening by Stake Holders to provide information					
Versatility of the Development Team to adapt to various technologies					

7.4 Results Obtained

After tabulating the data, Table 2 shows the results obtained taking into account the four general evaluation criteria.

Table 2. Results based on the general evaluation criteria.

Indicator	1	2	3	4	5	Final
Applicability	4.76%	4.76%	16.67%	42.86%	30.95%	73,81%
Integration of SCRUM and XP	0.00%	0.00%	10.71%	69.05%	20.24%	89,29%
Technical documentation	0.00%	0.00%	9.62%	60.58%	29.81%	90,38%
SCRUM team qualities	0.00%	0.00%	2.86%	62.86%	34.29%	97,14%

To interpret the results, the highest levels are considered according to the evaluation scale, adding level 4 with 5, and, as shown in Table 2, 73.81% of professionals consider the model presented applicable to projects with a medium to long-term duration, as well as medium to high complexity. When evaluating the integration of SCRUM and XP, an 89.29% acceptance is obtained; in the same way, when evaluating the minimum technical documentation required in each phase of the framework, 89.29% consider it relevant to use the proposed UML tools and artifacts. Finally, when evaluating the qualities of the SCRUM TEAM, 97.14% of the experts indicate the importance of the optimal attitude and aptitude of the team members to achieve the stated objectives. In Fig. 2, we can also see the importance of the last two evaluation levels 4 (Agree) and 5 (Totally agree) for the interpretation of the results obtained

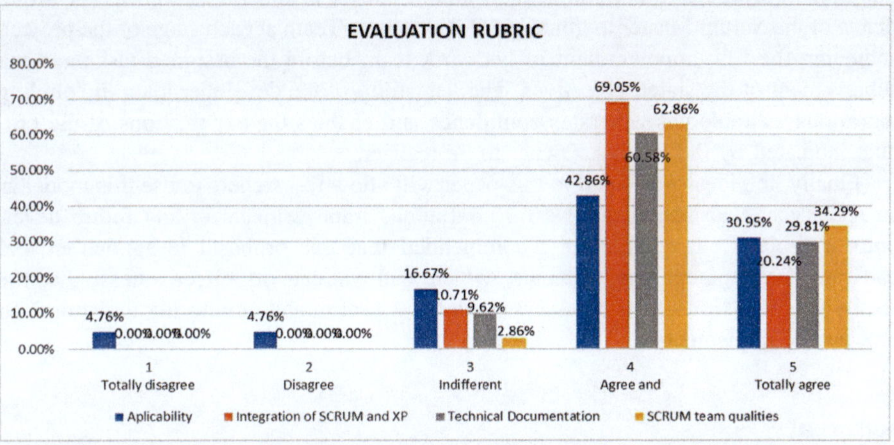

Fig. 2. Results of the publication for the evaluation of the proposed framework.

8 Conclusions

Based on the judgment of experts participating in the evaluation of the proposed framework, it is concluded that the integration of a project management process such as SCRUM and an agile development methodology such as XP, constitute a viable strategy for the development of software projects a medium (6 to 12 months) and long term (greater than 12 months), of medium and high complexity, that is, Apps that deploy APIs or services, integrated systems, expert systems, AI implementation, Big Data, ERP systems, etc.

The relevant aspects to be considered in the initial stages (Pregame Phase) include the importance of the training of the work team, the initial participation of the Stake Holders for the identification of requirements, and the leading role of the Product Owner as the spokesperson for the Stake Holders. In the Game Phase, the relevance of Sprint Planning stands out (using a dashboard and Daily Scrum of no more than 15 min) in which the development of each Sprint is planned (1 to 4 weeks), directly using the XP methodology (design, coding, testing, and deployment). And in the Postgame Phase, the participation of the Product Owner in the Sprint Review is essential; and, carry out the Sprint Retrospective for the timely identification of functional coding errors.

The minimum technical deliverables required for the Pregame phase are Product Backlog, functional requirements, non-functional requirements, and general use case diagram; In the Game phase, the domain model, the physical/logical architecture, user stories, process diagrams, prototype interfaces, progress board, source code, unit tests, integration tests, and error logs must be taken into account/defects; and in the Postgame phase, the errors/defects log must be updated, the acceptance tests must be carried out and the manuals must be documented.

When considering the qualities of SCRUM TEAM, it is necessary to highlight the importance of effective communication among all its members; as well as the experience of the Scrum Master to guide his Development Team at each stage of the project, achieving the full empowerment of his work team before the assigned tasks and the achievement of the stated objectives. The versatility of the developer team in coupling to various technologies generates confidence and enables the expectations of the product indicated by the Product Owner to be realized.

Finally, the ideas presented in this paper will allow researchers to use this model as an archetype; this method allows for continuous improvements in any future development projects. It is therefore recommended that this proposal is applied in real software development environments, which will validate its effectiveness, that the feasibility of UML artifacts be assessed, and that control and monitoring tools for each process are implemented.

References

1. Pons, C., Giandini, R., Pérez, G.: Desarrollo de Software dirigido por modelos: Conceptos teóricos y su aplicación práctica. McGraw Hill, New York (2010)
2. Pressman, R.S.: Ingeniería del software. Un enfoque práctico, Mexico (2010)
3. Canós, J., Letelier, P., Penad, M.C.: Repositorio institucional de la Universidad de Las Tunas: Métodologías Ágiles en el Desarrollo de Software. http://roa.ult.edu.cu/handle/123456789/476
4. Navarro, A., Fernández, J.: Revisión de metodologías ágiles para el desarrollo de software. Prospectiva [en linea] **12**, 30–39 (2013)

5. Galvan, S., Mora, M., O'Connor, R.V., Acosta, F., Alvarez, F.: A compliance analysis of agile methodologies with the ISO/IEC 29110 project management process. Procedia Comput. Sci. **64**, 188–195 (2015)
6. Menzinsky, A., López, G., Palacio, J.: Scrum Master. In: Scrum Manager - Troncal I v. 2.61 (2019). https://scrummanager.net/files/scrum_manager.pdf
7. Beck, K., Andres, C.: Extreme Programming Explained: Embrace Change. Addison-Wesley Professional, Boston (2004)
8. Ken, S., Jeff, S.: Software in 30 Days: How Agile Managers Beat the Odds, Delight Their Customers, and Leave Competitors in the Dust. Wiley, Toronto (2012)
9. Angarita, L.B., Guapachá, J.A.: Gamified system for learning of Scrum development process. In: IEEE Conference Publication, 14th Iberian Conference on Information Systems and Technologies (CISTI), Coimbra, Portugal. IEEE (2019)
10. Lei, H., Ganjeizadeh, F., Jayachandran, P.K., Ozcan, P.: A statistical analysis of the effects of Scrum and Kanban on software development projects. Robot. Comput. Integr. Manuf. **43**, 59–67 (2017). https://doi.org/10.1016/j.rcim.2015.12.001
11. Betta, J., Chlebus, T., Kuchta, D., Skomra, A.: Applying scrum in new product development process. In: Lecture Notes in Mechanical Engineering, pp. 190–200. Pleiades Publishing (2019)
12. Saleh, S.M., Huq, S.M., Rahman, M.A.: Comparative study within Scrum, Kanban, XP focused on their practices. In: 2nd International Conference on Electrical, Computer and Communication Engineering, ECCE 2019. Institute of Electrical and Electronics Engineers Inc. (2019)
13. Adnan, M., Afzal, M., Asif, K.H.: Ontology-oriented software effort estimation system for e-commerce applications based on extreme programming and Scrum methodologies. Comput. J. **62**, 1605–1624 (2019). https://doi.org/10.1093/comjnl/bxy141
14. Aftab, S., Nawaz, Z., Anwer, F., Salman, M., Ahmad, M., Anwar, M.: Empirical evaluation of modified agile models. Int. J. Adv. Comput. Sci. Appl. **9**, 284–290 (2018). https://doi.org/10.14569/IJACSA.2018.090641
15. Gracia Burgués, J.E.: Aprende a Modelar Aplicaciones con UML. IT Campus Academy (2014)
16. Organización Internacional de Normalización ISO, Comisión Electrotécnica Internacional IEC: ISO - ISO/IEC 19501:2005 - Information technology — Open Distributed Processing — Unified Modeling Language (UML) Version 1.4.2. (2005)
17. Larman, C.: UML y patrones. Una introducción al análisis y diseño orientado a objetos y al proceso unificado. Pearson Educación, Madrid, España (2003)
18. Carrasco, M.K., Ocampo, W.J., Ulloa, L.J., Azcona, J.: Metodología Híbrida de Desarrollo de Software combinando XP Y SCRUM. Mikarimin. Rev. Científica Multidiscip. **5**, 109–116 (2019)
19. Salazar, J., Tovar, Á., Linares, J., Lozano, A., Valbuena, L.: Scrum versus XP: similitudes y diferencias. Tecnología Investigación y Academia. TIA-Tecnología Investig. y Acad. **6**, 29–37 (2018)
20. Costa, G., Torres, R., Ruete, D.: Good Practices Guide in the use of SCRUM and XP Methologies oriented to inexperienced teams. Chile (2015)
21. Matas, A.: Diseño del formato de escalas tipo Likert: Un estado de la cuestión. Rev. electrónica Investig. Educ. versión On-line **20**, 38–47 (2018). https://doi.org/10.24320/redie.2018.20.1.1347

Disruptive Use of Spreadsheets in the Teaching-Learning Process of Technical Scientific Subjects

Wilson G. Simbaña L.[1](✉), Andrés E. Castillo R.[2], Edgar A. Bravo D.[1], Luis M. Guallasamin P.[1], and Rosa M. Feria G.[1]

[1] Instituto Superior Tecnológico Rumiñahui, Sangolqui, Ecuador
wilson.simbana@ister.edu.ec
[2] Universidad Politécnica Territorial de Aragua Federico Brito Figueroa, La Victoria, Venezuela

Abstract. This document proposes the use of spreadsheets in a disruptive way, putting them forward not only as a tool for budget calculations, but as a means to create interactive environments, based on mathematical models, graphs, GUI elements, databases to create learning and training environments, applied both as didactic tools in the teaching processes, and as an environment for developing learning situations for the student.

In addition, the possibilities of interoperation with other tools, such as calculation sheets, design applications such as Autocad, geographic applications such as MapInfo and many others, allow us to think about the development of an interconnected system of applications, which would form a digital ecosystem of learning, namely Personal Learning Environment.

A series of learning tools are presented by using the same instrument: spreadsheets. Due to the diversity of disciplines to which they are applied and the variety of used strategies, this work show that spreadsheets can be developed as a learn-by-doing tool in controlled environments.

Keywords: Spreadsheets · Education · Simulators

1 Introduction

The demands in engineering education are increasingly complex and multidimensional, that is why it is interesting to set forth some elements, from Tendencias en la Formación de Ingenieros en Iberoamérica (Trends in the Training of Engineers in Latin America).

> "This inability to understand each procedure requires that engineering address the complexity in all study programs, specifically in the area of modeling and simulation, without neglecting the implicit social language in current problems" [1]
>
> "Teacher training also comprises some issues that should be promptly addressed: the lack of clear training objectives aligned with the economic and social needs of the countries, which has misled the actual practices they need." [1]
>
> "…training engineers is more expensive than training professionals in other areas" [1]

M. Botto-Tobar et al. (Eds.): CI3 2020, AISC 1277, pp. 362–373, 2021.
https://doi.org/10.1007/978-3-030-60467-7_30

The three quotes stated above allow the researchers to come up with the following three reflections:

- Models and simulation are part of the engineer's training.
- It is important to focus on teacher training.
- Engineering career is expensive.

The following views arise from the document Competencias y Perfil del Ingeniero Iberoamericano, Formación de Profesores y Desarrollo Tecnológico e Innovación (Documentos Plan Estratégico ASIBEI):

> "It is widely accepted that engineers must not only know, but also know how to do. The know-how does not arise from the mere acquisition of knowledge, but is the result of a complex structure of knowledge, skills and abilities that work altogether. This structure needs to be explicitly incorporated in the learning process so that pedagogical proposals include activities to favor know-how" [2]

Besides, from the same document, these observations appear:

> "Ibero-American engineers should encompass, among others, the following desirable characteristics:..." [2]

Self-learning ability and commitment to continuous training, especially to the application and implementation of technological breakthroughs.

> "The ability to analyze, model, experiment and solve design issues, through accessible solutions and multidisciplinary approach." [2]
> "The ability to efficiently use the increasing development of telecommunications and computer tools"

In the paper Simulation-Based Learning for Conveying Soft-Skills to XL-Classes [3], Simulation Based Learning (SBL) stands out as an approach to put theoretical knowledge into practice.

Another lecture from the same event: *Shifting from Teaching to Learning: A Special Challenge for Large Classes,* [4] emphasizes that there is a shift from teaching to learning.

The foregoing arguments highlight the importance of simulation environments, but at the same time, the importance of developing solutions and self-learning, for both students and teachers, and also for low-cost solutions.

Thus, the spreadsheet is proposed here as a viable means for the development of simulation environments, with relative simplicity of development, low cost and accessible practically on any computer.

The worldwide popularization of spreadsheets and its broad use among companies is unquestionable. However, its use as a didactic support tool in the teaching-learning process and as a simulation tool is still unexplored. Their use in engineering as a calculation tool is also widespread, given its high calculation performance [5]. Another aspect is the complexity and extension of calculations, such as in applications for economic power supply solutions, where matrix calculation [6] is involved; or in tools for calculating used in transmission substations [7].

The publication La Hoja de Cálculo, poderosa Herramienta de Aprendizaje [8] issues a pioneering work by Professor Pamela Lewis (Magic SpreadSheet 2006), where she states the multiple characteristics of the spreadsheets that make them useful tools for developing the following skills:

a) to organize data (sort, categorize, generalize, compare and highlight key elements);
b) to plot information in a variety of graphs in a meaningful way, facilitating interpretation and analysis;
c) to use graphs to strengthen the concept of percentage;
d) to identify and interpret descriptive statistics such as mean, median, mode and range for a set of data;
e) to use concrete visual elements in order to explore abstract mathematical concepts (visual and spatial intelligence);
f) to discover patterns;
g) to understand basic mathematical concepts such as counting, addition and subtraction;
h) to stimulate higher order mental abilities through conditional formulas with IF and SO functions;
i) to troubleshoot and
j) to use formulas to work with numbers, explore how and which formulas can be used in a given problem and how to modify the variables that affect the result.

Pamela Lewis's work is basically oriented to mathematics learning.

In this work, we present the spreadsheet used in a disruptive way in the sense of unconventional, to create tools to shed light on concepts and knowledge based on mathematical models (using formulas, graphs, interactive user graphic elements (GUI), and databases. All of this, complemented by programming.

Spreadsheets, by means of programming and facilities provided by operating systems, allow interoperation of applications. Particularly, Windows Operating System encourages and calls for these applications which allows users to organize whole processes based on interconnection of applications to set a learning environment underpinned, for instance, by the strengths of Excel calculation capacity, Autocad three-dimensional graphing capacity, database development with Access, and creation of automatic reports with Word.

The following examples are taken from applications in telecommunications. This discipline is supported, among others, by geometry, antennas, cartography, measuring equipment, propagation, satellites, computer networks, etc.

2 Methods

This paper presents a series of learning tools developed with spreadsheets, and framed in engineering didactics. It is important to understand that engineering is learned by doing, given that it is a controlled space, where values can be changed without restrictions so as to see their effects.

This learning space requires a mathematical model, the ability to interact, a graphic design and a pedagogical objective that gauges the specific space of knowledge that needs to be delivered. These requirements are met in each of the experiences shown. Each experience presents the model, the graphic design and the pedagogical objective.

3 Applications

- Parabola, graphing and effects of changes in focal diameter, antenna types based on parabolic reflectors.
- Sinusoid applied in visualization of reflected waves and antenna radiation pattern.
- Digitalization of the radiation pattern of an antenna from the manufacturer's data.
- Obtaining an image in a spreadsheet by coloring the cells.
- Simulation of an electrocardiograph screen with signals.
- Simulation to explain how data from a computer graphics card is stored and plotted in graphs.
- Use of a spreadsheet as a graphic tool for geographic information.
- Visual comparison of dimensions of the earth and artificial satellites. Determination of the minimum size of the satellite to be visualized in graphs.

3.1 Parabola

This sheet shows the parabola and the fundamental property that every ray that strikes parallel to the axis of the parabola bounces off towards a point called the focus. The parabolic shape is the design basis of the satellite transmitting and receiving antennas.

Pedagogical Objective: The significant learning deals with the law of reflection, with ray direction determination where any ray incidence is directed to the focus, and with adequate feed position.

Without explaining the mathematical formulas and their implementation, both the student and the teacher approach knowledge on reflecting antennas, taking into account scientific law and a specific geometric shape.

Model: The laws of reflection are used. At the point of incidence where the ray reaches the mirror if a line is drawn perpendicular to the surface (N) that is known as normal. It turns out that incident, normal and reflected rays are in the same plane; in addition, the angle of the incident with the normal ray is the same as the one formed by the normal and the reflected ray. When the contact surface is curved, the plane tangent to the surface is used, at the point of incidence.

The model is implemented using concepts like equations of parabolas, linear equations and derivative of a parabola, thus the drawing is the graphic representation of a mathematical model, the sheet is part of a book, where more complex antennas are modeled, but the reflection model is used in all cases.

Graphic Design and Interactivity: It includes the image of a parabola, the focus, the focus of the antenna, the incident ray, the reflected ray, the tangent at the point of incidence, and the normal ray to the tangent that passes through the incidence point (see Fig. 1).

Spin buttons have been added to regulate the angle of incidence, the point of incidence and the position of the feed.

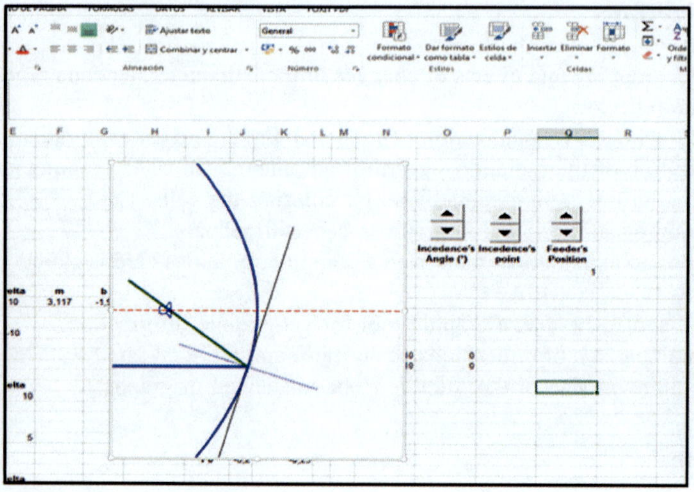

Fig. 1. The parabola, focus, incident ray, reflected ray, tangent, normal to the tangent and numerical control elements.

3.2 Sinusoid

Sinusoid is a fundamental mathematical structure that allows to understand many aspects of radiofrequency. Sinusoid properties favor to model phenomena of interest such as wave phenomenon, particularly radio waves.

In the example below, the spreadsheet models the wave that leaves a transmitter, which is directed towards the antenna, as well as the fraction of the wave that bounces from the antenna to the transmitter.

Pedagogical Objective: Considering that radiofrequency energy is spread as sine waves, students are to learn about the effects of superposition of these waves and also about the reflected effect that the proportion of waves have, namely the effect of reflection coefficient.

A major significant learning deals with the understanding of the behavior of the signal amplitude, along the line of propagation and its behavior over time.

Model: Travelling sine wave equation is used, which models it both in space (linear, in this case) and in time. The model encompasses direction of propagation and reflection coefficient.

As in a parametric amplifier, the highest value that an audio sample reaches for a given frequency is recorded graphically.

Graphic Design and Interactivity: Two graphics are used, one to visualize the incident and reflected signals and the other to see the combined signals (see Fig. 2). Users can use a fast-forward button to notice waves motion in space, and how they combine.

Furthermore, a button has been added to alter the fraction of the signal that is reflected (reflection coefficient) and a check box that records the highest value reached

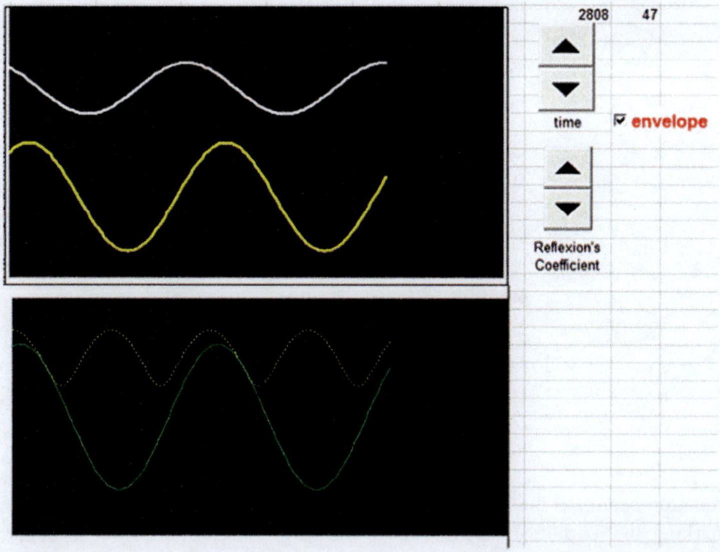

Fig. 2. Sinusoid graphs representing direct wave, reflected wave and their superposition and graphic controls.

by signal for a given value of distance, so it is possible to see how the maximum amplitude varies with distance.

3.3 Digitalization

Pedagogical Objective: The student is expected to digitize the antenna radiation patterns to do calculations and come up with graphs based on mathematical equations using the obtained values.

Model: An image of the antenna radiation pattern is cropped and inserted at the bottom of the plot area. The essential step is to match the limits of the X and Y axes, which match with the maximum values of the outermost circle of the diagram, so as to determine its value at any point that is plotted in the area, regarding the intrinsic scale of the graph.

Graphic Design and Interaction: The model uses a circle centered on the origin and a line that joints the origin and the outermost circle. The formula used is parametric and depends on the angle formed by the line segment and the X axis (see Fig. 3).

It contains an antenna radiation pattern graph with background, with a circle that shows the amplitude and a line designating the azimuth, graphic controls to change amplitude and azimuth, and value capture button.

Points of the radiation pattern can be chosen by the students and they can match the circle and the line, in a triple intersection, circle pattern, or line; to achieve this, the button captures values reflected on the spots and records them.

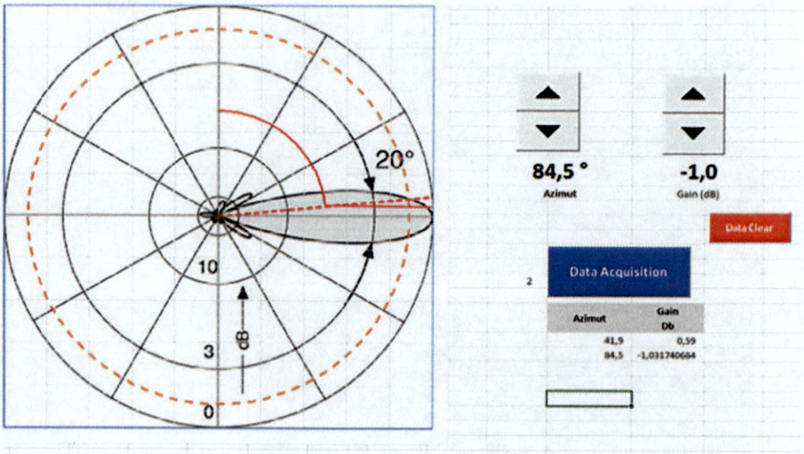

Fig. 3. Radiation pattern as the background of an XY graph, with a circle marking the amplitude and line marking the azimuth, and controls to vary the values and button to capture desired values.

3.4 Images

Pedagogical Objective: Have students know and understand basic principles of image formation on a screen and how data is stored in electronic memory.
Among the significant learnings are considered:

- The images are formed by dots that form lines and lines that form shapes.
- Each section of the image has a color that results from the combination of intensities of three primary colors: red, green, and blue.
- Colors are encoded numerically.

Model: The model is a matrix system, specifically three matrices that store red, green and blue values from each image. The matrix has as many rows and columns as the image may represent, and a region of cells of the spreadsheet contains the same number of rows and columns as the matrices where the image is formed. Colors of each matrix are represented in binary form in a region of three columns. Equations are used to map a matrix to vector and vice versa. Formulas reflect the color of each cell on the spreadsheet after the intensities of its components.

Graphic Design and Interactivity: an area of images is presented, which can be displayed in white or black through buttons, an area of 3 columns with more than 30,000 rows, each one, records binary values of the intensity of the red, green and blue that make up each point of the image (see Fig. 4).

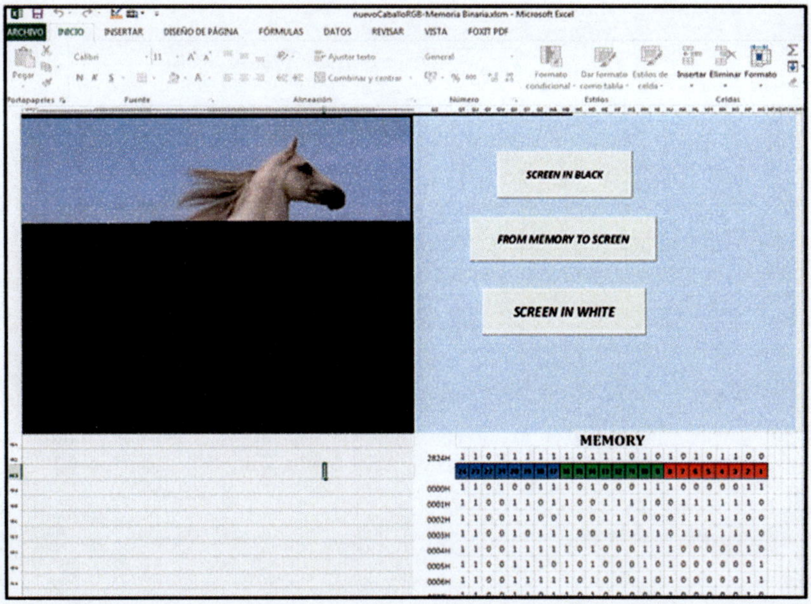

Fig. 4. Image formation control. Upper left part: image formation area, upper right part, buttons to put the screen in black and white or load the image, lower right part: matrix of 300000 rows, showing the value of the intensities of color components, that are reflected on the pixels of the image.

3.5 Digital Modulation

Pedagogical Objective: To show how a binary data modifies the shape of a sinusoidal signal, by modulation by phase-shift keying.

Model: A sinusoid that changes the phase in periods of fixed time; there is a set of phases related to specific combinations of ones and zeros, if the combination is 6 binary digits, there is a set of 16 values of different phases (Fig. 5).

Graphic Design and Interaction: Three elements are presented: the graph with two sinusoids, one modified in phase according to a specific combination of binary numbers, and the other unmodified to serve as a visual reference to visualize the changes that occur.

In addition, there is a button that generates a random binary number of 30 digits, and 5 numbers of 6 digits appear below, which is a 5-part decomposition of the generated number. Thus, when students press the button, a figure is generated and the modulated sinusoid automatically appears.

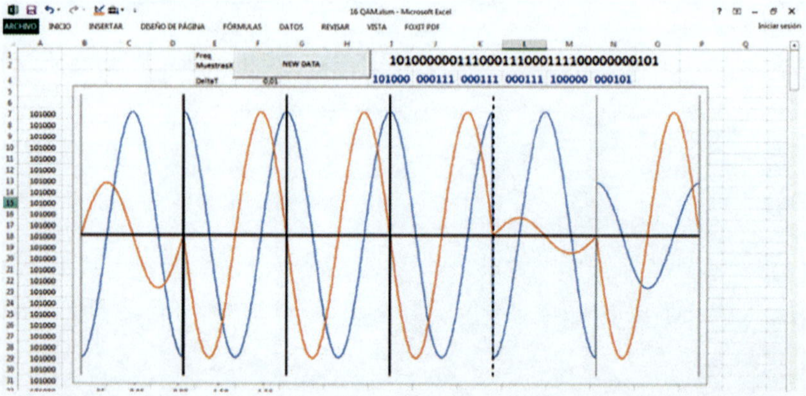

Fig. 5. Sinusoid in phase and in quadrature, modulated in phase. A generator button with a value of 30 binary digits and its coding in 5 digits of 6 binary digits is shown.

3.6 Applications of GPS

Pedagogical Objective: To determine the geographical coordinates of any point within a given territory.

Model: a region of geo-referenced space is used, where longitude is represented by the X axis, and latitude by the Y axis. Geographic images are used, to depict exactly the coordinates (Fig. 6).

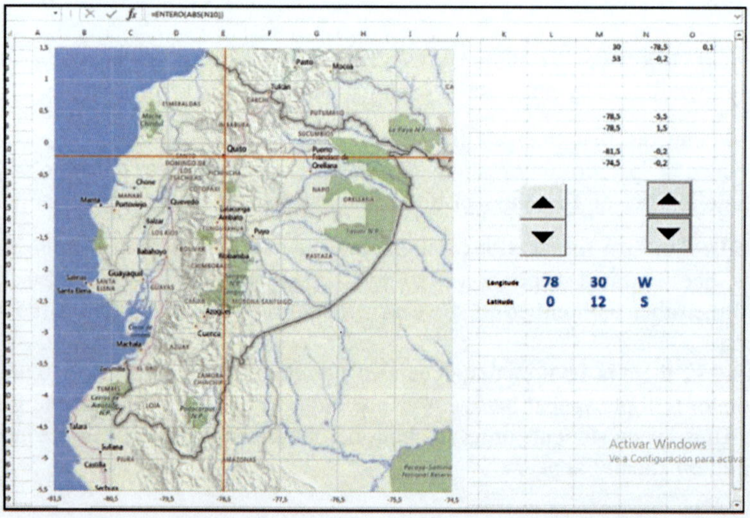

Fig. 6. Sample of a geographical tool

Graphic Design and Interaction: A graphic displays a region of space with two vertical and horizontal lines that represent the longitude and latitude, whose values appear in cells. The spreadsheet includes two sets of spin buttons. The one on the left controls the vertical line (marking the longitude), and the one on the right left controls the horizontal line (marking the latitude). Simultaneously, the values are updated with after each change of value.

3.7 Slotted Waveguide

Pedagogical Objective: Have students measure SWR parameters, using a simulated slotted waveguide, a voltage meter and a variable value load. These measurements are made in Radio Frequency and the student wants to do the same steps he would do with a real SWR measuring device.

Model: Wave theory and its mathematical equations are used. In this case, the voltage is calculated as a function of distance, frequency and reflection coefficient (Fig. 7).

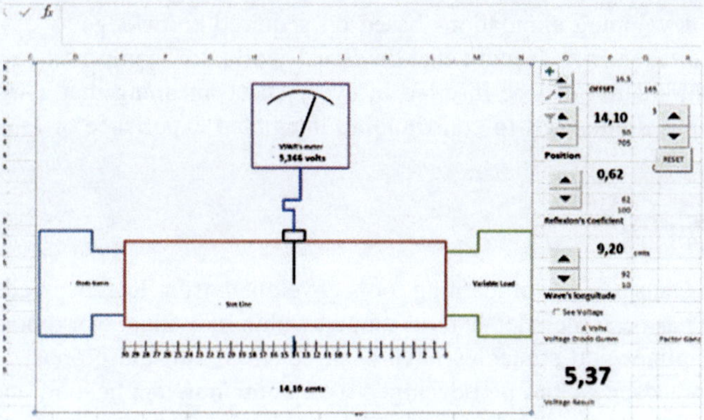

Fig. 7. Example of a slotted waveguide

Graphic Design and Interaction: A graphic containing the slotted waveguide, the voltage meter, the generator and the load is presented to the student. It also comprises spin buttons able to control the position of the guide, the value that the meter marks and the frequency that is chosen, next to the graph, the student has a series of numerical controls, in addition to all the numerical controls, there is a check box to show or not the voltage curve. By itself, this means a kind of augmented reality, which shows information that would remain unseen in real conditions.

Personal Learning Environments: The aim is to meet a continuous learning. After UNESCO, this is defined as lifelong learning where technology facilitates communications (namely Connectivism according to UNESCO); aside from the fact that it has generated the idea of Personal Learning Environment (PLE), which already encompasses trends and conceptions. Thus, a concept in concordance with the possibility of considering a spreadsheet, as a PLE component, determines that it:

> "…is the set of tools, sources of information, connections and activities that each person uses assiduously to learn" [9]

These authors point out that students' PLE is shaped by processes, experiences and strategies that they can - and should - start to learn within the current social and cultural conditions which are determined by the possibilities offered and enhanced by technology. Consequently, that implies that currently some of these processes, strategies and experiences are new, and that they have arisen alongside with information and communication technologies, but it also implies that they shall to be used frequently and serve to enrich the manner in which people learn both individually and collectively.

Developing tools such as those presented in this document is relatively simple on condition that the model supporting the tool is understood. The process of creating graphics, inserting user controls, editing presentation, is customary for spreadsheets users; conversely, the systematic exploitation of spreadsheets to create simulation environments remains unusual so far. At the beginning, teachers explain how the tools are created, then, besides editing these provided learning tools, it is desirable that students star developing simulations based on acquired knowledge.

It must be taken into account that an Excel workbook comprises several spreadsheets, so a simulation can be inserted in every sheet, meaning that a workbook can include multiple simulations to constitute an integrated experience of knowledge.

4 Conclusions

This work presents a series of learning tools developed using Excel spreadsheets. Each of these tools has been included and developed within individual workbooks. A diverse range of disciplines and strategies have been covered with the proposed tool, which shows that spreadsheets can be developed as a know-how application and applied in controlled learning environments, meaning actually that they are virtual mini-laboratories.

Even if specialized software already exists, the ubiquitous use of the spreadsheets, its low cost and flexibility and relative ease to develop both graphics and implement mathematical models, makes of it a tool that shall be part of the classroom and blended teaching strategies.

When students get involved in the creation of these scenarios, higher levels of cognition are achieved, by means of applying mathematical formulas that are part of their existing knowledge.

During the last cycles of the technical careers, students have appropriate knowledge, maturity and experiences, not only to replicate the proposed tools, but also to modify them and come up with other ones.

References

1. ASIBEI Asociación Iberoamericana de Ingenieria: Tendencias en la formación de ingenieros en Iberoamérica. Asibei, Puebla Mexico (2014)
2. Consejo Federal de Decanos de Ingeniería de Argentina - Confedi, Competencias y Perfil del Ingeniero Iberoamericano, Formación de Profesores y Desarrollo Tecnológico e Innovación, ASIBEI (2016)
3. Janßen, D., Borowski, E., Vossen, R.: Simulation-based learning for conveying. In: Education 4.0, Bochum, Alemania. Springer (2016)
4. Stechling, V., Bach, U., Jeschke, S.: Teaching professional knowledge to XL-Class. In: Engineering Education 4.0, Bochum, Alemania. Springer (2016)
5. Santamarta Cerezal, J.C., Tómas Jover, R., Rodriguez-Martin, J., Gutierrez Hernández, L.E., Cano Gonzalez, M., Riqueme Guill, M.: Optimización y Eficiencia en Los Cálculos de Ingeniería Mediante. In: de De la innovación imaginada a los procesos de cambio, Vicerrectorado de Docencia, Universidad de La Laguna; Servicio de Publicaciones de la Universidad de La Laguna (2018)
6. Alcázar-Ortega, M., Álvarez-Bela, C.: Utilización de Microsoft EXCEL en la enseñanza de sistemas eléctricos de potencia: desarrollo de un método matricial para la. In: de I-Red 2018 Congreso Nacional de Inonvación Educativa y de Docencia en Red. Universidad Politecnica de Valencia, Valencia (2018)
7. Cordova Saavedra, E.A.M.: Diseño de Subestaciones de Transmisión de Potencia Convencional Mediante la Elaboración de UNA Herramienta Computacional en Microsoft Excel. Universidad Católica Santo Toribio de Mogrovejo Facultad de Ingeniería Escuela de Ingeniería Mecánica Eléctrica, Chiclayo (2018)
8. Eduteka: La Hoja de Cálculo, poderosa herramienta de aprendizaje. Eduteka, Septiembre 2003. http://eduteka.icesi.edu.co/articulos/HojaCalculo2. Accessed 15 Enero 2020
9. Castañeda, L., Jordi, A.: Entornos Personales de Aprendizaje: claves para el ecosistema. Marfil, Alcoy (2013)

Technological Trends

Nitrate Characterization as Phase Change Materials to Evaluate Energy Storage Capacity

Marco Orozco[1,2], Francis Vásquez[1], Javier Martínez-Gómez[1,3(✉)], K. Acurio[2], and A. Chico-Proano[2,4]

[1] Instituto de Investigación Geológico y Energético (IIGE), Quito, Ecuador
marco.orozco@geoenergia.gob.ec,
javier.martinez@uisek.edu.ec
[2] Departamento de Ingeniería Química, Escuela Politécnica Nacional, Quito EC170525, Ecuador
[3] Universidad Internacional SEK Ecuador, Quito, Ecuador
[4] Department of Chemical Engineering, Centre for Process Systems Engineering, University College London, London WC1E 7JE, UK

Abstract. This research aims to characterize nitrates as phase change materials (PCM) for energy storage in renewable energy systems. Sodium Nitrate ($NaNO_3$), Sodium Nitrite ($NaNO_2$) and Potassium Nitrate (KNO_3) have been considered to be characterized by applying differential scanning calorimetry (DSC), scanning electron microscopy (SEM) and Thermogravimetric analysis (TGA). The heat capacity, thermal stability and microstructure of each material have been evaluated. $NaNO_2$ has the highest enthalpy value (221 $kJ.kg^{-1}$), has good thermal stability recording a maximum mass loss of 1.7% at 500 °C and its specific heat has been determined to be 1.8 $kJ.kg^{-1}.K^{-1}$. Generally, all the PCM studied have granular microstructure but have different grain sizes, in addition, they do not have cracks and porosity in their structure.

Keywords: Nitrates · PCM · Energy storage · Renewable energy systems

1 Introduction

Economic development of society and quality of life are directly linked to energy consumption. The continuous change in the oil price, the emergence of renewable energy systems, the growing energy demand, the need to optimize the use of fossil fuels, global warming and high greenhouse gas emissions (GHG), pose a challenge to the development of efficient energy storage systems [1]. With this background, it is imperative to find alternatives that optimize the use of fossil fuels in order to minimize environmental impact. One option to achieve this goal is to develop renewable energy storage systems [2–4]. Generally, it is known that renewable resources are intermittent (time-varying), therefore, it is necessary to ensure energy availability in the shortage of traditional resources. One way to achieve this purpose is to implement energy storage systems to provide stability to the system [5–7]. Energy storage is defined as any installation or method, usually with an independent control, which can store the energy generated by a system and eventually release it when required [2, 8–11]. Thermal

M. Botto-Tobar et al. (Eds.): CI3 2020, AISC 1277, pp. 377–389, 2021.
https://doi.org/10.1007/978-3-030-60467-7_31

energy storage systems, focused on the use of renewable resources, are classified into sensible heat storage (SHS), latent heat (LHS) and thermochemical storage (TQS) [1, 12–14]. In an SHS, energy is stored by changing the temperature of the storage medium. The performance of these systems depends mainly on the density and specific heat (Cp) of the storage medium, affecting the volume of the system [15, 16]. Therefore, a sensible heat storage system is larger than a latent heat storage system. The storage medium can be solid or liquid [17–19]. Also, thermochemical storage systems use the heat collected to stimulate an endothermic chemical reaction. This reaction is characterized by being reversible, in this way the stored heat can be recovered [20]. On the other hand, a promising alternative for thermal energy storage is to use phase change materials (PCM) which take advantage of the latent heat of phase change; that is, the amount of energy that a body absorbs or releases during its change of state, and then dispose of that energy according to a need or demand [20, 21]. It is possible to say that thermal storage is a promising alternative to guarantee energy security with efficiency and at the same time taking care of the environment. The efficiency of these systems reduces energy losses between supply and demand [22].

According to the application, there are several criteria to select a PCM. The main criteria are melting points within the operating range of the system, high melting enthalpy, high thermal conductivity, low weight variation during the phase transition, chemical stability, non-poisonous, non-flammable, available in large quantities and low cost [1]. Generally, there are three types of PCM: organic, inorganic and eutectic. This study is primarily focused on inorganic PMCs as nitrates. Inorganic PCMs have advantages such as high storage capacity per unit volume. However, no hydrated salt solidifies at the freezing point since they have sub-cooling [20].

Currently, there are several thermal analysis techniques to evaluate material properties such as differential scanning calorimetry (DSC) which is a calorimeter method, scanning electron microscopy (SEM) and Thermogravimetric analysis (TGA) which measures mass changes of a compound as a function of temperature. However, a standard methodology is not available. This statement explains that the results obtained in published analyzes differ from each other. In addition, factors such as measurement procedures, equipment calibration, sample preparation and crucibles used in the tests constitute variables that favor these differences [23]. In a study carried out by Bauer (2012) determined by DSC the appropriate thermal characteristics of $NaNO_3$ as PCM, finding the melting enthalpy 178 $kJ.kg^{-1}$ [24]. Kourkova et al. (2009) evaluated the thermal properties of $NaNO_2$ by two calorimeter methods where they observed three types of phase change corresponding to structural modifications and found that the melting temperature is 437.2 K with melting enthalpy 13.9 $kJ.mol^{-1}$ [25]. Graeter and Rheinlander (2001) determined the thermal characteristics of KNO3 by calorimetry and found that the melting temperature is 336 °C and melting enthalpy 116 $kJ.kg^{-1}$ [26]. In the research carried out by Bauer (2012), it shows the suitability of using $NaNO_3$ as a thermal storage medium, where thermal stability is examined by long-term heating tests and kinetic experiments. In addition, thermal diffusivity is determined by "laser-flash" technique, thermal conductivity by "hot wire" technique and heat capacity by DSC.

One of the important parameters to select PCMs is to evaluate the operating temperature. TGA analysis provides quantitative information on losses of water, crystallization water, residual solvent, pyrolysis processes; oxidation processes, ash percentage and thermal stability of substances. Generally, mass variations imply a reduction, however, in some cases an increase may occur [27]. Although the TGA method does not allow the study of phase change processes, it allows estimating the useful life of the material for energy storage. According to [28, 29], the maximum temperature or the stability limit of a material is usually defined as the temperature when the sample has lost 3% relative to the initial weight, this point is known as T_3. According to the research carried out by [30], the degradation temperature of the $NaNO_3$ sample starts at 400 °C; however, it is lower than the material decomposition temperature, around 450 °C. To avoid the instability of $NaNO_2$ that appears above 330 °C, the first isothermal process must be reduced to 300 °C, ensuring that there is no weight loss before the dynamic segment [31]. And, according to Chaozhen et al. (2017), KNO_3 is stable at 180 °C, losing only 0.044% of the total weight [32].

Scanning electron microscopy is an appropriate technique for visualization and analysis of morphological characteristics of solid samples. The technique (SEM), has great depth of field that allows observing samples with varied surface topography, especially granulated materials can be observed. The main disadvantage of this technique is the quantitative analysis of the mixture due to processes associated with sample preparation [21].

With this background, this research work is aimed at determining thermo-physical properties of nitrate salts with low melting points. In this context, nitrates were characterized as PMC through these techniques to establish their feasibility of use in renewable energy resource applications. The present study shows the analysis of thermal properties of compounds such as sodium nitrate, potassium nitrate and sodium nitrite. These substances were studied by Differential Scanning Calorimetry (DSC), Scanning Electron Microscopy (SEM) and Thermogravimetric analysis (TGA).

2 Materials and Equipment

Thermal analysis consists of using various techniques to determine physical or chemical properties when a substance is heated, cooled or kept at a constant temperature. Thermo analytical techniques and procedures are used for the evaluation and interpretation of the measured values. The data obtained in the thermal analysis are shown as a curve in a thermo analytical diagram [33]. Sodium nitrate ($NaNO_3$), sodium nitrite ($NaNO_2$) and potassium nitrate (KNO_3) were selected as materials. Table 1 shows the chemical characteristics of materials such as purity and melting temperature.

Table 1. Materials

Molecular formula	CAS	Purity (%)	Melting point (°C)
KNO_3	7757-79-1	99	334
$NaNO_3$	7631-99-4	99	306
$NaNO_2$	7632-00-0	98	271

From these materials, specific heat in solid and liquid state, melting temperature, enthalpy of phase change was measured and the microstructural state was evaluated. These materials were selected due to the melting point suitable for thermal storage applications whose operating temperature does not exceed 350 °C, for example in solar thermal systems [1]. These materials were characterized using DSC, TGA and SEM, to determine their thermo-physical and chemical properties compared to data in previous studies, see Table 2.

Table 2. Experimental data of thermo-physical properties of PCM

Property	KNO_3	$NaNO_3$	$NaNO_2$
Melting temperature (°C)	333 [34]	308 [26]	297.8 [37]
	336 [26]	310 [1]	271 [25]
	330 [35]		
Melting enthalpy (kJ/kg)	266 [36]	174 [26]	172 [1]
	116 [26]	199 [36]	
	266 [1]	172 [1]	
Specific heat	1.22 [1]	1.82 [1]	1.78 [38]
	1.29 [38]	1.44 [39]	

In DSC analysis, material samples were 5 to 7 mg. The sample was heated at 10 °C. min^{-1} to achieve total fusion of each material. Finally, the resulting mixture was cooled to 25 °C to ensure good thermal contact between the sample and the container thus achieving a uniform composition [33]. The TGA samples have been taken between 2 and 4 mg; have been heated between 25 and 600 °C at a rate of 10 °C.min^{-1} and under a protective atmosphere of N_2 with flow 20 ml min^{-1} [33]. The preparation of the sample for SEM analysis involves removing moisture through the progressive heating of the compound from room temperature to 110 °C in a muffle oven, approximately for one hour. To verify the absence of moisture, the sample is weighed in time intervals every 5 min until the absence of variations between consecutive measurements is verified.

2.1 Differential Scanning Calorimetry (DSC)

For DSC analysis, a Meter Toledo calorimeter model HP DSC 1 was used which consists of DSC chamber cooling equipment. According to [33] the enthalpy of phase change is obtained from Eq. (1).

$$\Delta H = \int_{t_1}^{t_2} \phi dt = \int_{t_1}^{t_2} \frac{dH}{dt} dt \tag{1}$$

Where: $ø$ is the heat flux in mW, t_1 initial test time, t_2 final test time. The specific heat is defined by Eq. (2).

$$C_p = \frac{dH}{dT}\frac{1}{m} \tag{2}$$

Where, C_p is the specific heat of the sample, m is the mass of the sample, $\frac{dH}{dt}$ is the enthalpy increase when the temperature increases. To calculate this parameter, the equation is derived as a function of time and Eq. (3) is obtained. Where: C_p is the specific heat, $\varnothing$ is the heat flux, β_S is the heating rate dT/dt and m is the mass of the compound.

$$C_p = \frac{dH/dt}{dT/dt}\frac{1}{m} = \frac{\emptyset}{\beta_s m} \tag{3}$$

2.2 Thermogravimetric Analysis (TGA)

In this case, a SHIMADZU model TGA50 Thermogravimetric analysis device will be used, which has tolerance to vibrations and to ambient temperature fluctuations. It is suitable for detecting mass changes in order of micrograms. The methodology establishes that a sample of material between 2 and 4 mg must be prepared. It will be heated in a range between 25 and 600 °C at a heating rate of 10 °C/min and under a protective atmosphere of N_2 with a volumetric flow 20 ml/min.

2.3 Scanning Electron Microscopy (SEM)

This technique is based on the information contained in the electrons that cross or bounce on the surface of the material when a coherent and high-speed electronic beam is struck [21]. The equipment used is a TESCAN VEGA 3 SEM electronic microscope.

3 Results and Discussion

3.1 Potassium Nitrate (KNO3)

Figure 1 shows the thermogram obtained from a sample of KNO_3. The vertical axis shows the heat flux in mW and the vertical axis the time in minutes. Two characteristic peaks generated from two endothermic processes are observed, the first process starts at 130 ° C and ends at 135 °C.

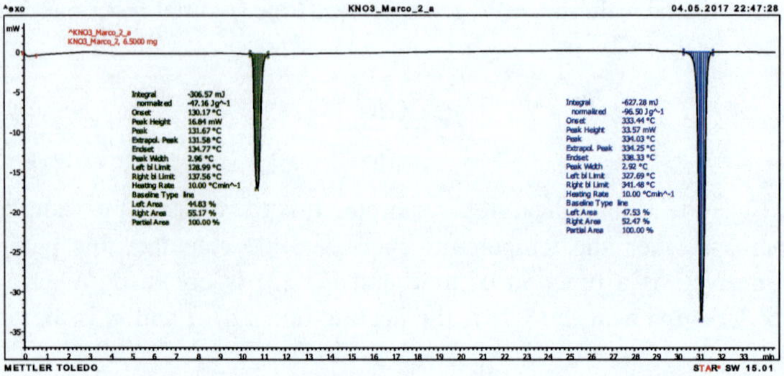

Fig. 1. DSC curve for Potassium Nitrate (KNO_3)

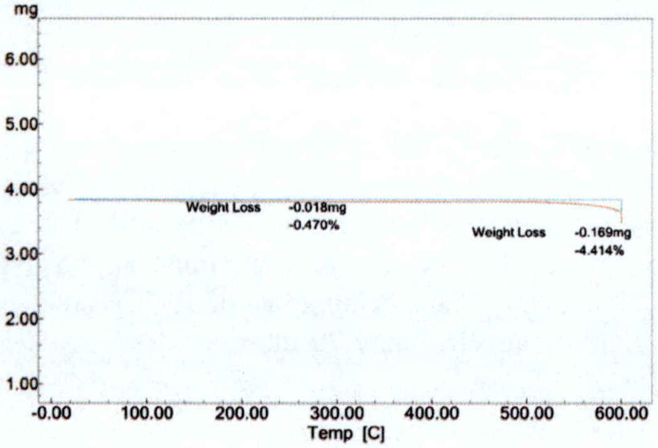

Fig. 2. Curva TGA para el Nitrato de Potasio (KNO_3)

The area under the curve represents the enthalpy variation corresponding to a solid-solid transition state (S-S) and has a value of 47 $kJ.kg^{-1}$. These processes are evidenced due to increased heat flow. Likewise, a new endothermic process is generated at a higher temperature which starts at 333 ° C and ends at 338 °C with a maximum heat flow of 34 mW. This process corresponds to the phase change from solid to liquid (S-L) and is due to the correspondence between the process start temperature and the melting point that occurs at the peak of the highest heat flux (334 °C) (Fig. 2).

In this case, the melting enthalpy reaches 97 $kJ.kg^{-1}$, approximately twice the number found during the first endothermic process. The melting enthalpy found in this study differs significantly from two authors [1, 36] who found values of 266 $kJ.kg^{-1}$ but less error is obtained with the author [26] who found a value of 116 $kJ.kg^{-1}$, approximately 20%.

Fig. 3. SEM microscopy for a) KNO_3, b) $NaNO_3$, c) $NaNO_2$

Figure 3 shows on the vertical axis the mass in mg and on the horizontal axis the temperature in °C and two curves are constructed; one shows the mass as a function of the temperature and the other represents the loss of mass as a function of the temperature. The heating process starts at room temperature and ends at 600 °C. The mass remains stable throughout the heating process, however certain changes are observed. At 250 °C the sample has undergone a slight change in mass (0.018 mg) which corresponds to approximately 0.5% of the initial mass. At the end of the process (600 ° C) there has been a 4.5% loss of mass (0.0169). The details of KNO3 morphology using SEM are shown in Fig. 3a. The morphological characteristics of a granulated material can be observed at a microscopic level, with grain sizes that vary from approximately 100 μm to 300 μm and have no porosity or cracking.

3.2 Sodium Nitrate ($NaNO_3$)

Figure 4 shows the thermogram (DSC) obtained from the $NaNO_3$ compound, shows an endothermic process between 271 °C and 277 °C which generates an inflection point at 276 °C corresponding to the solid-solid transition state of the sample, generating an enthalpy change of 107 kJ.kg^{-1}.

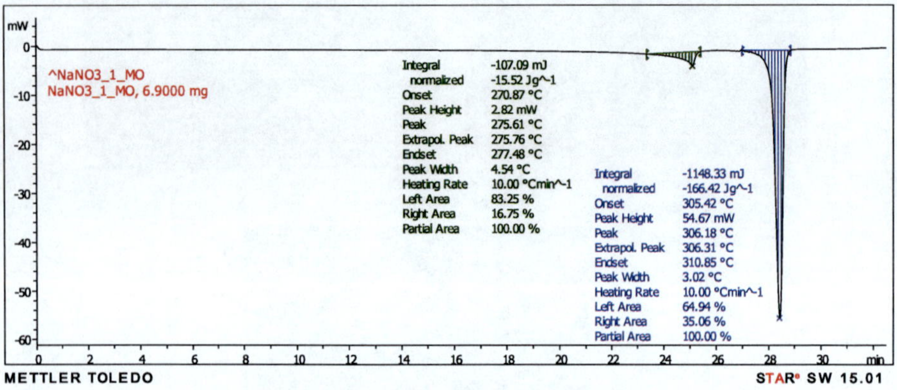

Fig. 4. Curva DSC para el Nitrato de Sodio ($NaNO_3$)

Subsequently, the temperature increase generates a second endothermic process where the phase change from solid to liquid is evidenced (Fig. 5). This process starts at 305 °C and ends at 311 °C. The phase change occurs at 306 °C, the resulting melting enthalpy is 166 kJ/kg. If these results are compared with [1, 26, 36] there are errors of approximately 3.6%, 4.8% and 20%, respectively shows the Thermogravimetric (TGA) diagram for the $NaNO_3$ compound between room temperature (approx. 18 °C) and 600 °C.

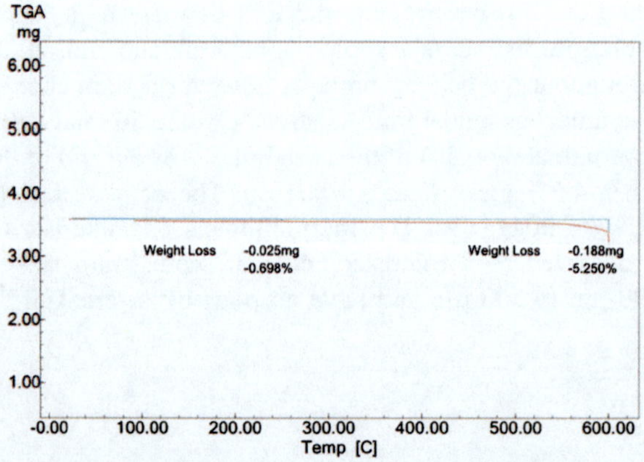

Fig. 5. TGA curve for Sodium Nitrate ($NaNO_3$)

The curves do not show pronounced negative slopes up to 600 °C. Although a slight weight loss (0.7%) is observed at approximately 250 °C and this percentage increases to 5.3% when it reaches 600 °C. In Fig. 3b the morphological characteristics of $NaNO_3$ can be observed, which has characteristics of granulated material, do not have porosity or cracking at the microscopic level, with grain sizes between 250 µm to approximately 500 µm.

3.3 Sodium Nitrite ($NaNO_2$)

Figure 6. shows the thermogram for $NaNO_2$ and similar characteristics of sodium nitrate are observed. The difference between $NANO_3$ and $NaNO_2$ lies in the first endothermic process which for sodium nitrite starts at 163 °C, achieving a temperature peak at 165 °C.

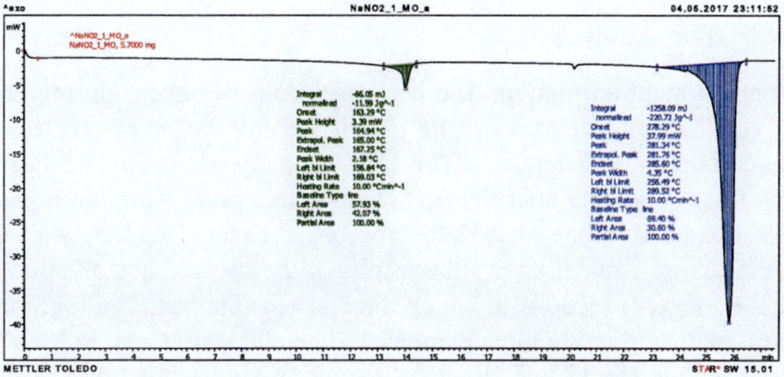

Fig. 6. DSC curve for Sodium Nitrite ($NaNO_2$)

The process finishes at 167 °C, has a heat flux of 3.37 mW and an enthalpy change of 11 $kJ.kg^{-1}$. The change of state (S-L) occurs at 281 °C and the measured melting enthalpy is 221 $kJ.kg^{-1}$. The error in this compound compared to [1] is 3.6%, however in this case a higher value has been found in relation to the previous compounds. The change in mass of $NaNO_2$ as a function of temperature is shown in Fig. 7. Sodium nitrite does not show considerable mass changes up to 500 °C. At 250 °C it has undergone a mass change of 1.5% and at 500 °C it has not increased considerably (1.7%). On the other hand, the microscopic morphology of sodium nitrite is shown in Fig. 3 c. Grains with a size ranging from approximately 50 to 250 µm can be observed and some porosity is observed but the presence of cracks is not visible.

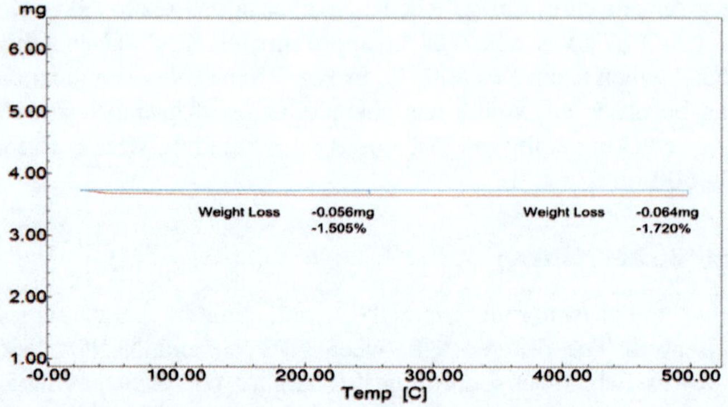

Fig. 7. TGA curve for Sodium Nitrite (NaNO2)

3.4 Specific Heat Analysis

It is important to mention that specific heat (*Cp*) does not occur during endothermic processes. To calculate this parameter, Eq. (3) that relates the heat flux, heating rate and mass of the compound will be used. The solid state Cp has been calculated from the average heat fluxes obtained after the first endothermic process. In the liquid state, the average heat fluxes have been used after the phase change, that is, after the second endothermic process. Table 3 shows the results of the specific heat in solid and liquid state in the thermogram sections in which there is sensible heat. Sodium nitrite shows higher values compared to the three elements. The solid state *Cp* is 1848 $J * kg^{-1}.K^{-1}$ and the liquid state is 1600 $J.kg^{-1}.K^{-1}$. Potassium nitrate has the lowest values for the solid and liquid state of 0.508 and 0.1 $kJ.kg^{-1}.K^{-1}$. Sodium nitrate has intermediate values of 1,321 and 0.74 $kJ\text{-}kg^{-1}.K^{-1}$ in solid and liquid state respectively.

Table 3. Valores de calor específico

	Mass (mg)	Weight (kg/mol)	Heat flux S-S (mW)	Heat flux S-L (mW)	Cp-S (J/kg K)	Cp-L (J/kg K)
KNO_3	6,5	0,10	0,55	0,11	507,69	99,69
$NaNO_3$	6,9	0,08	1,52	0,85	1.321,74	739,13
$NaNO_2$	5,7	0,07	1,76	1,52	1.848,42	1.600,00

It is important to mention that the difference between solid and liquid *Cp* for $NaNO_2$ is smaller than the other compounds which have much wider differences. When comparing KNO_3 with previous research, a considerable high error has been found according to [1]. However, for sodium nitrate and sodium nitrite the errors do not exceed 10% according to [33].

4 Conclusions

The latent heat thermal storage has its maximum representation in solid-liquid transition systems (S-L). The melting enthalpy in all cases is higher than the variation of enthalpy during the solid-solid transition (S-S). In the S-L transition, a minimal variation in the volume of the storage medium also occurs. The TGA results showed that $NaNO_3$ and KNO_3 exceed the T_3 point of weight loss in an analysis performed up to 600 °C. This aspect determines that nitrates alone would not be suitable for high temperature energy storage applications. Of the results obtained as a whole, particularly, emphasis should be placed on the higher melting enthalpy observed in $NaNO_2$ with a value of 220.72 J/kg, however, by itself this parameter does not indicate the suitability of use in applications determined, it is also important to consider the operating temperature of the system to select the storage medium. Thus, this material alone will not work for systems whose operating range is less than 278 °C. The values shown in this research work are intended to make available several results of thermal parameters of phase change materials that obtained under a validated methodology, are suitable for specific applications in thermal energy storage systems.

Acknowledgments. This research takes part of the project Selection, characterization and simulation of phase change materials for thermal comfort, cooling and energy storage. This project is part of the INEDITA call for R&D research projects in the field of energy and materials. This research takes part of the project P121819, Parque de Energias Renovables founded by Universidad International SEK.

References

1. Agyenim, F., Hewitt, N., Eames, P., Smyth, M.: A review of materials, heat transfer and phase change problem formulation for latent heat thermal energy storage systems (LHTESS). Renew. Sustain. Energy Rev. **14**, 615–626 (2010)
2. Kousksou, T., Bruel, P., Jamil, A., El Rhafiki, T., Zeraouli, Y.: Energy storage: applications and challenges. Sol. Energy Mater. Sol. Cells **120**(PART A), 59–80 (2014)
3. Acurio, K., Chico-Proano, A., Martínez-Gómez, J., Ávila, C.F., Ávila, Á., Orozco, M.: Thermal performance enhancement of organic phase change materials using spent diatomite from the palm oil bleaching process as support. Constr. Build. Mater. **192**, 633–642 (2018)
4. Beltrán, R.D., Martínez-Gómez, J.: Analysis of phase change materials (PCM) for building wallboards based on the effect of environment. J. Build. Eng. **24**, 100726 (2019)
5. Forrester, J.: The value of CSP with thermal energy storage in providing grid stability. Energy Procedia **49**, 1632–1641 (2014)
6. Aldás, P.S.D., Constante, J., Tapia, G.C., Martínez-Gómez, J.: Monohull ship hydrodynamic simulation using CFD. Int. J. Math. Oper. Res. **15**(4), 417–433 (2019)
7. Espinoza, V.S., Guayanlema, V., Martínez-Gómez, J.: Energy efficiency plan benefits in Ecuador: long-range energy alternative planning model. Int. J. Energy Econ. Policy **8**(4), 52–54 (2018)
8. Kastillo, J.P., Martínez, J., Riofrio, A.J., Villacis, S.P., Orozco, M.A.: Computational fluid dynamic analysis of olive oil in different induction pots. In: 1st Pan-American Congress on Computational Mechanics–PANACM 2015, pp. 729–741 (2015)

9. Kastillo, J.P., Martínez-Gómez, J., Villacis, S.P., Riofrio, A.J.: Thermal natural convection analysis of olive oil in different cookware materials for induction stoves. Int. J. Food Eng. **13** (3) (2017)
10. Rodríguez, D., Martínez-Gómez, J., Guerrón, G., Riofrio, A.: Impact of induction stoves penetration over power quality in Ecuadorian households. Revista ESPACIOS **40**(13) (2019)
11. Martínez-Gómez, J., Ibarra, D., Villacis, S., Cuji, P., Cruz, P.R.: Analysis of LPG, electric and induction cookers during cooking typical Ecuadorian dishes into the national efficient cooking program. Food Policy **59**, 88–102 (2016)
12. Martínez-Gómez, J.: Material selection for multi-tubular fixed bed reactor Fischer-Tropsch reactor. Int. J. Math. Oper. Res. **13**(1), 1–29 (2018)
13. Villacís, S., Martínez, J., Riofrío, A.J., Carrión, D.F., Orozco, M.A., Vaca, D.: Energy efficiency analysis of different materials for cookware commonly used in induction cookers. Energy Procedia **75**, 925–930 (2015)
14. Martínez-Gómez, J., Guerrón, G., Riofrio, A.J.: Analysis of the "Plan Fronteras" for clean cooking in Ecuador. Int. J. Energy Econ. Policy **7**(1), 135–145 (2017)
15. Villacreses, G., Martínez-Gómez, J., Quintana, P.: Geolocation of electric bikes recharging stations: city of Quito study case. Int. J. Math. Oper. Res. **14**(4), 495–516 (2019)
16. Segarra, M., Martorell, I., Cabeza, L.F., Fernandez, A.I., Martínez, M.: Selection of materials with potential in sensible thermal energy storage. Sol. Energy Mater. Sol. Cells **94**, 1723–1729 (2010)
17. Villacreses, G., Gaona, G., Martínez-Gómez, J., Jijón, D.J.: Wind farms suitability location using geographical information system (GIS), based on multi-criteria decision making (MCDM) methods: the case of continental Ecuador. Renew. Energy **109**, 275–286 (2017)
18. Martínez, J., Martí-Herrero, J., Villacís, S., Riofrio, A.J., Vaca, D.: Analysis of energy, CO2 emissions and economy of the technological migration for clean cooking in Ecuador. Energy Policy **107**, 182–187 (2017)
19. Kousksou, T., Bruel, P., Jamil, A., El Rha, T., Zeraouli, Y.: Energy storage: applications and challenges. Sol. Energy Mater. Sol. Cells **120**, 59–80 (2014)
20. Oliver, A., Neila, F.J., García-Santos, A.: Clasificación y selección de materiales de cambio de fase según sus características para su aplicación en sistemas de almacenamiento de energía térmica. Mater. Construcción **62**, 131–140 (2012)
21. Juárez Varón, D., Ferrándiz Bou, S., Balart Gimeno, R.A., García Sanoguera, D.: Estudio de materiales con cambio de fase (PCM) y análisis SEM de micro PCM. 3c Tecnol. **3**, 54–77 (2012)
22. Gaona, D., Urresta, E., Martínez, J., Guerrón, G.: Medium temperature phase change materials thermal characterization by the T-History method and differential scanning calorimetry. Exp. Heat Transf. **30**(5), 463–474 (2017)
23. Lazaro, A., Peñalosa, C., Solé, A., Diarce, G., Haussmann, T., Fois, M., Zalba, B., Gshwander, S., Cabeza, L.F.: Intercomparative tests on phase change materials characterisation with differential scanning calorimeter. Appl. Energy **109**, 415–420 (2013)
24. Bauer, T., Laing, D., Tamme, R.: Characterization of sodium nitrate as phase change material. Int. J. Thermophys. **33**(1), 91–104 (2012). https://doi.org/10.1007/s10765-011-1113-9
25. Kourkova, L., Svoboda, R., Sadovska, G., Podzemna, V., Kohutova, A.: Heat capacity of NaNO2. Thermochim. Acta **491**(1–2), 80–83 (2009). https://doi.org/10.1016/j.tca.2009.03.005
26. Graeter, F., Rheinlander, J.: Thermische Energiespeicherung mit Phasenwechsel im Bereich von 150 bis 400 °C, 65–75 (2001)
27. Chango, J.I.: Instrucciones de uso TGA 50 Shimadzu. Quito (2015)

28. Bauer, T., Breidenbach, N., Eck, M.: Overview of molten salt storage systems and material development for solar thermal power plants, pp. 1–8 (2012)
29. Raade, J.W., Padowitz, D.: Development of molten salt heat transfer fluid with low melting point and high thermal stability. Sol. Energy Eng. **133**, 031013 (2011)
30. Hoshino, Y., Utsunomiya, T., Abe, O.: Thermal decomposition of sodium nitrate and the effects of several oxides on the decomposition. Bull. Chem. Soc. Jpn. **54**(5), 1385–1391 (1981)
31. Gimenez, P., Fereres, S.: Effect of heating rates and composition on the thermal decomposition of nitrate based molten salts. Energy Procedia **69**, 654–662 (2015)
32. Li, C., Yan, N., Ye, Y., Lv, Z., He, X., Huang, J., Zhang, N.: Thermal analysis and stability of boron/potassium nitrate pyrotechnic composition at 180 °C. Appl. Sci. (Switzerland) **9** (17) (2019). https://doi.org/10.3390/app9173630
33. Wagner, M., Widmann, G.: Thermal Analysis in Practice. Mettler-Toledo (2009)
34. Granados, Y.A.: Importancia De Los Ensayos TGA y DSC en el Estudio de las Propiedades Térmicas de Mezclas Asfálticas, Universidad Distrital Francisco José De Caldas (2015)
35. Solar, P., Gmbh, I.: Survey of thermal storage for parabolic trough power plants period of performance, September 2000
36. Zalba, B.: Review on thermal energy storage with phase change: materials, heat transfer analysis and applications. Appl. Therm. Eng. **23**(3), 251–283 (2003)
37. Svoboda, R., Sadovska, G., Podzemna, V., Kohutova, A., Organické, Ú., Av, B.: Heat capacity of NaNO2, July 2009
38. National Center for Biotechnology Information, Pubchem Compound Database (2017)
39. Kawakami, M., Suzuki, K., Yokoyama, S., Takenaka, T.: Heat capacity measurement of molten NaNO 3 -NaNO 2 -KNO 3 by drop calorimetry, pp. 201–208 (2004)
40. Jeong, S.G., Chung, O., Yu, S., Kim, S., Kim, S.: Improvement of the thermal properties of Bio-based PCM using exfoliated graphite nanoplatelets. Sol. Energy Mater. Sol. Cells **117**, 87–92 (2013)

Thermal Comfort Evaluation in a Building with Phase Change Materials in Different Ecuadorian Climatic Zones

Hugo Sebastián Romero Espinosa[1], E. Catalina Vallejo-Coral[2], Miguel Darío Ortega López[1], and Javier Martínez-Gómez[3](✉)

[1] Escuela Politécnica Nacional, Quito, Ecuador
[2] Instituto de Investigación Geológico y Energético, Quito, Ecuador
catalina.vallejo@geoenergia.gob.ec
[3] Universidad Internacional SEK, Quito, Ecuador
javiermtnezg@gmail.com

Abstract. Nowadays, the building sector aims to improve the thermal comfort and to reduce the energy consumption implementing new alternative systems. For this purpose, a new phase change material (PCM) is being studied to analyze the performance of a building in three different climates of Ecuador by using a simulation tool. This work assesses the PCM performance through a virtual model in Design Builder, when this material is used as part of a building envelope. To improve the reliability of simulation results, the model is calibrated with experimental previous existing data of a monitoring room without occupancy. The PCM is located in the roof and walls of the virtual calibrated model. The prime matter of PCM is found in Ecuador and thermal and chemical characteristics were product of a previous study. On the other hand, this study assesses the thermal comfort by using the predicted mean vote (PMV). The original model and the PCM model results are compared by the total time of thermal comfort during a year in three Ecuadorian cities.

The main conclusions include that the use of a PCM in a building located in Quito increases the thermal comfort time. Since the PCM decreases, the air temperature fluctuation decreases throughout the year. However, the use of PCM in Zumbahua represents a disadvantage because the climate conditions do not allow energy to be stored by the PCM. In Guayaquil, the PCM performance could be improved by using air-conditioned systems, reaching both, the change phase temperatures and energy savings.

Keywords: Phase change material (PCM) · Building thermal comfort · Energy simulation · Ecuador climatic zones

1 Introduction

The thermal environment quality, inside a building, is a fundamental requirement for the occupants, since it has a great influence on satisfaction level with the general interior environment [1]. Under this context, on 2015, the United Nations agreed on a set of objectives for sustainable development, one of them is related to buildings and

M. Botto-Tobar et al. (Eds.): CI3 2020, AISC 1277, pp. 390–402, 2021.
https://doi.org/10.1007/978-3-030-60467-7_32

living conditions [2]. Currently, urban areas are inhabited by approximately 130 million families in Latin America, of which 42 million live under the minimum living conditions, affecting their comfort and health [3]. Ecuador has an estimated population of 17 million, which constitute 3.8 million homes. Additionally, 45% of Ecuadorian households have a housing deficit; 36% and 9% of households suffer from qualitative and quantitative deficit respectively. The qualitative deficit refers to the buildings with low thermal performance envelope, presenting an inadequate conservation condition. In Ecuador, 1.37 million households live in these conditions [4].

The building envelope has an important impact on energy conservation [5] since it regulates the influence of outdoor weather conditions on the internal environment. Currently, in the construction industry, technologies and materials are emerging to increase the envelope energy performance. Phase change materials (PCM), due to their ability to absorb thermal energy when it exists in abundance and reject it in periods of deficiency, have presented considerable development in the construction market through research [6].

The scientific interest, for the application of PCM in the building envelope, is reflected in of Lagou′s work [6]. Which, through a finite element method (FEM), studied the operation of PCM considering influential aspects such as: its location in the building envelope and melting temperature, and the external and interior environmental conditions. The analysis was performed to six European locations, under diverse climatic conditions, to non-conditioned vernacular buildings. The results show that the PCM optimal melting and freezing temperatures depend on the external climatic conditions to which the material is exposed. Furthermore, Lagou defined that the wall internal side is the optimal place to position the PCM in all cities evaluated, since the inner wall surface temperature presents lower variability, throughout a day, than external surface. Several research have shown an optimal performance of passive envelope systems with PCM when the melting temperature is in accordance with the climatic conditions. In this regard, Saffari [5] developed an optimization method through energy simulation. The methodology selects an optimal melting temperature to enhance energy performance of heating and cooling systems in different climate conditions defined on Köppen-Geiger classification. The results show that, a PCM temperature range of 24–28 °C is the best material for a cooling dominant climate to reduce annual energy consumption; while to heating dominate climates the best PCM temperature range is 18–22 °C.

Markarian [6] used a multi-objective optimization technique, based on energy simulation in DesignBuilder software, to determine the PCM type and location inside the envelope. The study carried out lets to select a suitable configuration to reduce air conditioning loads in five cities in Iran with different climates. The results show that a PCM with a melting temperature of 25 °C has excellent performance to reduce cooling loads. To minimize the heating load, a PCM with a melting temperature of 21 °C is required. Moreover, it is possible to reach energy savings between 4.5–5.5% in all locations and carbon footprint reduction of 2040 kg of CO_2.

Previous studies show evidence about the advantages of using envelope materials according to their thermo-physical properties and climatic conditions. However, the architectural design of buildings in Ecuador generally does not consider those features. This is despite the fact that the Ecuadorian territory covers different climatic zones due to the influence of the Andean mountains, sea currents, location, and altitude [8]. Considering the importance of the proper use of building materials depending on local weather conditions and applications of the PCMs in comfort inside a dwelling, Beltran [9], selects PCMs to be used in wallboard (walls and roof) to enhance a generic social dwelling designed by MIDUVI and built in all Ecuadorian climatic zones. A multi-criteria analysis was carry out [10] followed by energy simulations to a thermal performance comparison of PCM energy performance in three cities located at Coast, Highlands and Amazon of Ecuador.

On the other hand, Acurio [11] characterized the thermal and chemical behavior of a PCM obtained from different mixtures of organic components, palm oil esters, commercial stearic acid and diatomite like support material. Diatomite is a residue of the palm oil industry in Ecuador. The results show phase change temperature between thermal comfort range, long-term thermal and chemical stability. For that reasons, the material can be used as building thermal energy storage.

The literature review shows global focus to PCM thermal behavior assessment. The effect over occupant thermal comfort, in regions where HVAC system are not widely used, it is not prioritized. This study aims to assess the influence of a new PCM in occupant thermal comfort. The assessment is carry out at different Ecuadorian climatic zones, and the PCM studied is a mixture of palm oil and commercial stearic acid derived from the production of palm oil in the Ecuador. PMV method is used with data gets from energy simulation with DesingBuilder.

2 Methodology

This section details the methodology used for: i) the calibration of the model, ii) PCM simulation in different climatic zones of Ecuador, and iii) thermal comfort assessment. In Ecuador, 63.7% of the buildings use concrete block for walls, while concrete is used in 82.6% of the buildings for foundations, 90.1% for the structure and 51.2% for the roofs. It means that the concrete block and reinforced concrete are the predominant building materials in Ecuador [12]. Due to, the aim of this project is to assesses the PCM thermal performance respect to Ecuador predominant envelope materials under different climatic conditions a building 2.6 meters high,7.2 m^2 area, concrete block walls, and solid concrete roof was selected. The building selected is located in Monterrey – Mexico and the superficial temperature of the envelope elements were measured while air conditioning system remained off. The model is part of the facilities of the ITESM (Instituto Tecnológico y de Estudios superiores de Monterrey) - Campus Monterrey. The photograph of the front façade of the building is presented in Fig. 1.

2.1 Calibration

The calibration of energy simulation increases the reliability of the results. The method used in this work is based on manual and iterative steps described by Reddy [13]. The construction of the model is based on the information available in Vallejo's work [14]. For the simulation, a model is generated considering the building geometry, materials properties, climatic data for the calibration period and other model that are mentioned below. Occupancy schedules are not used for calibration because during the collection of experimental data there was no occupancy.

Geometric and Geographic Data: The geometry was performed at the DesingBuilder program interface with the dimensions provided by Vallejo [15]. Considering the geographical coordinates, the orientation angle of the building, respect to the magnetic north, was determined.

Fig. 1. Building studied

Climate: Weather available data were obtained from a weather station near the monitoring building. The used climatic variables are outside temperature, relative humidity, solar radiation, precipitation, speed and wind direction. In order to DesignBuilder can read these variables, it needs an EPW file. Therefore, weather data were organized and transformed by some auxiliary programs like Berkley mentions [16].

Envelope Materials: A 3-part structure was created for the wall. Two of plaster 0.01905 m thick and an intermediate concrete block 0.1016 m thick. The roof has a concrete layer 0.15241 m thick exposed to the weather and an internal plaster layer. Firstly, the first model uses the thermal properties (without surface properties) and dimensions based on Vallejo's work [14] and show in Table 1. The surface properties were the adjustment variables to calibrate the model. These values were taken as a starting point from ASHRAE Handbook Fundamentals [17] and they were modifying them until that they were selected the best adjust values without exceeding a variation of 10%.

Table 1. Thermal properties of envelope materials.

Element	Conductivity [W/m-K]	Specific heat [J/kg-K]	Density [kg/m^3]
Plaster	0.7	840	2778
Concrete block	0.9	840	977
Concrete	1.73	840	2242

Table 2. Experimental measurement instruments

Parameter/Datalogger	Instrument	Range	Accuracy
Wall/roof surface temperature	K-Type TC	−270 to 1372 °C	±0,5 °C
Inside air temperature (with radiation shield)	K-Type TC	−270 to 1372 °C	±0,5 °C
Datalogger	OMEGA OMB-DAQ-55	Input ± 15 V	0.015% of reading

Parameters of Comparison: In order to validate the thermal simulation model, the internal surface temperature of the roof and walls, and the indoor air temperature obtained from the model and the experimentally recorded values are used as parameters of comparison. The experimental data was collected by the instruments of Table 2

Statistical Indicators: The indicators compute the simulation results variation respect to the experimental ones. The statistical indicators RMSE (root mean square error) and MBE (mean bias error) are used. When using experimental temperature data, it is recommended to use additional indicators [18, 19] like correlation coefficient (R), maximum and minimum error of all hourly calibration.

2.2 Simulation of the Building Using PCM

Based on the calibrated model and the climatic data of three cities in Ecuador, the building simulation is carried out by implementing a PCM layer 0.01 [m] thick. On the roof, the PCM is located between the layers of plaster and solid concrete, while on the wall, the PCM is between the concrete block and the internal plaster. To evaluate the envelope behavior, the temperatures of the air and the inner surface of the roof are compared with and without the PCM use. To assess the impact of the PCM use on occupant comfort, the model is simulated as an office occupied by a person who is sitting typing in front of a computer from 8:00 to 18:00, 365 days a year. Clothing depends on the climatic zone.

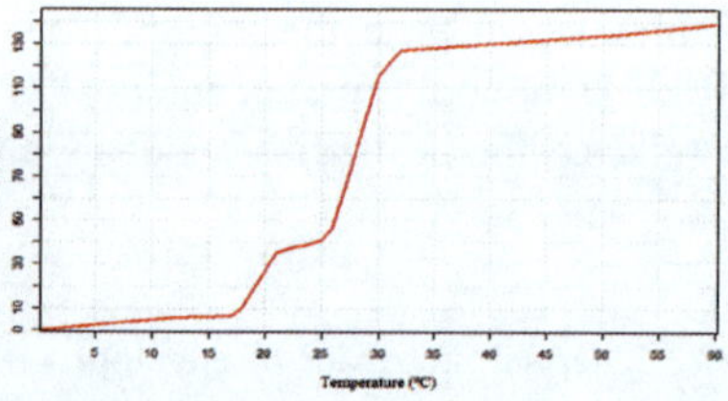

Fig. 2. PCM enthalpy - temperature curve

Phase Change Material (PCM): In this study, the PCM support material is new diatomite soil. It is a porous material and can be purchased from companies that extract oil in Ecuador. The PCM organic compound is a mixture of palm oil (matter over-produced in Ecuador) and commercial stearic acid derived from the production of palm oil in the country. The thermal characteristics of this material are found in the results of Acurio's study [11]. In this study, PCM with 100% stearic acid is used as an organic compound and new diatomite as support material.

The results of the DSC test (Differential Scanning Calorimetry) conducted by Acurio [11] show that this type of PCM has a greater capacity for latent heat storage, and its melting temperature range is close to the fluctuations in the ambient temperature. The PCM stabilizes at 360 cycles in the DSC test. For this reason, the enthalpy versus temperature curve, required by the DesignBuilder, was obtained from the test data at 360 cycles. The results obtained are presented in Fig. 2. which shows that the melting temperature range is between 17 °C and 32 °C. The material melting process occurs when the enthalpy abruptly increases in a small temperature range.

Climatic Zones: The cities, Quito and Guayaquil, are selected because they have the highest population density at national level. Zumbahua represents towns that exceed 3000 [m.a.s.l]. The climatic parameters considered are: dry bulb temperature, spray temperature, direct normal solar radiation, diffuse horizontal solar radiation, wind speed, wind direction, and atmospheric pressure. The weather data used for the simulations of each climate zone was generated by the Meteonorm software, which is a weather database, generating hourly data for a typical weather year. These data are generated for a point located in the centroid of each city.

2.3 Comfort Evaluation

The thermal comfort analysis of the study is based on the ANSI/ASHRAE standard 55, and the PMV (predicted mean vote) method applicable for specific environmental conditions is used [20]. The metabolism and clothing are determined depending on the activity and the climatic conditions of the study respectively [21]. Insulation values for clothing are obtained from ASHRAE Handbook Fundamentals [17]. The method uses, as evaluation parameters, air and radiant temperature, and relative humidity of the environment. They are obtained from the simulated model, resulting in the values of the hourly PMV. The acceptance comfort range is $-0.5 < \text{PMV} < +0.5$. Once the PMV values have been calculated, frequency histograms are performed at 0.1 PMV intervals, and statistical parameters of position and dispersion are determined. In addition, the percentage of comfort hours is calculated with respect to the hours that were occupied during a year.

3 Results

This section details the calibration results with experimental air temperature data and statistical indicators. In addition, the results of thermal comfort are presented with and without PCM use for a representative week in three climatic zones of Ecuador.

3.1 Calibration of the Simulation

In order to perform the calibration analysis, comparison charts between the experimental and calibration parameters and statistical indicators (RMSE, MBE, R, maximum and minimum error) were used. The emissivity and solar absorbance of the plaster were modifying until reach the lower values of RMSE and MBE. As result of calibration the final value of emissivity and solar absorptance were 0.99 and 0.12 respectively. The calibration period is from September 16 to 20, 2016, corresponding to 120 h. In Fig. 3, the air temperature, from the calibrated model and experimental data, are presented. A similar trend is observed, and the statistical indicators are within the ranges established by the M&V Guidelines (Measurement and Verification for Performance) and in ASHRAE 14: 2014. [17, 18] as is shown in Fig. 3. The differences between calculated and measured values can be generated because the properties of the materials were not obtained experimentally by altering the simulation results with respect to reality.

The RMSE and MBE statistical indicators, regarding the comparison of calibration parameters, are within the acceptable range according to the regulations [22, 23]. The R value is close to 1 for all calibration parameters, demonstrating that simulated and experimental temperatures have an acceptable direct relationship as described in the references [24] (Table 3).

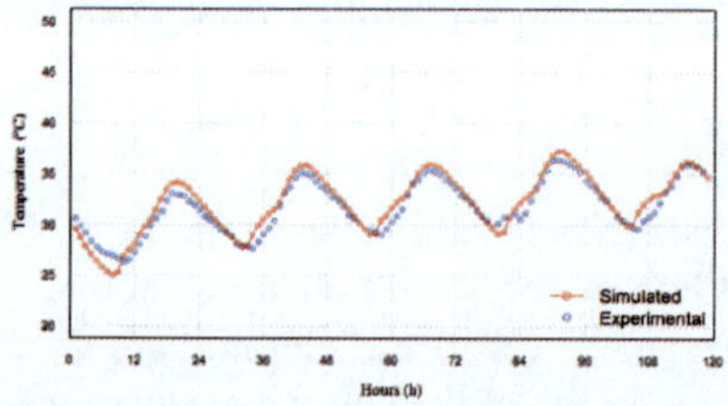

Fig. 3. Experimental and simulated data of the indoor air temperature from September 16 to September 20, 2016, in Monterrey.

Table 3. Calculated calibration indicators values.

Parameters of comparison - Temperature	RMSE (hourly) [%]	MBE (hourly) [%]	Max error [°C]	Min error [°C]	R
Indoor air temperature	2.85	−1.59	1.8	−1.8	0.97
Wall internal surface	3.2	1.85	1.5	−1.8	0.97
Roof internal surface	3.99	1.37	3.44	−1.62	0.98

Considering that, the experimental data presents uncertainty and, according to the relevant standards, the statistical indicators are within ranges defined as acceptable, an acceptable correspondence between the simulated calibration parameters and the experimental measurements is observed.

3.2 Building Thermal Behavior Evaluation in Ecuador

In order to know the influence of the PCM use in occupant thermal comfort, the thermal behavior of the building was evaluated. It considering 3 different climatic zones of Ecuador based on the calibrated model. The evaluation is carried out by comparing the internal surface temperature of the roof and the internal air. The analyzed period, from December 25 to 29, corresponds to the five continuous days of a typical meteorological year that present the average temperature of the outside air within the most frequent range in Quito, Guayaquil, and Zumbahua as shown in Fig. 4.

Quito has two seasons, a rainy cold season and temperate dry season with mean temperatures that oscillate between 19 and 10 °C. Figure 5a shows the internal surface temperatures of the roof and two behaviors are observed. In the first 60 h, the roof temperature is not higher than the lower limit of the PCM melting range due to the climatic conditions. It behavior is similar to the roof without PCM. From the 61st hour, during the day, the temperature of the roof rises and exceeds the lower melting limit of the material. PCM stores latent heat by phase change and the air temperature decreases. At night, the roof temperature decreases and it produces heat rejection to the indoor air, increasing its temperature. Therefore, it can be concluded that the PCM use in Quito decreases the fluctuation range of the indoor air temperatures.

Zumbahua is the highest city studied. It has a cold climate with a lot of cloudy dates. It has same seasons of Quito with temperatures that oscillate between 14 and 6 °C. Figure 5b illustrates the comparison of the roof internal surface temperature and the air temperature behavior with and without the use of the PCM in Zumbahua.

Guayaquil has a tropical climate with dry and rainy seasons. The temperatures oscillate between 30 and 20 °C. Figure 5c shows the comparison of the roof internal surface temperature and the air temperature with and without the PCM use, in Guayaquil. The weather conditions keep the roof temperature close to the upper limit of the PCM melting range. It allows that PCM is as liquid or dissolved in a liquid-solid mixture.

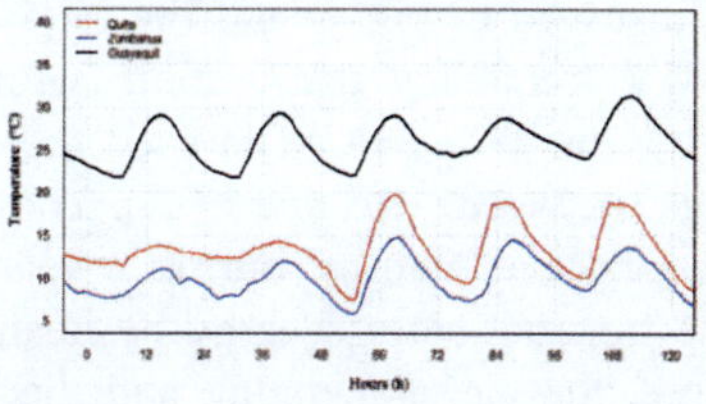

Fig. 4. Environmental temperature of the studied cities from December 25 to 29 of a typical meteorological year.

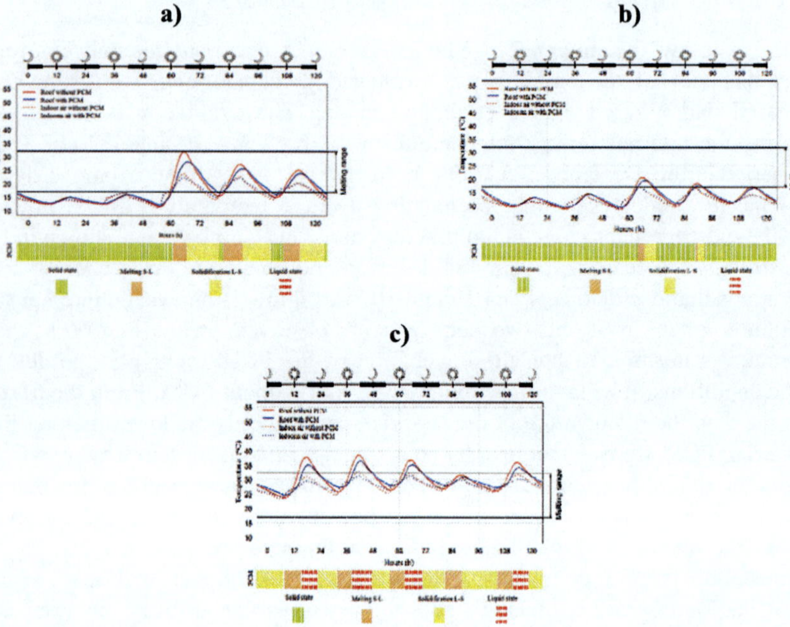

Fig. 5. Roof internal surface and air temperature with and without the PCM use, from December 25 to 29 of a typical meteorological year, in a) Quito; b) Zumbahua and c) Guayaquil

3.3 Evaluation of Thermal Comfort

The PMV method was used to assess the thermal comfort. The building is considered as office and the occupation schedule is from 8:00 to 18:00, resulting 3650 h per year. For the comfort analysis in each city, the comfort and discomfort percentage hours caused by low or high temperature, are determined respect to the total occupation hours in a year. The results are illustrated in Fig. 4 and Fig. 5.

Figure 5 shows the PMV frequency distribution, without the PCM use and it presents a positive bias in Quito. When PCM is used the PMV distribution is symmetrical to the comfort range (between −0.5 and −0.4). The mode range change shows that the PCM use increases thermal comfort during a year inside the building. Furthermore, in Guayaquil, the material stores energy by change of state in the greatest thermal load hours; therefore, the air temperature and the roof internal surface temperature decrease. For this reason, the PMV range mode increases, and the frequency histogram distribution changes from positive to negative bias. Finally, in Zumbahua there is a symmetrical distribution with and without PCM use. Since, PCM does not accumulate or reject heat most days, it does not generate a significant change in thermal comfort (Table 4).

Table 4. Results of thermal comfort with and without the use of PCM in Quito, Guayaquil and Zumbahua.

Comfort state	Quito		Guayaquil		Zumbahua	
	PCM	No PCM	PCM	No PCM	PCM	No PCM
Comfort [%]	53.91	48.35	12.02	15.42	17.69	21.34
Discomfort by high temperature [%]	0.69	4.01	87.75	84.19	0	0.1
Discomfort by low temperature [%]	45.4	47.64	0.23	0.39	82.31	78.56

In Guayaquil, the climatic conditions keep the roof and the walls temperatures close to the upper limit of PCM melting. It decreases the air temperature fluctuations range. Therefore, thermal comfort decreases by 3.4%, which represents 124 h, due to the PCM use. It means a decrease of 0.16% (6 h) of discomfort due to low temperatures, and an increase of 3.56% (130 h) of discomfort due to high temperatures. In addition, it was observed that PCM change its thermal state more days in Guayaquil than in Quito and Zumbahua.

4 Discussion

The phase change material in Quito increases the comfort hours because it increases the roof internal surface and the air temperatures during the coldest day hours, while it decreases the temperatures in the hottest hours. The behavior of the PCM for Quito is similar to Beltrán's work in which it shows that the thermal behavior of a PCM is a response to weather conditions. It mentions that in Quito the PCM exhibits good thermal behavior during the day and especially at night [10]. The PCM with best performance for a homologous climate similar to Quito (Cwb in the Kôppen-Geiger classification [25]) has a similar melting temperature to the PCM used in this study [5]. Therefore, the PCM used in this study presents the best behavior in the climatic conditions of Quito respect to Guayaquil and Zumbahua. In Zumbahua, the PCM does not generate a significant change in the behavior of the air temperature. In cold climatic zones PCM studied remains in solid state and it is not able to store or reject heat. It could be observed in Park´s work [26] that shows since PCMs does not reach its melting temperature (Fig. 6).

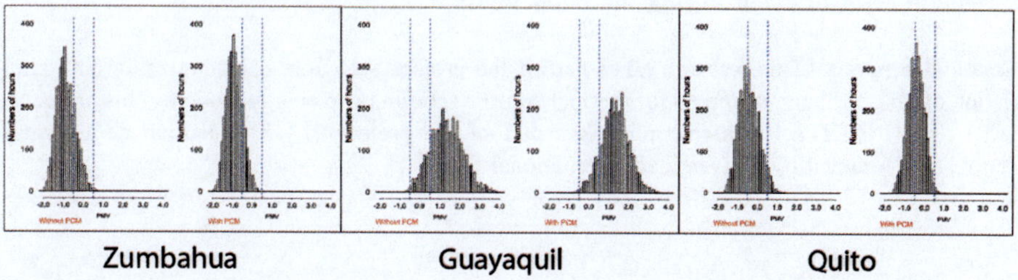

Fig. 6. PMV histograms with and without PCM in Quito, Guayaquil, and Zumbhua.

In Guayaquil, the indoor air temperature behavior when PCM is used it is similar to Beltrán's results [10]. From 0:00 to 12:00 the air temperature increases while from 12:00 to 24:00 it decreases respect to the model without PCM increasing discomfort due to high temperatures. During occupation hours, PCM modifies meaningfully the thermal environment because it presents more phase change cycles throughout the year which results in PMV values distribution change from a positive to negative bias. However, in Guayaquil′s case could be relevant when talking about energy savings. According Esbati′s work, PCM can decrease 8% of cooling energy [25]. Furthermore, PCMs could be a practical passive solution for higher outdoor temperatures [26]

When the PCM reaches temperatures within its melting range, it modifies the indoor environment otherwise it does not generate a considerable change in air temperature. In Quito and Guayaquil, the PCM stores energy due to weather conditions. In Guayaquil, the PCM stores and rejects heat more days, in a year, than in Quito as it is clearly observed by comparing Fig. 5a and Fig. 5c. By comparing the roof behavior of the 3 cities, Quito stores more heat because the weather conditions allow the roof temperatures to be within the PCM melting range. While, in Guayaquil, the roof temperature is close to the upper limit and the PCM remains as liquid, at the hottest day temperatures and it does not store latent heat by phase change.

5 Conclusion

The surface properties of the materials are important to the simulation validation process. The properties values modify the air and internal surface temperatures which are simulation results. The errors between the simulated results and the experimental data are reduced by selecting values, of the materials surface properties, of the close to those found in literature.

Generally, the MBE and RMSE statistical indicators are used to assess the calibration of energy simulations taking, as a comparison parameter, the building energy consumption over a period of time. To evaluate the calibration of a simulation using experimental temperature data, it is necessary to use additional statistical indicators which complements those recommended. This decreases the calibration time and improves the results. Precipitation is an influential parameter in the calibration process. The correct precipitation value may decrease the difference between calibration results and experimental data. A higher value than real in the precipitation generates an error by decreasing the indoor temperature of the roof and the air, causing 1.8 [°C] of maximum error to occur during the rainy hours.

Acknowledgment. This research takes part of the project Selection, characterization and simulation of phase change materials for thermal comfort, cooling and energy storage. This project is part of the INEDITA. This research takes part of the project P121819, Parque de Energias Renovables founded by Universidad International SEK.

References

1. Mora, R., Bean, R.: Thermal comfort: designing for people. ASHRAE J. **60**(2), 40–46 (2018)
2. Villacreses, G., Gaona, G., Martínez-Gómez, J., Jijón, D.J.: Wind farms suitability location using geographical information system (GIS), based on multi-criteria decision making (MCDM) methods: the case of cont. Ecuador. Renew. Energy **109**, 275–286 (2017)
3. Espinoza, V.S., Guayanlema, V., Martínez-Gómez, J.: Energy efficiency plan benefits in Ecuador: long-range energy alternative planning model. Int. J. Energy Econ. Policy **8**(4), 52–54 (2018)
4. Aldás, P.S.D., Constante, J., Tapia, G.C., Martínez-Gómez, J.: Monohull ship hydrodynamic simulation using CFD. Int. J. Math. Oper. Res. **15**(4), 417–433 (2019)
5. Saffari, M., de Gracia, A., Fernández, C., Cabeza, L.F.: Simulation-based optimization of PCM melting temperature to improve the energy performance in buildings. Appl. Energy **202**, 420–434 (2017)
6. Lagou, A., Kylili, A., Šadauskienė, J., Fokaides, P.A.: Numerical investigation of phase change materials (PCM) optimal melting properties and position in building elements under diverse conditions. Constr. Build. Mater. **225**, 452–464 (2019)
7. Markarian, E., Fazelpour, F.: Multi-objective optimization of energy performance of a building considering different configurations and types of PCM. Sol. Energy **191**(May), 481–496 (2019)
8. Martínez-Gómez, J., Guerrón, G., Riofrio, A.J.: Analysis of the "Plan Fronteras" for clean cooking in Ec. Int. J. Energy Econ. Policy **7**(1), 135–145 (2017)
9. Martínez, J., Martí-Herrero, J., Villacís, S., Riofrio, A.J., Vaca, D.: Analysis of energy, CO2 emissions and economy of the technological migration for clean cooking in Ecuador. Energy Policy **107**, 182–187 (2017)
10. Beltrán, R.D., Martínez-Gómez, J.: Analysis of phase change materials (PCM) for building wallboards based on the effect of environment. J. Build. Eng. **24**, 100726 (2019)
11. Acurio, K., Chico-Proano, A., Martínez-Gómez, J., Ávila, C.F., Ávila, Á., Orozco, M.: Thermal performance enhancement of organic phase change materials using spent diatomite from the palm oil bleaching process as support. Constr. Build. Mater. **192**, 633–642 (2018)
12. Martínez-Gómez, J., Ibarra, D., Villacis, S., Cuji, P., Cruz, P.R.: Analysis of LPG, electric and induction cookers during cooking typical Ecuadorian dishes into the national efficient cooking program. Food Policy **59**, 88–102 (2016)
13. Reddy, T.A.: Literature review on calibration of building energy simulation programs: uses, problems, procedures. ASHRAE Trans. **112**(1), 226–240 (2006)
14. Vallejo, E.: Determinación de CLTD para cargas de enfriamiento de edificaciones ubicadas en ciudades de clima cálido en México, no. c, pp. 1–4 (2017)
15. Vallejo-Coral, E.C., Rivera-Solorio, C.I., Gijón-Rivera, M., Zúñiga-Puebla, H.F.: Theoretical and experimental development of cooling load temperature difference factors to calculate cooling loads for buildings in warm climates. Appl. Therm. Eng. **150**, 576–590 (2018)
16. Berkeley, L., et al.: Energy Plus: Auxiliary Programs (2018)
17. ASHRAE.: Handbook - Fundamentals (SI Edition), American Society of Heating, Refrigerating and Air-Conditioning Engineers, Inc. (2017)
18. Roberti, F., Oberegger, U.F., Gasparella, A.: Calibrating historic building energy models to hourly indoor air and surface temp: meth. Energy Build. **108**, 236–243 (2015)
19. Gaona, D., Urresta, E., Marínez, J., Guerrón, G.: Medium-temperature phase-change materials thermal characterization by the T-History method and differential scanning calorimetry. Exp. Heat Transf. **30**(5), 463–474 (2017)

20. ASHRAE.: Standard 55. Thermal Environmental Conditions for Human Occupancy (2017)
21. Kastillo, J.P., Martínez-Gómez, J., Villacis, S.P., Riofrio, A.J.: Thermal natural convection analysis of olive oil in different cookware materials for induction stoves. Int. J. Food Eng. **13**(3) (2017)
22. ASHRAE.: ASHRAE Guideline Project Committee 14P, vol. 8400 (2002)
23. US Department of Energy Federal Energy Management Program: M & V Guidelines : Measurement and Verification for Contracts, no. November (2015)
24. Martínez-Gómez, J.: Material selection for multi-tubular fixed bed reactor Fischer-Tropsch reactor. Int. J. Math. Oper. Res. **13**(1), 1–29 (2018)
25. Esbati, S., Amooie, M.A., Sadeghzadeh, M., Ahmadi, M.H., Pourfayaz, F., Ming, T.: Investigating the effect of using PCM in building materials for energy saving: case study of Sharif Energy Research Institute. Energy Sci. Eng. **8**(4), 959–972 (2019)
26. Auzeby, M., Wei, S., Underwood, C., Chen, C., Ling, H.: Using phase change materials to reduce overheating issues in UK. Energy Procedia **105**, 4072–4077 (2017)

Simulation of a Phase Change Material for an Automotive Rooftop Thermal Insulation System

Andrés Mendez[1], Javier Martínez-Gómez[1,2(✉)], and Juan Francisco Nicolalde[1]

[1] Universidad Internacional SEK, Quito Albert Einstein s/n and 5th, Quito, Ecuador
javier.martinez@uisek.edu.ec
[2] Instituto de Investigación Geológico y Energético (IIGE), Quito, Ecuador

Abstract. The temperature that a car can reach inside is between 25 °C to 40 °C in certain environmental circumstances, for this reason studies have been carried out in the field of materials science that allow to control the interior environment of a car. This study proposes the thermal simulation of paraffin a as PCM, the simulation is performed with the NX software, the material selected was chosen through multicriteria given its melting point that is 37 °C, the result of simulating this material as a heat insulator is that in a 50 °C of incidence in the roof of a vehicle, the exterior of the material shows 48 °C and the internal temperature decreases through the thickness of the part until is kept in 20 °C, also the simulation presents the temperature gradient, the non-existing heat flow what means that phase change process of the material is taking place, it stores thermal energy efficiently as expected, finally the simulation of the rooftop show how the paraffin allows maintaining an indoor temperature close to the thermal comfort range which is between 18 °C and 22 °C.

Keywords: Thermal simulation · PCM · Automotive

1 Introduction

The material to be use in this simulation was selected by multi-criteria means, as a first step taking in consideration the needs of the application [1], in these case, among these characteristics the most important are: density, heat of fusion, melting point and latent heat, where paraffin result as the most apt for the task.

The use of latent heat of phase change materials to cool the environment is the oldest concept of air conditioning that exists. These materials have the capacity of store latent heat, high fusion heat and a transition phase point in the operative temperature [2], high thermal conductivity, congruent fusion, low volume transition phase [3] these systems store thermal energy to release them when it's needed, there are some materials that have this properties, one of the is the paraffin [4]. Phase change materials have a wide range of application in the industry, construction, factories, foods, medical, clothing and automotive [5, 6].

M. Botto-Tobar et al. (Eds.): CI3 2020, AISC 1277, pp. 403–415, 2021.
https://doi.org/10.1007/978-3-030-60467-7_33

In the construction industry it is where its use is most popular, they are already used in thermal insulation systems, which improves the comfort of buildings and constructions, since the PCMs being incorporated with traditional construction materials such as plaster, improve the properties of this and add their own characteristics, which create new compounds capable of isolating the outside temperature in the buildings and maintain the internal temperature of the rooms [7]. The biggest advantage of the phase change materials is the thermal range in which they can work, this range goes from below −10 °C to above 90 °C, which results in the applications of the PCM have a wide range of possibilities, within the automotive industry we can highlight the following applications for these materials.

In the logistics and transport sector, PCMs are able to guarantee the temperature stabilization of products and goods. Especially in the field of pharmaceutical products logistics, which meet high demands. In addition, they have already been tested for the transport of machines, works of art and food; sectors that have special temperature sensitivity. The PCMs, with thermal windows between 2–8 °C, are poured into plastic containers and act as cooling accumulators inside the containers and transport trucks. What preserves the quality and integrity of the products that are being stored in conditions without other ventilation or air conditioning [7]. In the automotive sector, PCMs have a variety of applications. While driving, the PCM stores the cold from the air conditioner. During the waiting periods, this stored cold is released by a fan, keeping the temperature inside the cabin at a comfortable level.

In the automotive industry, phase change materials have a somewhat limited application capacity, since their main function is to store thermal energy, however, a PCM could store cold while driving to prevent the use of air conditioner, during long waiting periods this store cold could be released by the fan, keeping de temperature at a comfortable level, also, the PCM could store the heat of the engine in order for the next cold start to short the heating phase, by this means minimizing its degradation and reducing the fuel consumption [8]. Paraffin that is an organic PCM which is a stable, safe and has a high latent fusion heat (180 kJ/kg) [9], also is very stable to the different freezing and fusion cycles, has low reactivity, low hysteresis, it's not toxic but is inflammable [10], for this reason, as an insulating material, that can reduce the heat transfer, protecting from heat and cold and contributing to the energy efficiency [11], the paraffin can be used in part of the car's upholstery which improves comfort inside the vehicle.

The paraffin can be found in a micro encapsulation state, this could have spherical shape with a wall around the core or asymmetric with little drops of material along the micro capsule [12], the intentions of the micro encapsulation is that the base material could be inside the capsule for specific time period, also the material can be encapsulated in order to be release part by parte through the walls, this event is known as controlled liberation or diffusion, when the conditions are extreme the capsule could break, melt or dissolve [2].

One of the most favorable results in this automotive section for phase change materials is meet thanks to its performance as thermoregulators, this serves to reduce the thermal oscillations of a space, such as the interior of a car if the PCM is incorporated in the upholstery of the same as an insulator, this taking into account that the material always works around the temperature of change of the material's own phase.

The physics of the automotive and their parts is very complex, there are many energetic flow exchanges between de interior and the exterior of the vehicle and those define the thermal behavior. The energy that enters in a zone is called thermal gain and the potential that has to be introduced or extract from a zone to maintain the comfort conditions of heat and humidity is called thermal load [13].

The best way of validate a result is by the simulation means, especially in thermal simulations specialized software are needed [14] because the simulation of thermal conditions is a useful tool for the evaluation of different alternatives for the design, or thermal properties of construction materials. Thermal management is a fundamental aspect to consider in a wide variety of products, such as vehicles and consumer electronics. The objective of any thermal simulation is to keep the temperature of the product within an optimum range for its performance. However, achieving this may involve the addition or extraction of heat, either actively or passively, something that can be evaluated using thermal simulation software such as the NX program [15].

By using the NX10 software, from Siemens, a material can be created and assigned the actual physical and chemical properties that it possesses, and with this data the software gives us the possibility to perform tests on the material, in this case in particular a Thermal load simulation, solved by the finite element method incorporated in the solver of the same program, allows to visualize the behavior that the material will have in a real environment when it is subjected to the given temperature conditions, and observe its reaction.

Being a simulation on the thermal behavior of a material, the NX10 program gives us a series of results that must be interpreted for later analysis,

In this research the simulation allows us to know the behavior of the paraffin as an insulator and will tell us how the paraffin will keep a comfortable environmental temperature by using it as the upholstery of the rooftop in an automotive.

2 Simulation

The method of validation of the result obtained through a selection process based on multicriteria methods will be carried out through software simulation, the program used for this purpose will be NX10, from Siemens, since this software allows simulation of thermal conditions in a way relatively simple, in order to know the behavior that phase change materials will have once they are subjected to different temperature conditions.

The simulation will be carried out on the roof of the car, since it is in this area where the outside temperature affects directly and the PCM would be incorporated into its panels in the form of an insulating shield to incorporate the phase change material inside the vehicles and thus take advantage of its properties as a thermal insulator in order to improve the comfort of a car.

2.1 Boundary Conditions

Table 1, presented below, indicates the thermal conditions in which the simulation will be performed, as well as the data that the program requires on the material in order to work on its behavior.

In this particular case the software library does not have the paraffin, which is the material necessary for the simulation, so the material is created with all the necessary data according to the chemical and physical properties of the material.

Table 1. Simulation conditions

Condition/characteristic	Values
Environmental temperature	22 °C
Temperature of incidence on the material	50 °C
Melting point of paraffin	37 °C
Latent heat of paraffin	70 kJ/kg
Specific paraffin heat	2.5 kJ/kg.k
Paraffin density	900 kg/m^3
Total simulation time	720 g

2.2 Starting Simulation

To carry out the simulation it is necessary to follow a series of steps within the program.

First it we start with the creation of a body that will be used in the aforementioned application with the data on the chosen material, in the event that the software library itself does not own this material then the boundary conditions are given, the variables that will intervene in the process are known and the basis for the simulation, with the data entered this software has a solver that allows to simulate the storage conditions and thermal behavior that the material will have.

2.3 Body, Material and Meshing

With the initial set to start you proceed to create the 3-dimensional model of the material, starting by entering the environment called "Model" of the program where you can apply a material and the boundary conditions necessary for the simulation, then you proceed to change the environment to the own advanced simulation environment.

Next the material that was created is chosen in order to assign it to the body. Once this process is completed the body changes color indicating that it has an assigned material and that from this point that body behaves according to the physical and chemical characteristics of the material.

Next it must be done a meshing of the body, which will allow the program to use its solver by means of the finite element method for the solution of the simulation, and it allows to know the nodal behavior of the material as well as its total behavior.

To mesh the body it is configured in automatic and since the body has a thickness of 3 mm, the assigned size of the mesh is 1 mm.

Once the mesh is applied on the body it changes color again and the mesh divisions can be seen as thin black lines on the body.

2.4 Loads

Next, the body with which to work and also the material that was created is chosen, in order to assign it to the body. Once this process is completed, the body changes color, indicating that it has an assigned material and that from this point that body behaves according to the physical and chemical characteristics of the material.

Next it must be done is a meshing of the body, which will allow two things, first that the program uses its solver by means of the finite element method for the solution of the simulation, and as a second point it allows to know the nodal behavior of the material, so as well as its total behavior.

To mesh the body it is configured in automatic and since the body has a thickness of 3 mm, the assigned size of the mesh is 1 mm.

Once the mesh is applied on the body it changes color again and the mesh divisions can be seen as thin black lines on the body.

Next, the environment must be changed within the program in order to assign the thermal loads with which to work in the simulation by placing the type of restriction that the simulation will have, in this case the option of type of restriction and the option of "thermal restrictions" are selected, within this select the "fixed temperature" option and the body, which will be the object on which the restriction is applied.

Then the body is rotated 180°, to be able to apply the following restriction on the opposite side, this will be a convection restriction, which indicates to the program that the temperature on this face is different than that applied on the previous face. This allows the software to simulate what will happen inside the material when the heat source begins to act on it.

It is also necessary to complete the ambient temperature values that will be on this side of the body and its convection coefficient. This data is obtained from the properties of the material.

Next, the loads are applied to the material, in this case the applied load will be a thermal, heat generating load.

2.5 Solver

Once you have these material conditions assigned, as well as all the restrictions given, and the thermal loads applied, what is needed is to solve the simulation in order to know the results of it.

The simulation solver, by clicking on Accept starts the simulation process and problem solving, this is done according to the parameters entered above, which are the data with which the program determines the behavior of the material and thus find the results according to the processing.

3 Results and Discussion

Table number 2 shows the results of the values obtained in the simulation, after applying the given conditions and values, these values indicate that the material behaves correctly, that is to say that the simulation gives a result that is expected, showing that The material absorbs thermal energy, or heat, from outside preventing it from entering the car, which can be said that the PCM fulfills its function.

Table 2. Temperature results

Temperature condition	Values
Outside temperature	50 °C
PCM indoor temperature	42 °C
PCM outside temperature	18 °C
Temperature inside the car	22 °C

3.1 Temperature Through Thickness

Figure 1 and 2 show the result of the temperature accumulated in the material, when a thermal load of 50 °C is applied on one side and on the opposite side there is a condition of room temperature at 22 °C, this images shows how the temperature changes through de thickness of the material.

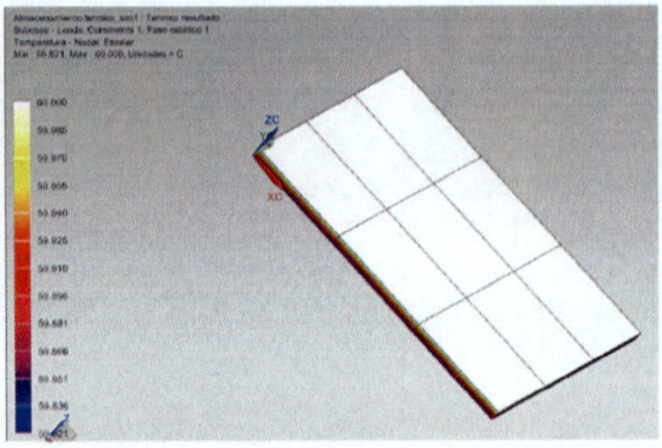

Fig. 1. Front face of paraffin sheet

Figure 1 shows the material on the external side, on which the temperature coming from the outside is affecting, which begins to accumulate inside the paraffin, thus fulfilling the function for which it is intended to provide better comfort thermal. Figure 2 indicates the rear face of the paraffin, according to the color scale that indicates the temperature, this side is colder, because the material is accumulating energy in the form of heat internally.

As can be seen in the two previous figures, the temperature scale is given by colors on the left side of the image, the white color being the highest temperature, which decreases until it reaches the blue color, which represents the highest temperature low that is obtained after the material does its job that is to absorb the heat coming from the opposite side of the material. What this indicates is that paraffin behaves as expected when subjected to a thermal load as an energy storage. The temperature applied is on

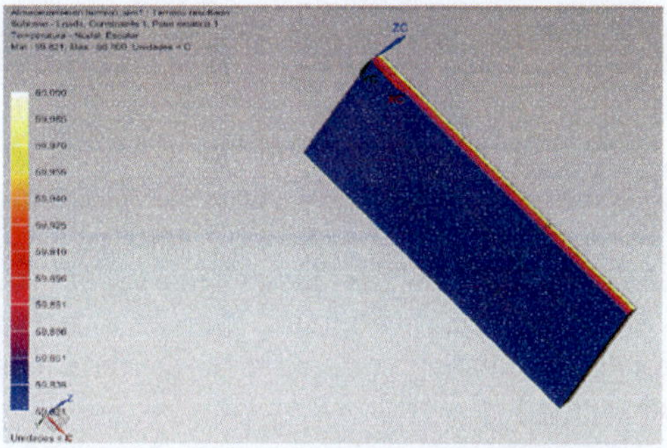

Fig. 2. Back side of paraffin sheet

the white side of the object, which stores this energy in its phase change process, thus preventing heat dispersion, which is why the opposite side of the iron remains blue, indicating that the temperature on that face is lower, the temperature obtained on this side of the material is 18 °C.

Figure 3 shows the interior of the paraffin, in which the process of dispersion and storage of thermal energy within the material can be observed through the thickness of the plate. In this image the side of the material that receives the temperature or the thermal load is remarkably heated, while the opposite side remains colder thanks to the internal process that occurs in the material.

Fig. 3. Cross section of the paraffin sheet

This phenomenon occurs thanks to the characteristics of this material, which allows it to behave in this way, which indicates the simulation is that, basically, the temperature or heat coming from the outside is trapped in the material, which uses this energy so that its phase change process occurs, in which the heat coming from the outside melts the paraffin, which results in the material changing phase, so that this thermal energy is retained in the material and heat does not cross the entire element, which does not get to disperse inside the car, thereby improving comfort inside.

3.2 Temperature Gradient

Next, Fig. 4 shows the result on the temperature gradient, elementally and as a total magnitude. In an elementary way you can see the behavior of the material divided into nodes, both on the X, Y and Z axis. The magnitude is observed completely for the paraffin plate, this temperature variation occurs over the time required by the material for its change of state.

The result of the elementary temperature gradient, according to the simulation can be seen in Fig. 4, here you can also see the division in nodes of the plate and the same color code for the applied and resulting temperature. As it can be seen, the temperature is always higher next to the application of the thermal load, while on the opposite side there is a lower temperature, which indicates that the material behaves as expected and fulfils its function to store thermal energy.

Fig. 4. Nodal paraffin division

Likewise, Fig. 5 shows the temperature gradient, but in this case the dispersion of it on the Y axis is observed within the analysis. This change of axis indicates that the temperature is absorbed by the material evenly throughout its surface. This is mainly due to two factors, the first that the applied thermal load is not punctual, but is distributed over the entire surface of the piece, and the second factor corresponds to the way in which the material behaves, the paraffin maintains its total temperature while the phase change occurs, this means that a part of the material cannot have a temperature higher than the rest of the body during the phase change process.

The total magnitude of the temperature gradient is observed in Fig. 6, here you can see that the accumulation of energy occurs inside the material, which confirms that it behaves as theoretically expected, absorbing the temperature from the side of its incidence causing the phase change inside and keeping the opposite side at a lower temperature. With this it is achieved that the temperature of the inner side is lower than the outside temperature.

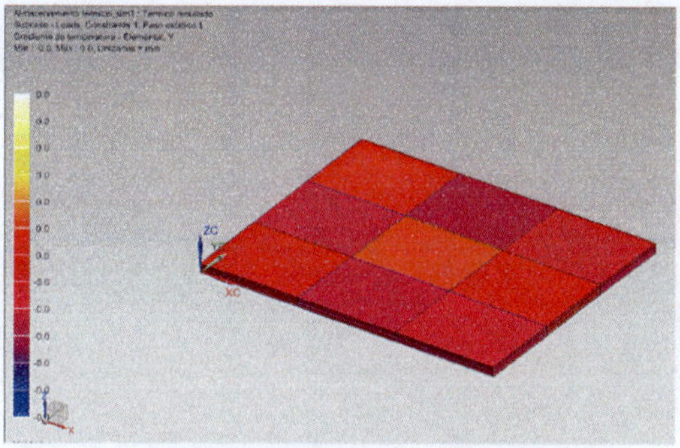

Fig. 5. Nodal temperature gradient

Fig. 6. Thermal energy accumulation

3.3 Heat Flow

Figure 7 shows the result obtained on the behavior of heat flow within the paraffin. This image shows the flow of heat in the different senses within the material, and in the same way as previously indicated, the thermal load was applied on the face of the material, and on the opposite side is where the result is seen, that is the behavior of paraffin.

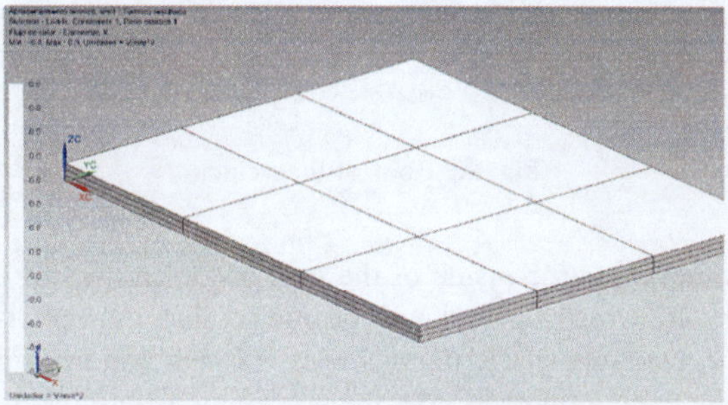

Fig. 7. Heat flow inside the material

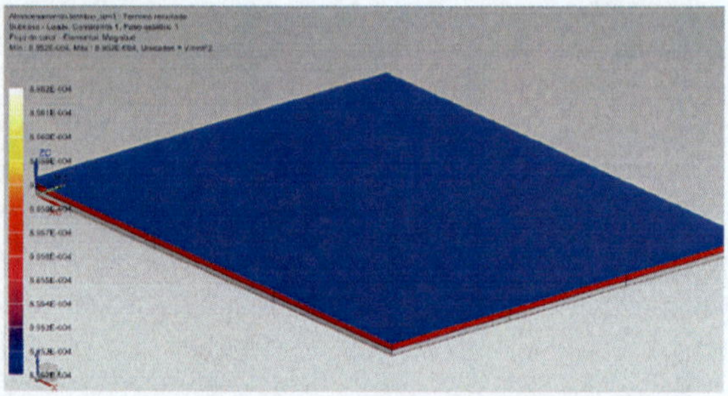

Fig. 8. Temperature distribution in PCM

In this case, the response provided by the simulation, indicates that there is no heat flow, which is why there is no temperature variation, the heat flow is maintained at 0. This is because the material is designed to accumulate thermal energy, which is why it should not allow heat flow, but rather the accumulation of heat, which the software represents as a constant total temperature. The total magnitude of the heat flux can be seen in Fig. 8, in this figure it is seen how the heat flux occurs uniformly over the entire surface of the material.

3.4 Rooftop Simulation

Figure 9 shows the simulation result on the roof panel of a car that has our PCM incorporated. Also shows the color temperature that is applied to the metal part of the simulated roof, this temperature is 50 °C to simulate the direct impact that would have on the car outdoor temperature to which it is subjected.

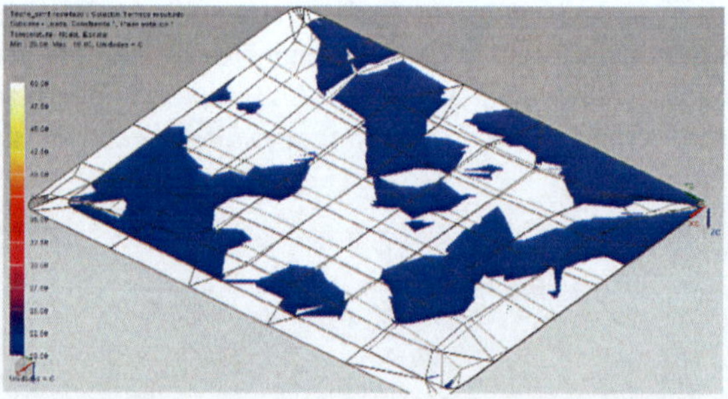

Fig. 9. Roof of the vehicle

Figure 10 then indicates the inside of the roof of the car, which is applied paraffin as an insulating layer, which could be incorporated into the upholstery, so that this material can fill the available space, its thickness is 6 mm and measures of this plate is 10 mm less than the ceiling on each side, the size of the material is the one that can be found in the market.. The temperature reached by the PCM in the inner part of the car is 20 °C, which indicates that the material fulfils its function and the heat is trapped inside, thereby applying thermal comfort is acceptable.

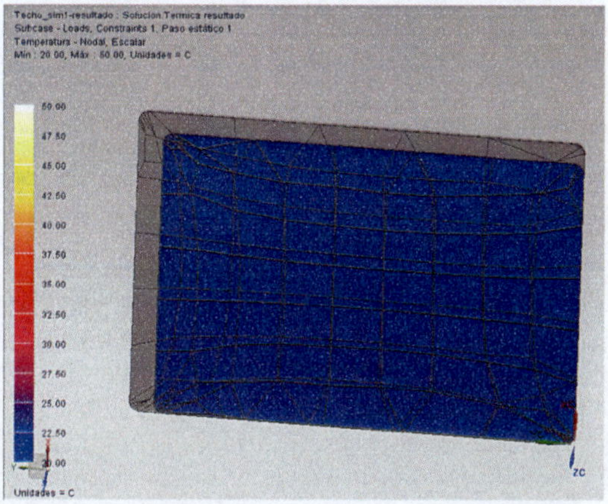

Fig. 10. Interior part of the roof of the vehicle

Table number 3 presented below shows a summary of the values resulting from this simulation and the characteristics of the materials used.

Table 3. Results of the simulation on the roof of the vehicle

Parameter	Value
Car roof measurements	1800 mm * 1500 mm * 1.8 mm
Material of the roof	Commercial steel
PCM plate dimensions	1700 mm * 1400 mm * 6 mm
Material of PCM	Paraffin
Exterior temperature	50 °C
Interior temperature	20 °C

4 Conclusions

When performing the thermal simulation in the NX10 software, it can be verified that the paraffin behaves as ideally expected; which means that while its phase change process is taking place, it stores thermal energy efficiently. The simulation shows how the paraffin, with the size of the plate that can be found in the market, would behave in the real world, with which it can be established that its application in vehicles is acceptable, in search of having constant control over its internal temperature. The implementation of phase change materials, such as paraffin, has demonstrated its contribution to free cooling and thermal comfort control by reducing temperature fluctuations.

Acknowledgment. This research takes part of the project Selection, characterization and simulation of phase change materials for thermal comfort, cooling and energy storage. This project is part of the INEDITA call for R&D research projects in the field of energy and materials. This research takes part of the project P121819, Parque de Energias Renovables founded by Universidad International SEK.

References

1. Acurio, K., Chico-Proano, A., Martínez-Gómez, J., Ávila, C.F., Ávila, Á., Orozco, M.: Thermal performance enhancement of organic phase change materials using spent diatomite from the palm oil bleaching process as support. Constr. Build. Mater. **192**, 633–642 (2018)
2. Aldás, P.S.D., Constante, J., Tapia, G.C., Martínez-Gómez, J.: Monohull ship hydrodynamic simulation using CFD. Int. J. Math. Oper. Res. **15**(4), 417–433 (2019)
3. Beltrán, R.D., Martínez-Gómez, J.: Analysis of phase change materials (PCM) for building wallboards based on the effect of environment. J. Build. Eng. **24**, 100726 (2019)
4. Espinoza, V.S., Guayanlema, V., Martínez-Gómez, J.: Energy efficiency plan benefits in Ecuador: long-range energy alternative planning model. Int. J. Energy Econ. Policy **8**(4), 52–54 (2018)
5. Gaona, D., Urresta, E., Marínez, J., Guerrón, G.: Medium-temperature phase-change materials thermal characterization by the T-History method and differential scanning calorimetry. Exp. Heat Transf. **30**(5), 463–474 (2017)
6. Kastillo, J.P., Martínez, J., Riofrio, A.J., Villacis, S.P., Orozco, M.A.: Computational fluid dynamic analysis of olive oil in different induction pots. In: 1st Pan-American Congress on Computational Mechanics–PANACM 2015, pp. 729–741 (2015)
7. Kastillo, J.P., Martínez-Gómez, J., Villacis, S.P., Riofrio, A.J.: Thermal natural convection analysis of olive oil in different cookware materials for induction stoves. Int. J. Food Eng. **13**(3) (2017)
8. Rodríguez, D., Martínez-Gómez, J., Guerrón, G., Riofrio, A.: Impact of induction stoves penetration over power quality in Ecuadorian households. Revista ESPACIOS **40**(13) (2019)
9. Martínez-Gómez, J., Ibarra, D., Villacis, S., Cuji, P., Cruz, P.R.: Analysis of LPG, electric and induction cookers during cooking typical Ecuadorian dishes into the national efficient cooking program. Food Policy **59**, 88–102 (2016)
10. Martínez-Gómez, J., Guerrón, G., Riofrio, A.J.: Analysis of the "Plan Fronteras" for clean cooking in Ecuador. Int. J. Energy Econ. Policy **7**(1), 135–145 (2017)
11. Martínez-Gómez, J.: Material selection for multi-tubular fixed bed reactor Fischer-Tropsch reactor. Int. J. Math. Oper. Res. **13**(1), 1–29 (2018)
12. Villacís, S., Martínez, J., Riofrío, A.J., Carrión, D.F., Orozco, M.A., Vaca, D.: Energy efficiency analysis of different materials for cookware commonly used in induction cookers. Energy Procedia **75**, 925–930 (2015)
13. Villacreses, G., Martínez-Gómez, J., Quintana, P.: Geolocation of electric bikes recharging stations: City of Quito study case. Int. J. Math. Oper. Res. **14**(4), 495–516 (2019)

14. Villacreses, G., Gaona, G., Martínez-Gómez, J., Jijón, D.J.: Wind farms suitability location using geographical information system (GIS), based on multi-criteria decision making (MCDM) methods: the case of continental Ecuador. Renew. Energy **109**, 275–286 (2017)
15. Martínez, J., Martí-Herrero, J., Villacís, S., Riofrio, A.J., Vaca, D.: Analysis of energy, CO2 emissions and economy of the technological migration for clean cooking in Ecuador. Energy Policy **107**, 182–187 (2017)

QA/QC Validation of the GMAW Welding Process, Used in the Construction of Body Bodies in the Ecuadorian Industry

Alfredo Icaza LLuglla[1], Javier Martínez-Gómez[1,2(✉)], and V. Diego F. Bustamante[1]

[1] Universidad Internacional SEK (UISEK), Calle Alberto einstein s/n y 5ta. Transversal, Quito, Ecuador
javier.martinez@uisek.edu.ec

[2] Instituto de Investigación Geológico y Energético (IIGE), Quito, Ecuador

Abstract. The Ecuadorian bodybuilding sector uses several types of steels to configure its structural system, among them the ASTM A36, ASTM A514 Gr. B and A517; application that is based on structural calculations and various production criteria; all these materials can be shaped into rectangular, square, omega, flat profiles, among others; with thicknesses that oscillate between 3 mm, same that can be welded according to their transversal section and other technical characteristics, with joints of the square type, fillet, etc.; using positions that go from Flat grooves "1G" to Overhead "4G". It should be noted that these welded joints are of vital importance, since more than allowing the union of the different profiles must meet the requirements of mechanical integrity established in the structural design, so it becomes absolutely necessary to execute a control and assurance of their quality. In the present study, technical characteristics, mechanical behavior and the effectiveness of running welding cords were analyzed through the GMAW process with ER706S-6 wire and CO2 protective gas; incorporated design parameters, non-destructive tests and mechanical tests as indicated in the AWS and ISO standards, allowing to check defects in each stage of the process, to later develop an experimental methodology under QA/QC standards to validate its correct execution; The proposed standardization contains highly relevant components to optimize production and ensure the quality implicit in the construction of metal bodies.

Keywords: Public transport · Metallurgy · Structures · Quality control

1 Introduction

For the construction of metal bus coachworks, the main material used is ASTM A36 steel and it is proposed that ASTM 514 Gr. B steel be used for improvement, given the structural calculations and corresponding validations; these coachworks are made of rectangular, square, angular, U-shaped, omega, T-shaped, platen profiles, among others. Said profiles may be welded using corner, fillet, double edge joints, etc. (corner, bevel, V, J, U joints) as per their cross section and other technical features, using

M. Botto-Tobar et al. (Eds.): CI3 2020, AISC 1277, pp. 416–427, 2021.
https://doi.org/10.1007/978-3-030-60467-7_34

welding positions that range from Flat or Down Hand (1G) to Overhead (4G) or from 1F to 4F for fillet [1].

There are many definitions of the QA/QC methodology, but it can ultimately be understood as a set of management and measurement processes and tools used to assure the quality of a product and/or service in terms of consumers' expectations. Quality Assurance focuses on management systems and planned preventive activities prior to executing a process that assure the quality of a product and/or service, and Quality Control relates to planned measurement and control activities that indicate the effectiveness that QA has on a product or service according to previously defined criteria [2]. In metallurgy, a welded joint is not only beads that look well. In order to reduce risk and to assure a definite metallurgic quality of the joints, it is important to safeguard, among other things, joint design, applied processes (WPS), technology implanted in order to characterize the materials used for welding, T° electrode, humidity control, chemical dilution, personnel qualifications and skills, etc. [1].

The present study which uses Quality Assurance and Quality Control aims to analyze joints welded using the GMAW process through nondestructive testing, mechanical testing, and metallographic analysis, linking them to regulations such as ISO, ANSI, AWS D1.1, AWS D1.3, ASTM, ASME, among others, in order to verify the existence of defects or errors in the different execution stages of the process, in existing metallographic structures in the base material, in the mechanical properties of the welded joints and a correct applicability of the welding process [3–11].

Subsequently, it aims to develop an experimental methodology under standards that not only consist of Quality Control but also incorporate technical foundations for its Assurance, allowing for a correct execution of the welding process, in addition to maintaining good practices currently executed in the construction and repair of metal coachworks for buses in the Ecuadorian industry, optimizing productivity and assuring the implicit quality of the welding manufacturing process.

2 Materials and Method

The study was carried out on the front section of metal coachworks and/or structural auto parts made of ASTM A 36, ASTM A514 Gr. B, and ASTM A517 steel, welded using the GMAW process with ER70S-6 as filler material and CO2 as protective gas. "According to the report issued by CANFAC in the year 2015 there are 54 coachbuilders authorized by the National Transit Agency (ANT) within this Chamber" [3–11].

2.1 Procedure for the Execution of Welded Joints

The present welding procedure qualification record (PQR) will be carried out under the standards stipulated in the AWS D1.3/D1.3M: 2008 code for a GMAW welding process applied to 3 mm thick (1/8 inch) steel sheets that may be welded together (Table 1).

Table 1. Code D1.3 application matrix, based on plate thickness.

Espesor del material	$t_2 < 1/8$ plg (3mm)	1/8 plg $\leq t_2 \leq$ 3/16 plg (3 mm $\leq t_2 \leq$ 5 mm)	$t_2 > 3/16$ plg (5mm)
$t_1 < 1/8$ plg (3mm)	D1.3	D1.3 con Anexo A	Anexo A
1/8 plg $\leq t_2 \leq$ 3/16 plg (3 mm $\leq t_2 \leq$ 5 mm)	D1.3 con Anexo A	D1.3 con Anexo A y D1.1	Anexo A y D1.1
$t_1 > 3/16$ plg (5mm)	Anexo A	Anexo A y D1.1	D1.1

Type of Joint to Be Used: The type of joint that is frequently used to make bus coachworks is a "butt joint with a square groove", because the average steel thickness used is 3 mm. It is important to note that this particular is recommended in many welding manuals [3–11].

Welding Position: For the manufacture of the different samples, as contemplated in AWS D1.a/D1.1 M:2010, the 3G qualification position is selected; its approval also serves to ratify the 1G and 2G positions. It should be noted that these considerations are frequently used according to the structural arrangement applied to the coachwork.

SSPC Cleaning Levels: The cleaning process will be carried out with the use of electric or pneumatic tools with metal brushes (Wire Brush) under SSPC-SP3 specifications, a method used in the coachwork industry to remove rust, peeling mill scale, loose rust and peeling paint.

Type and Size of Electrode: Given that the materials to be tested belong to the group of low carbon steel, and that the ER70S-6 electrode is frequently used in this industry, the specimens to be tested will use the following filler material (Table 2):

Table 2. Characteristics of electrode for GMAW welding processes

Metal base	Diámetro del electrodo		Amperaje
	Pulgadas	Milímetros	Rango (A)
Acero al Carbono	0.035	0.90	80-220
	0.045	1.20	125-380
	1/16	1.60	75-40

Technical Conditions of Equipment: Fort the execution of the welding beads according to the manufacturers of the different systems implemented in the welding equipment, and given that the base material has a thickness between 1 to 3, the following parameters were selected:

- Mix of 75% Argon + 25% CO2, with a flow rate of 8 to 16 L per minute.
- Amperages from 125 to 380 A; with a voltage range of 16 to 22 V.
- Direct current electrode positive (DCEP) polarity.

It should be noted that the protective gas mixture, the flow rate and the current described above may all be modified during the welding process, depending on the requirement of the operator, the welding bead and/or the equipment.

Preheating. Preheating is not required in steel of less than 0.2% carbon and with thickness of less than 1 inch. In steel from 0.2% to 0.4% carbon, preheating is necessary in ½ inch pieces. In steel of more than 0.4% carbon, heating is generally required with all thicknesses. In this scheme, steels such as ASTM A-517, A543, A654, etc. are not considered.

Using the Seferian formula described in the article "Selection of the Most Appropriate Technology of Reparatory Hard Facing of Working Parts on Universal Construction Machinery", it was determined that the optimal preheating temperature mainly in the process involving ASTM 514 Gr. B steels is 54.22 °C [3–11].

Heat Input and Heat Treatment
According to the Welding Handbook, Vol 1. American Welding Society (1991) [11], heat input is one of the factors that influence the welding process significantly, which is why it is important to know the approximate source of heat of the electric arc. Therefore, this analysis was performed based on the critical experimental data obtained during the GMAW welded joint in the front structure of a transport bus. Using Eq. 1, we get:

$$Q = P/V = (E \times I)/V \tag{1}$$

$$Q = (20\,V \times 215\,A)/48.6\,mm/s \tag{2}$$

$$Q = 88.47\,J/mm \tag{3}$$

A post welding treatment will not be considered, since this process has not been demonstrated in the coachwork industry. It should also be noted that given the thickness of the sample, the welding bead will be executed in a single pass, which is why it is not considered important to reduce the level of residual stress produced (Table 3).

2.2 Sample Parameterization of the Front Section of a Coachwork

To manufacture it, segments of varying dimensions will be used, which will be consistent with the requirements of each of the tests to be performed and in accordance with the standards stipulated in the corresponding guidelines.

Sample for tensile test, in accordance with ISO 5178-1. Sample for rotating bending test, in accordance with ISO 5173. Sample for hardness test, in accordance with ISO 9015-1. Sample for Microstructure analysis, in accordance with ISO 17639 e ISO 17640 [3–11].

Table 3. Welding plan for ASTM A 36, A 514, and A 517 steel

Variables	ASTM A36	ASTM 514 Gr. B	ASTM 517
Espesor del material Base	3m	3m	3m
Limpieza superficial	SSPC- SPC3	SSPC- SPC3	SSPC- SPC3
Tipo de Junta	Junta a tope con ranura cuadrada	Junta a tope con ranura cuadrada	Junta a tope con ranura cuadrada
Material de aporte	ER70S-6 Θ1.20 mm (0.045 plg).	ER70S-6 Θ1.20 mm (0.045 plg).	ER70S-6 Θ1.20 mm (0.045 plg).
Posición de la soldadura	3G, Vertical Descendente		
Gas de protección	Mezcla 75% Argón + 25 % CO_2	Mezcla 75% Argón + 25 % CO_2	Mezcla 75% Argón + 25 % CO_2
Equipo	XMT 350 CC / CV Auto-Line, marca MILLER y 420 Lincoln Electric	XMT 350 CC / CV Auto-Line, marca MILLER y 420 Lincoln Electric	XMT 350 CC / CV Auto-Line, marca MILLER y 420 Lincoln Electric
Características eléctricas	Amperaje 125 a 380 A.	Amperaje 125 a 380 A.	Amperaje 125 a 380 A.
	Voltaje de arco (16 a 22 V)	Voltaje de arco (16 a 22 V)	Voltaje de arco (16 a 22 V)
Polaridad	DCEP	DCEP	DCEP
Precalentamiento	No aplica	54.22 °C	54.22 °C
Tratamiento Térmico	No aplica	No aplica	No aplica
Cantidad de calor aportado en el proceso	88.47 Joules / mm	88.47 Joules / mm	88.47 Joules / mm
Cantidad de soldadura	4.71 kg/m	4.71 kg/m	4.71 kg/m

According to the references described in ISO 6892-1:2009 Annex B, the technical specifications for sheets, backing strips and flat products between 0.1 mm and 3 mm thick indicate the standard values for test samples that will be subjected to different tensile tests [3–11] (Table 4).

Every test to be executed will begin with visual inspection of the welding bead and the corresponding analysis by penetrating inks to rule out convergence of possible discontinuities. Subsequently, the specimens to be used in mechanical and metallographic tests will be traced, cut and identified.

Welding tests will be executed under the predetermined "Welding Plan" in a 300.00 mm × 216.46 mm specimen, which will provide a minimum of 8 specimens for tests, in addition to the corresponding sections not to be tested and the contingency sections (Fig. 1).

Table 4. Sample dimensions for tensile test

Tipo de probeta	Anchura	Longitud inicial entre puntos	Longitud de la parte calibrada		Longitud libre entre las mordazas (probetas de lados paralelos)	Tolerancia del mecanizado	Tolerancia en la forma
N#	b_o	L_o	L_c (Mínima)	L_c (Máxima)	L_t		
2	20 (+/- 1)	80	90	120	140	+/- 0.10	0.12

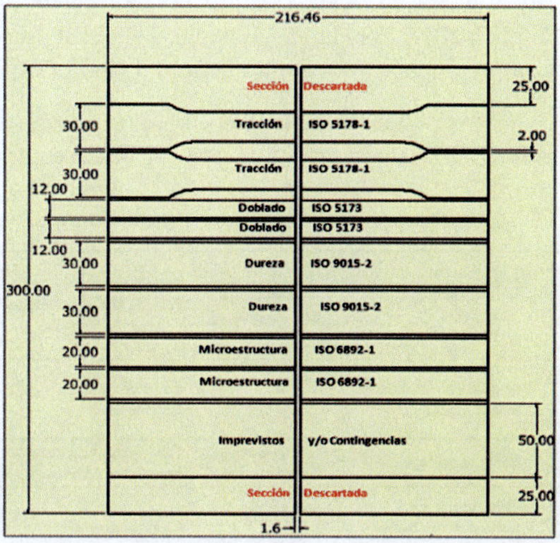

Fig. 1. Sample dimensions for mechanical and study tests (ISO Standards)

In accordance to ISO 9016, Welding Impact tests are omitted given that the thickness of the base material to be used does not meet the dimension requirements in said standard [3–11].

Execution of Welding Beads: According to the parameters established in the "Welding Plan", it is necessary to consider the following sequence required for the qualification and quality control of the welded joint prior to execution.

WPS. Execution of Welding Bead. Visual Inspection. Execution of Non-destructive tests. Location, identification, marking and stamping of samples. Execution of Destructive tests. Results Analysis (Criteria of considered code). Qualification or Rejection report (PQR y WPQ).

Welding Procedure Specification (WPS): The basic requirement to execute a GMAW welding process is to have a WPS and qualify it according to the requirements indicated in AWS D1.1/D1.1 M: 2015 and AWS D1.3/D1.3 M: 2018 – Structural Welding Code-Steel standards. Therefore, it must be in accordance with QW-402, QW-403, QW-405, and QW-406 parameters [3–11].

Preparation and Welding of samples: The main stages in the manufacturing of the samples are:

- Sectioning (square groove) and cleaning of parts to be welded. Fixing of plate to be welded; measurement verification and welding opening. Execution of welding bead in 3G position (Descending recommended). Cleaning prior to executing welding bead SSPC- SPC3. Execution of welding bead according to Welding Plan and WPS
- Visual Inspection: Inspection of the welding beads was carried out in the 20 beads of the different specimens formed in the ASTM A 36, ASTM A514, and ASTM A517 steels, welded using the GMAW process, executed under the standards described in ISO 5817-2014 Welding – Fusion – Welded Joints in Steel, nickel, titanium and their alloys Quality for imperfections [3–11] (Table 5).

Table 5. Visualized defects of GMAW bead samples

Probeta	Defecto visualizado	Posible Consecuencia	Imperfección según Norma ISO 6520-1 (Niveles de calidad de imperfecciones tipo B).	
A36-GMAW-001A A36-GMAW-010L A36-GMAW-014R A36-GMAW-016C A517-GMAW-012A	Cordón estrecho y poca penetración	Bajo amperaje Falta de pericia	Poro de superficie, Falta de fusión, Incompleta penetración de la raíz	
A36-GMAW-004D A514-GMAW-005C A514-GMAW-012A	Condón muy ancho con mucha salpicadura	Alto amperaje	Fisura de cráter, Superposición	
A517-GMAW-008F A517-GMAW-012A	Poca penetración	Arco corto Falta de pericia	Falta de fusión, Incompleta penetración de la raíz	**Múltiple imperfecciones en cualquier cruz sección**
A36-GMAW-007X A517-GMAW-012A A514-GMAW-005C A514-GMAW-012A	Poca penetración y mordeduras en ambos lados	Arco largo Falta de pericia	Corte continuo	
A36-GMAW-010L A514-GMAW-012A A514-GMAW- 003S A517-GMAW-008F	Cordón ancho y muy acumulado con ondas casi rectas	Baja velocidad de avance Falta de pericia	Exceso de soldadura metal (tope-soldar), Concavidad de la raíz	
A36-GMAW-001A A36-GMAW-014R A36-GMAW-016C A514-GMAW- 003S A517-GMAW-012A	Cordón estrecho con ondas puntiagudas	Baja velocidad de avance Falta de pericia	Exceso de penetración	

3 Tests Results

Tensile Tests Results

These tests were executed under the criteria defined in ISO 6892-1, Base material Tensile test (Table 6).

Macrographic Test Results: These tests were executed according to the methodology implemented in 17639:2003 Destructive Test on Welds in Metallic Materials – Macroscopic and Microscopic Examination of Welds. This test was executed with a composition of 4% nitric acid Nital etchant, to visualize the zones to be used for the Brinell Hardness test.

Table 6. ASTM A36, A514, and A517 Steel Tensile Tests, with GMAW welding bead

	ASTM A36		ASTM A514		ASTM A517	
	Probeta 1	Probeta 2	Probeta 1	Probeta 2	Probeta 1	Probeta 2
Carga Máxima (MPa)	363.33	559.17	400.83	456.67	514.71	550.00
Carga Rotura (MPa)	280.83	275.00	310.00	371.67	420.83	427.50
Fuerza Máxima (KN)	21.80	33.55	24.05	22.30	30.85	33.00

Brinell Hardness Tests: The methodology used for the present test has been developed as stipulated in the NTE INEN ISO 6506-1 standard, Metal materials – Brinell Hardness Tests – Part 1, in the Metallographic Analysis Laboratory of the HGPT Metallurgical Coachwork Productive Development Center (Tables 7 and 8).

Table 7. Macrograph of ASTM A36, A514, and A517 Specimens

Acero	Macrografía: Probeta 1 / Probeta 2
ASTM A 36	
ASTM A 514 GrB	
ASTM A 517	

Table 8. Macrograph of the ASTM A36, A514, and A517 Brinell Tests

ACERO	Macrografía: Probeta 1 (MB, ZAT, ZF/ZAT, ZF) / Probeta 2 (MB, ZAT, ZF/ZAT, ZF)
ASTM A 36	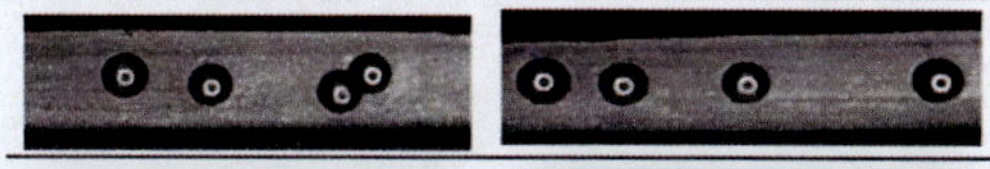
ASTM A 514	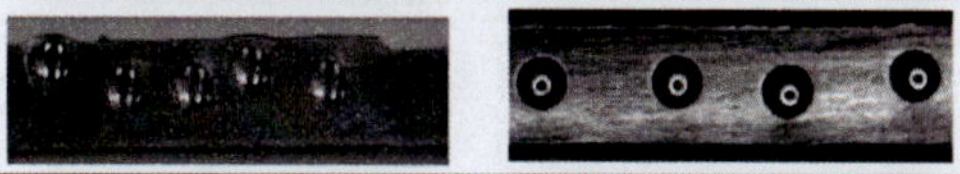
ASTM A 517	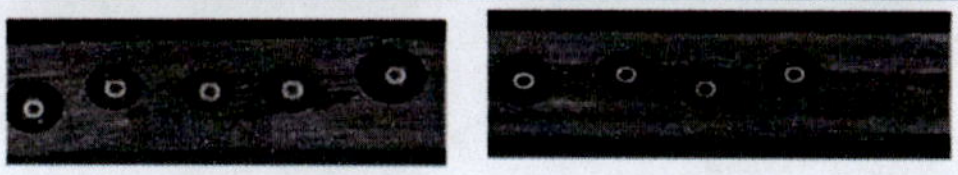

After obtaining the results of the hardness test, hardness graphs were made according to study areas, comparing the application of the GMAW process in ASTM A 36, A 514, and A 517 steel, obtaining the following values (Fig. 2):

Acero	Probetas	Mediciones			
		ZF	ZF/ZAT	ZAT	MB
ASTM A 36	1	127.00	136.00	121.00	112.00
	2	129.00	137.00	118.00	115.00
	Promedio	128.00	136.50	119.50	113.50
ASTM A 514	1	136.00	128.00	126.00	114.00
	2	135.00	129.00	127.00	115.00
	Promedio	135.50	128.50	126.50	114.50
ASTM A 517	1.00	158.00	133.00	89.00	82.00
	2	156.00	132.00	90.00	83.00
	Promedio	157.00	132.50	89.50	82.50

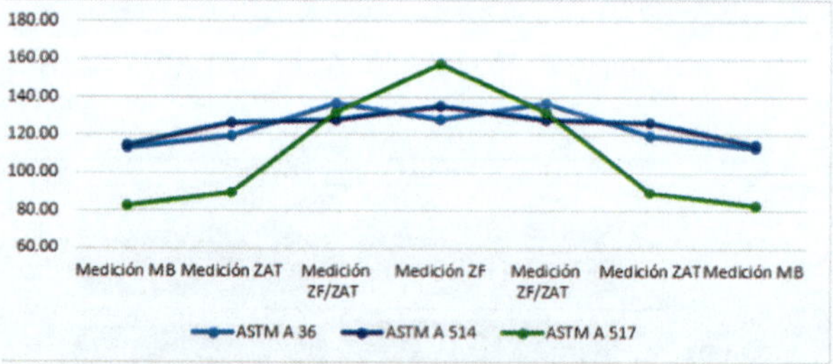

Fig. 2. Profile of ASTM A36, A514, and A517 (CFPMC) Brinell Hardness Test.

Rotating Bending Test: The present test was developed with the methodology detailed in ISO 5173:2009, Destructive Testing of Metal Material Welding, Rotating Bending Tests, by means of a universal Microtexte machine of 1500KN at a speed of 50 mm/min and 2000 N pre- load, with micro polished and rounded specimen edges (Table 9).

Table 9. Steel samples subjected to Rotating and Bending

	Ensayo de Flexión		Evidencia del Ensayo	Defecto Abierto encontrado	Evaluación del ensayo
ASTM A 36	TRBB 180°	Probeta 1		Ninguno	Conforme
		Probeta 2		Ninguno	Conforme
ASTM A 514	TRBB 180°	Probeta 1		1.40mm	Conforme
		Probeta 2		Ninguno	Conforme
ASTM A 517	TRBB 180°	Probeta 1	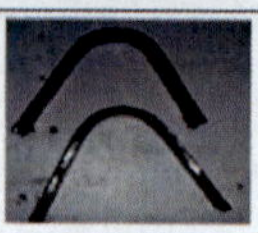	Ninguno	Conforme
		Probeta 2		Ninguno	Conforme

4 Discussion

The steels subjected to testing in the present article are ASTM A36, ASTM A514, and ASTM A517, which are used in the national coachwork building industry with 3 mm thickness regularly. Since they offer mechanical properties that supply different design requirements, it should be noted that these steels are also applied to the construction of auto parts [3–11].

Table 10. Comparative chart among ASTM A36, A514 Gr.B, and A-517 steels.

Table 10. Steel samples subjected to tensile test

	ASTM A36	*ASTM A514*	*ASTM A517*
Resistencia a la tracción	58 - 80 ksi 400 - 500 MPa	450 - 650 MPa	115 - 135 ksi 795 - 930 MPa
Límites de elasticidad	36 ksi 250 MPa	490 MPa	100 ksi 690 MPa

One of the absent documents used to contrast the results previously presented was the Certified Material Test Report (CMTR), especially regarding ASTM A514 and ASTM A517 steels. In order to implement a standard process for the qualification and certification of ANSI/AWS QC1 Welding Inspector in the coachwork industry, the following criteria must be met:

Through a correct execution of NDT tests, 17 out of 20 specimens have been discarded, equivalent to an 85% of manufactured specimens, thus reducing the spectrum to be studied and, therefore, the inherent costs. If this filter is incorporated in the bus construction process, it will encourage a substantial reduction in time and investment, since the welded joint could be repaired or manufactured again before undergoing any treatment.

With the execution of Brinell Hardness tests carried out in accordance with ISO 9015-1, it is evident that ASTM A 514 and A517 steels have the greatest hardness concentrated in their Molten Zone (MZ) and gradually decreases in their Heat Affected Zone (HAZ) and their Base Material (BM). This phenomenon is directly related to the deformation of the grain structure resulting from the welding fusion and the cooling rate, which allows a formation of dendritic structures possessing a greater degree of hardness than the rest of the analyzed areas [3–11].

5 Conclusion

The parameters involved in the GMAW welding process are: base material, structural profiles, thicknesses, types of joints, welding positions, technical characteristics required for the execution of processes, etc., in order to relate the essential and complementary variables in a Welding Plan and a WPS according to the AWS code, complying with QW-402, QW-403, QW-405, and QW-406 standards, which could be partially or totally implemented in the different coachwork industries as a measure to guarantee the correct execution of welding beads. To assure the correct execution of the process, it is required that the welder have WPQ stamping in GMAW welding processes, a duly approved SCWI/CWI Inspector, and that the process be monitored by a CWS Supervisor according to the standards outlined in the AWS D1.1/D1.1 M – AWS D1.3/D1.3 M code.

In order to obtain the standardization and the desired improvement of welding processes, ISO standards are a quality reference that will allow the coachwork industry to incorporate QA/QC systems, starting their implementation with the factors raised in the present research for the execution of welded joints in ASTM A36 steels which obtained satisfactory results in non-destructive and mechanical tests. It should be noted that no Quality Assurance (QA) system can avoid application errors that may occur due to an incorrect selection of GMAW welding processes, especially in regards to the execution of welding beads in steel structures such as A514 and A517, which have microstructural arrangements and mechanical characteristics different to those of A2S ER 70S-6 filler material used for the construction of specimens. This type of steels has a high mechanical resistance, which not only requires preheating, but also a welding process with low diffusible hydrogen content and low alloy electrodes with AWS A5.5/ASME-SFA 5.5 classification, such as E-11018-M, E7018-1 H4R, E11018M-H4R, etc., which use short arcs.

This process, if applied correctly in ASTM A 514 and ASTM A 517, can provide a tensile strength of 700Mpa, a creep limit of 717Mpa and a 2 inch elongation at 23%, remarkably exceeding the results obtained in the mechanical tests carried out to the GMAW process with AWS ER 70S-6 filler material.

Acknowledgments. This research takes part of the project P121819, Parque de Energias Renovables founded by Universidad International SEK.

References

1. García, G.V., Reyes, C.F., et al.: Optimization of experimental temperature measurement in GTAW process by means of DoE technique and computational modeling. Metal-Mechanical Engineering Department, Technological Institute of Morelia, Mexico (2016)
2. Lazić, V., Mutavdži, M., et al.: Selection of the most appropriate technology of reparatory hard facing hard facing of working parts on universal construction mechinery. Tribol. Ind. **33**, 18–27 (2011)
3. ISO 5178-1: 2014 Welding – Fusion-Welded Joints in Steel, nickel, titanium and their alloys (beam welding excluded) Quality Levels For Imperfections (2014)
4. ISO 5173:2009: Ensayos Destructivos de Soldaduras de Materiales Metálicos Ensayos de Flexión (2009)
5. ISO 17639 2003: Destructive Test on Welds in Metallic Materials – Macroscopic and Microscopic Examination of Welds (2003)
6. ISO 6892-1:2009: Metallic materials-Tesile testing Par 1. Anexo B (2009)
7. ISO 9016:2012: Destructive test son welds in metallic materials – Impact tests – Test specimen location, notch orientation an examination (2012)

8. ISO 5817-2014: Welding – Fusion – Welded Joints in Steel, nickel, titanium and their alloys Quality for imperfections (2014)
9. ISO 23277:2006: Non-Destructive Testing of Welds – Acceptance Levels. Tabla 1 (2006)
10. NTE INEN ISO 6506-1: Materiales metálicos – Ensayos de dureza brinell – Parte 1 (2014)
11. AWS QC1: Standard for AWS Certification of Welding Inspectors. American Welding Society (2007)

Material Selection Based on Multicrieria Decision Methods for Brake Disc Manufacture

Mario Cherrez[1], Javier Martìnez-Gomez[1,2(✉)], Juan Francisco Nicolalde[1], and Augusto Riofrio[3]

[1] Universidad internacional SEK, Quito Albert Einstein s/n and 5th, Quito, Ecuador
javier.martinez@uisek.edu.ec
[2] Instituto de Investigación Geológico y Energético (IIGE), Quito, Ecuador
[3] Budapest University of Technology and Economics (BME), Magyar Tudósok Körútja 2, Budapest 1117, Hungary

Abstract. This study evaluates an alternative material in the manufacture of a brake disc in light SUV vehicles, using multicriteria methods; for which five candidate materials are taken into consideration. The multicriteria methods used are: multidisciplinary optimization and compromise solution (VIKOR), elimination and options that reflect reality (ELECTRE I), proportional complex evaluation (COPRAS), additive relationship evaluation (ARAS), multi-objective optimization is based on the radius analysis (MOORA) and the entropy method used for weighting the criteria. The best alternative is ASTM A536 according to the COPRAS, ELECTRE I and ARAS methods, due to its low density, good compressive strength and good thermal conductivity.

Keywords: Brake disc · Multi-criteria methods · MCDM · Manufacture

1 Introduction

In the development of the automotive industry, brakes constitute one of the main safety devices. Therefore, the materials to be selected must have the appropriate physical and mechanical properties for optimum brake disc performance [1, 2]. The importance of carrying out this study lies in the increase of the automotive fleet, especially light vehicles. For this reason, it is necessary to select an alternative material for manufacturing this device, considering geometry, weight, material, mechanical resistance, thermal deformation, among other parameters [3].

The selection of materials plays a recognized role in engineering design. Also, design and development of products as well as competitiveness of manufacturing organizations depends on selected materials [4]. An inaccurate selection of one material could negatively affect productivity, profitability, and reputation of an organization due to growing demands for extended producer responsibility [5].

The objective is to select a material for the construction of a brake disc, through multicriteria methods to improve the operation and safety conditions of the vehicle [6, 7]. The selected material will be able to dissipate heat quickly according to its thermal properties and the analysis of the mechanical characteristics of the materials will allow

M. Botto-Tobar et al. (Eds.): CI3 2020, AISC 1277, pp. 428–439, 2021.
https://doi.org/10.1007/978-3-030-60467-7_35

the good performance of this device. One of the most common problems related to ventilated brake discs is the formation of cracks, especially under high loads and stresses caused by braking. [8]. Hwang and Wu [9] affirm that during breaking the kinetic and potential energy is converted into thermal energy, therefore it is necessary to know the temperature and thermal tension in braking [9]. It is important to select an alternative material to cast iron for a lightweight material to reduce fuel consumption, depending on its specific weight [10]. Chatterjee and Chakraborty [11] states that a systematic and efficient approach to material selection is necessary in order to select the best alternative for a given engineering application.

2 Selection of Materials and Multicriteria Methods

Among the important properties that the selected material must contain, thermal conductivity is the most important because a high value will allow heat to dissipate quickly [12]. Likewise, it is necessary a high thermal expansion coefficient will allow a good thermal expansion when the brake disc is exposed to a variation of temperature [13, 14]. In addition, a good elastic limit, Young's modulus and a Poisson coefficient, will allow to withstand high tensions without suffering permanent deformations in the disc. A high value of resistance to compression, traction and Brinell hardness, will prevent the material from fracturing due to the efforts produced by the jaws at the time of braking [15, 16] To reduce the consumption of the vehicle it is necessary to reduce the weight of the vehicle, for this reason the brake disc must have a low density [17].

The material used in the automotive industry for the design and manufacture of brake discs is gray cast iron, however, different types of alloys with superior technological characteristics have been developed, because they must meet extremely high parameters, because they are working at high degrees of wear and temperature [18]. Gray cast iron discs have better wear resistance than alloys or Ti compounds. Nevertheless, the addition of hard particles to a Ti-based compound can substantially improve wear resistance [18]. The analysis of the mechanical properties between an Aluminum alloy, cast iron, Titanium alloy, ceramic materials and composites resulted in the most appro- priate material for the manufacture of a brake disc to an Aluminum alloy [19]. Taking into account the aforementioned properties, the candidate materials for the manufacture of brake discs in Ecuador are the following: Ti 6Al 4 V (Titanium alloy, number 1), Al10Si C (Aluminum or Duralcan alloy, number 2), AISI 304L (stainless steel, number 3), ASTM A536 (nodular gray cast iron, number 4) and ASTM A48 (pearl gray cast iron, number 5).

The MCDM applied in this study for the selection of the material for the manufacture of the brake disc are: COPRAS, VIKOR, ELECTRE I, ARAS and MOORA. The ENTROPIA method is used to calculate the weights of the criteria for the candidate materials, in order to obtain objective results. The procedure and mathematical model of the MCDM is described below.

2.1 Entropy Method

Entropy measures the uncertainty in the information formulated using the theory of probability. It indicates that a wide distribution represents more uncertainty than that of a distribution with pronounced peaks. The Entropy method is explained in [20].

2.2 COPRAS Method

The COPRAS method selects the best decision alternatives considering the ideal and the worst-ideal solutions, in a step-by-step classification and evaluation of the alternatives in terms of their importance and degree of utility. The algorithm of the COPRAS method explained in [20].

2.3 VIKOR Method

The basic concept of VIKOR is to first define the positive and negative ideal solutions. The positive ideal solution indicates the alternative with the highest value (score of 100) while the negative ideal solution indicates the alternative with the lowest value (score of 0). The VIKOR engagement algorithm explained in [22].

2.4 ELECTRE I Method

The ELECTRE method has the ability to handle discrete quantitative and qualitative criteria and provides a complete order of alternatives. The limitation is replaced by the agreement and discordance of the matrix index and the procedure of the ELECTRE I method is explained in [20].

2.5 ARAS Method

The ARAS method determines the complex relative efficiency of a feasible alternative is directly proportional to the relative effect of the values and weights of the main criteria considered. Based on the theory of utility and the quantitative method. The steps of this method are explained in [20].

2.6 MOORA Method

The MOORA method starts from reference points, which will be the largest evaluation of the alternative radius vector with respect to each criterion, whether maximum or minimum. The steps of this method are described in [22].

3 Results of the Multicriteria Methods

The candidate materials and the criteria under analysis are shown in Table 1, these values represent the initial decision matrix. The physical, mechanical and thermal proper- ties of the candidate materials are: density (A), price (B), Young's modulus (C), elastic limit (D), Poisson radius (E), tensile strength (F), compressive strength (G), Brinell hardness (H), thermal conductivity (I), and thermal expansion coefficient (J). The En- tropy method is applied for the weighting criteria, in order to obtain objective weights at the time of the evaluation, since it is based on defined mathematical models; unlike the AHP method that is based on expert criteria applied by (Anojkumar, Ilangkumaran and Sasirekha and 2014). Then each proposed method is detailed with the respective results of the mathematical models.

Table 1. Evaluation matrix

MAT.	Density (kg/m3)	Price (USD/ kg)	Young Mod ulus (GPa)	Elastic limit (MPa)	Poisso n's ratio	Tensile Strength (MPa)	Compressive Strength (MPa)	Brinell Har dness (HV)	Thermal conductivity (W/m°C)	Thermal expansion coefficient (µstrain/°C)
	1	2	3	4	5	6	7	8	9	10
Tit-6Alu-4Va	4430	27.5	115	898	0.349	620	848	347	8.91	9.1
Alu-10Sil-Ca	2770	8.29	88	358	0.32	372	358	118	148	18
AISI-304L	7980	4.53	205	310	0.275	620	310	210	16	18
AST M-A 536	7150	0.67	173	339	0.28	500	351	217	41	12.5
AST M-A48	7200	0.67	120	149	0.265	250	170	252	46	13

Table 2 shows the standard decision matrix of the Entropy method, which is cal- culated according to Eq. (1). The entropy values (Ex) of each variable, the diversity of criteria (Dj) and the normalized weights of each criterion (Wj) are indicated in Table 3.

Table 2. Normalized decision matrix

1	2	3	4	5	6	7	8	9	10
0.151	0.661	0.167	0.441	0.234	0.262	0.421	0.299	0.029	0.131
0.094	0.201	0.131	0.169	0.209	0.161	0.180	0.099	0.571	0.261
0.271	0.111	0.289	0.149	0.179	0.259	0.149	0.179	0.059	0.261
0.243	0.018	0.251	0.157	0.191	0.209	0.169	0.191	0.161	0.181
0.245	0.018	0.169	0.069	0.181	0.099	0.079	0.221	0.181	0.179

Table 3. Calculations of Entropy method.

Criteria	A	B	C
1	0.959	0.041	0.041
2	0.599	0.397	0.397
3	0.969	0.031	0.031
4	0.901	0.111	0.111
5	0.994	0.004	0.004
6	0.971	0.029	0.029
7	0.909	0.091	0.091
8	0.971	0.029	0.029
9	0.751	0.248	0.248
10	0.979	0.021	0.021

3.1 COPRAS

The matrix normalized by weights (Dij) it was calculated and represented in Table 4. The sum of the weighted normalized values (Si+), (Si−) the relative importance (Qi) shows the degree of satisfaction of an alternative and the performance index (Pi) which determines the ranking of the candidate materials for the manufacture of a disc of brake, are calculated and all these calculations are indicated in Table 5, in this method the best material is material 4 (ASTM A536) due to the evaluation of the best decision criteria related to Young's modulus (C), elastic limit (D), Poisson's radius (E), tensile-compression resistance (F and G), hardness (H) and thermal conductivity (I) (Tables 6, 7 and 8).

Table 4. Normalized decision matrix Fij by the method VIKOR.

Material	1	2	3	4	5	6	7	8	9	10
1	0.319	0.939	0.349	0.829	0.519	0.559	0.809	0.651	0.061	0.279
2	0.201	0.279	0.271	0.329	0.481	0.341	0.329	0.221	0.921	0.549
3	0.569	0.161	0.619	0.291	0.409	0.559	0.301	0.389	0.101	0.549
4	0.509	0.019	0.531	0.309	0.421	0.449	0.341	0.399	0.261	0.379
5	0.521	0.019	0.371	0.141	0.401	0.231	0.159	0.471	0.291	0.399

Table 5. Ideal and non-ideal solution according to VIKOR

	1	2	3	4	5	6	7	8	9	10
A	0.569	0.951	0.631	0.829	0.519	0.559	0.809	0.639	0.921	0.549
B^i	0.201	0.019	0.271	0.137	0.395	0.0226	0.162	0.219	0.055	0.280

Table 6. Calculations

Material	A	B	C
1	0.321		
2	0.551		
3	0.761	0.854	0.320
4	0.794		
5	0.865		

Table 7. Calculations

Material	D	E	F	G
1	0.249		0.000	
2	0.284		0.000	
3	0.339	0.400	0.000	0.000
4	0.400		0.003	
5	0.400		0.004	

Table 8. Calculation of VIKOR Ranking

Material	H	Ranking
1	0.316	5
2	0.569	4
3	0.829	3
4	0.935	2
5	1.000	1

3.2 ELECTRE I

The data of the initial decision matrix is tabulated in Table 1 and the weighted standardized decision matrix (Vij), these values are indicated in Table 9. The matrix of concordance intervals (Cab) and is represented in Table 10. The matrix values of discrepancy intervals are calculated (Dab), which are tabulated in Table 11. The maximum threshold ($\bar{c}$) for the concordance index, and the dominant concordance matrix (*cdij*) is represented in Table 12. While the maximum threshold for the index of disagreement $\bar{()}$, tabulated in Table 13 and the discordant matrix (*ddij*) is shown in Table 14. Finally, the upper and lower net worth (Ca) and (Cb), respectively, these values are indicated in Table 15. The best rated material is ASTM A536 (material 4). The materials with the highest score are Al 10 Si C (number 2) and ASTM A536 (number 4), with an impact on thermal conductivity (I), elastic limit (D) and tensile-compression resistance (F and G).

Table 9. Vij Weighted Standard Decision Matrix according to ELECTRE I.

Material	1	2	3	4	5	6	7	8	9	10
1	0.027	0.019	0.010	0.091	0.002	0.018	0.072	0.022	0.014	0.005
2	0.031	0.286	0.007	0.036	0.002	0.011	0.031	0.008	0.230	0.011
3	0.018	0.338	0.018	0.031	0.002	0.018	0.027	0.014	0.026	0.011
4	0.020	0.391	0.016	0.032	0.002	0.015	0.030	0.014	0.065	0.008
5	0.020	0.391	0.009	0.015	0.002	0.008	0.015	0.016	0.070	0.008

Table 10. C_{ab} matrix of concordance intervals.

	Alt. 1	Alt. 2	Alt. 3	Alt. 4	Alt. 5
Alt. 1	0.000	0.289	0.284	0.302	0.302
Alt. 2	0.709	0.000	0.501	0.511	0.541
Alt. 3	0.712	0.499	0.000	0.081	0.274
Alt. 4	0.701	0.489	0.919	0.000	0.495
Alt. 5	0.701	0.459	0.723	0.503	0.000

Table 11. D_{ab} matrix of discrepancy intervals.

	Alt. 1	Alt. 2	Alt. 3	Alt. 4	Alt. 5
Alt. 1	0.000	0.219	0.179	0.151	0.201
Alt. 2	1.000	0.000	1.000	1.000	1.000
Alt. 3	1.000	0.249	0.000	0.059	0.341
Alt. 4	1.000	0.631	1.000	0.000	1.000
Alt. 5	1.000	0.661	1.000	0.421	0.000

Table 12. cdij matrix of dominant Concordancey concordance threshold c

	Alt. 1	Alt. 2	Alt. 3	Alt. 4	Alt. 5	$\bar{c}$
Alt. 1	0.000	0.000	0.000	0.000	0.000	
Alt. 2	1.000	0.000	0.000	1.000	1.000	
Alt. 3	1.000	1.000	0.000	0.000	0.000	0.5
Alt. 4	1.000	0.000	1.000	0.000	0.000	
Alt. 5	1.000	0.000	1.000	1.000	0.000	

Table 13. ddij matrix of dominant discordance and threshold of disagreement $\overline{d}$

	Alt. 1	Alt. 2	Alt. 3	Alt.4	Alt. 5	$\overline{d}$
Alt. 1	1.000	1.000	1.000	1.000	1.000	
Alt. 2	0.000	1.000	0.000	0.000	0.000	
Alt. 3	0.000	1.000	1.000	1.000	1.000	0.654
Alt. 4	0.000	1.000	0.000	1.000	0.000	
Alt. 5	0.000	0.000	0.000	1.000	1.000	

Table 14. acd_{ij} matrix of aggregate dominance (concordance-discordant)

	Alt. 1	Alt. 2	Alt. 3	Alt.4	Alt. 5
Alt. 1	0.000	0.000	0.000	0.000	0.000
Alt. 2	0.000	0.000	0.000	0.000	0.000
Alt. 3	0.000	1.000	0.000	0.000	0.000
Alt. 4	0.000	0.000	0.000	0.000	0.000
Alt. 5	0.000	0.000	0.000	1.000	0.000

Table 15. Calculation of the upper and lower net worth and ELECTRE I ranking

Materials	C_{ai}	D_{ai}	Ranking
1	0.0000	0.000	2
2	1.0000	−1.000	1
3	−1.0000	1.000	3
4	1.0000	−1.000	1
5	−1.0000	1.000	3

3.3 ARAS

The normalized decision matrix is calculated ($\overline{X}ij$), taking into account the calculation of non-beneficial. Subsequently, the decision matrix normalized by pesos ($\hat{i}$j) whose values are presented in Table 16. To calculate the optimization function values (Si) of each of the alternatives, the degree of utility (Ki), which determines the ranking of the alternatives for the application under study.

These values are represented in Table 17, showing that ASTM A536 material (number 4) is the best alternative due to the relative effect of the values of thermal conductivity (I), elastic limit (D) and compressive strength (G).

Table 16. $X\hat{}ij$ weighted standard decision matrix, of the ARAS method.

Material	1	2	3	4	5	6	7	8	9	10
1	0.009	0.006	0.006	0.049	0.0011	0.009	0.039	0.008	0.009	0.004
2	0.017	0.013	0.004	0.021	0.0009	0.006	0.014	0.004	0.143	0.005
3	0.006	0.027	0.009	0.018	0.0008	0.010	0.012	0.007	0.015	0.005
4	0.006	0.181	0.008	0.019	0.0009	0.008	0.014	0.007	0.039	0.004
5	0.006	0.181	0.005	0.011	0.0008	0.004	0.006	0.008	0.044	0.004

Table 17. Calculation Si, Ki and ranking.

Material	S_i	K_i	Ranking
1	0.129	0.469	4
2	0.219	0.799	3
3	0.106	0.369	5
4	0.281	1.000	1
5	0.259	0.941	2

3.4 MOORA

Table 18 shows the weighted standardized decision matrix. Then you get the aggregation function S(xi) which evaluates each alternative, this calculation also determines the preference ranking of each alternative. The values are shown in Table 19, showing that the material Al 10Si C (number 2) is the best because its thermal conductivity (I) and coefficient of thermal expansion (J) are high compared to the rest of the materials experienced.

Table 18. Xij weighted standard decision matrix, by MOORA method.

Material	1	2	3	4	5	6	7	8	9	10
1	0.009	0.381	0.010	0.091	0.002	0.021	0.069	0.019	0.009	0.004
2	0.011	0.109	0.006	0.041	0.002	0.009	0.031	0.010	0.231	0.011
3	0.019	0.059	0.016	0.029	0.001	0.021	0.027	0.015	0.026	0.010
4	0.021	0.010	0.014	0.029	0.001	0.009	0.030	0.015	0.065	0.006
5	0.021	0.010	0.009	0.009	0.001	0.009	0.011	0.017	0.069	0.006

Table 19. Aggregation function S(xi) and ranking MOORA

Material	S(xi)	Ranking
1	−0.161	5
2	0.209	1
3	0.061	4
4	0.151	2
5	0.109	3

3.5 Discussion Results

MCDM classify a finite number of decision alternatives, each explicitly described in terms of different decision criteria that must be taken into account simultaneously. For this reason, these methods are used in the selection of the material for the construction of a brake disc. The analysis of the properties of the candidate materials, using qualitative and quantitative criteria of the MCDM, yields ASTM A536 as the best material are shown in Fig. 1. Through the development of the proposed MCDM, it can be said that the best material in the COPRAS, ELECTRE I, and ARAS methods is ASTM A536, due to its low density (A), a high elastic limit (D) and a good resistance to compression (G), the MOORA method and VIKOR place it as a second alternative. The second-best option evaluated is Al 10Si C and ASTM A48 by the criteria of ELECTRE I, MOORA and VIKOR, since it has good thermal conductivity (I), low density (A) and an accessible price (B). These results are aligned with the materials used in the study conducted by [13]. In addition, Kharate and Chaudhari [18] study the effect of material properties on noise and brake disc performance through FEM and EMA approach, for which they experiment with gray cast iron, ceramic coal and steel, resulting in the gray cast iron having a lower natural frequency than the rest of the materials experienced.

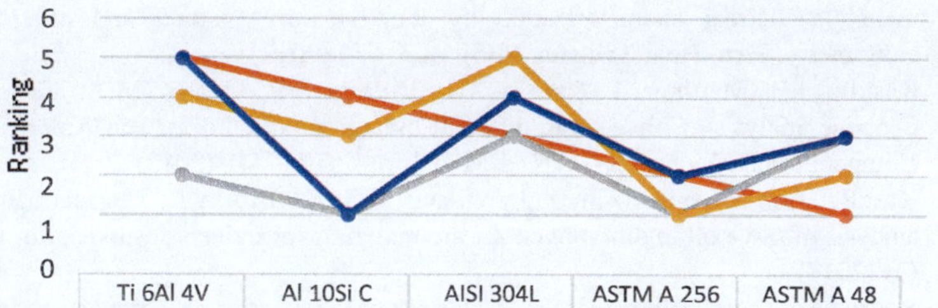

	Ti 6Al 4V	Al 10Si C	AISI 304L	ASTM A 256	ASTM A 48
COPRAS	4	3	5	1	2
VIKOR	5	4	3	2	1
ELECTRE 1	2	1	3	1	3
ARAS	4	3	5	1	2
MOORA	5	1	4	2	3

Alternative

Fig. 1. Ranking of candidate materials according to the MCDM.

4 Conclusions

The application of MCDM methods in this research allows the selection of brake disc material, incorporating quantitative and qualitative criteria. The weighting of the properties in candidate materials for the construction of a brake disc was obtained by the ENTROPY method. The achieved results, according to the COPRAS, ELECTRE I and ARAS methods, select ASTM A536 as the best option because it has better thermal and mechanical properties. Also, the paper presents the use of MCDM techniques with a different approach that allows to solve complex problems, which will be adapted to any type of need. Therefore, MCDMs can be applied to different areas of engineering and material selection.

Acknowledgment. This research takes part of the project P121819, Parque de Energias Renovables founded by Universidad International SEK.

References

1. Acurio, K., Chico-Proano, A., Martínez-Gómez, J., Ávila, C.F., Ávila, Á., Orozco, M.: Thermal performance enhancement of organic phase change materials using spent diatomite from the palm oil bleaching process as support. Constr. Build. Mater. **192**, 633–642 (2018)
2. Aldás, P.S.D., Constante, J., Tapia, G.C., Martínez-Gómez, J.: Monohull ship hydrodynamic simulation using CFD. Int. J. Math. Oper. Res. **15**(4), 417–433 (2019)
3. Beltrán, R.D., Martínez-Gómez, J.: Analysis of phase change materials (PCM) for building wallboards based on the effect of environment. J. Build. Eng. **24**, 100726 (2019)
4. Espinoza, V.S., Guayanlema, V., Martínez-Gómez, J.: Energy efficiency plan benefits in ecuador: long-range energy alternative planning model. Int. J. Energy Econ. Policy **8**(4), 52–54 (2018)
5. Gaona, D., Urresta, E., Marínez, J., Guerrón, G.: Medium-temperature phase- change materials thermal characterization by the T-history method and differential scanning calorimetry. Exp. Heat Transfer **30**(5), 463–474 (2017)
6. Kastillo, J.P., Martínez, J., Riofrio, A.J., Villacis, S.P., Orozco, M.A.: Computational fluid dynamic analysis of olive oil in different induction pots. In: 1st Pan-American Congress on Computational Mechanics–PANACM 2015, pp. 729–741 (2015)
7. Kastillo, J.P., Martínez-Gómez, J., Villacis, S.P., Riofrio, A.J.: Thermal natural convection analysis of olive oil in different cookware materials for induction stoves. Int. J. Food Eng. **13** (3) (2017)
8. Rodríguez, D., Martínez-Gómez, J., Guerrón, G., Riofrio, A.: Impact of induction stoves penetration over power quality in Ecuadorian households. Revista ESPACIOS **40**(13) (2019)
9. Hwang, P., Wu, X.: Investigation of temperature and thermal stress in ventilated disc brake based on 3D thermo-mechanical coupling model. J. Mech. Sci. Technol. **24**(1), 81–84 (2010). https://doi.org/10.1007/s12206-009-1116-7
10. Chatterjee, P., Chakraborty, S.: Material selection using preferential ranking methods. Mater. Des. **35**, 384–393 (2012). https://doi.org/10.1016/j.matdes.2011.09.027
11. Martínez-Gómez, J., Ibarra, D., Villacis, S., Cuji, P., Cruz, P.R.: Analysis of LPG, electric and induction cookers during cooking typical Ecuadorian dishes into the national efficient cooking program. Food Policy **59**, 88–102 (2016)
12. Martínez-Gómez, J., Guerrón, G., Riofrio, A.J.: Analysis of the Plan Fronteras for clean cooking in Ecuador. Int. J. Energy Econ. Policy **7**(1), 135–145 (2017)
13. Martínez-Gómez, J.: Material selection for multi-tubular fixed bed reactor Fischer-Tropsch reactor. Int. J. Math. Oper. Res. **13**(1), 1–29 (2018)
14. Villacís, S., Martínez, J., Riofrío, A.J., Carrión, D.F., Orozco, M.A., Vaca, D.: Energy efficiency analysis of different materials for cookware commonly used in induction cookers. Energy Procedia **75**, 925–930 (2015)
15. Villacreses, G., Martínez-Gómez, J., Quintana, P.: Geolocation of electric bikes recharging stations: City of Quito study case. Int. J. Math. Oper. Res. **14**(4), 495–516 (2019)

16. Martínez, J., Martí-Herrero, J., Villacís, S., Riofrio, A.J., Vaca, D.: Analysis of energy, CO2 emissions and economy of the technological migration for clean cooking in Ecuador. Energy Policy **107**, 182–187 (2017)
17. Villacreses, G., Salinas, S.S., Ortiz, W.D., Villacís, S., Martínez-Gómez, J.: Environmental impact assessment of internal combustion and electric engines for maritime transport. Environ. Process. **4**(4), 907–922 (2017)
18. Kharate, N.K., Chaudhari, S.S.: Effect of material properties on disc brake squeal and performance using FEM and EMA approach. Mater. Today Proc. **5**(2), 4986–4994 (2018)
19. Nicolalde, J.F., Martinez-Gomez, J., Maiguashca, J.: Characterization of the Mocora leaf for mechanical purposes. Revista ESPACIOS **41**(08) (2020)
20. Villacreses, G., Gaona, G., Martínez-Gómez, J., Jijón, D.J.: Wind farms suitability location using geographical information system (GIS), based on multi-criteria decision mak- ing (MCDM) methods: The case of continental Ecuador. Renew. Energy **109**, 275–286 (2017)
21. Chingo, C., Martínez-Gomez, J.: Material selection using multi-criteria decision making methods for geomembranes. Int. J. Math. Oper. Res. **16**(1), 24–52 (2020)
22. Loor, R.B.S., Martínez-Gómez, J., Rocha-Hoyos, J.C., Cedeño, E.A.L.: Selection of materials by multi-criteria methods applied to the side of a self-supporting structure for light vehicles. Int. J. Math. Oper. Res. **16**(2), 139–158 (2020)

fPCM Selection for Latent Heat Storage by MCDM

Javier Martínez-Gomez[1(✉)], Gonzalo Guerrón[2], C. Ricardo A. Narváez[2], Francis Vasquez[2], Luis Godoy-Vaca[2], E. Catalina Vallejo-Coral[2], and Marco Orozco[2]

[1] Universidad internacional SEK, Quito Albert Einstein s/n and 5th, Quito, Ecuador
javier.martinez@uisek.edu
[2] Instituto de Investigación Geológico y Energético (IIGE), Quito, Ecuador

Abstract. A common complication in industry is the difference between available thermal energy and its application period. Using thermal energy storage with phase change materials (PCMs) will increase considerably the energy efficiency in industry and will solve the gap between energy supply and consumption. This research aims to select a PCM which better accomplish the solution of the TES between 200–400 °C and reduce the cost of production.

MCDM have been developed to resolve the problem. The MCMD methods were complex proportional assessment of alternatives such as COPRAS-G, TOPSIS and VIKOR methods. The criteria weighting was performed by AHP and Entropy method. The correlation of the results between three ranking methods has been developed by the Spearman's correlation coefficient. The results illustrated the best choices of the three MCDM were NaOH and KNO_3, due to the highest values of the most important criteria. Furthermore, Spearman's correlation between both methods exceeds 0,714.

Keywords: Phase change material (PCM) · Latent heat thermal energy storage (LHTES) · Material selection · Multi-criteria decision making (MCDM)

1 Introduction

The production of electricity on solar thermal power plants demands to develop thermal energy storage (TES) with high accuracy, energy efficiency and low cost [1, 2]. This TES solve the gap between the available solar source and energy demand [1, 3]. It is necessary that they hold the thermal energy during high solar irradiation periods and maintained for until the periods of its application [4–6].

LHS systems with solid-liquid change of PCM are an alternative to SHS systems, particularly in solar energy applications of middle-temperature-range (200–400 °C) [6, 7]. Molten salts have been used as conducive LHS resources in solar power systems [8, 9]. This research tries to resolve the material selection for a PCM which better accomplish the solution of the TES between 200–400 °C.

Selecting the most proper material for a specific purpose is a vital component for designing and developing products. Materials selection has become an imperative

M. Botto-Tobar et al. (Eds.): CI3 2020, AISC 1277, pp. 440–449, 2021.
https://doi.org/10.1007/978-3-030-60467-7_36

source at engineering processes due to technological, economic, and environmental parameters. The objectives and criteria in material selection process are often in conflicts, such as chosen properties, production process, operating environment, costs, market values, product performance and available supplying sources [10, 11].

Multi-criteria decision-making methods (MCDM) applications for material selection has been conducted in many fields. In 2006, Shanian and Savadogo [12] developed a material selection model based on multiple attribute decision making. In 2008, Ho [13] made a review of the AHP model integrated applications with other methods. In 2014, Anojkumar et al. [14] compared the MCDM methodologies to select pipe material in sugar industry.

Many researches about applications of MCDM have been developed in the energy storage field. In 2009, Bartin et al. [15] considered MCDM for managing storage energy technologies on renewable hybrid systems, analytic hierarchy process and fuzzy logic, which studies five kinds of SES and are examined giving six criteria, in order to find the most suitable SES that will be used in costs and environment scenarios. In 2010, Cavallaro [16] developed Fuzzy TOPSIS approach for evaluating TES in concentrated solar power systems (CSP). In this research, a fuzzy logic methodology compares diverse heat transfer fluids between 400–500 °C in a CSP based on costs and benefits, also investigate the viability of employing molten salts. Fernández et al. [17] presented the materials selection of SES in the range of 150–200 °C by the CES Selector software. Additionally, Khare et al. (2013) [18] studied the material selection for high temperature SES. In this research, a materials selection software package Granta Design's CES Selector was used to assess SHS between 500–750 °C. However, a study related to the selection of LHS material for TES between 200–400 °C has not been accomplished yet.

This study has developed three preferences ranking based MCDM methods for ranking accuracy the alternative materials for PCMs. The compromised weights have been performed by AHP and Entropy methods. The Spearman's rank correlation has been used to measure the linear relationship strength between results. All the methods have been extensively studied and refined in several articles [19–28]. A more extensive explanation of these methods can be observed in Sect. 2. For these methods, a list of possible in ascending order of importance of suitable materials has been obtained considering diverse material selection criteria.

2 Materials and Methods

2.1 Decision-Making Problem Definition

Figure 1 illustrates the scheme of a commercial PCM. The most important criteria are heat of fusion (λ)and specific heat (C_p). High values are desired for the TES. Also, it is necessary a high thermal conductivity (k). Low costs (C) could improve the sails and competitive. The demanded melting temperature (T_m) provides a window of working temperatures. Density (ρ) reduce the downsize of the HSU.

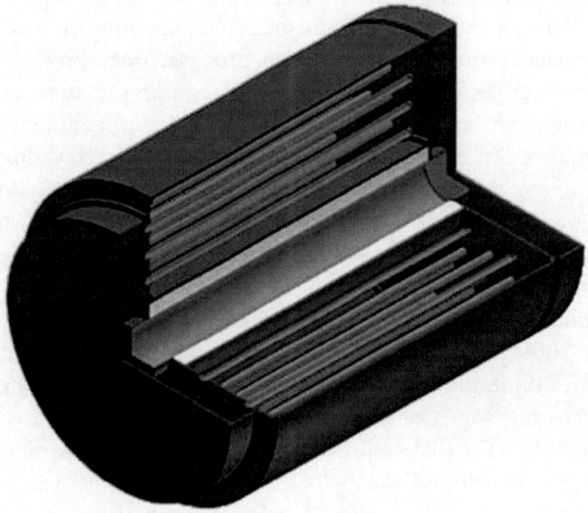

Fig. 1. Schema of a TES unit that uses PCM as the thermal storage media

Seven alternatives for a PCM were taken into consideration: $NaNO_3$, KNO_3, NaOH, KOH, $ZnCl_2$, $NaNO_3/KNO_3$ (0,5/0,5), $ZnCl_2/KCl$ (0,319/0,681). The properties of the alternatives are represented in Table 1.

Table 1. Material properties for a PCM [2–13].

Material	(A) Heat of fusion, $\left[\frac{KJ}{Kg}\right]$ (λ)	(B) Specific Heat $\left[\frac{J}{g°C}\right]$ (C_p)	(C) Thermal conductivity $\left[\frac{W}{m \cdot K}\right]$ (k)	(D) Cost $\left[\frac{\$}{kg}\right]$ (C)	(E) Melting Temp. [°C] (T_m)	(F) Density, $\left[\frac{kg}{m^3}\right]$ (ρ)
(1) $NaNO_3$	172	1,82	0,5	0,53	310	2260
(2) KNO_3	266	1,22	0,5	0,78	330	2110
(3) NaOH	165	2,08	0,92	0,34	318	2100
(4) KOH	149,7	1,47	0,5	1,05	380	2044
(5) $ZnCl_2$	75	0,74	0,5	0,9	280	2907
(6) $NaNO_3$/ KNO_3 (0,5/0,5)	100,7	1,35	0,56	0,64	220	1920
(7) $ZnCl_2$/KCl (0,319/0,681)	198	0,67	0,8	0,59	235	2480

2.2 MCDM Methods

Criteria Weighting
The criteria weights are calculated using a AHP and Entropy methods. The synthesis weight for the *jth* criteria is:

$$w_j = \frac{\alpha_j x \beta_j}{\sum_{j=1}^{n} \alpha_j x \beta_j} \quad \mathrm{j} = 1, \ldots, \mathrm{n} \tag{1}$$

where α_j is the weight of *jth* criteria obtained via AHP method, and β_j is the weight of *jth* criteria obtained through Entropy method.

AHP Method
It was developed by Saaty in [19], more information about the method could be found in [19].

Entropy Method The development of the method could be found in [21].

COPRAS-G Method
The development of the method could be found in [23].

TOPSIS Method
The development of the method could be found in [24].

VIKOR Method
The development of the method could be found in [25].

Spearman's Rank Correlation Coefficient The development of the method could be found in [26].

3 Results

3.1 Criteria Weighting Results

In Table 2 shows the relative importance criteria scale for AHP method.

Table 2. Criteria scale.

Definition	Intensity of importance
Equal importance	1
Moderate importance	3
Strong importance	5
Very strong importance	7
Extreme importance	9
Intermediate importance	2, 4, 6, 8

In Table 3 is illustrated the criteria comparision. The results have a consistency index (CI = 0,052) and the consistency ratio (CR = 0,041). The entropy method decision matrix appears in Table 4. The weights results are shown in Table 5.

Table 3. Criteria comparasion

(λ)	(C_p)	(k)	(C)	(T_m)	(ρ)
1	3	5	5	7	9
0,333	1	3	3	5	7
0,2	0,333	1	1	3	5
0,2	0,333	1	1	3	5
0,143	0,2	0,333	0,333	1	3
0,111	0,143	0,2	0,2	0,333	1

Table 4. Entropy method decision matrix.

Material	(λ)	(C_p)	(k)	(C)	(T_m)	(ρ)
1	0,380	0,485	0,299	0,276	0,390	0,374
2	0,588	0,325	0,299	0,407	0,415	0,350
3	0,364	0,554	0,550	0,177	0,400	0,348
4	0,331	0,391	0,299	0,548	0,478	0,339
5	0,166	0,197	0,299	0,469	0,352	0,482
6	0,222	0,359	0,335	0,334	0,277	0,318
7	0,437	0,178	0,478	0,308	0,295	0,411

Table 5. Weights results.

(λ)	(C_p)	(k)	(C)	(T_m)	(ρ)
0,461	0,240	0,109	0,109	0,051	0,028
0,146	0,146	0,173	0,157	0,187	0,192
0,440	0,228	0,123	0,112	0,063	0,035

3.2 COPRAS-G Method Results

The matrix of decision are presented in Table 6. Matrix with normalized are show in Table 7. The results of this MCDM method are presented in Table 8. The results in order are 2–3–1–7–4–6–5. NaOH and KNO_3 are the best alternatives.

Table 6. Matrix of decision

Material	(λ)		(C_p)		(*k*)		(*C*)		(T_m)		(ρ)	
1	154,8	189,2	1,638	2,002	0,45	0,55	0,477	0,583	279	341	2034	2486
2	239,4	292,6	1,098	1,342	0,45	0,55	0,702	0,858	297	363	1899	2321
3	148,5	181,5	1,872	2,288	0,828	1,012	0,306	0,374	286,2	349,8	1890	2310
4	134,73	164,67	1,323	1,617	0,45	0,55	0,945	1,155	342	418	1839,6	2248,4
5	67,5	82,5	0,666	0,814	0,45	0,55	0,81	0,99	252	308	2616,3	3197,7
6	90,63	110,77	1,215	1,485	0,504	0,616	0,576	0,704	198	242	1728	2112
7	178,2	217,8	0,603	0,737	0,72	0,88	0,531	0,649	211,5	258,5	2232	2728

Table 7. Matrix normalized

Material	(λ)		(C_p)		(*k*)		(*C*)		(T_m)		(ρ)	
1	0,060	0,074	0,040	0,049	0,013	0,016	0,011	0,014	0,008	0,010	0,004	0,005
2	0,093	0,114	0,027	0,033	0,013	0,016	0,016	0,020	0,009	0,011	0,004	0,005
3	0,058	0,071	0,046	0,056	0,024	0,029	0,007	0,009	0,009	0,011	0,004	0,005
4	0,053	0,064	0,032	0,039	0,013	0,016	0,022	0,027	0,010	0,013	0,004	0,005
5	0,026	0,032	0,016	0,020	0,013	0,016	0,019	0,023	0,008	0,009	0,006	0,007
6	0,035	0,043	0,030	0,036	0,015	0,018	0,013	0,016	0,006	0,007	0,004	0,005
7	0,070	0,085	0,015	0,018	0,021	0,007	0,012	0,015	0,006	0,008	0,005	0,006

Table 8. Results of the method

Material	Pi	Ri	Qi	Ui	Rank
1	0,134	0,015	0,153	83,823	3
2	0,156	0,019	0,172	94,076	2
3	0,157	0,011	0,183	100,000	1
4	0,122	0,023	0,135	73,938	4
5	0,078	0,023	0,090	49,510	7
6	0,098	0,016	0,117	64,076	6
7	0,115	0,017	0,133	72,745	5

3.3 TOPSIS Method Results

Table 9 shows the TOPSIS weights. The ideal and nadir ideal solutions are show in Table 10. The best PCM alternatives are shown in Table 11. The results are 2–3–1–7–4–6–5. For TOPSIS method KNO_3 and NaOH are the best alternatives.

Table 9. TOPSIS weights.

Material	(λ)	(C_p)	(k)	(C)	(T_m)	(ρ)
1	0,167	0,110	0,037	0,031	0,024	0,013
2	0,258	0,074	0,037	0,046	0,026	0,012
3	0,160	0,126	0,068	0,020	0,025	0,012
4	0,145	0,089	0,037	0,061	0,030	0,012
5	0,073	0,045	0,037	0,053	0,022	0,017
6	0,098	0,082	0,041	0,037	0,017	0,011
7	0,192	0,041	0,059	0,034	0,018	0,014

Table 10. TOPSIS method solutions.

	(λ)	(C_p)	(k)	(C)	(T_m)	(ρ)
V^+	0,258	0,126	0,068	0,020	0,030	0,011
V^-	0,073	0,041	0,037	0,061	0,017	0,017

Table 11. TOPSIS method results.

Material	S_i^+	S_i^-	C_i	Rank
1	0,098	0,121	0,552	3
2	0,066	0,189	0,741	1
3	0,098	0,133	0,575	2
4	0,130	0,088	0,405	5
5	0,208	0,011	0,050	7
6	0,170	0,054	0,242	6
7	0,110	0,124	0,531	4

3.4 VIKOR Method Results

The values of E_i, F_i and P_i were shown in Table 12. The best selection materials were 2–3–1–7–4–6–5, which indicates that KNO_3 and NaOH obtain the best solution.

Table 12. VIKOR method results.

Material	E_i	F_i	P_i	Rank
1	0,514	0,216	0,293	3
2	0,352	0,139	0,000	1
3	0,397	0,232	0,201	2
4	0,520	0,268	0,385	5
5	0,842	0,440	1,000	7
6	0,766	0,380	0,824	6
7	0,564	0,228	0,364	4

3.5 Spearman's Correlation Results

The magnitude of this parameter for a PCM exceeds 0,714 for the relation of the results between the methods COPRAS-G, TOPSIS and VIKOR. In case of the magnitude of the parameter between TOPSIS and VIKOR it has a value of 1, which indicates that all the results have the same rank.

4 Discussion

In Fig. 2 are presented results of MCDM methods for PCMs. NaOH and KNO_3 were the best alternatives for COPRAS-G, VIKOR, and TOPSIS methods.

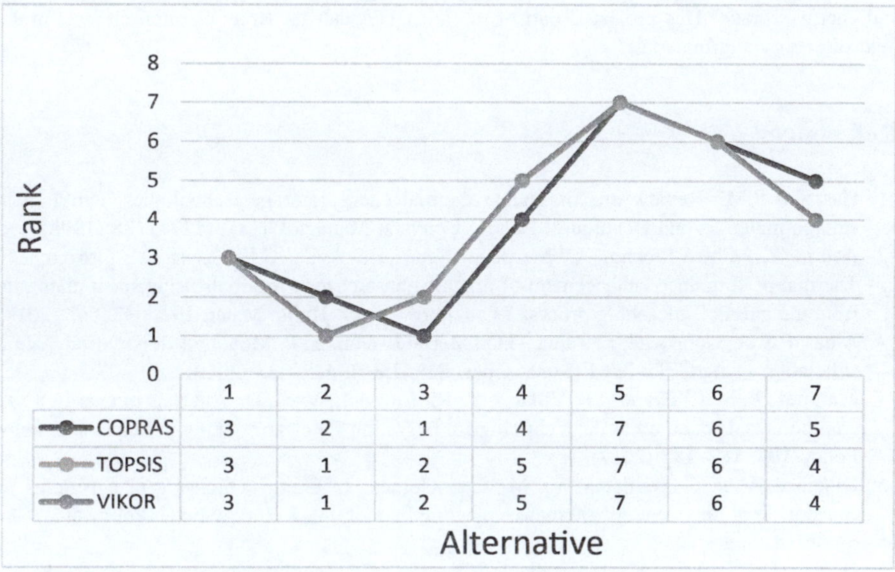

Fig. 2. Results of MCDM methods for PCMs

Similar properties have been used in the studies of Khare et al. [18] in the "selection of materials for high temperature sensible energy storage" and Fernández et al. [17] in the "Selection of materials with potential in sensible thermal energy storage between 150–200 °C", which have been taken into account how the most important properties the high heat of fusion (λ), specific heat (C_p) and thermal conductivity (k), with cost (C) and density (ρ) are the most important properties for materials in TES applications. Moreover, Fernández et al. [17] considered the fracture toughness, but this is relevant for the structure which contains PCM and not for the materials of the LHS.

5 Conclusions

PCM selection at middle temperature between 200–400 °C for LHS has been solved.

The results show that make a TES with NaOH and KNO_3, could reduce the manufacturing cost with a high heat of fusion (λ), specific heat (C_p) and thermal conductivity (k).

These properties should improve the energy efficiency of the TES. In addition, it should consider that to obtain all the heat of fusion for the TES, the operation temperature should be above 310 °C for NaOH and above 330 °C KNO_3.

Acknowledgment. This research takes part of the project P121819, Parque de Energias Renovables founded by Universidad International SEK. This research takes part of the project Selection, characterization and simulation of phase change materials for thermal comfort, cooling and energy storage. This project is part of the INEDITA call for R&D research projects in the field of energy and materials.

References

1. Hasnain, S.M.: Review on sustainable thermal energy storage technologies, Part I: heat storage materials and techniques. Energy Convers. Manag. **39**(11), 1127–1138 (1998)
2. Acurio, K., Chico-Proano, A., Martínez-Gómez, J., Ávila, C.F., Ávila, Á., Orozco, M.: Thermal performance enhancement of organic phase change materials using spent diatomite from the palm oil bleaching process as support. Constr. Build. Mater. **192**, 633–642 (2018)
3. Aldás, P.S.D., Constante, J., Tapia, G.C., Martínez-Gómez, J.: Monohull ship hydrodynamic simulation using CFD. Int. J. Math. Oper. Res. **15**(4), 417–433 (2019)
4. Martínez, J., Martí-Herrero, J., Villacís, S., Riofrio, A.J., Vaca, D.: Analysis of energy, CO2 emissions and economy of the technological migration for clean cooking in Ecuador. Energy Policy **107**, 182–187 (2017)
5. Espinoza, V.S., Guayanlema, V., Martínez-Gómez, J.: Energy efficiency plan benefits in ecuador: long-range energy alternative planning model. Int. J. Energy Econ. Policy **8**(4), 52–54 (2018)
6. Gaona, D., Urresta, E., Marínez, J., Guerrón, G.: Medium-temperature phase-change materials thermal characterization by the T-History method and differential scanning calorimetry. Exper. Heat Transfer **30**(5), 463–474 (2017)
7. Kastillo, J.P., Martínez, J., Riofrio, A.J., Villacis, S. P., Orozco, M.A.: Computational fluid dynamic analysis of olive oil in different induction pots. In: 1st Pan-American Congress on Computational Mechanics–PANACM 2015, pp. 729–741 (2015)
8. Kastillo, J.P., Martínez-Gómez, J., Villacis, S.P., Riofrio, A.J.: Thermal natural convection analysis of olive oil in different cookware materials for induction stoves. Int. J. Food Eng. **13** (3) (2017)
9. Rodríguez, D., Martínez-Gómez, J., Guerrón, G., Riofrio, A.: Impact of induction stoves penetration over power quality in Ecuadorian households. Revista ESPACIOS **40**(13) (2019)
10. Tawancy, H.M., Ul-Hamid, A., Mohammed, A.I., Abbas, N.M.: Effect of materials selection and design on the performance of an engineering product–an example from petrochemical industry. Mater. Des. **28**(2), 686–703 (2007)
11. Jee, D.H., Kang, K.J.: A method for optimal material selection aided with decision making theory. Mater. Des. **21**(3), 199–206 (2000)

12. Shanian, A., Savadogo, O.: A material selection model based on the concept of multiple attribute decision making. Mater. Des. **27**(4), 329–337 (2006)
13. Ho, W.: Integrated analytic hierarchy process and its applications–a literature review. Eur. J. Oper. Res. **186**(1), 211–228 (2008)
14. Anojkumar, L., Ilangkumaran, M., Sasirekha, V.: Comparative analysis of MCDM methods for pipe material selection in sugar industry. Expert Syst. Appl. **41**(6), 2964–2980 (2014)
15. Barin, A., Canha, L., Magnago, L.K., Abaide, A., Wottrich, B.: Multicriteria decision making for management of storage energy technologies on renewable hybrid systems-the analytic hierarchy process and the fuzzy logic. In: Energy Market. EEM 2009. 6th International Conference on the European, pp. 1–6 (2009)
16. Cavallaro, F.: Fuzzy TOPSIS approach for assessing thermal-energy storage in concentrated solar power (CSP) systems. Appl. Energy **87**(2), 496–503 (2010)
17. Fernandez, A.I., Martínez, M., Segarra, M., Martorell, I., Cabeza, L.F.: Selection of materials with potential in sensible thermal energy storage. Sol. Energy Mater. Sol. Cells **94**(10), 1723–1729 (2010)
18. Khare, S., Dell'Amico, M., Knight, C., McGarry, S.: Selection of materials for high temperature sensible energy storage. Sol. Energy Mater. Sol. Cells **115**, 114–122 (2013)
19. Martínez-Gómez, J., Ibarra, D., Villacis, S., Cuji, P., Cruz, P.R.: Analysis of LPG, electric and induction cookers during cooking typical Ecuadorian dishes into the national efficient cooking program. Food Policy **59**, 88–102 (2016)
20. Martínez-Gómez, J., Guerrón, G., Riofrio, A.J.: Analysis of the "Plan Fronteras" for clean cooking in Ecuador. Int. J. Energy Econ. Policy **7**(1), 135–145 (2017)
21. Martínez-Gómez, J.: Material selection for multi-tubular fixed bed reactor Fischer-Tropsch reactor. Int. J. Math. Oper. Res. **13**(1), 1–29 (2018)
22. Villacreses, G., Gaona, G., Martínez-Gómez, J., Jijón, D.J.: Wind farms suitability location using geographical information system (GIS), based on multi-criteria decision making (MCDM) methods: the case of continental Ecuador. Renew. Energy **109**, 275–286 (2017)
23. Villacreses, G., Martínez-Gómez, J., Quintana, P.: Geolocation of electric bikes recharging stations: City of Quito study case. Int. J. Math. Oper. Res. **14**(4), 495–516 (2019)
24. Beltrán, R.D., Martínez-Gómez, J.: Analysis of phase change materials (PCM) for building wallboards based on the effect of environment. J. Build. Eng. **24**, 100726 (2019)
25. Godoy-Vaca, L., Almaguer, M., Martínez-Gómez, J., Lobato, A., Palme, M.: Analysis of solar chimneys in different climate zones-case of social housing in Ecuador. In: IOP Conference Series: Materials Science and Engineering. vol. 245, No. 7, p. 072045 (2017)
26. Villacreses, G., Salinas, S.S., Ortiz, W.D., Villacís, S., Martínez-Gómez, J.: Environmental impact assessment of internal combustion and electric engines for maritime transport. Environ. Process. **4**(4), 907–922 (2017)
27. Chingo, C., Martínez-Gomez, J.: Material selection using multi-criteria decision-making methods for geomembranes. Int. J. Math. Oper. Res. **16**(1), 24–52 (2020)
28. Acurio, K., Chico-Proano, A., Martínez-Gómez, J., Orozco, M.: Regeneration of waste diatomite from palm oil production process as a support material for pcms in thermal energy storage in buildings. Adv. Mat. Research **1151**, 29–33 (2019)

Phase Change Materials. Material Selection Based on Better Thermal Properties: A Literature Review

E. Reyes-Cueva[1], Javier Martínez-Gómez[1,2](✉), and Mónica Delgado Yánez[1]

[1] Universidad Internacional SEK Ecuador, Quito, Ecuador
javier.martinez@uisek.edu.ec
[2] Instituto de Investigación Geológico y Energético (IIGE), Quito, Ecuador

Abstract. This study aims to make the optimal selection of Phase Change Materials (PCM) based on thermal properties and according to industrial application, in addition to publicize the most used methods to facilitate this selection. Although there are several types of PCM, depending on the composition and characteristics they can be used in different applications. The most used characterization methods, throughout their study, are the DSC and the T-history, but new methods that replicate or exceed the expected are still being studied, which contribute to make an adequate material selection and their components. The bibliography and study of these materials has aroused curiosity and over the years new materials and new ways of obtaining them are studied. Also, to improve the environment in which they work, wider applications and those currently used, it is possible to obtain greater energy efficiency. Abstract should summarize the contents of the paper in short terms, i.e. 150–250 words.

Keywords: Phase change materials · Thermal properties · Material selection · Energy storage

1 Introduction

Phase change materials (PCM) are known as those whose main characteristic is to store or transfer energy in large quantities while maintaining constant temperature during the change of state [1–3]. PCMs are also known as latent heat storage materials [4].

It is usual to find studies on phase change materials (PCM), so they take off around the 70s [5–9], growing exponentially up to the present date according to data obtained from SCOPUS as seen in Fig. 1, with little development in the countries of America, except for the USA and Canada indicated in Fig. 2 and Fig. 3, which makes it important to carry out at least a bibliographical investigation on this subject in Ecuador.

When talking about materials for storage and conservation of latent heat energy. Phase change materials are the best candidates [10–13] taking into account that they provide the following advantages in the following applications. They can store large latent heat in a small volume [10, 14, 15]; heat losses when applied in a system are

M. Botto-Tobar et al. (Eds.): CI3 2020, AISC 1277, pp. 450–463, 2021.
https://doi.org/10.1007/978-3-030-60467-7_37

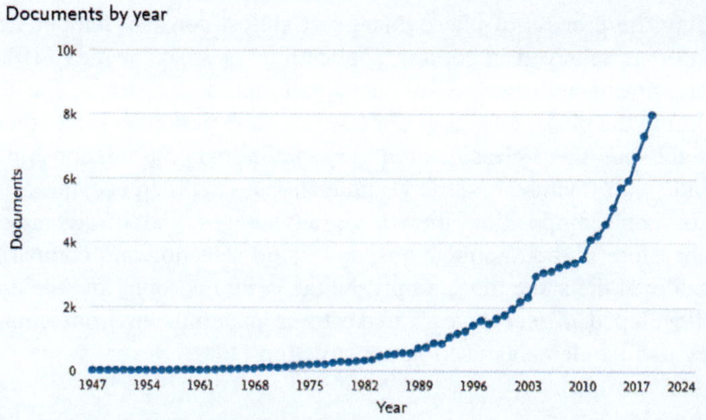

Fig. 1. Studies of phase change materials by year. (SCOPUS, 2020)

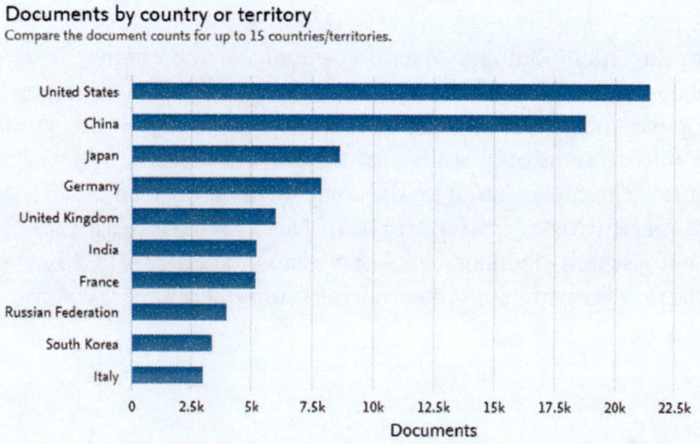

Fig. 2. Studies of phase change materials by country in the world. (SCOPUS, 2020)

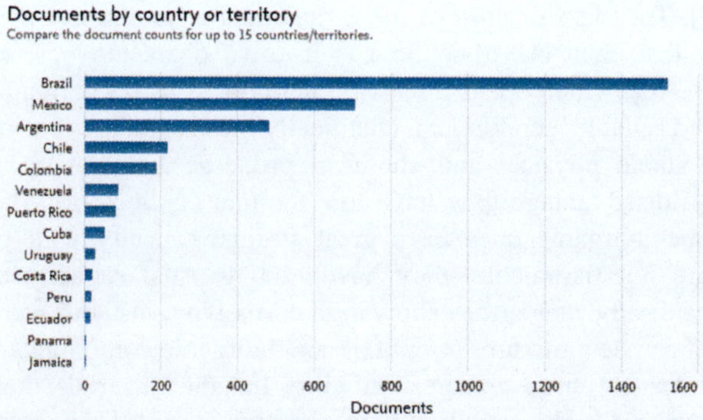

Fig. 3. Studies of phase change materials by country in Latin America. (SCOPUS, 2020)

minimal, being the process of phase change at almost constant temperature excellent thermal comfort is achieved in cooling applications or space heating [16].

The development and research of these materials is important due to their high latent heat during the phase transition, for this reason their correct selection is necessary taking into account their classification (organic/inorganic), relationship with other materials and with themselves; and applications in different areas, besides that according to their composition they have advantages and disadvantages in their properties, therefore in their applications, so a good selection and comparison method is indispensable; at the same time, it must be taken into account that the encapsulation techniques developed in recent years make these materials environmentally friendly, and that they can be elements used in construction [17].

2 Materials and Methods

For this study, the main methodology used is the search for existing bibliography about PCM.

To develop this methodology, several academic search engines were used such as: Google Scholar, Scielo, World Wide Science, Google Scholar, Springer Link, Science Gov, and Scopus. And several academic articles and books on the subject have been found, taking into account only studies of the last 20 years.

The literature search was based on different words used such as: PHASE CHANGE MATERIALS, THERMAL PROPERTIES, MATERIAL SELECTION, ENERGY STORAGE, with search operators and expressions such as «PCM», «phase change material», «thermal properties», «material selection» and «energy storage».

3 Results

As a result of the literature review, there are several scientific articles and books. By relevance and bibliographic utility have been used that refer to this article. The main classification of PCMs according to chemical composition is: organic, inorganic and eutectic [17–19]. The classification of these materials is shown in Fig. 4.

Considering that each classification has its own characteristics, advantages and disadvantages [6, 20], which opens up a wide field for its use, so organic products they can be considered reliable, economical, chemically inert, safe, non-corrosive and stable during casting, stable physical and chemical properties, almost no overcooling or segregation, as a disadvantage they have low thermal conductivity, flammability and low enthalpy; the inorganic ones have great storage capacity with respect to their volume, they are not flammable, they have high thermal capacity and low vapor pressure, but as a disadvantage, they show high corrosivity, instability and overcooling; the eutectics that can be a mixture of organic and inorganic compounds which presents a fixed temperature at phase change, knowing that the materials that make up the eutectic PCM have the highest melting point than when mixed, freeze or melt without segregation in addition to having a high volumetric storage density, unfortunately they are very limited with respect to their availability [18, 20–23].

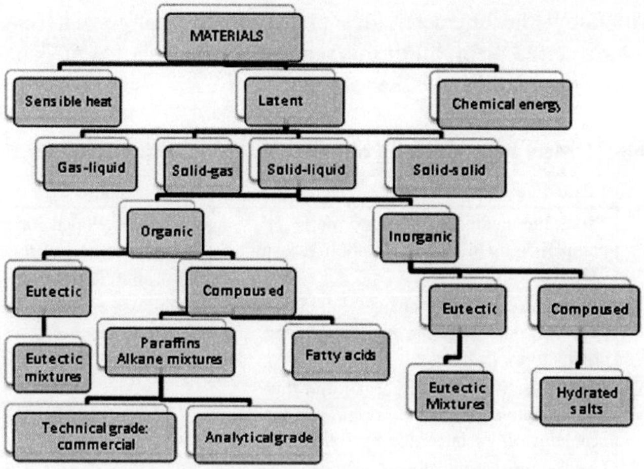

Fig. 4. Classification of phase change materials [10].

Phase change materials also have mostly features such as (see Table 1).

Table 1. Phase change materials features

Thermal	Physical	Kinetics	Chemical	Economic
- A good working temperature according to applicability	- Favorable phase equilibrium	- Avoid hysteresis problems	- Long-term chemical stability	- Abundant
- Defined melting temperature	- High density	No supercooling or lag when solidifying	- Compatibility with building materials, contact materials and encapsulation to avoid oxidation, corrosion, thermal decomposition or hydrolysis	-Available
- High latent transition heat	- Small volumetric variation	- Enough crystallization rate	- Non toxic	- Profitable
- High enthalpy at phase change	- Low vapor pressure		- Non flammable	- Technically an economically viable
- Good heat transfer			- Reversible phase change and temperature depend only	- Reasonable life cycle based on cost
- High thermal conductivity				

Note: Prepared on the basis of [23, 24].

In order to know the characteristics of the phase change materials, it has been necessary to characterize them, all of this using technique shown in Table 2, with their respective advantages and disadvantages.

Table 2. Main methodologies of phase change materials characterization

Methodology	Advantage	Disadvantages
T-History	Economic methodology economic in comparision with others metohologies studied [25, 26]	It must developed for each case with the specification of its range of application [27]
	Determines the phase change enthalpy, the specific caloric capacity and the thermal conductivity [28]	Requires a controlled temperature chamber, which can result in high initial costs [31]
	Solve effectively the trouble of find the representativeness of the sample and the problems of crystalization by size shows [29]	
	Lets analyze curves the subcooling on the enthalpy-temperature curves of the samples [30]	
Zhang's Method	Transfer of hot unidirectional in radial direction [32]	It present control problems since it requieres methodical vigilance
	Records the evolution of temperatures in two tubes and even It allows microencapsulation of materials [33]	To get the properties thermophysical temperatura curves must be obtained previously [34]
	Heat transfer is by natural convention [27]	
Improvements of the Zhang's Method	Allows variation in specific heats of the substances	Minor errors in natural convection
	It supposes a coefficient of convection constant in differenvial temperature less than 0,5 K [28]	There must be methodical controlk of the amount of sensible and latent heat absorbed by the susbtance [28]
		Lack of control of the symmetrical geometry with respect to the axis causes errors systematic serious [35]
Adiabatic Calorimetry	Study the energy delivery as input or heat transfer	The sample size must be considered, It must always be stable for several test and in large quantites
	Accurate for large periods of time in the study [36]	One must work in very low speed to measure the thermal balance
		High costs
Thermograms	Heat capacity determines by the energy exchanged	Equipment must be available, it must have a period reasonable working to avoid distortion in the curves
Self built methods	It is built according to the study material, and the object of development	They present problems in relation to conventional methods in relaion to reliability
Two bath methods	Comparison with two measurements made	Good results are no obtained in relation to standard substance measurements
Kousksou	Determines the specific heat capacity without correction of empty cells [28]	Depends on exact speed measurement
Differential Scanning Calorimetry (DSC)	Wide and flexible temperature ranges	The treated samples are very small, therefore errors can be generated
	Easy drive	

3.1 Uses and Applications of Phase Change Materials

PCM applications cover all those where energy storage or temperature maintenance is required [13, 37, 38], although, there are several types of PCM, depending on their composition and characteristics can be used in different applications and uses. Among these uses, use in "solar thermal systems, water heating systems, in low enthalpy energy systems, in thermoregulatory systems for food preservation, in thermoelectric plants" can be highlighted. The main applications developed to date are for thermal comfort for buildings, protection of electronic devices and storage for cooling systems [20]. Other uses of PCMs are in domestic hot water tanks, heating and air conditioning in buildings, refrigeration, transportation and storage of food, beverages, pharmaceuticals, blood products [39], box type solar cookers, helmet cooling systems for motorcyclist and athletes [40]. The latest technological advances and deeper studies of this type of materials have found new uses such as: pocket heaters, fibers and fabrics, transport containers, solar energy systems for electricity storage, drug administration, information storage and memory devices, cancer biomarker detection, barcodes [41], solar thermal energy storage, bioclimatic construction, temperature regulation and thermal protection of electronic devices, drying, desalination of water, greenhouse temperature conservation, heat and cold therapies for medical treatments, solar power plants, aerospace terms systems. In chemical reactions for softening of exothermic temperature peaks [42].

3.2 Phase Change Materials and Energy Efficiency

The PCMs have aroused curiosity, that over the years, new materials and new ways of obtaining research are being investigated, as well as improving the working environment, greater scope of applications and the current ones used are more efficient.

Based on the knowledge and development of PCM properties, studies have been carried out to optimize thermal energy storage by improving production costs, taking into account these qualities and properties, to choose the most recommended materials by selection methods [43].

Energy efficiency can be improved using PCM in air conditioning installations, in addition it should be taken into account to popularize its use, by studying new PCMs regarding the heat transmission and economic viability of these applications, as they are low risk materials and large benefits, considering that globally it is necessary to improve energy costs [44].

3.3 Phase Change Materials, Enthalpy and Phase Change Temperature

The study of the PCM, as has been clearly indicated, has to do with two important points such as the enthalpy and the temperature of the phase change, this in order to be clear about the type of applicability that can be given to the material, in this way and As shown in Fig. 5, a classification can be given to the PCM according to the implementation of their thermal accumulation, so they can be declared high, medium and low temperature [45], in this way, those whose phase change is below 10° C can be considered low temperatures, having in this case the saline solutions, water and a small part of clathrates, hydrate salts, eutectic mixtures and paraffins, those of average temperatures would consider those whose phase change occurs in the range between 10° C and 80° C where most of the materials studied are found, among which are hydrated salts, eutectic mixtures, paraffins, fatty acids, polyethylene glycols and finally the high temperature materials whose phase change temperatures are greater than 80° C where salts and eutectics (mixtures of salts), sugars and alcohols, and some hydrate salts, eutectic mixtures and paraffins are found.

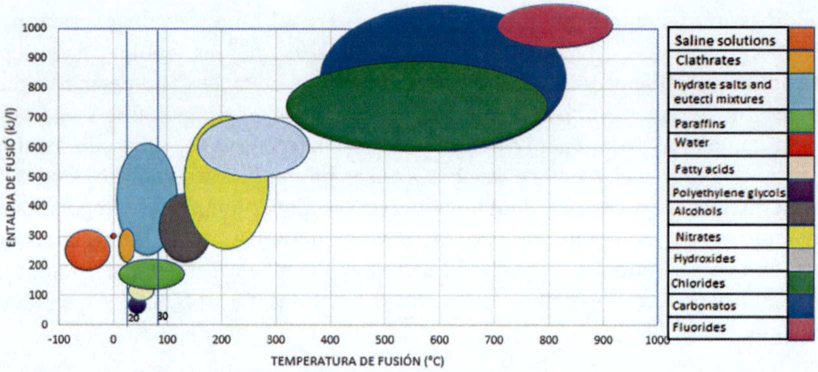

Fig. 5. Phase change materials by temperature and enthalpy, (Adapted from [46, 47]).

3.4 Phase Change Materials for Low Temperature Use

The main characteristics of PCMs for this type of application would be: high latent transition heat, no overcooling, high crystallization rate, negligible volume change, compatibility with container materials, no corrosion, availability, low cost, recyclability [18].

Complete a reversible freeze cycle, in other words that may melt, lack of degradation after several freeze and melt cycles, do not produce corrosion in building and encapsulation materials [39].

Among the main applications of PCM for low temperatures are transport containers used for various applications, such as food handling, beverages, pharmaceuticals derived from blood, biomedicines, electronic circuits [48].

Studies with nanofluids such as PCM have been carried out, resulting in materials suitable for low temperature energy storage uses [49].

3.5 Phase Change Materials and Costs

Although the applications of these materials are varied, the cost of obtaining these materials is quite high, which makes the discovery of new PCMs or their mixtures and their encapsulation attractive, which is why low-bio-PCMs have been developed cost to be able to commercialize it, this through the use of waste oils or low-cost fats from hydrodeoxygenation for the production of paraffins [50].

For the cost of thermal storage systems it is important to know the price of the phase change materials, which can be indicated in $/kg, the price of some of these materials are tabulated in Table 3, based on This table can understand the costs that are granted in reference to the properties that they present, so that when the latent heat of fusion and the specific heat are greater, it results in a reduction in the dimensioning of the container of these because the density of the substances is greater, directly affecting the final cost of the project since the cost of said containers decreases, in the same way when the thermal conductivity of the material is low, encapsulations must be performed, and this process increases the final cost [45].

3.6 Selection Methods

For the following study, the main methods of selecting PCMs are divided in:

Existing Materials with Better Energy Efficiency
Select PCMs for use in thermal energy storage (TES) around 200–410° C and reduce production costs, using the multi-criteria decision making method (MCDM), based on thermal characteristics, qualitative criteria, so that the best material for this purpose is obtained, having as methods used: COPARA-G, TOPSIS, VIKOR [20].

Another form of PCM is to use in (construction boards) and compare two methodologies, using MCDM and Construction Energy Simulation (BES), considering that participation with BES allows a contrast to the MCDM methodology. At the same time, it allows a numerical evaluation of the thermal behavior of the material, likewise, an energy consumption can be structured by integrating the PCM. For this type of experimentation, the environment variables are important but create a type of ambiguity between the methods [51].

In addition to using fatty acids (lauric, myristic, palmitic, stearic and acetamide) as a single component, they can be used as a binary eutectic mixture and selected for drying applications. The selection is made using the differential scanning calorimetry thermal analysis (DSC) technique with heat flow or power compensation techniques [42].

Among the techniques for selecting PCMs by means of characteristics analyzed such as DSC and T-history, a technique called the Peltier method based on elements of adiabatic scanning calorimetry (pASC) has also been developed. Good results were found using PCMs in materials containing alkanes, with the advantage that this methodology exceeds the previous ones due to the use of smaller mixtures and also when amounts of tests are greater and faster results are required [41].

New Materials Developed and Application Improvement
In order to improve thermal performance, PCMs are being developed in nanofluids, which demonstrate greater efficiency characteristics for cold storage applications, hoping that conventional PCMs will be replaced in these applications [52].

For technical characteristics and applications, for example, selection made for application of energy saving in greenhouses, based on salt hydrates, paraffins and polyethylene, with variations in the types of heat exchangers, type of storage and quantities of PCM used per m^2 of greenhouse soil [53].

Another way to select PCMs is by microencapsulation method, among these physical, chemical and physiochemical. For production of micro-encapsulated PCMs,

Table 3. Cost per unit of mass of different phase change materials [45]

Category	Storage material	Cost per unidad of mass ($/kg)	Melting Temperature (°C)	Density (kg/m3)	Latent heat of fusion kJ/kg	Specific heat capacity (kJ/kg°K)	Thermal conductivity W/(m °K)
Nitrate Salts	NaNO3	0,4	306	2261	172	1655	0,514
	KNO3	0,9	335	2109	266	0,953	0,5
Hydroxide Salts	NaOH	0,35	318	2100	165	0,92	
	KOH	1	380	2040	150	1,34	0,5
Chloridre Salts	NaCl	0,1	802	2160	420	5	5
	KCl	0,5	771	1980	353	-	-
	MgCl	0,1	714	2320	452	-	-
	LiCl	10	610	2070	441	-	-
	CaCl2	0,15	772	2150	253	-	-
Fluorine Salts	LiF	0,5	850	2640	1044	-	-
	NaF	0,9	996	2558	794	-	-
	KF	0,1	858	2370	468	-	-
	CaF2	0,35	1418	3180	391	-	-
Solid Materials	Sand oil mineral and rocks	0,15		1700	-	1,3	1
	Cast steel	5		7800		0,6	40
	Concrete	0,05		2200		0,85	1,5
Paraffins	Paraffin wax	1	64	916	173,6		0,346
	N-Pentadecane	0,59	10		-	-	206
	N-Hexadecane	4	20	773	-	-	236
Fatty acids	Caprylic acid	5	16	981	148,5	-	0,149
	Capric acid	4	32	1004	152,7	-	0,153
	Lauric acid	1	42	870	171	-	-
	Myristic acid	1,89	54	860	190	-	-
	Palmitic acid	1	64	989	185,4	-	0,162
Stearate	Palmitate methyl	100	27	-	163,2	-	-
	Stearate methyl	10	39	-	160,7	-	-
Glycols	Polietilen glycol	2	4,2		117,6		
	PEG600	2	12,5	-	129,1	-	-
	PEG1000	2	40	-	168,6	-	-

the common methods are by artificial polymerization, suspension polymerization, coacervation, emulsion polymerization, spray drying. From the study, it is obtained that nanoencapsulation would drastically reduce the over-cooling problem of PCMs, which is one of the biggest concerns with these materials. This method uses paraffin-type PCM, therefore, for other PCMs it is excluded [54].

Through the creation of new PCMs from nitrate processing and characterization, since being tested and obtaining relevant results, they contribute mostly to solve the problem of energy storage, and can help different industrial applications [20].

The selection based on the creation of sustainable PCM versus the existing PCM, an analysis of thermal characteristics and properties with materials obtained from palm oil residues is carried out. For its evaluation, different mixtures with organic components were made and those with the highest latent melting heat and phase change temperature range closest to the thermal comfort in each support material can be selected, resulting in these materials being able to easily compete with existing ones being important for the selection criteria. In addition to being potential energy saving materials, they are sustainable [55, 56].

Select new PCMs created from composite materials, such as based on paraffin and high density polyethylene and expanded tap aggregates, taking into account thermal properties and varying the amount of the compounds, thus achieving selection criteria not only based to characteristics and storage capacity but costs when encapsulation is not necessary [57].

4 Discussion

Although several PCMs are known, it is necessary to focus the study on new products. It is much better if the PCMs were sustainable, economical and easily accessible. It could contribute with new applications or improvement of the current ones, mainly the study should be further extended towards the characterization of new materials for low temperature applications. This is because the study of PCM is focused on energy conservation for thermal comfort applications and for medium and high temperatures. With the search for new PCMs or their mixtures for use in energy efficiency technology that advances exponentially with the passing of days, these new forms of energy storage will be highlighted in the near future.

5 Conclusions

A systematic review for the optimal selection of Phase Change Materials (PCM) is carried out, with the table of thermal properties and methodology being the reference when deciding on their use. The use of PCM to improve energy storage efficiency, especially in the form of heat, has been developing and increasing in recent years.

Each time it seeks to improve the thermal properties of these materials. Depending on the application, these materials have been highly used and studied, especially in the construction of buildings to maintain the air conditioning of the environments, taking into account the storage capacity of solar thermal energy. Although there are techniques to measure the characteristics of PCM in some way, new techniques continue to expand that facilitate the comparison between materials, and thus help in the selection of materials for their properties and application. It seeks to generate new PCMs that are sustainable, easy to use and reduced in costs, which will generally be compared with those already existing, and chosen by methods such as these multi-criteria, simulation, among others.

References

1. Lee, B.C., Ipek, E., Mutlu, O., Burger, D.: Architecting phase change memory as a scalable dram alternative. In: Paper presented at the Proceedings of the 36th Annual International Symposium on Computer Architecture (2009)
2. Tan, F., Hosseinizadeh, S., Khodadadi, J., Fan, L.: Experimental and computational study of constrained melting of phase change materials (PCM) inside a spherical capsule. Int. J. Heat Mass Transf. **52**(15–16), 3464–3472 (2009)
3. Yinping, Z., Yi, J.: A simple method, the-history method, of determining the heat of fusion, specific heat and thermal conductivity of phase-change materials. Meas. Sci. Technol. **10**(3), 201 (1999)
4. Wong, H.-S.P., Raoux, S., Kim, S., Liang, J., Reifenberg, J.P., Rajendran, B., Goodson, K. E.: Phase change memory. Proc. IEEE **98**(12), 2201–2227 (2010)
5. Sharma, A., Tyagi, V.V., Chen, C.R., Buddhi, D.: Review on thermal energy storage with phase change materials and applications. Renew. Sustain. Energy Rev. **13**(2), 318–345 (2009)
6. Bohorquez, R.D.B.: Proceso De Recuperacion De Revestimientos Viejos Mediante La Tecnica De Mapeo De Corriente (Pipeline Current Mapper-Pcm) Y Aplicacion De Pinturas Epoxicas, Efectuados En En Tramo De Dina-Guasimal Del Gasoducto Centroriente De Ecogas. Universidad Industrial de Santander, Escuela De Ingeniería de Petróleos (2005)
7. Estrella, A.L., Martínez, J.H.A., Sevilla, E.E.: Revisión bibliográfica. Los PCM para el ahorro energético en edificación (2016)
8. Carmona, J.M.: Caracterización de materiales de cambio de fase para su aplicación en revestimientos y acabados. Universitat Politècnica de Catalunya (2013)
9. Padilla, P.C.M., del Castillo, G.P.F., Arnau, R.G.: Visiones de Estados Unidos y Canadá desde América Latina a principios del siglo XXI, FLACSO Mexico, vol. 1 (2001)
10. Sanoguera, D.G., Varón, D.J., Gimeno, R.A.B., Bou, S.F.: Estudio de materiales con cambio de fase (PCM) y análisis SEM de micro pcm. 3C Tecnología: glosas de innovación aplicadas a la pyme **1**(3), 4 (2012)
11. Acurio, K., Chico-Proano, A., Martínez-Gómez, J., Ávila, C.F., Ávila, Á., Orozco, M.: Thermal performance enhancement of organic phase change materials using spent diatomite from the palm oil bleaching process as support. Constr. Build. Mater. **192**, 633–642 (2018)
12. Acurio, K., Chico-Proano, A., Martínez-Gómez, J., Orozco, M.: Regeneration of waste diatomite from palm oil production process as a support material for PCMs in thermal energy storage in buildings. In: Advanced Materials Research, vol. 1151, pp. 29–33. Trans Tech Publications (2019)
13. Ortega, A.Z.R.: Evaluación analítica y experimental de un sistema de calentamiento solar de agua a baja temperatura con tecnología de almacenamiento térmico con material de cambio de fase (PCM) en términos energéticos y exergéticos. Universidad del Norte (2019)

14. Román, E.B., Calle, D.A.V.: Evaluación De Modelos Termofísicos De Cambio De Fase En Una Aplicación De Almacenamiento De Energía Térmica. Mecánica Computacional **13**(1) (2015)
15. Kalternbach, F.: PCM: Acumuladores de calor latente para calefacción y refrigeración. Detail: revista de arquitectura y detalles constructivos (4), 482 (2005)
16. Jiménez, A.Z.: Diseño de una instalación solar térmica para el abastecimiento energético de acumuladores de calor latente de tipo PCM (2019)
17. Li, Y., Liu, S., Zhang, Y.: Progress in Sustainable Energy Technologies, vol II, II (2014)
18. Hyun, D.C., Levinson, N.S., Jeong, U., Xia, Y.: Emerging applications of phase-change materials (PCMs): Teaching an old dog new tricks. Angewandte Chemie - International Edition (2014)
19. Martínez-Gómez, J.: Material selection for multi-tubular fixed bed reactor Fischer-Tropsch reactor. Int. J. Math. Oper. Res. **13**(1), 1–29 (2018)
20. Gaona, D., Urresta, E., Marínez, J., Guerrón, G.: Medium-temperature phase- change materials thermal characterization by the T-History method and differential scanning calorimetry. Exp. Heat Transf. **30**(5), 463–474 (2017)
21. Orozco, M., Martínez, J., Chico, A., Acurio, K.: Procesado y Caracterización de Nitratos como Materiales de Cambio de Fase para Evaluar su Capacidad de Almacenamiento Energético. Revista Técnica Energía, 224–235 (2018)
22. Moscoso, M.: Aplicación de los materiales de cambio de fase en el mobiliario interior como reguladores de temperatura, p. 81 2016
23. Wi, S., Seo, J., Jeong, S.G., Chang, S.J., Kang, Y., Kim, S.: Thermal properties of shape-stabilized phase change materials using fatty acid ester and exfoliated graphite nanoplatelets for saving energy in buildings. Sol. Energy Mater. Sol. Cells **143**, 168–173 (2015)
24. He, M., Yang, L., Zhang, Z.: Experimental studies on cycling stable characteristics of inorganic phase change material CaCl2 6H2O-MgCl2 6H2O modified with SrCl2 6H2O and CMC. In: IOP Conference Series: Earth and Environmental Science, vol. 108, No. 2, p. 022058. IOP Publishing (2018)
25. Gallardo, R.P.: Optimización de la temperatura de cambio de fase en PCM para refrigeración de módulos fotovoltaicos en función de las condiciones climáticas (2018)
26. Peck, J.H., Kim, J.-J., Kang, C., Hong, H.: A study of accurate latent heat measurement for a PCM with a low melting temperature using T-history method. Int. J. Refrig **29**(7), 1225–1232 (2006)
27. Solé, A., Miró, L., Barreneche, C., Martorell, I., Cabeza, L.F.: Review of the T-history method to determine thermophysical properties of phase change materials (PCM). Renew. Sustain. Energy Rev. **26**, 425–436 (2013)
28. Hasan, A., McCormack, S., Huang, M., Norton, B.: Characterization of phase change materials for thermal control of photovoltaics using differential scanning calorimetry and temperature history method. Energy Convers. Manag. **81**, 322–329 (2014)
29. García, M.C.P. :Avances en determinación de propiedades termofísicas de materiales de cambio de fase: búsqueda y análisis de nuevos materiales PCM-TES de bajo coste. Universidad de Zaragoza (2015)
30. García, J.O.: Título del trabajo: Metodología T-History para la determinación de curvas entalpía-temperatura de materiales compuestos para aplicaciones de construcción. Ensayos y análisis de resultados. Universidad de Zaragoza (2016)
31. Martínez, M.J.V.: Título del trabajo: Adaptación de la metodología T-History para determinación de curvas entalpía-temperatura de materiales compuestos para aplicaciones de construcción. Universidad de Zaragoza (2015)

32. Gallardo, R.P.: Optimización de la temperatura de cambio de fase en PCM para refrigeración de módulos fotovoltaicos en función de las condiciones climáticas. Universidad de Sevilla (2018)
33. Pujante, A.G.: Evaluación de las propiedades en estado fresco y endurecido (mecánicas, termofísicas y reacción al fuego) del yeso con PCM mediente tres métodos diferentes de incldioe liex y exclusiones de PCM. Universitat Politècnica de Catalunya (2014)
34. Figueira, A.J.: Potência nuclear aproximada para o caso de reatividade linear e métodos alternativos de solução da equação da cinética pontual inversa. Universidade Federal do Rio de Janeiro (2009)
35. Varón, D.J., Bou, S.F., Gimeno, R.A.B., Sanoguera, D.G.: Estudio de materiales con cambio de fase (PCM) y análisis SEM de micro PCM. 3C Tecnologia **3**, 54–77 (2012)
36. Kastillo, J.P., Martínez-Gómez, J., Villacis, S.P., Riofrio, A.J.: Thermal natural convection analysis of olive oil in different cookware materials for induction stoves. Int. J. Food Eng. **13** (3) (2017)
37. Almiñanaa, D.G., Castellb, A., Cabezab, L.F.: Estudio sobre el sistema de almacenamiento de agua caliente sanitaria en un sistema solar térmico (2006). Espacio de reflexión y comunicación en Desarrollo Sostenible **3**(14) (2012)
38. Pérez, A.L.: Evaluación de las propiedades físicas y térmicas de yeso combinado con materiales cambio de fase para su utilización en construcciones en la República Dominicana. Universitat Politècnica de Catalunya (2018)
39. Oró, E., De Gracia, A., Castell, A., Farid, M.M., Cabeza, L.F.: Review on phase change materials (PCMs) for cold thermal energy storage applications **99**, 513–533 (2012)
40. Shukla, A.: Latent heat storage through phase change materials. Resonance **20**(6), 532–541 (2015)
41. Leys, J., Duponchel, B., Longuemart, S., Glorieux, C., Thoen, J.: A new calorimetric technique for phase change materials and its application to alkane-based PCMs. Mater. Renew. Sustain. Energy **5**(1) (2016)
42. Jain, A., Kumar, A., Shukla, A., Sharma, A.: Solar Drying Technology (2017)
43. Martínez-Gómez, J., Urresta, E., Gaona, D., Guerrón, G.: Selection of phase change materials for latent heat storage. Revista Técnica Energía 179–190 (2017)
44. Domínguez, M., Csic, F., Novais, C.A.: Aprovechamiento de los Materiales de Cambio de Fase (PCM) en la Climatización Utilization of the Phase Change Materials (PCM) in the Air Conditioning **20**, 107–115 (2009)
45. Hidalgo, D.B., González, J., Martínez, Y.: Costos de las tecnologías de almacenamiento de energía térmica **44**, 67–76 (2017)
46. Bielsa, D.P.: Almacenamiento térmico de energía mediante cambio de fase. Diseño y modelización de equipos de almacenamiento para intercambio de calor con aire, 1–318 (2011)
47. Mayayo, J.G.: Materiales de Cambio de Fase. Diseño de una instalación para la caracterización de PCM a altas (2012)
48. Mehling, H., Cabeza, L.F.: Applications in transport and storage containers. In: Heat and cold storage with PCM. Heat and Mass Transfer. Springer, Berlín, Heidelberg (200)
49. He, Q., Wang, S., Tong, M., Liu, Y.: Experimental study on thermophysical properties of nanofluids as phase-change material (PCM) in low temperature cool storage. Energy Convers. Manag. **64**, 199–205 (2012)
50. Biswas, K., Abhari, R.: Low-cost phase change material as an energy storage medium in building envelopes: experimental and numerical analyses. Energy Convers. Manag. **88**, 1020–1031 (2014)
51. David, B.R., Martínez-Gómez, J.: Analysis of phase change materials (PCM) for building wallboards based on the effect of environment. J. Build. Eng. **24**, 100726 (2019)

52. Liu, Y.D., Zhou, Y.G., Tong, M.W., Zhou, X.S.: Experimental study of thermal conductivity and phase change performance of nanofluids PCMs. Microfluidics and Nanofluidics (2009)
53. Kurklo, A.: Energy storage applications in greenhouses by means of phase change materials (PCMs) : a review. Pergamon Renew. Energy **13**(1), 89–103 (1998)
54. Jamekhorshid, A., Sadrameli, S.M., Farid, M.: A review of microencapsulation methods of phase change materials (PCMs) as a thermal energy storage (TES) medium. Renew. Sustain. Energy Rev. (2014)
55. Acurio, K., et al.: Thermal performance enhancement of organic phase change materials using spent diatomite from the palm oil bleaching process as support. Constr. Build. Mater. **192**, 633–642 (2018)
56. Rozanna, D., Chuah, T.G., Salmiah, A., Choong, T.S.Y., Sari, M.: Fatty acids as phase change materials (PCMs) for thermal energy storage: a review. Int. J. Green Energy (2005)
57. Sari, A.: Form-stable paraffin/high density polyethylene composites as solid-liquid phase change material for thermal energy storage: preparation and thermal properties. Energy Convers. Manage. (2004)

Technological Innovation for the Sustainability of Knowledge and Natural Resources: Case of the Choco Andino Biosphere Reserve

Marco Heredia-R[1,2(✉)], Verónica Falconí[3], Jamil H-Silva[5], Katherine Amores[6], Carla A. Endara[6], and Karina F-Ausay[3,4]

[1] Department of Life Sciences, Universidad Estatal Amazónica, Puyo 160101, Pastaza, Ecuador
mheredia@uea.edu.ec
[2] Natural Resources Economics and Business Development Program, Universidad Estatal Amazónica, Puyo 160101, Pastaza, Ecuador
[3] Faculty of Economic and Administrative Sciences, Universidad de las Américas, Quito 170125, Ecuador
[4] Distance Education Unit, Universidad de las Fuerzas Armadas, ESPE, Quito, Ecuador
[5] Ciencias económicas y administrativas, Universidad International de la Rioja, Logroño, La Rioja, Spain
[6] Department of Economic, Administrative and Commercial Sciences, Universidad de las Fuerzas Armadas, ESPE, Latacunga 050102, Ecuador

Abstract. The management of natural resources has multiple challenges, such as covering the needs of a population that is estimated to exceed nine billion people by 2050, Therefore, the need for a digital evolution in agriculture that meets the needs of a world population in the future is identified The objective was: evaluate the sustainability of natural resources as a contribution towards the technological innovation of agriculture; the snowball sampling technique was used, in four productive systems: mixed, agroecological, indigenous y conventional, located in the transition and buffer zones of the Choco Andino Biosphere Reserve, SAFA evaluation framework (FAO) was used. The data were collected through a semistructured interview based on a questionnaire of 117 questions; the interview lasted 75–85 min conducted to the heads of household In the results were evidence of the 12 case of studies and the different sustainability dynamics in the dimensions: good governance, economic residence, environmental integrity and social welfare. The similar characteristic between case studies was the precarious capacity to govern natural resources, which strengthens the importance of intervention with a digital literacy process to improve management capacity.

Keywords: Digital technology · Rural territories · SAFA

M. Botto-Tobar et al. (Eds.): CI3 2020, AISC 1277, pp. 464–476, 2021.
https://doi.org/10.1007/978-3-030-60467-7_38

1 Introduction

1.1 The Technological Challenge of the Agriculture and Food

The agriculture and food sector faces multiple challenges such as: demographic growth, urbanization, aging, economic growth, investments, food prices, increased competition for natural resources for food and energy production, climate change, transboundary pets [1], crises, natural disasters, conflicts, inequality, poverty, food insecurity, health problems, structural changes, employment, migration and feminization of agriculture, changing food systems, food loss and waste, development financing, governance for nutrition and food security, agricultural productivity and innovation [2].

As global trends and challenges affect food security and the sustainability of food systems, several uncertainties arise as: 1) food and agricultural systems will be able to meet the needs of a population that is estimated to exceed nine billion people by 2050 [2] and 2) the necessary increases in production will be achieved in a context of climate change, even if doing so means putting even more pressure on the land, water resources and in general to natural resources already depleted [4], To meet the growing demand, agricultural productivity has to be increased, a potential solution to sustainable intensification. [5]. Productivity from 2012 should increase 50% by 2050 [6].

Agriculture has undergone several revolutionary processes to increase efficiency, performance and profitability to successful levels; it is evident that in the next decades, there must be a digital technological revolution in agriculture in the markets, establishing itself as one of the strategies to help agriculture satisfy the needs of a growing population [1]. For the use of digital technologies in the agri-food sector, there are certain basic conditions such as: 1) connectivity and infrastructure, 2) affordability, degree of instruction and 3) institutional support.

In terms of infrastructure and information technology networks in rural areas, it is a challenge [7]. Computers and cell phones (information and communication technologies, ICT) have changed the direction of people who have access to knowledge and information, despite this, there are considerable digital gaps within and between countries [8]. The degree of instruction, digital literacy and use of digital technologies in rural contexts affected literacy [9] and knowledge of arithmetic, as well as knowledge and spatial technical skills, levels of instruction are often lower in rural areas than in urban areas [10], digital literacy is low particularly among women [11, 12]. Government policies and programs are the force for digitalization, creating an environment of digital markets and electronic services, for the transformation to a digital government, designed and administered a high level program of capacity is required, which some countries have been successful due to the complexity of the subject [13, 14].

The technological - digital transformation for the management of natural resources [15] especially the agriculture sector has the potential to produce benefits in terms of sustainability: economic, social, environmental and governance. For this reason, the objective was to evaluate the sustainability of natural resources as a contribution towards the technological innovation of agriculture.

2 Methodologic

The case studies were carried out in the buffer and transition zone located northeast of the CABR (parish San José de Minas), area that was categorized as the seventh Biosphere Reserve of Ecuador in 2018 (Fig. 1). It has an area of 286,000 ha; it belongs to the humid forest of Choco - Darién, considered a biodiversity hotspot [16].

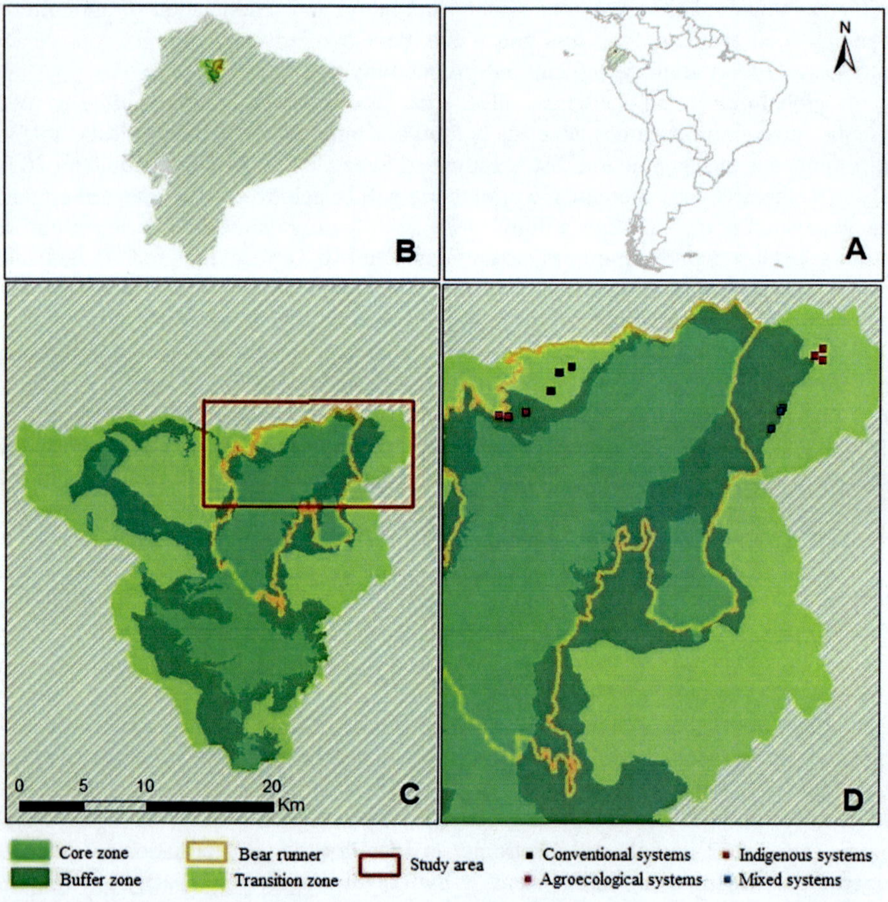

Fig. 1. A) Ecuador in South America, B) Andean Choco Biosphere Reserve (ACBR) located in northern Ecuador, C) Study area overlapping the northeast of the ACBR y D) Case studies.

In the sampling of case studies (Table 1) the non-probabilistic "snowball" technique was used, which starts from individuals not known to reach others and thus increase the size of the population, this approach is used in small populations that are difficult to access due to its closed nature or difficult access [17] and have special features [18],

Table 1. Case studies selected from the snowball technique at the ACBR

Zoning of the ACBR	Coding (indexing)	Study cases (Productive Systems)
Transition zone	AA.1 – AA.2 – AA-3	Mixed
	BB.1 – BB.2 – BB.3	Agroecological
Buffer zone	CC.1 – CC.2 – CC.*3*	Indigenous
	DD.1 – DD.2 – DD.3	Conventional

Three methodological processes were used: 1) methodology for technological innovation SAFA Tool. "Sustainability assessment for food and agriculture", 2) field annotation also called "Marginal Notes" and 3) Indexing using the OCM method.

The definition of each case study is as follows: **Mixed Systems:** are the food production systems that use synthetic products pesticides, herbicides and fertilizers [19] and in another proportion it uses ecological processes and cycles; crop rotation is performed; organic matter is added from the soil, biodiversity is taken care of and biological control of pests is by natural means [20]. **Agroecological Systems:** productive systems that take more advantage of natural processes and beneficial interactions in the production unit to reduce the use of inputs outside the farm, hacienda etc., and seeks to improve the efficiency of the systems [21]. The technologies used are: cover crops, green manures, intercalated crops, agroforestry and the mixing of crops and livestock [22]. **Conventional Systems:** it is agriculture that only uses chemical inputs, such as herbicides, fertilizers and pesticides throughout the production process [19]. **Indigenous Systems:** also called indigenous farming systems where production is traditional, is related to the conservation of natural resources to produce food or livelihoods, these are a subsistence alternative that uses family labor [23].

2.1 SAFA Tool Version 2.4.1

The SAFA version 2.4.1 program was used (Fig. 2) developed by the United Nations Organization for Agriculture and Nutrition (FAO) in 2012 as an innovative strategy and a contribution to digital technologies in agriculture and rural areas and used as a good ICT practice in Education defined as such, when better and/ or new learning is achieved, generates a pedagogical change (or innovation) and produces an organizational change [25].

Data were collected in December 2019 through a semi-structured interview (ethnographic interview) [25, 26], from a questionnaire of 117 questions based on the indicators of the SAFA methodology [27]. The interview lasted 75–85 min was made to the heads of household, the questions were translated from English to Spanish and Kichwa (language spoken by the indigenous Kichwa Otavalo).

The SAFA methodology has hierarchical levels: dimensions, themes, sub-themes and indicators, the dimensions are: good governance, environmental integrity, economic resilience and social welfare. It includes 21 sustainability issues, established by 58 subtopics, each subtopic includes several indicators, in total there are 116, which are measured based on a performance scale of 1 to 5 [28]. With sustainability thresholds:

better (dark green), good (light green), moderate (yellow), limited (orange), unacceptable (red).

The sustainability assessment through the SAFA methodology is developed in four stages: mapping, contextualization, indicators and final report or report (Fig. 2). Each of the stages can be re-executed or evaluated throughout the process, being a dynamic methodology that is fed back with the information obtained in each of the stages.

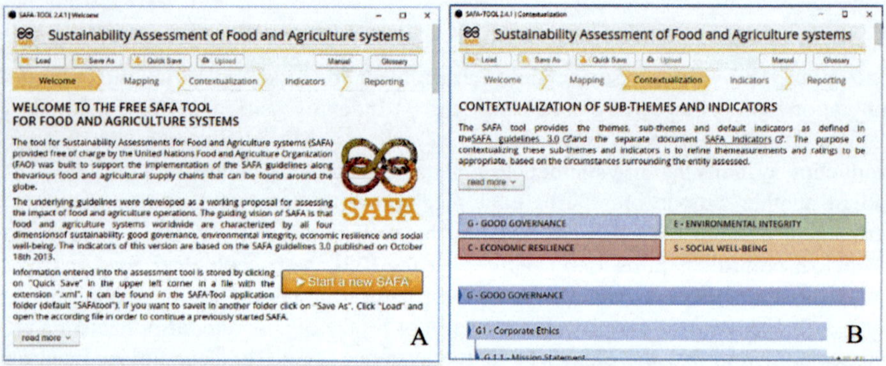

Fig. 2. SAFA program - version 2.4.1. created by FAO: A) Program interface - B) Visualization of the four stages mapping, contextualization, indicators and report and Evaluation dimensions.

2.2 Field Annotation also Called "Marginal Notes"

The marginal notes made were 15: three for each case study except for the indigenous productive systems that were carried out nine, due to the difficulty in conducting interviews due to the cultural context and language, field notes are a technique for qualitative research [29]. Field notes contextualize the study and provide a perspective on the lives of participants that may be useful when analyzing data in the future or examining perceptions over time [30].

2.3 OCM Method

For the coding or indexing of the information, the Outline of Cultural Materials, or OCM tool was used, developed by George Peter Murdock (1971), the codes and number are described in Table 1; the capital letter indicates the code as the designation for each production system and the number is the sequence of the case studies [31].

3 Results

The transition zone is the area where programs are carried out to promote economic and human development, which are sustainable from the sociocultural and ecological point of view [32]. The area is located north of the ACBR, six case studies were evaluated, then the degree of sustainability by productive systems is detailed: Mixed and Agroecologists Systems.

3.1 Mixed Systems

Production systems coded as AA1, AA2, AA3 (Fig. 3) coincide in unacceptable issues are: responsibility, participation, rule of law and holistic management, moderate theme: materials and energy and as good: safety and human health.

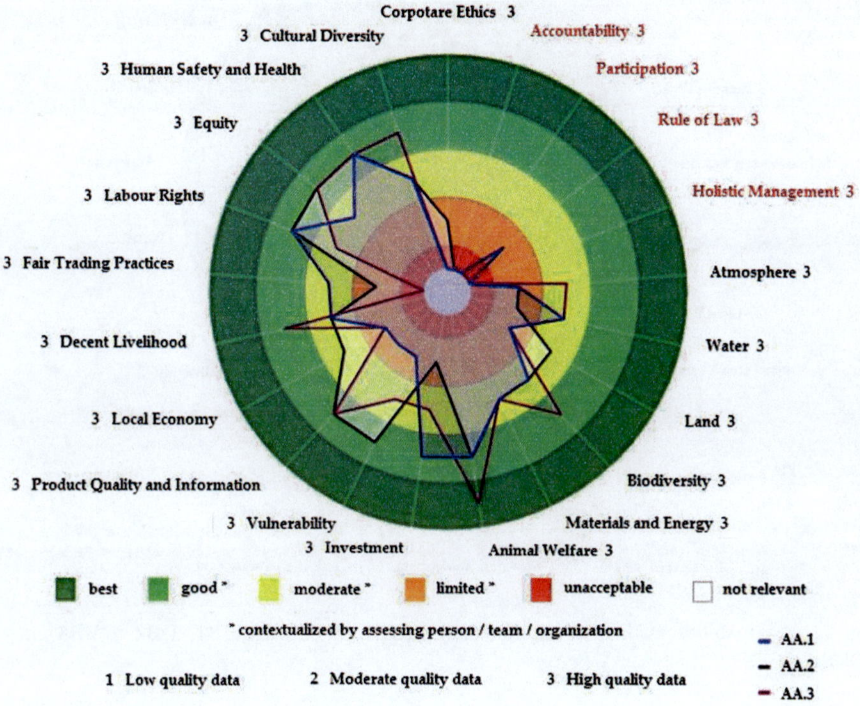

Fig. 3. Sustainability evaluation of mixed systems (coded as: AA1, AA2 y AA3) in the ACBR.

The topic Best is animal welfare (AA3), the good themes by productive system are: biodiversity, decent livehood and cultural diversity (AA3), animal welfare and labour rights (AA1; AA2), invesment (AA1), vulnerability (AA2), equity and product quality and information (AA2, AA3).

3.2 Agroecologists Systems

The valuation of production systems BB1, BB2, BB3 (Fig. 4) classified as unacceptable in the following topics: corporate ethics, accountability, participation, rule of law, holistic management. The topics with the best score per production system are: animal wealfare (BB2) and labour rights, equity cultural diversity (BB3). The topics with a good score for productive systems are: biodiversity, human safety and health (BB2), materials and energy (BB2, BB3), animal welfare, investment (BB3), decent livelihood, labourt rights (BB1), equity and cultural diversity (BB1, BB2).

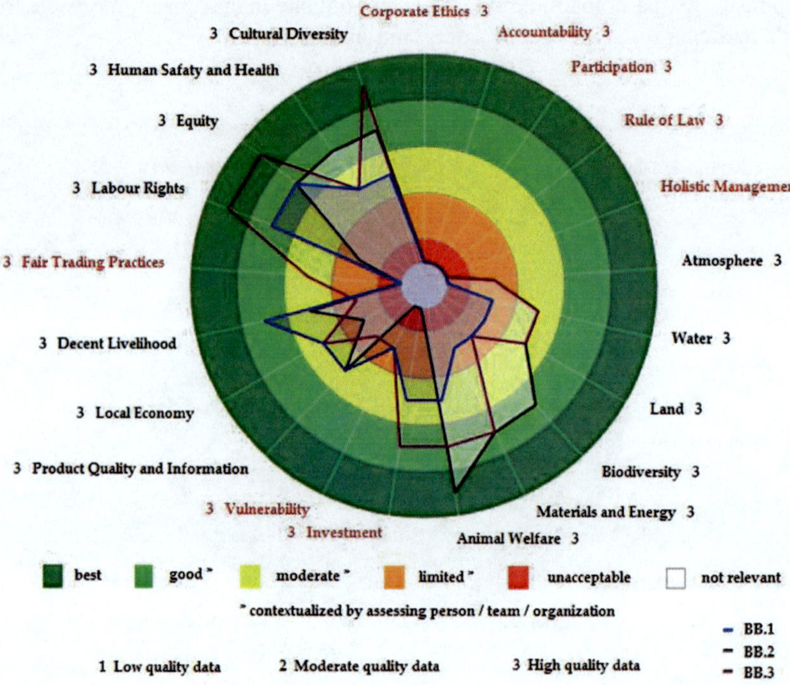

Fig. 4. Sustainability evaluation of mixed systems (encoded as: BB1, BB2 y BB3) in the ACBR.

The buffer zone is the area adjacent to the central areas (nuclei), the permitted activities are ecological practices, scientific research, educational programs, monitoring plans, and training [32]. The area are located to the north of the ACBR were six studies of In the case evaluated below, the degree of sustainability of the productive systems is detailed: Indigenous and conventional systems.

3.3 Indigenous Systems

In the production systems CC1, CC2, CC3 (Fig. 5), corporate ethics and holistic management issues are unacceptable. The topic with the best score is cultural diversity (CC1), animal welfare (CC1, CC2), fair trading practices and equity (CC2) are classified as good.

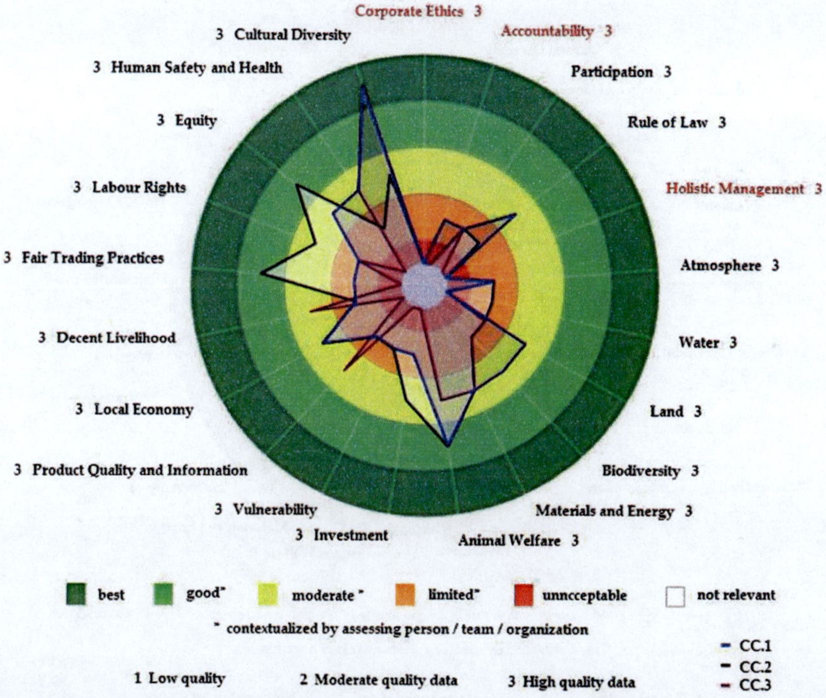

Fig. 5. Sustainability assessment of indigenous systems (coded as: CC1, CC2 y CC3) in the ACBR.

The topics with a limited score are: rule of law (CC1); land, biodiversity, materials and energy (CC1, CC2), animal welfare, decent livelihood, product quality and information (CC3), investment (CC2), local economy, human safety and health (CC1, CC3), equity and cultural diversity (CC1).

3.4 Conventional Systems

In the production systems DD1, DD2, DD3 (Fig. 6) the themes: corporate ethics, accountability, participation, rule of law, holistic management and the themes atmosphere (DD2, DD3), vulnerability and fair trading practices (DD2) are classified as unacceptable. The themes good are: land (DD2), animal welfare (DD1, DD3), investment, product quality and information, local economy, decent economy (DD3), labour rights and equity (DD1, DD3), human safety and health (DD2, DD3).

The moderated themes are: biodiversity, human safety and health (DD1), land, animal welfare, local economy, labour rights (DD2), water (DD3), product quality and information and decent livelihood (DD1, DD2), fair trading practices (DD1, DD3) and cultural diversity (DD1, DD2, DD3).

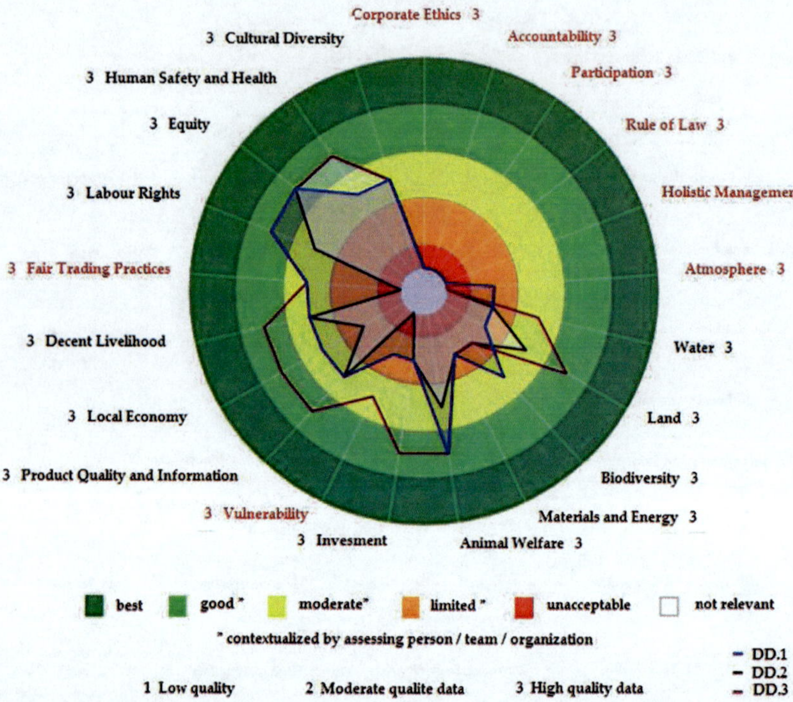

Fig. 6. Sustainability assessment of indigenous systems (coded as: DD1, DD2 y DD3) in the ACBR.

4 General Discussion

The 12 case studies carried out in the transition and buffer zones in the RBCA, do not exceed the value of 3.10 and their classification is moderate, which evidences the unsustainability of the productive systems (Fig. 7) and the precarious management of resources natural in the environmental, social, economic and governance dimensions, conditioning human well-being [33] that depends intimately on the living components of natural systems and are well managed [34].

Unacceptable governance in productive systems is corroborated by the challenges of improving agriculture and food globally by 2050 [35], influenced by the precarious innovation in holistic management in systems; to increase the values of governance, citizens and communities are required to be responsible for the sustainability of resources [36], efficient governance frameworks are necessary to achieve the sustainable development goals.

In the environmental integrity dimension, three case studies do not exceed the value of 2,49, qualifying them as moderate only the conventional system is limited, showing that the most compromised areas are the quality of air, water and land. Air pollution affects global, local temperature changes and rainfall patterns [37] has a direct

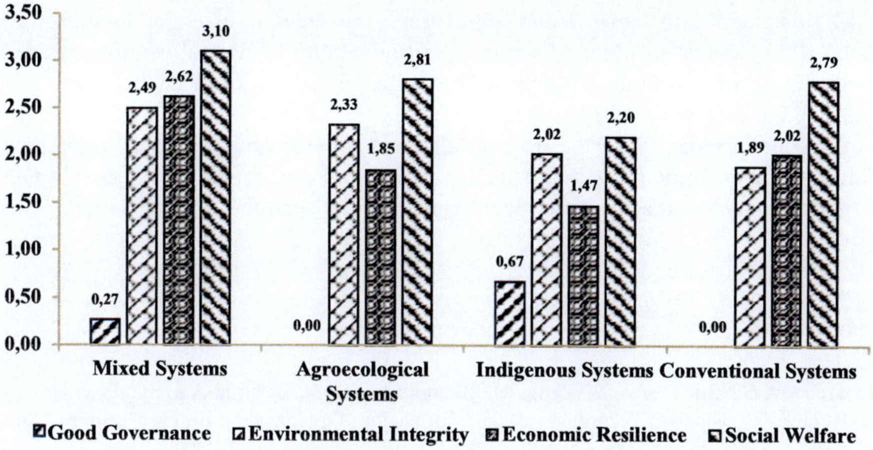

Fig. 7. Values by sustainability dimension on the case studies evaluated in the ACBR.

relationship with climate change; Air pollution by agriculture imposes a great health and economic burden on society, reducing ammonia can generate economic and social benefits [38]. The quantity and quality of water is a threat to food security and sovereignty, in agriculture it is important to fill out an adequate management and monitoring plan with the help of digital technology [39]. In order to reduce production costs, stop soil degradation and improve its quality, it is important to manage crops based on the minimum disturbance of the soil, the use of vegetable crops with residues or other crops, this procedure is called conservation agriculture [40], These conservation systems can be incorporated by improving instructional levels with digital literacy processes and use of digital technologies [9].

The economic resilience and social welfare scores in the case studies can be improved by increasing economic vulnerability with crop diversification or improving livelihoods, the benefits obtained from crop diversity include reduced nutrient loss in the soil, greater resource efficiency, greater absorption of resources by plants and greater productivity, stability of the production and economic system [41]. It is an option to obtain more stable agricultural income while improving the governance of natural resources and sustainability in all its dimensions [42] and fair trade practices allow the growth of assets for farmers [43].

5 Conclusions

The use of the SAFA program as a technological and innovative tool makes it possible to identify the critical points and strengths of productive systems in areas of high biodiversity, such as the precariousness of governance in the systems evaluated and facilitate alternatives to improve their management capacity until formulation of government ordinances.

Digital literacy in isolated and vulnerable rural areas is essential for the use of digital technologies, in order to achieve a sustainability of natural resources and the strengthening of capacities to achieve the sustainability of knowledge in precarious rural population's political interference.

Data management with the use of ICTs can provide new ways for a profitable, socially accepted agriculture that benefits the environment, driven by the development of public policies that support legal and market architecture to achieve smart agriculture.

References

1. Trendov, S.N.M., Varas, S., Zeng, M.: Tecnologías Digitales en la Agricultura y las Zonas Rurales. División de Tecnología de la Información. Organización de las Naciones Unidas para la Alimentación y la Agricultura (FAO), pp. 1–26, Roma (2019)
2. Calicioglu, O., Flammini, A., Bracco, S., Bellù, L., Sims, R.: The future challenges of food and agriculture: an integrated analysis of trends and solutions. Sustainability **11**(1), 222 (2019)
3. FAO, El futuro de la alimentación y la agricultura: Tendencias y desafíos la agricultura. Organización de las Naciones Unidas para la Alimentación y la Agricultura (FAO), pp. 1–44, Roma (2017)
4. Calicioglu, O., Flammini, A., Bracco, S., Bellù, L., Sims, R.: The future challenges of food and agriculture: an integrated analysis of trends and solutions. Sustainability **11**(1), 222
5. Garnett, T., Appleby, M.C., Balmford, A., Bateman, I.J., Benton, T.G., Bloomer, P., Burlin-Game, B., Dawkins, M., Dolan, L., Fraser, D., et al.: Sustainable intensification in agriculture: premises and policies. Science **341**, 33–34 (2013)
6. World Bank Group. Global Monitoring Report 2015/2016: Development Goals in an Era of Demographic Change; World Bank: Washington, DC, USA (2016)
7. Kapur, R.: Science and technology for rural development. Acta Sci. Agric. **3**(7), 167–173 (2019)
8. European Parliament: ICT in the developing world. Brussels: European Parliamentary Research Service (2015)
9. Castilla, D., Botella, C., Miralles, I., Bretón-López, J., Dragomir-Davis, A.M., Zaragoza, I., Garcia-Palacios, A.: Teaching digital literacy skills to the elderly using a social network with linear navigation: a case study in a rural area. Int. J. Hum. Comp. Stud. **118**, 24–37 (2018)
10. Nallusamy, A., Balasubramaniam, S., Chellappan, S.K.: Use of information and communication technology (ICT) to achieve information literacy in agriculture. Int. J. Agric. Extension **3**(2), 111–122 (2015)
11. Guloba, M., Ssewanyana, S., Birabwa, E.: Low literacy levels likely to impede ruralwomen's success in business (No. 675-2017-1126) (2017)
12. UNESCO: Reading the past, writing the future fifty years of promoting literacy. Paris: UNESCO (2017)
13. Fakhoury, R.: Digital government isn't working in the developing world. Here's why. The Conversation (2018). https://theconversation.com/digital-government-isntwork-ing-in-the-developing-world-heres-why-94737
14. McKinsey & Co.: Offline and falling behind: Barriers to Internet adoption. New York: McKinsey and Company (2014)

15. Glynn, P.D., Voinov, A.A., Shapiro, C.D., White, P.A.: From data to decisions: Processing information, biases, and beliefs for improved management of natural resources and environments. Earth's Future **5**(4), 356–378 (2017)
16. Sorgato, V.: Conoce el Chocó Andino, la séptima reserva de la biósfera de Ecuador. Mongabay.com 11 noviemde de 2019 (2018). https://es.mongabay.com/2018/08/ecuador-choco-andino-reserva-de-la-biosfera/
17. Breweton, P., Millward, L.: Organizational Research Methods. SAGE, London (2001)
18. Shaghaghi, A., Bhopal, R.S., Sheikh, A.: Approaches to recruiting 'hard-to-reach' populations into research: a review of the literature. Health Promot. Perspect. **1**(2), 86–94 (2011)
19. National Research Council: Toward Sustainable Agricultural Systems in the 21st Century. National Academies of Sciences, Engineering, and Medicine, Washington, DC (2010)
20. Reganold, J., Wachter, J.: Organic agriculture in the twenty-first century. Nat. Plants **2**, 15221 (2016)
21. Reintjes, C., Haverkort, B., Waters-Bayer, A.: Farming for the future: an introduction to low-external input and sustainable agriculture (1992)
22. Francis, C., Lieblein, G., Gliessman, S., Breland, T.A., Creamer, N., Harwood, R., Salomonsson, L., Helenius, J., Rickerl, D., Salvador, R., Wiedenhoeft, M., Simmons, S., Allen, P., Altieri, M., Flora, C., Poincelot, R.: Agroecology: the ecology of food systems. J. sustain. Agric. **22**(3), 99–118 (2003)
23. Suárez-Torres, J., Suárez-López, J.R., López-Paredes, D., Morocho, H., CachiguangoCachiguango, L.E., Dellai, W.: Agroecology and health: lessons from indigenous populations. Current Environ. Health Rep. **4**(2), 244–251 (2017)
24. Claro, M.: La incorporación de tecnologías digitales en educación. Modelos de identificación de buenas prácticas (LC/W.328), Santiago de Chile, Comisión Económica para América Latina y el Caribe (CEPAL) (2010)
25. Ahlin, E.M.: Semi-Structured Interviews with Expert Practitioners: Their Validity and Significant Contribution to Translational Research. SAGE Publications Ltd. (2019)
26. Kallio, H., Pietilä, A.M., Johnson, M., Kangasniemi, M.: Systematic methodological review: developing a framework for a qualitative semi-structured interview guide. J. Adv. Nurs. **72**(12), 2954–2965 (2016)
27. FAO: SAFA Sustainability Assessment of Food and Agriculture Systems: Tool User Manual Version 2.4.1; FAO Food and Agriculture Organization of the United Nations: Roma, Italy, p. 20 (2014)
28. FAO: SAFA Sustainability Assessment of Food and Agriculture Systems: Indicators Food and Agriculture Organization of the United Nations; Roma, Italy, p. 271 (2013)
29. Emerson, R., Fretz, R., Shaw, L.: Writing Ethnographic Field Notes, 2nd edn. University of Chicago Press, Chicago (2011)
30. Phillippi, J., Lauderdale, J.: A guide to field notes for qualitative research: context and conversation. Qual. Health Res. **28**(3), 381–388 (2018)
31. Bernard, H.R.: Research Methods in Anthropology: Qualitative and Quantitative Approaches. Rowman & Littlefield (2017)
32. UNESCO. World Network of Biosphere Reserves homepage. Spring (2000). http://www.unesco.org/mab/brfaq-3.htm
33. Ajayi, C.O., Oguntade, A.E.: Natural resource management, food security and violent conflicts in Nigeria: concepts, issues and policy considerations. Asian J. Agric. Extension Econ. Sociol. **22**(4), 1–11 (2018)
34. Pecl, G.T., Araújo, M.B., Bell, J.D., Blanchard, J., Bonebrake, T.C., Chen, I.C., Falconi, L.: Biodiversity redistribution under climate change: impacts on ecosystems and human well-being. Science **355**(6332), eaai9214 (2017)

35. Davies, J.: Enabling Governance for Sustainable Land Management. In: Land Restoration, pp. 67–76. Academic Press (2016)
36. Mustalahti, I., Gutiérrez-Zamora, V., Hyle, M., Devkota, B.P., Tokola, N.: Responsibilization in natural resources governance: a romantic doxa? Forest Policy Econ. **111**, 102033 (2020)
37. Samset, B.H.: How cleaner air changes the climate. Science **360**, 148–150 (2018)
38. Giannadaki, D., Giannakis, E., Pozzer, A., Lelieveld, J.: Estimating health and economic benefits of reductions in air pollution from agriculture. Sci. Total Environ. **622**, 1304–1316 (2018)
39. Murtaza, G., Zia-ur-Rehman, M., Rashid, I., Qadir, M.: Use of poor-quality water for agricultural production. In: Research Developments in Saline Agriculture, pp. 769–783. Springer, Singapore (2019)
40. Kassam, A., Derpsch, R., Friedrich, T.: Global achievements in soil and water conservation: the case of conservation agriculture. Int. Soil Water Conserv. Res. **2**, 5–13 (2014)
41. Reich, P.B., Knops, J., Tilman, D., Craine, J., Ellsworth, D., Tjoelker, M., Lee, T., Wedin, D.: Plant diversity enhances ecosystem responses to elevated CO2 and nitrogen deposition. Nature **410,** 809–810 (2001)
42. Bobojonov, I., et al.: Crop Diversification in Support of Sustainable Agriculture in Khorezm. In: Martius, C., Rudenko, I., Lamers, J., Vlek, P. (eds.) Cotton, Water, Salts and Soums. Springer, Dordrecht (2012)
43. Khan, M.I., Khan, S., Haleem, A.: Compensating Impact of globalization through fairtrade practices. In: Globalization and Development, pp. 269–283. Springer, Cham (2019)

Prediction of Trophic State of San Marcos Lagoon Based on AQUATOX Eutrophication Model

Juan Gabriel Mollocana Lara[1(✉)],
Erika Samantha Quezada Espinosa[2],
and Joselyn Magaly Vizcaino Angamarca[2]

[1] Grupo de Investigación Ambiental en el Desarrollo Sustentable GIADES, Carrera de Ingeniería Ambiental, Universidad Politécnica Salesiana, Quito 170702, Ecuador
jmollocana@ups.edu.ec

[2] Carrera de Ingeniería Ambiental, Universidad Politécnica Salesiana, Quito 170702, Ecuador

Abstract. Eutrophication of lakes is related to the decrease in water quality. It generates problems that are particularly important if the water is used for irrigation and human consumption. This research assesses and predicts the trophic state of San Marcos Lagoon using the AQUATOX model. Six sampling campaigns were carried out, for three months and at intervals of two weeks. Simple and compound water samples were taken at three different points in the lagoon to measure physicochemical and biological parameters such as chlorophyll-a, phosphates, nitrates, cyanobacteria, dissolved oxygen, among others. Additionally, a sensitivity analysis was performed to determine the critical parameters of the model, which will be used to improve the model fit. The comparison between the measured and simulated data indicates that the current trophic state of the San Marcos Lagoon varies between eutrophic and hypereutrophic. Three error indexes used at calibration and validation stages to calculate the goodness of fit (R2, RMSE, and ER). The obtained R2 for chlorophyll-a, total phosphorus and Secchi disc depth validation were 0.89, 0.97 and 0.93, respectively. Once the model is validated, the trophic state prediction is made for the next six months, obtaining 1.56 ug/L for chlorophyll-a, 114 ug/L for total phosphorus and 1.21 m for Secchi disc depth. These values correspond to a hypereutrophic state. Also, the simulation of a mitigation scenario is presented, where a mesotrophic state is obtained through the renewal of 25% of water volume and detritus removal.

Keywords: Eutrophication · AQUATOX · Water quality

1 Introduction

Water quality in water bodies could deteriorate due to excessive nutrients increasing, mainly phosphorus and nitrogen, causing eutrophication processes. Eutrophication occurs with the excessive proliferation of aquatic organisms such as algae,

M. Botto-Tobar et al. (Eds.): CI3 2020, AISC 1277, pp. 477–490, 2021.
https://doi.org/10.1007/978-3-030-60467-7_39

phytoplankton, cyanobacteria and others [1]; which when decomposing cause turbidity, biodiversity impoverishment and dissolve decreased oxygen in deep areas of the water body [2]. According to [3], the main factors for the increase of nutrients are wind speed, temperature, evaporation, precipitation, and anthropogenic actions.

In the other hand, to determinate the trophic state of a water body, the relation between the amount of nutrients and the amount of organic matter must be estimated [4]. One way to assess the trophic state of a lagoon is based on the measurement of three variables: chlorophyll-a, (Chl), transparency measure by Secchi disc depth (SD) and total phosphorus (TP) concentration; if Chl and TP increases, and SD decreases, then the trophic state increases [5].

Modeling of eutrophication process in different bodies of water has been extensively studied. Paper [6] presents a one-dimensional model for evaluating eutrophication in wetlands. The model was calibrated, validated, and used to simulate mitigation actions to decrease eutrophication; being the most effective the implementation of a treatment unit to remove phosphorus and nitrogen. Paper [7] presents a water quality model for the Cau River basin, using the integration of the SWAT hydrological model, used to predict lateral inflows and discharges of ungauged tributaries, with the quality water model QUAL2K. Paper [8], presents a model for water quality management (WQM) based on Interval quadratic programming (IQP). The IQP-WQM model was solved using three algorithms, including a piecewise linear approximation (PLA) method, derivative algorithm (DEA) and a duality-based algorithm (DUA). Paper [9], uses various one-dimensional and two-dimensional models to calculate water level and water quality of rivers and coastal estuaries on the Ca Mau peninsula. Another model used to simulate water quality and aquatic life is AQUATOX. This model addresses ecotoxicological processes, water quality and food network; it has been used to assess ecological risks in aquatic ecosystems and lakes [10, 11].

This paper seeks to predict the trophic state of the San Marcos Lagoon reservoir, based on the simulation of the state variables: total phosphorus, chlorophyll-a and transparency by Secchi disk depth. For this, the AQUATOX model will be used with physicochemical (e.g. total phosphorus, nitrates, CO2, detritus), biological (e.g. cyanobacteria, phytoplankton, and diatoms) and meteorological parameters (e.g. evaporation, wind speed and solar radiation).

2 Materials and Methods

2.1 Study Zone

The current study was carried out in the San Marcos Lagoon located between Pichincha and Sucumbíos provinces, within the Cayambe-Coca National Park, Ecuador, at the points of 0° 7′0.88″N-77′57′55.91″O, at 3400 m.a.s.l. It originates from the thaws of the Cayambe volcano. It has a surface area of 81.2 km^2 and a length of 2.097 km. It has an average evaporation of 3.5 mm/d, average wind speed of 0.7 m/s and average solar radiation of 750.26 Ly/d. San Marcos Lagoon is part of the "Cayambe-Pedro Moncayo Irrigation Canal" project, which seeks to provide water for irrigation and human consumption to communities in Cayambe city.

2.2 Model Description

AQUATOX is an ecosystem simulation model that can assess past, present and future interactions of biomass, nutrients and dissolved oxygen, from the incoming loads to the water mass [6]. The modeled food network includes nutrients, algae, phytoplankton, macrophytes, fish and invertebrates. Interaction with organic toxics may also be included. The amount of nutrients is determined by considering the decomposition of organic matter, bacterium nitrification/denitrification, atmospheric deposition, detrital decomposition, excretion of biota and algae assimilation. On the other hand, algae biomass consider mortality, sinking, animal ingestion, algae metabolism and photosynthesis process, where the amount of light varies depending of suspended sediment. Dissolved oxygen increase with wind reaeration and photosynthesis; and decrease with biota respiration and detritus decomposition, this factors directly influence the Biochemical Oxygen Demand (BOD) (Fig. 1).

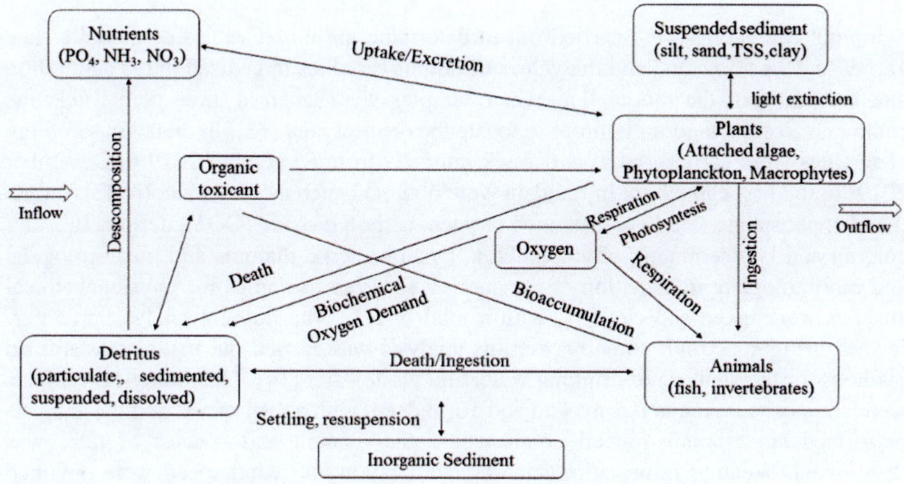

Fig. 1. AQUATOX model

2.3 Trophic State Assessment

This paper applied the international accepted criteria for trophic state classification propose by the Organization for Economic Co-operation and Development (OECD) [4]. These criteria is based on chlorophyll-a maximum concentration, total phosphorus concentration and maximum transparency measure by Secchi disc depth (Table 1).

Table 1. OECD trophic state classification criteria

Trophic level	Chlorophyll-a (ug/L)	Secchi disc depth (m)	Total phosphorus (ug/L)
Ultra-Oligotrophic	<1	>12	<4
Oligotrophic	1 to 2.5	6 to 12	4 to 10
Mesotrophic	2.5 to 7.9	3 to 6	10 to 35
Eutrophic	8 to 25	1.5 to 3	35 to 100
Hypereutrophic	>25	<1.5	>100

This classification will be used to compare both measured and simulated trophic states. In addition, they will be used for predicting the trophic state over the next six months and to assess the effectiveness of eutrophication mitigation actions.

2.4 Sampling Campaigns and Model Parameters

Sampling campaigns were carried out to determine the initial values of model's state variables, its parameters, and the values of output variables to be used in the calibration and validation of the model. The water samples were taken at three points near the central area of the lagoon, in order to locate the deepest area [13]. In total, six sampling campaigns were carried out at two-week intervals from September 2019 to December 2019. In the first campaign, initial data were obtained such as: ammonia (NH3), nitrate (NO3-), phosphate (PO43-), dissolved oxygen, carbon dioxide (CO2), detritus in water column and bed sediment, phytoplankton, cyanobacteria, diatoms and meteorological and morphological information of the study area. Water samples for physicochemical analysis were taken superficially, with a total of 6 L. The nutrient analysis was performed using spectrophotometry; detritus analysis was carried out using standardized methods for the analysis of drinking water and wastewater [14]. The biological samples were composite, taken at 0.5 m, 1 m and 10 m deep, with a total of 6 L and the analysis performed corresponds to cell count where the amount and species of flora was determined. Weather information such as evaporation and wind speed were obtained from data collected from 2008 to 2018 by the nearest weather stations (Tomalón-Tabacundo, Ibarra and Lumbaquí). Solar radiation data were downloaded from the NASA POWER database [15]. The area and length were estimated using ArcMap 10.5 and SENTINEL 2 satellite images from August 30 of 2019. Finally, the volume is the product of the area and the depth measured at each of the points.

Total phosphorus, chlorophyll-a and Secchi disc depth were considered as output model variables. As being the variables used in the trophic state classification of OECD. These variables were measured during September, October, November, and December 2019, and used for model calibration and validation.

2.5 Sensibility Analysis and Calibration

In order to improve the model based on the experimental data, the calibration is performed, considering as parameters to calibrate the results of the statistical sensitivity analyses tool of AQUATOX software. In addition, the calibration recommendations described in AQUATOX documentation were applied, where the first step is the stabilization of the food network [13].

2.6 Validation

In order to validate the model, the data of total phosphorus, Secchi disc depth and chlorophyll-a from the sampling campaigns were compared with the simulated data. In addition, the following three goodness-of-fit measures are used to quantify this difference:

Determination Coefficient. It is defined as the proportion of the total variance of a response variable, explained by an explanatory variable. Oscillates between 0 and 1. The closer to 1 is its value, the better the setting.

$$R^2 = \frac{\sum_{i=1}^{n}(\hat{y}_i - \bar{y})^2}{\sum_{i=1}^{n}(y_i - \bar{y})^2} \tag{1}$$

Where $\hat{y}_i$ are the predicted (simulated) values, y_i are the measured values, $\bar{y}$ is the mean of measured values and $\bar{n}$ is the total number of observations.

Mean Quadratic Error (RMSE). This index quantifies the error between two datasets. That is, it compares a predicted value and an observed value, the closer it approach to zero, the better the goodness of fit is.

$$RMSE = \sqrt{\frac{\sum_{i=1}^{n}(A_i - A_i')^2}{n}} \tag{2}$$

Where Ai are the observed values, $A'i$ are the simulated values and n is the total number of observations.

Medium Relative Error (ER). This index quantifies the proximity between the simulated and observed values. The smaller the indicator, the closer the measured values are to the simulated values.

$$ER = \frac{\sum_{i=1}^{n}\frac{|A_i - A_i'|}{A_i}}{n} \tag{3}$$

Where Ai are the observed values, $A'i$ are the simulated values and n is the total number of observations.

3 Results and Discussion

3.1 Sampling Campaigns and Model Parameters

The first sampling campaign compiled the initial values soured by the AQUATOX model. In addition, it determined the plants with the greatest presence in the lagoon and its concentrations (Synechococcus sp., Navicula sp. and Oscillatoria sp.). These results are summarized in Table 2. The output variables, used to assess the trophic state, were obtained in six sampling campaigns (Table 3).

Table 2. Results of first sampling campaign

Nutrients			
Parameter	Unit	Value	Date
Ammonia as N	mg/L	0.02	1/9/2019
Nitrate as N	mg/L	3.3	1/9/2019
Phosphate as P	mg/L	0.126	1/9/2019
Carbon dioxide	mg/L	6.69	1/9/2019
Dissolved oxygen	mg/L	10	1/9/2019
Detritus			
Parameter	Unit	Value	Date
Labile detritus	g/m^2	44.16	1/9/2019
Refractory detritus	g/m^2	24	1/9/2019
Sedimentable solids	mg/L	1	1/9/2019
Percentage of particulate detritus	%	10	1/9/2019
Refractory detritus percentage	%	20	1/9/2019
Plants			
Parameter	Unit	Value	Date
Synechococcus sp.	mg/L	0.0082	1/9/2019
Navicula sp.	g/m^2	0.599568	1/9/2019
Oscillatoria sp.	g/m^2	0.17286	1/9/2019
Water temperature			
Parameter	Unit	Value	Date
Epilimnion	°C	12.3	1/9/2019
Hypolimnion		10.5	1/9/2019
pH of water			
Parameter	Unit	Value	Date
Ph	unit pH	7.85	1/9/2019

Table 3. Output variables sampling campaigns

Date	Total phosphorus	Chlorophyll-a	Secchi disc depth
1/9/2019	0.126	0.6512	1.46
22/9/2019	0.323	1.48	1.87
20/10/2019	0.1	1.68	1.73
3/11/2019	0.1	1.48	1.73
17/11/2019	0.13	1.66	1.46
3/12/2019	0.1	2.072	1.7

On the other hand, the results of the values obtained from weather stations, satellite images and NASA POWER data base are summarized in Table 4.

Table 4. Values obtained from weather stations, satellite imagery and NASA POWER data base

Feature of the study area			
Parameter	Unit	Value	Date
Long body of the water	Km	2.097	1/9/2019
Surface area	m^2	811809	1/9/2019
Average depth	m	0.5	1/9/2019
Maximum depth	m	43	1/9/2019
Average evaporation	in.a.a.	50.3	1/9/2019
Latitude	Degrees	0.116911	1/9/2019
Volume			
Parameter	Unit	Value	Date
Volume	m^3	32472373	1/9/2019
Wind speed			
Parameter	Unit	Value	Date
Wind speed	m/s	0.5	1/9/2019
Solar radiation			
Parameter	Unit	Value	Date
Solar radiation	Ly/d	733	1/9/2019

3.2 Sensitivity Analysis and Calibration

As a result of the sensitivity analysis, the Synechococcus sp. specie and the initial concentration of nitrogen were very influential in the output of the model (Table 5), so these parameters were varied to improve the goodness of fit of the simulation with the calibration data. The new values are summarized in Table 6. While errors indexes are shown in Table 7.

Table 5. Sensitivity analysis results

10% Sensitivity test	Parameter value	Tot. Pho. (mg/L)	Secchi (m)	(ug/L)
Value is base		0.126	0.811	4.652
Synechococcus: Optimal temperature (deg. C) + (10%)	19.8	0.133	0.831	3.903
Synechococcus: Optimal temperature (deg. C) − (10%)	16.2	0.12	0.794	5.294
Synechococcus: Max photosynthetic rate (1/d) + (10%)	5.5	0.119	0.793	5.349
Synechococcus: Max photosynthetic rate (1/d) − (10%)	4.5	0.132	0.829	3.937
Extinction coeff water (1/m) + (10%)	0.605	0.123	0.775	4.913
Extinction coeff water (1/m) − (10%)	0.495	0.128	0.850	4.395
NH3 & NH4+: Initial condition (mg/L) + (10%)	0.011	0.126	0.810	4.654
NH3 & NH4+: Initial condition (mg/L) − (10%)	0.009	0.126	0.810	4.650
NO3: Initial condition (mg/L) + (10%)	0.55	0.125	0.808	4.752
NO3: Initial condition (mg/L) − (10%)	0.45	0.127	0.814	4.546

Table 6. Calibrated parameters

Parameter	Original value	Adjusted value
Synechococcus: Optimal temp (deg. C)	**32**	**9**
Synechococcus: Max photosynthetic rate (1/d)	**3**	**11**
Extinction coeff water (1/m)	**0.2**	**0.6**
NH3 & NH4+: Initial cond (mg/L)	**0.02**	**0.01**
NO3: Initial condition (mg/L)	**3.3**	**0.5**

3.3 Validation

Model validation was carried out by comparing the observed data of the output variables with the simulated data generate with the AQUATOX model. The validation plots for the total phosphorus, Secchi disc depth and chlorophyll-a are shown in figures Fig. 2 and Fig. 3, respectively. The values of error related with calibration (Cal) and validation (Val) were all less than 40% (Table 7) and match the errors reported in similar studies [14]. On the other hand, values of RMSE and R2 errors also match similar studies [11, 15]. Although the differences between the simulations and the observed TP and DS data were small, the model did not capture certain peaks, while for Chl the simulated results have the same trends as the observed results.

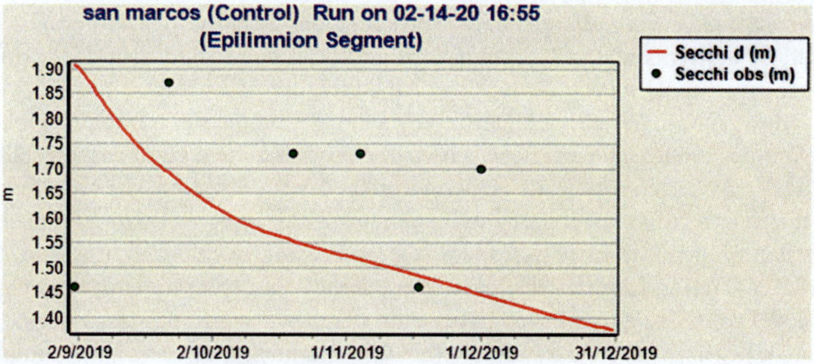

Fig. 2. Secchi disc depth validation

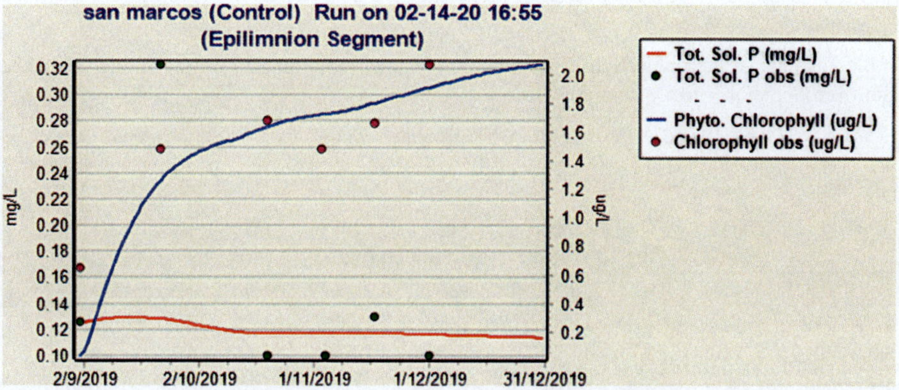

Fig. 3. Total phosphorus and Chlorophyll-a validation

Table 7. Errors with calibration data, validation data and both

	Variable	R2	RMSE	ER
Cal	TP	0.98	0.19	29%
	Chl	0.86	0.27	9%
	DS	0.87	0.25	14%
Val	TP	0.97	0.023	11%
	Chl	0.89	0.68	39%
	DS	0.93	0.56	19%
Both	*TP*	0.677	0.198	20.3%
	Chl	0.89	0.73	24%
	DS	0.069	0.613	13.3%

The results of the simulations between September 2019 and December 2019 show high load of nutrients and chlorophyll-a. In addition, the Secchi disc depth decrease, which indicates transparency decreasing. At estimation of trophic status using OECD classification criteria, not all variables have values within the same trophic level, so the more common trophic level is chosen. It can be seen that the lagoon has a trophic state that varies between eutrophic to Hypereutrophic level (Table 8). On the other hand, during the months of October 2019 to November 2020, there is it strong evidence that as the water temperature increases, there is an increasing in chlorophyll-a, due to algae proliferation. These processes decrease transparency and, consequently, increases the trophic level from this eutrophic to hyper-eutrophic. Starting from December 2019, the temperature begins to decrease and therefore also gradually decreases the optimal range for algae growth during the coldest months of the year. It should be noted that the estimated trophic status based on simulations matches the estimate based on the measured data.

Table 8. State of eutrophication of Laguna San Marcos according to observed and simulated data from September to December 2019.

Date	Observed value			Simulated value			Observed and simulated trophic state
	TP (ug/L)	Chl (ug/L)	DS (m)	TP (ug/L)	Chl (ug/L)	DS (m)	OECD
1/9/2019	126	0.6512	1.46	124	0.0345	1.90	Eutrophic
22/9/2019	323	1.48	1.87	128	1.2602	1.69	Eutrophic
20/10/2019	100	1.68	1.73	117	1.6405	1.55	Hypereutrophic
4/11/2019	100	1.48	1.73	117	1.728	1.52	Hypereutrophic
17/11/2019	130	1.66	1.46	118	1.8042	1.48	Hypereutrophic
3/12/2019	100	2.072	1.7	116	1.9277	1.44	Eutrophic

3.4 Trophic State Prediction

The model simulation period was extended to six months after the last sampling campaign, it means from December 3, 2019 to June 30, 2020 (Fig. 4 and Fig. 5). It can be seen that the TP tends to stabilize over time, while the chlorophyll-a reach its highest peak in January 2020, coinciding with the lowest TP value. In the other hand, the Secchi disc depth maintains an accelerated decreasing that matches chlorophyll-a simulation and indicates that the water body will reach higher trophic states. The classification of the trophic status based on the OECD index shows that the lagoon will have a hypereutrophic state which far exceeds the minimum value for this classification (>25 ug/L for chlorophyll-a, and >100 ug/L for phosphorus and <1.5 m for Secchi disc depth) (Table 9).

Table 10. Mitigation actions simulation – trophic state prediction

Date	TP (ug/L)	Chl (ug/L)	DS (m)	Trophic state
1/1/2020	35.4	1.13	1.69	Hypereutrophic
1/2/2020	23.1	1.50	1.64	Eutrophic
1/3/2020	15.9	1.33	1.64	Eutrophic
1/4/2020	11.1	1.15	1.65	Mesotrophic
1/5/2020	7.8	0.97	1.66	Mesotrophic
1/6/2020	5.3	0.68	1.68	Mesotrophic

4 Conclusions

This research used AQUATOX model to assess and predict the trophic stated of San Marcos lagoon. Focusing on three main variables used for trophic state classification, these are: total phosphorus, disc Secchi depth and chlorophyll-a.

Observed data of total phosphorus and Secchi disk depth indicate that the trophic level of San Marcos Lagoon is between eutrophic and Hypereutrophic, since concentrations of both, total phosphorus and Secchi disk depth are high, with values between 0.1–0.3 mg/L and 1.46–1.7 m respectively. These values exceed the limits of 0.01 mg/L for total phosphorus and 3-1.5 m deep in the hypereutrophic state. In the other hand, chlorophyll-a presents values between 0.65–2.072 ug/L, that positions it in the oligotrophic state, which has a range between 1 to 2.5 ug/L, according to OECD classification. As noted in Table 8, the higher the water temperature is, the higher the degree of eutrophication. This indicates that, during the summer season, an excessive proliferation of algae can be experienced, resulting in high levels of chlorophyll-a, and a considerable decrease in the transparency of the water column. The simulated data matches the observed data favorably, indicating the model is successfully represented.

The mitigation measures proposed have positive effects on the water body, decreasing nutrients which consequently results in the decrease of photosynthetic organisms and the improvement of the transparency of water. The more the renewal water volume increase, the better the water quality and is and the conservation of aquatic biodiversity will be maintained.

References

1. Du, H., Chen, Z.: Evaluation of eutrophication in lakes: a new non-equilibrium statistical approach. Ecol. Indic. **102**, 686–692 (2019). LNCS Homepage. http://www.springer.com/lncs. Accessed 21 Nov 2016
2. Dou, M., Ma, X., Zhang, Y., Zhang, Y., Shi, Y.: Modeling the interaction of light and nutrients as factors driving lake eutrophication. Ecol. Model. **400**, 41–52 (2019)
3. Zhang, Y., Liang, J., Zeng, G.: How climate change and eutrophication interact with microplastic pollution and sediment resuspension in shallow lakes: a review. Sci. Total Environ. **705**, 135979 (2020)

4. López, M., Madroñero, S.: A trophic state of tropical lake high mountain: case of La Cocha Lagoon. Sci. Eng. Neogranadina, 21–42 (2015)
5. Zou, W., Zhu, G., Cai, Y.: Relationships between nutrient, chlorophyll-a and Secchi depth in lakes of the Chinese Eastern Plains ecoregion: Implications for Eutrophication management. J. Environ. Manage. **260**, 109923 (2020)
6. Dash, S., Swapna, S., Kalamdhad, A.S.: Study of the limnology of wetlands through a one-dimensional model for assessing the eutrophication levels induced by various pollution sources. Ecol. Model. **416**, 108907 (2020)
7. Bui, H.H., Hien, N., Nhu Dinh, T., Thao, A.: Integration of SWAT and QUAL2K for water quality modeling in a data scarce basin of Cau River basin in Vietnam. Ecologhydrol. Hydrobiol. **19**, 210–223 (2019)
8. Zhang, Q., Li, Z.: Development of an interval quadratic programming water quality management model and its solution algorithms. J. Clean. Prod. **249**, 119319 (2020)
9. Quang Tri, D., Hong Thai, T., Kandasamy, J.: Application of 1D–2D coupled modeling in water quality assessment: a case study in Ca Mau Peninsula, Vietnam. Phys. Chem. Earth Parts A/B/C **113**, 83–99 (2019)
10. Lombardo, A., Franco, A., Pivato, A., Barausse, A.: Food web modeling of a river ecosystem for risk assessment of down-the-drain chemicals: a case study with AQUATOX. Sci. Total Environ. **508**, 214–227 (2015)
11. Niu, Z., Gou, Q., Wang, X., Zhang, Y.: Simulation of a water ecosystem in a landscape lake in Tianjin with AQUATOX: sensitivity, calibration, validation and ecosystem prognosis. Ecol. Model. **335**, 54–63 (2016)
12. Yao, W.: AQUATOX application of clear lake California model calibration on blue-green algae bloom and nutrients. Northeastern University, Boston (2015)
13. Hudson, H.: Monitoring Lake Quality. Northeastern Illinois Planning Commission, Chicago (1998)
14. APHA, AWWA, WPCF: Normalized methods for the analysis of drinking water and wastewater. Diaz de Santos S.A., Madrid (1989)
15. Environmental Protection Agency: Modeling environmental fate and ecological effects in aquatic ecosystems. Office of Research and Development (ORD), Washington, D.C. (2018)
16. Wang, Y., Hu, W., Peng, Z., Zeng, Y., Rinke, K.: Predicting lake eutrophication responses to multiple scenarios of lake restoration: a three-dimensional modeling approach. Water **10**(8), 994 (2018)
17. Chen, L., Yang, Z., Liu, H.: Assessing the eutrophication risk of the Danjiangkou reservoir based on the EFDC model. Ecol. Eng. **96**, 117–127 (2016)
18. Chislock, M.F., Doster, E., Zitomer, R.A., Wilson, A.E.: Eutrophication: causes, consequences, and controls in aquatic ecosystems. Nat. Educ. Knowl. **4**(4), 10 (2013)
19. Environmental Protection Agency: Guidance in AQUATOX Setup and Application. EPA, United States (2014)
20. Xu, M.J., Yu, L., Zhao, Y.W., Li, M.: The simulation of shallow reservoir eutrophication based on MIKE21: a case study of Douhe Reservoir in North China. Procedia Environ. Sci. **13**, 1975–1988 (2012)
21. Reilly, J., Horne, A., Miller, C.: Nitrogen removal in large-scale free-surface constructed wetlands used for pre-treatment to artificial recharge of groundwater, pp. 33–47. Elsevier (2003)
22. Yang, C., Ma, J.: Research progress of ecological restoration and aquatic plant community construction for urban lake, pp. 13–20. Elsevier (2003)

Design of an Aerial Cable Transport Cabin

Villarreal Pamela(✉), Caza Paúl, Macha Vinicio, and López Víctor

Superior Technology Major in Industrial Mechanics, Instituto Superior Tecnológico Vida Nueva, Quito, Ecuador
pamela.villareal@istvidanueva.edu.ec

Abstract. This document contains the detailed engineering of the design of a cabin for the cable transport system. The cabin has a capacity for ten people that can be transported. After carrying out a study using design alternatives, analysis and evaluation, the calculations carried out for the design process of the cabin are presented below. These are complemented with a simulation of the system elements by finite elements in ANSYS and structural analysis such as SAP2000. Also, the assembly and workshop plans for the construction and assembly process of the cabin and a cost analysis of the unit production of the cabin. The result is a cabin with a modern design that meets national and international standards for cable transport systems.

Keywords: Design of machines · Simulation · Ropeways · Machine elements · Finite elements analysis

1 Introduction

Over the years, cable transport systems have been positioned as a viable alternative to the problems of mobilization in large cities. While it is true that initially cable transport systems were created with the objective of reaching areas of difficult access, where people could not walk to their destination. Today, developed cities use this type of transport to avoid traffic congestion.

1.1 Components of Cable Transport

Cable transport systems are made up of three basic systems:

Loading System
This system is made up of the cabin, holding arm, cabin-arm coupling system (known as "H") and drive or support cable attachment clamp. Its function is to allow the transport of freight or passengers. Generally, these devices are made of steel with a galvanized surface coating because they are installed outdoors to avoid corrosion and oxidation effects. In order to lighten the structure of the cabin, it is built with high-strength aluminum alloys and with a fiberglass coating.

The cabin is coupled with a connection system that works as a support and stabilization for load variations due to the entry or exit of people. The holding arm has two functions:

M. Botto-Tobar et al. (Eds.): CI3 2020, AISC 1277, pp. 491–504, 2021.
https://doi.org/10.1007/978-3-030-60467-7_40

It attaches the cab to the cable by means of the clamp and establishes an operating distance between the cargo vehicle and the bollards or towers in such a way as to ensure collisions between these elements.

Support System
The support system is made up of bollards which are intermediate structures, whose main function is to support the conveyor cable along the route, the anchors which are special types of foundation that are designed to support the lateral or upward tensioning force transmitted to the ground, and the foundation is an important component that allows the transmission of the forces that a structure supports to the ground, so that it does not exceed the bearing capacity of the ground, and also that the deformations produced in the ground are acceptable for the structure (Fig. 1).

Fig. 1. Support system Source: (POMA 2016)

Traction System
It is made up of a conveyor cable that generally consists of a certain number of strands, placed in a helical manner around a support core, the pulleys trains that allow support in the towers along the route and in the stations, and by two elements that are located in the start and return station: the drive motor and the return pulley, respectively (About Ropeways, s.f.)

The pulley trains are an essential part of this system as they serve to align the cable and hold it in place during operation. They can be of three types:

- When there is contact between the cable and the top of the rollers, these are called support rockers.
- When there is contact between the cable and the lower part of the rollers, it is called compression rockers.
- When there is contact of the roller both at the top and at the bottom, due to stresses caused in the tower, they are called mixed rockers.

1.2 Cabin

The cabin is a substantial part of the cable transport system, it is responsible for transporting people or cargo from one place to another. For the design of the cabins, several factors that affect their performance and structural safety are taken into account, such as: weather conditions (wind, rain, hail, etc.), weight of the structure, materials used for its construction, among others The current aerodynamic design of the cabins demonstrates the technological development that has been perfected over the years in order to avoid the negative effect of wind in the operation stage.

The cabins are attached to a series of cables by means of a special clamp with a damping system included, they advance through a series of fixed stations with control towers known as bollards that have the capacity to support the weight of the cabins and the transported load. The structure of the upper part of the cabin is designed in such a way that when fastened with the clamp and the cable, it complies with the appropriate safety factor according to the manufacturer's specifications (Miravete et al. 1998).

2 Methodology

For the selection of the parameters that define the construction of the system, the following methodology is used.

2.1 Design Parameters

Wind Load
According to the hydrological and meteorological study (EMMOP 2016) for the investigation, reviews of the weather stations adjacent to the sector of implementation were used, in addition a statistical analysis was made with the data taken from 1981 to 2012 in order to obtain the values of the minimum wind speed (37.1 km/h), average (53.9 km/h) and maximum (70.4 km/h).

$$q_{red} = \beta x q$$

q_{red}: Reduced pressure, kN/m^2
β: Reduction Coefficient
q: Wind pressure

The reduction coefficient is chosen according to the maximum length of the span (light = l*), for which the following measurement assignment is established $l^* = 0 \text{ m} \rightarrow 1,00$, $l^* = 600 \text{ m} \rightarrow 0,65$ and $l^* \geq 2000 \text{ m} \rightarrow 0,50$.

Hail Load
Making an analogy for the calculation of the hail load considered in buildings, established according to the NEC standard (NEC 2014), the hail load is calculated with the following equation:

$$S = \rho_S \, x \, H_S$$

ρ_S: Specific weight of hail (generally 1000 kg/m^3)
H_S: Accumulation height, m

Reaction Force

The reaction force according to the UNE-EN 13796 (AENOR, Spanish standard UNE-EN 12930: Safety requirements for cableway installations designed to carry persons, 2015) indicates that it should be considered a force caused by the stirrup which acts on the cabin when it is in movement when entering the station as well as allowing for transverse stability when passengers enter and leave.

The calculation of the reaction force is simple, a linear motion quantity analysis must be performed to obtain the load acting on the stirrup, which is calculated with the following equation:

$$P = mV$$

$$I = F_{impacto} x \Delta t$$

$$F_{impacto} = R = \frac{\Delta P}{\Delta t}$$

m: Fully loaded cabin weight, kg
V: Speed, m/s
P: Amount of linear movement, kg m/s
R: Reaction force, N
I: Variation in the amount of movement
$F_{impacto}$: Impact force, N
Δt: Time variation, s

Dimensioning of the Cabin

Based on the technical specifications for cable transport (Miravete et al. 1998), the following parameters are used to calculate the floor area:

$$Superficie = 0,5 + a \times n$$

Where:

a: 0.15 to 0.25 m^2
n: Number of passengers

Figure 2 shows the layout of the cabin floor, the space occupied by the seats and the person with reduced capacity.

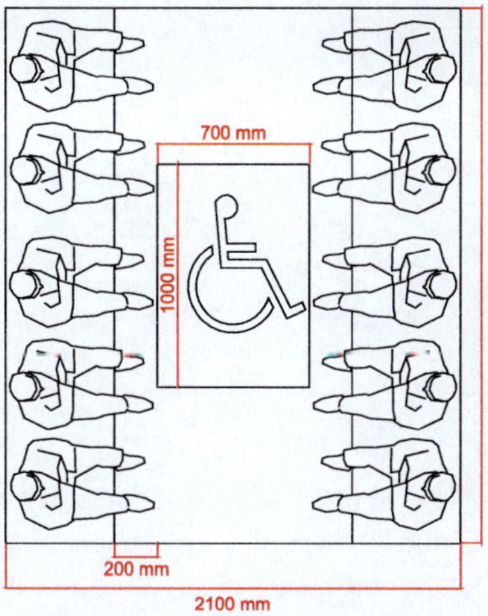

Fig. 2. Distribution of passengers. Source: Own elaboration

2.2 Loads and Load Combinations

For the structural analysis using the program SAP2000 all loads, and combinations thereof are evaluated to determine a critical operation condition. The following are the combinations used in the simulation.

Combination 1: $G + G_1 + Q + F_{Wx(+)} + S$
Combination 2: $G + G_1 + Q + F_{Wy(+)} + S$
Combination 3: $G + G_1 + Q + F_{Wy(-)} + S$
Combination 4: $G + G_1 + Q + S$
Combination 5: $G + G_1 + 0,5Q + R + S$
Combination 6: $G + G_1 + Q_P$

2.3 Calculation of the Damping System

Four springs are used that are capable of supporting the weight of the passengers, in addition to the weight of the cabin itself. The equivalent elastic constant of the springs is calculated as follows:

$$k_{eq} = k_1 + k_2 + k_3 + k_4$$

$$k_{eq} = 4k$$

k: Spring constant, N/m

$$m_c g = \delta k_{eq}$$

$$k_{eq} = \frac{mg}{\delta}$$

Static design

$$\tau_{xy} = k_s \times \frac{8 \times F \times D}{\pi \times d^3}$$

τ_{xy}: Shear stress, MPa
k_s: Shear Multiplication Factor
F: Spring load, N
D: Average spring diameter, mm
d: Spring wire diameter, mm

Dynamic design

$$D = d_{eje} + d + holgura$$

D: Average spring diameter, mm
d: Spring wire diameter, mm

2.4 Calculation of Bolted Joints

The different bolts that form part of the cabin are designed based on the different stress, cut and crush analyses to which they are subjected.

The calculating prestressing force of a bolt is given by:

$$F_p C_d = 0,7 fub \times A_s$$

A_s: Tensile area of the screw, m^2
fub: Ultimate screw capacity

Crush resistance

The ultimate stress of the base metal must be taken into account, this stress must be less than or equal to 1.2 times the stress of the base metal.

$$F_p \leq 1,2 S_u$$

However, it is rare for seals to fail because the permissible crushing stress has been exceeded. Therefore, the mentioned Fp is considered as an efficiency index of the bolted joint and not as a permissible stress in reality. (Mc Cormac & Csernak, 2013).

Tensile Strength

The axial tensile strength of a bolt is based on the tensile strength area (As)

$$F_t = fub \times A_S$$

However, after a statistical study based on the experimental method, it is concluded that it is necessary to apply the formula:

$$F_t = 0,9fub \times A_S$$

When the line of action of the applied force is eccentric to the screw shaft, additional traction is induced in the screw due to the effect.

Checking joints

$$F_{s,Rd} = \frac{k_s \times n \times \mu}{\gamma_{M3}} \times F_{p,Cb}$$

k_s: Coefficient of value 1, for holes with normal clearances.
n: It is the number of surfaces in contact with each other that produce the frictio force. 1 if one inner surface is in contact, 2 if two inner surfaces are in contact.

2.5 Photovoltaic System Calculation

The following equation is used to calculate the solar radiation on the photovoltaic solar panel:

$$\overline{H_T} = k \times \bar{H}\ (kWh/m^2/día)$$

k: Correction factor chosen from the tables according to the latitude and tilt angle of the panel.
$\bar{H}$: Average daily solar radiation, kWh/m^2/day

$$HS = \frac{H_T}{1000}(h/día)$$

$$T_c = T_a + 1,25(NOCT - 20)$$

T_c: Cell temperature at 1000 W/m^2 irradiation
T_a: Ambient temperature, °C
$NOCT$: Normal operating temperature of the cell is specified by the supplier, otherwise the value of 45 °C is assumed.

$$\eta_{temp} = 1 - C_T(T_C - 25)$$

n_{temp}: Temperature efficiency
C_T: Coefficient of power variation at temperatures different from test standards, or 0.005.
T_C: Cell temperature at 1000 W/m^2 irradiation, °C

Once the temperature efficiency has been calculated, the performance of the system is calculated with the following equation according to the Ecuadorian construction standard (NEC-11 2011):

$$PR = \eta_{temp} \times \eta_{suc} \times \eta_{reg} \times \eta_{cab} \times \eta_{bat} \times \eta_{out}$$

η_{temp}: Temperature efficiency
η_{suc}: Efficiency by dirt (0.98)
η_{reg}: Efficiency of the regulator (battery and inverter) (0.92)
η_{cab}: Loss efficiency of cables (0.97)
η_{bat}: Battery efficiency (0.98)
η_{out}: Efficiency due to self-discharge losses in batteries of 0.98

Battery sizing
The capacity of the batteries is calculated with the following equation:

$$C_{bat} = \frac{Pot_p}{V_{bat}}$$

Pot_p: Peak power, Wh
V_{bat}: Battery voltage, 12 V

3 Results and Discussion

3.1 Simulation of the Cabin Structure

The structure of the cabin is simulated in the SAP2000 program. To determine the safety factors, it is indicated that it must be evaluated under the criterion of maximum deformation energy or equivalent stress of Von Mises and since it is a design with aluminium in relation to the apparent limit of elasticity of the material (Fig. 3).

All cabin profiles support the weight of the passengers and the cabin without any problem (Fig. 4).

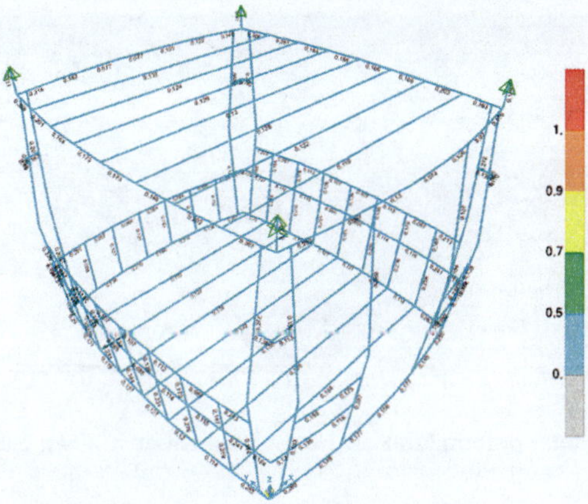

Fig. 3. Cabin structure simulation Source: Own elaboration (SAP2000)

3.2 Floor Plate Simulation

The calculation of the floor plate by means of the ANSYS program is carried out by taking the real stress produced and the apparent yield stress, so the result is:

$$n = \frac{\sigma_{adm}}{\sigma_{VM}}$$

The value of the safety factor present is 6.3 so the elements selected to form part of the cabin structure are fully reliable and meet the minimum safety requirements of the various rules governing the transport of passengers by air.

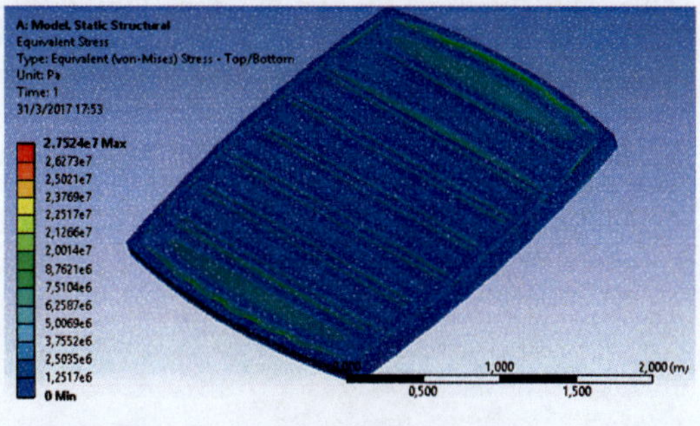

Fig. 4. Von Mises equivalent stress for the cabin floor Source: Own elaboration (ANSYS)

Once the structural resistance is guaranteed, it is important to verify the maximum deformation of the cabin floor as a design parameter. The total deformation of the element is a very small value, so the values presented can be widely accepted and do not affect the proper functioning of the cabin. The maximum total deformation of the cabin is 0.814 mm (Fig. 5).

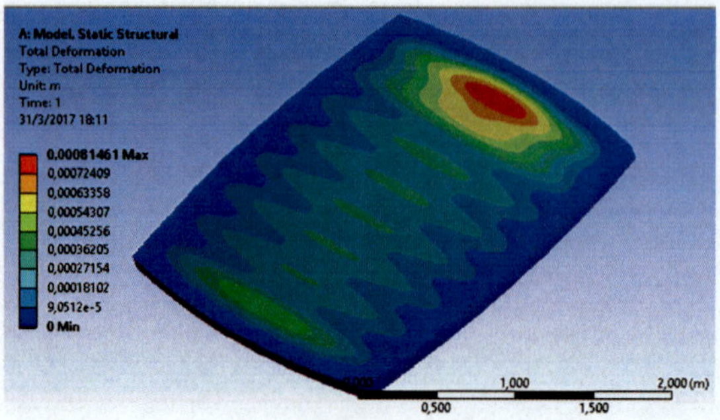

Fig. 5. Maximum total deformation of the cab floor Source: Own elaboration (ANSYS)

3.3 Simulation of the Coupling Element Cabin - Support "H"

The cab support element is located in the upper corner of the cab, so that the entire weight of the cab and accessories is supported, in addition to the load produced by the weight of the people. This element allows the coupling of the cabin with the support system called "H". The acting load corresponds to a quarter of the total load considering the weight of the cabin fully loaded (12740 N), equivalent to 3185 N. This element is made up of an aluminium plate 10 mm thick and a rectangular profile 95.3 × 76.2 × 6.6 in size, stiffened by two 8 mm thick plates.

Von Mises equivalent stress has a maximum value of 55.01 MPa at the base of the rectangular profile, associating this maximum stress with the reduced yield stress established by the AENOR UNE-EN 13796-1 standard gives a safety factor of 3.16, which is acceptable since, with respect to the actual yield stress, it represents a safety factor of 4.39. On the other hand, the maximum deformation for this state of load represents a small value, equivalent to 0.127 mm. Therefore, the design is validated with respect to the strength and deformation (Figs. 6 and 7).

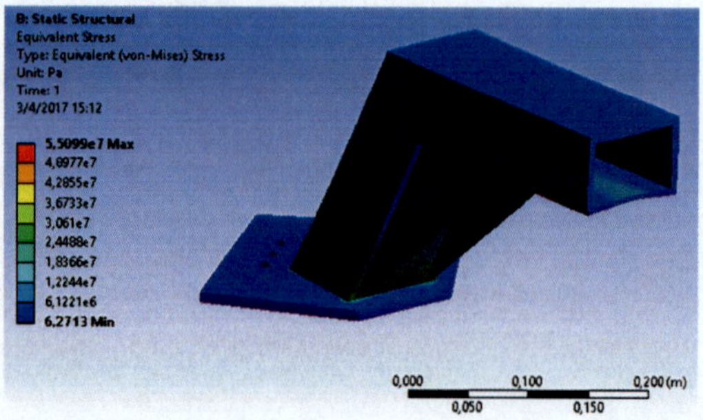

Fig. 6. Von Mises equivalent stress for the cab support element Source: Own elaboration (ANSYS)

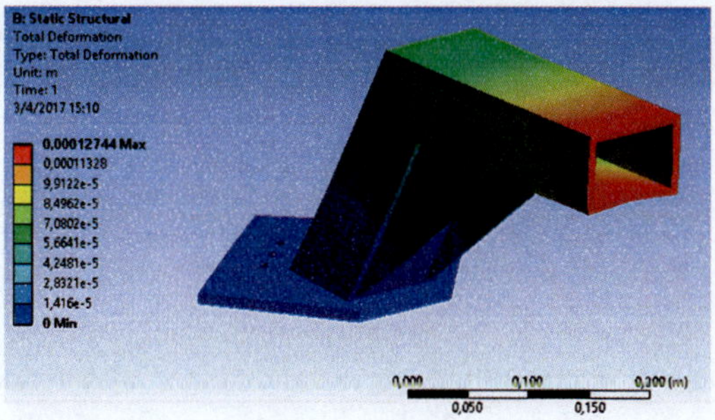

Fig. 7. Maximum total deformation of the cab support element Source: Own elaboration (ANSYS)

3.4 Design of the "H" Bracket

The design of this element consists of determining the sections of the profiles to be used in its construction. It consists of a central element and two lateral elements that give it the shape of a letter "H", from which its name comes. For its construction, ASTM A572-Gr50 steel profiles are used, since a material with good mechanical resistance, high resistance to corrosion and larger dimensions than those available in the national market for aluminum profiles is needed. In addition, unlike the cabin, the weight would not be an impediment due to its dimensional characteristics. Any load that produces a turning effect on the cabin is absorbed by the shock absorber systems that are strategically located on the coupling arm and disengageable clamp. However, in this structural system the damping device is installed to dissipate loads from the ascent and descent of passengers and additionally a critical effect produced by the incidence of hail load on the cabin roof.

As shown in Fig. 8, the point load applied at each end of the side elements is equivalent to 4576.6 N (a quarter of the total load generated by the combination of

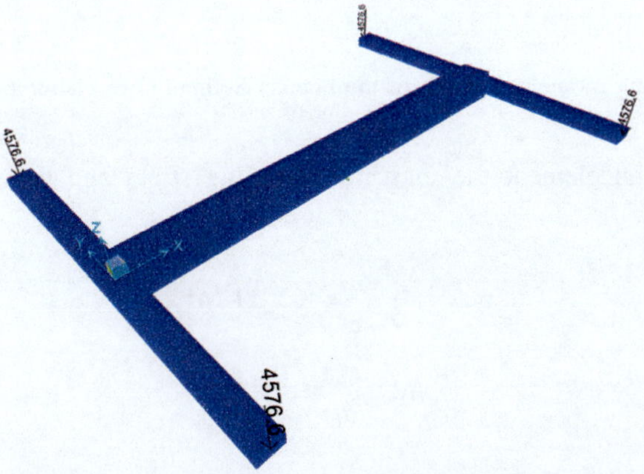

Fig. 8. Load status of support system "H" Source: Own elaboration (SAP2000)

loads in the cab). This value takes into account a possible critical load for the own weight of the cab including all its auxiliary systems, the load of the persons at their maximum capacity and the hail load. The selected geometrical and cross-sectional properties for the support profiles are shown in Table 1.

Table 1. Geometric characteristics of "H" bracket profiles

Element	Dimesions [mm]	Length [m]	Area [mm2]	Section module [mm3]	
				Axis 2–2	Axis 3–3
Central profile	150 × 150 × 6	2200	3456	159528,96	159528,96
Lateral profile	100 × 70 × 6	1700	1656	40786,06	50791,04

Source: Own elaboration (SAP2000)

Once the critical load state has been established, it is determined that the support elements work only in bending, with the corresponding bending moment diagrams shown in Fig. 8 Load state of support system "H", as the maximum moment is located in the middle of the profiles. For this study, a joint restricting translation in all axes is located at the midpoint of the central profile (Fig. 9).

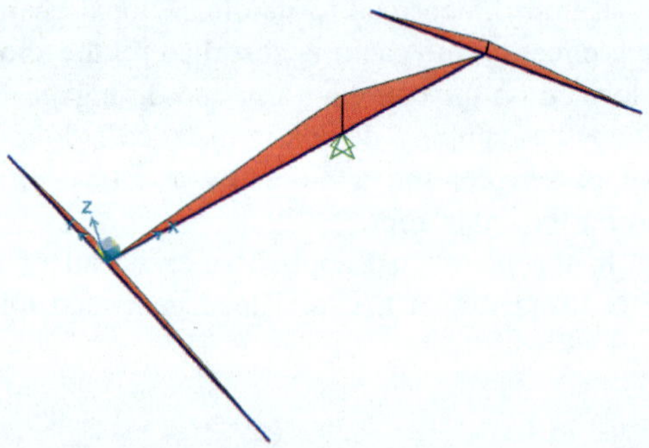

Fig. 9. Bending moment diagram of the bracket Source: Own elaboration (SAP2000)

For the central element the maximum bending stress and the safety factor are considered:

$$\sigma_b = \frac{M_{max}}{S_{2-2}} = 63,11 \text{ MPa}$$

$$n = \frac{\sigma_{adn}}{\sigma_b} = 5,46$$

For the side elements are considered:

$$\sigma_b = \frac{M_{max}}{S_{2-2}} = 95,38 \text{ MPa}$$

$$n = \frac{\sigma_{adn}}{\sigma_b} = 3,62$$

The calculated safety factors are satisfactory for all elements under the defined critical operating condition. The deflection values produced in the support structure "H" are satisfactory, so the design is validated with the determined geometry.

4 Conclusions

- A cabin with a capacity for 10 people was designed using all the criteria established by national and international regulations and codes in force used in the development of cable transport technology.
- The dimensioning of the cabin complies with the necessary space requirements for the comfort of the people inside a cubicle such as the cabin, in addition the current design contemplates the inclusion of people with reduced capacities.
- With regard to the structural design, all the scenarios that can generate a critical state of operation were used, i.e. when the system is exposed to overloads due to wind, hail and passenger overweight, based on the provisions of Spanish Standard UNE 13796-1: Safety requirements for cableway installations designed to carry persons, which determines the minimum safety factors for these operating conditions. All the structural elements of the cabin showed values above those established by the standard, which guarantee that the design carried out and validated by means of computer packages is adequate. Likewise, all the elements have been designed to generate joints that are easy to build and assemble. Additionally, all the selected profiles are available in the national market.
- Additional elements to the structural system of the cabin were designed, based on parameters established by the theory for the design of machine elements, and likewise the finite element method (FEM) was used for the simulation of certain elements of complex geometry.
- The welding procedures specified for the type joints identified in this cabin have been designed under the AWS D1.2 standard. This analysis is important since these procedures can be pre-qualified, thus knowing all the requirements and deficiencies that may arise in the process and thus guiding us to a greater development of knowledge in the qualification of welding processes in aluminium and its alloys.
- The construction processes implemented for aluminum welding can mean a great impediment for the construction of this device at a national level, since it must be submitted to the requirements established in the AWS D1.2 standard, being these an obstacle since there are not qualified personnel as well as high technology equipment to achieve an optimal welding procedure.

- From the investigation carried out, the total production of the cabins of a cableway would be limited by the unavailability of all the indispensable materials for its conformation, as it is the case of the acrylics, these must fulfill very strict requirements of ultraviolet protection, high mechanical resistance, low weight, reason why a 100% national production would be hindered.
- A system of electrical supply by alternative energies was dimensioned based on the requirements established by the Ecuadorian Norm of the Construction NEC-11, this aspect is important since resources are used for the generation of clean energy and to avoid the environmental pollution and where another type of power supply is impossible.
- The final budget to produce a cabin is high, however if the production will be made in greater quantity, the unit costs will decrease. The production process of the cabins would boost a large percentage of the country's industry by providing more knowledge about this type of technology.

References

Aenor, A.E.: Norma Española UNE 13796-1: Requisitos de seguridad de las instalaciones de transporte por cable destinadas a personas. AENOR, Madrid (2007)

Aenor, A.E.: Norma española UNE-EN 12930: Requisistos de seguridad de las instalaciones de transporte por cable destinadas a personas. AENOR, Madrid (2015)

Code-Aluminum, A.S.: ANSI/AWS D1.2/D1.2 M: 2014. American Welding Society (2013)

Emmop, E.M.: Estudio Hidrológico y Meterológico para el Proyecto: "Estudios Complementarios para el Detalle de Ingenierías de las Línea Roldós - Ofelia". Quito (2016). Obtenido de http://www.epmmop.gob.ec/doc_quito_cables/PRODUCTO%201/CAP.%20C%20-%20HIDROLOG%cdA%20Y%20METEOROLOG%cdA/Informe%20Meteorologico%20e%20Hidrologico.pdf

EMSD, E. a.: Code of practice on the design, manufacture and installation of aerial ropeways. Hong Kong (2002)

Euro, I.: (2010). bonnet.es

Miravete, A., Larrodé, E., Castejón, L., Cuartero, J.: Los Transpotes en la Ingeniería Industrial (teoría). Reverté, Barcelona (1998)

NEC, N. E.: Cargas (No Sísmicas). Quito: Código: NEC-SE-CG (2014)

NEC-11, N. E.: Capítulo 14: Energías Renovables. Quito (2011)

Norton, R.: Diseño de Máquinas: Un enfoque integrado. Prentice Hall, México (2011)

Valera, F. (s.f.). scribd.com. Recuperado el 20 de Febrero de 2017. de https://es.scribd.com/doc/314529830/Analisis-de-Precios-Unitarios

Alstom: Citadis on the move. The tram solution for each city. Paris: France, pp. 1–16 (2014)

Arboleda, M.A., Parra, I.F., Aristizábal, I., Sabogal, H.: Estudio dinámico de la movilidad en la ciudad de Santiago de Cali - Colombia. X Congreso Latinoamericano de Dinámica de Sistemas, III Congreso Brasileño de Dinámica de Sistemas, I Congreso Argentino de Dinámica de Sistemas (2015)

Nina Carolina Izabal: Metrocable de Medellín: transformando la dinámica social. Distintas Latitudes, Revista de Reflexión Latinoamericana (2014)

Environmental Protection Agency EPA: Guide to sustainable transportation performance measures. EPA 231-K-10-004. Southwestern Pennsylvania Commission (in Pittsburgh, Pennsylvania) and the Mid-America Regional Council (in Kansas City, Missouri), 3, 55 p. (2011)

Contribution Process for Producing Biofuel from Ripe Plantain Utilizing a HZSM-5 Catalyst

William Oñate[1(✉)], Luis Hernández[2], Sebastián Taco[3], and Gustavo Caiza[1]

[1] Universidad Politécnica Salesiana, UPS, 170702 Quito, Ecuador
wonate@ups.edu.ec

[2] Instituto Superior Tecnológico Central Técnico, ISTCT, 170502 Quito, Ecuador

[3] Escuela Politécnica Nacional, EPN, 170146 Quito, Ecuador

Abstract. This project describes the processes that were implemented during the experimental tests to obtain ethanol from the ripe plantain fruit, going through the treatment, fermentation and distillation stages. Afterwards, the alcoholic grades of the different obtained ethanol samples were measured, to select the samples with the greater alcoholic grade and inject them to the HZSM-5 packed bed reactor, to determine the amount of ethanol that oligomerizes to obtain fuel. Through the experimental analysis, appropriate conditions in the distillation temperature and fermentation time were established for the process of obtaining bioethanol, as well as the temperature of the reactor and the spatial velocity (WVHS) in the biotransformation process of ethanol to high octane gasoline, using a catalytic bed heterogeneous reactor. Grades of 50 resulted in the process of obtaining the bioethanol, and a selective performance of 23.5% of the biofuel was yielded. The gasoline was characterized by means of INEN (935:2016) standardized analyses for commercial gasoline, with which an octane number of 96.5 was determined, thus obtaining a gasoline of greater octane rating than the standard (super) gasoline commercialized in Ecuador.

Keywords: Biofuel · Bioethanol · Reactor · Spatial velocity · Zeolite

1 Introduction

Worldwide, the technologically developed countries are looking for alternatives to obtain security in energy supplies, seeking for a diversification of energy sources as a key part for the diversification of the productive matrix [1, 2], among which there is the lignocellulosic biomass. One of the more feasible alternatives than oil, coal and nuclear reactors of the developing countries, is the direct and indirect utilization of energy from vegetable and fruit residues; it is a renewable energy, abundant, decentralized and clean [3].

At a regional level, most Latin American countries seek different strategies for the reconversion of their productive technologies, due to the reduction of the reserves of fossil fuels and the increase in their usage (Molina, Martíne y Rudnick s.f.). The search

M. Botto-Tobar et al. (Eds.): CI3 2020, AISC 1277, pp. 505–514, 2021.
https://doi.org/10.1007/978-3-030-60467-7_41

for a renewable alternative raw material to replace oil derivatives has intensified, such as the production of ethanol using lignocellulosic material [3]; even though these processes currently have high cost, the advances in biotechnology should lead to a reduction in the costs of converting these materials to ethanol [1].

Since Ecuador is a country of agricultural nature, the biomass would be one of the more abundant renewable resources, [3, 4]. The lignocellulosic raw material (forage species, agricultural waste, etc.), namely the agricultural waste, are in this category and, in fact, have lower cost and are more abundant. However, according to studies of the Ecuadorian energy research corporation, these wastes are more difficult to process to obtain fuel; as a consequence, this alternative for transforming lignocellulosic biomass to fuel is in a research phase worldwide [5].

The production of musaceas is one of the most common and abundant crop in countries with tropical climates. For this reason, they are exported, and approximately 8% of the total production is sold [6]. The rejection in exportations does not exceed 2% of the annual total, and these excesses can be used for human, animal and industrial consumption; besides, 10% of the production is accumulated outdoors, causing problems due to inadequate handling, and generating residues and gases that harm the ecosystem [3]. Due to their bromatological composition both when unripe and ripe, the musaceas may be utilized for obtaining ethylic alcohol due to the amount of sugars and starches they possess [6]. As a consequence, this overproduction can be utilized in the generation of ethanol, contributing in part to the environmental issue and generating income sources [7].

The national demand of gasoline and the potential of Ecuador due to its geographic location, would enable producing biofuels and implementing technologies for obtaining hydrocarbons. This project studies the process of producing biogasoline from alcohols obtained by the enzymatic process, which will be activated on the ripe plantain fruit. This bioethanol is transformed to biogasoline by means of a catalytic reaction using a packed bed reactor for hydrocarbons [8].

The rest of the paper is organized as follows: Sect. 2 describes the utilized methodology; Sect. 3 shows the results and discussion and, finally, Sect. 4 presents the conclusions.

2 Methodology

2.1 Materials, Treatment and Reacting Substance

Among the physical properties of the ripe plantain (peeled) used in this study it can be mentioned: the variety (Musa AAB), weight of 215 $\pm$ 57.9 [g], length of 21.3 $\pm$ 2.9 [cm] and diameter of 4.14.1 $\pm$ 0.4 [cm].

[9], states that the chemical properties of the ripe plantain (Musa AAB) utilized in thus study are shown in use (see Table 1).

When unripe, this variety of plantain has between 20 and 22% of starch, and when in the ripe stage such starch becomes simple sugars, among which the most important are sucrose, fructose and glucose, components that facilitate the assimilation of sugars for the fermentation process [6].

Table 1. Chemical composition of the ripe plantain.

Components	Ripe plantain [%]
Humidity	69
Starch	3.4
Cellulose	1.33
Sucrose	66
Glucose	20
Dextrins	1.01
Gum	5.7
Protein	2
Raw Fiber	0.5
Fat	0.2
Fructose	14
Ashes	0.8
Tannins	–

Bioethanol from the Plantain Ripe Fruit. The process employed for the biotransformation of the fruit to ethanol comprises four main stages: mechanical fragmentation of the fruit, which produces physical and chemical changes, thus improving the mass transfer characteristics during the treatment, fermentation and distillation stages. The bioethanol was obtained from the ripe plantain fruit through a physical treatment, according to the block diagram shown in Fig. 1.

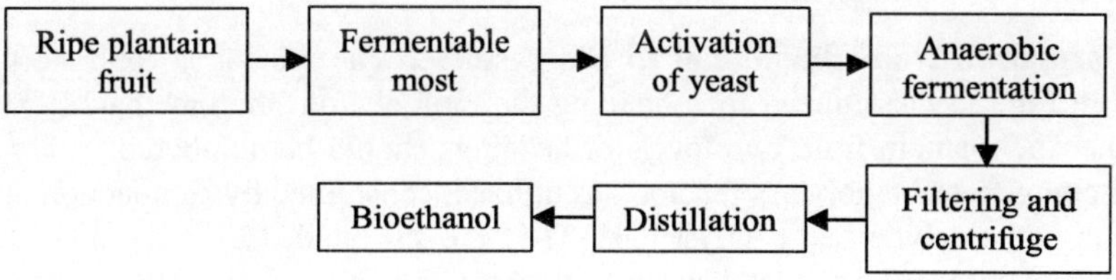

Fig. 1. Block diagram of the process for obtaining the bioethanol.

Treatment Stage. The fruit is triturated using a mill, to obtain small pieces with an approximate size between 0.2 and 3 [mm].

Fermentation Stage. An industrial yeast of type Saccharomyces cerevisiae [10], was utilized for the biotransformation of the glucose in ethanol; it takes sugar from the medium and initiates a series of reactions known as glycolysis route or Embden-Meyerhrda, where the glucose is catabolized to obtain final products such as: heat, ethanol carbon dioxide and pyruvate, that take different routes [11], as shown in Fig. 2.

The yeast was activated according the amount of samples to be taken; the relationship employed was 10 g of yeast per pound of ripe fruit. Afterwards the sample was stored and sealed in a container with air locks, which prevents oxidation of the

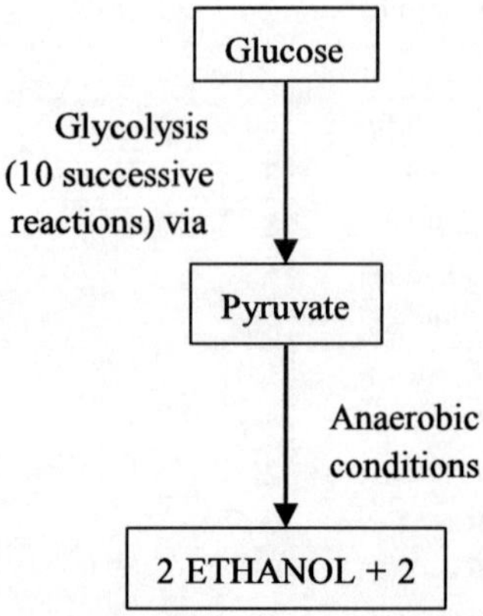

Fig. 2. Catabolic route of the pyruvate obtained by means of glycolysis (Fermentation).

mixture and at the same time releases carbon dioxide during the fermentation in a period of 3 to 9 days [12].

Sifting Stage. Due to the density of the sample, a centrifugal sifting was employed to separate the solids, thus obtaining the fermented liquid.

Distillation Process. A handcrafted distillation equipment was utilized. This equipment is constituted by a pressure cooker, a temperature sensor, a condenser and a water pump. The fermented liquid was distilled at a temperature established through experimental analysis between 77 and 88 [°C], thus obtaining bioethanol of different grades as a function of the distillation temperature.

Transformation from Bioethanol to Bio-gasoline. The reactor at laboratory scale shown in Fig. 3, was utilized for obtaining the biofuel from the bioethanol extracted from the ripe plantain fruit. two levels of headings should be numbered.

A homogeneous laboratory reactor was utilized, constituted by an injection pump, a preheater and catalytic bed catalyst [13]. They are surrounded by a thermal isolation made of mineral wool, and also have a container with cooling coil for the condensation of the biofuel [4].

The process for obtaining the biofuel from ethanol, is illustrated by the block diagram shown in Fig. 4.

The process for obtaining the gasoline starts with the injection of a constant flow of ethanol at the inlet of the preheater and at a previously established temperature, such that the ethanol (reactant) evaporates and enters the reactor, improving the contact with the surface of the catalyst (packed bed) [14, 15]. This catalyst is the microporous crystalline aluminosilicate (zeolite), which has as main physical characteristic a system of channels of sinusoidal path and almost spherical section, and a system of channels of linear path and an elliptical section. Due to its isotropic structure, this compound may exchange ions and the reversible desorption of water; as a consequence, this product is utilized in applications such as: heterogeneous catalysis, petrochemistry, fine chemistry

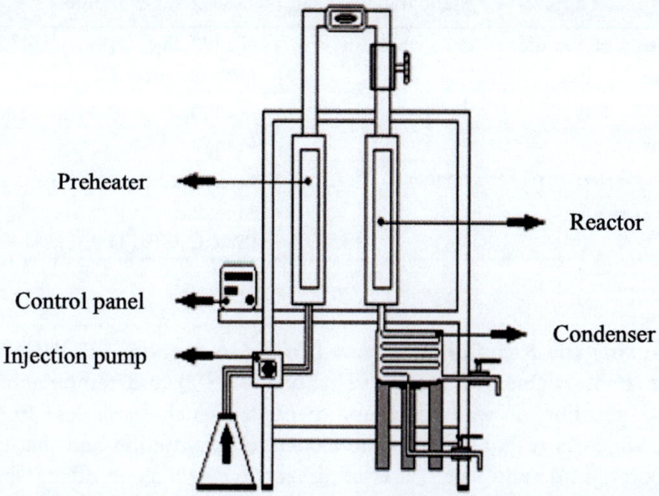

Fig. 3. Schematic diagram of the reactor at laboratory scale.

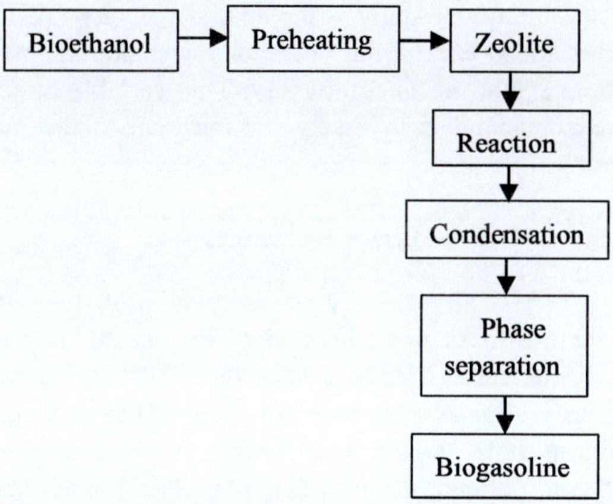

Fig. 4. Block diagram of the process for obtaining de biofuel.

and environmental processes of separation and purification [16]. Table 2 shows the crystallographic detail of the structure of the zeolite utilized in this study.

As a consequence of the previous, the products outgoing from the reactor are exposed to a low temperature (approximately 0°) by means of a coolant constituted by ice and water, and separated by a funnel at ambient temperature, with the organic phase of the gasoline located at the top and the aqueous phase at the bottom of the separator; thus, the liquid fuel is obtained in this manner [8].

Table 2. Crystallographic detail of the HZSM-5 zeolite.

Composition of the unitary cell; where (n < 27)	$Na_n[Al_nSi_{96n}O192] \sim 16H_2O$
Crystalline System	Orthorhombic
Parameter of cell	$a = 20,\ b = 19.9\,y\,c = 13.4\,[\dot{A}]$
Spatial group	Pnma
Shape and diameter of the channels	Oval, $5.3\,x\,5.6\,[\dot{A}]$
Isotropic structure	Silicalite ($Si_{96n}O_{192}F_4(TPA)_4$), Al – TS – 1

Variables During the Reaction of Ethanol to Hydrocarbons (T, WHSV). In order to determine stable values of spatial velocity (WHSV) and temperature [T] when obtaining the gasoline, it was necessary to control such variables. In the case of temperature, which is responsible for the alcoholic dehydration and that the reactants become carbon dioxide and water, it is necessary to establish an effective temperature that enables developing the process [17]. Due to the energy of reaction (oligomerization) and the difficulty for controlling the heat transfer, the temperature at the center of the reactor has a variation between ±3 [°C] with respect to the value established with the experimental analysis, guaranteeing the molecular traffic during the reaction [18].

The following variable under study is the spatial velocity (WHSV), which should be controlled in order not to exceed the upper or lower bounds established by Eq. 1, guaranteeing that the reactant reacts completely. This variable directly depends of the mass flow of the reactant, and is inversely proportional to the mass of the catalyst stocked in the reactor [17].

$$WHSV = \frac{m_{reactant}}{m_{catalyst}} \quad (1)$$

Where $m_{reactant}$, is the mass flow directed to the reactor expressed in [g/h] and $m_{catalyst}$ is the mass of the catalyst expressed in [g].

The utilized catalyst was a zeolite of type HZSM-5 composed by 280 silicon mole/Aluminum mole, which was broken to obtain pieces approximately between 0.1 and 0.5 cm (pellets), material that was placed inside the reactor such that the reactant makes contact with the zeolite, producing the decomposition of complex molecules in others simpler (cracking) [13, 16]; as a consequence, there is a decomposition in the oligomerization reactions when the temperature exceeds 300 [°C] (see Eq. (2))

$$\frac{1}{2}C_2H_4 \leftrightarrow \frac{1}{3}C_2H_9 \leftrightarrow \frac{1}{4}C_2H_8 \leftrightarrow \frac{1}{2N}C_nH_n \quad (2)$$

In addition, the reactor has crystal packing in the extremes, thus avoiding vapor escapes during the process. Figure 5 shows the internal structure of the reactor [17].

At last, optometry tests were carried out by means of parameters that determine the octane number RON in the biofuel.

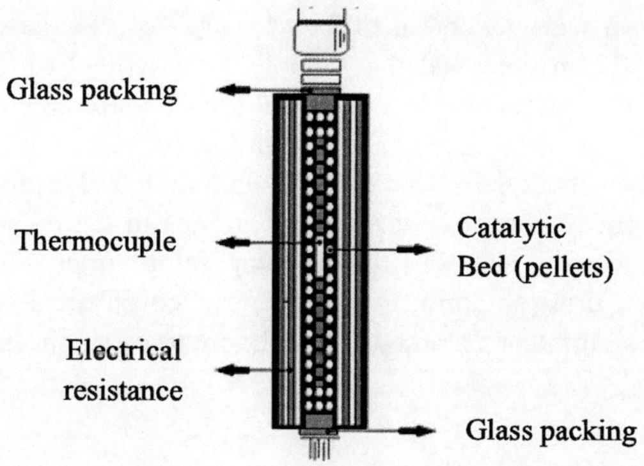

Fig. 5. Block diagram of the process for obtaining de biofuel.

3 Results and Discussion

3.1 Stage for Obtaining Bioethanol

Due to the physicochemical composition of the ripe fruit of plantain, such as a high content of glucose of 20%, sucrose 66% and fructose 14%, the sugars present in this fruit are easily assimilable and fermentable. As a consequence, the fermentable liquid was obtained between the third and the ninth day; on the contrary, microorganisms were generated that destroy the cellulose and the glucose in the samples, totally eliminating the alcoholic grades of the ethanol. The data gathered was divided in three samples from a total fruit weight of 47 [kg], obtaining the bioethanol results shown in Fig. 6.

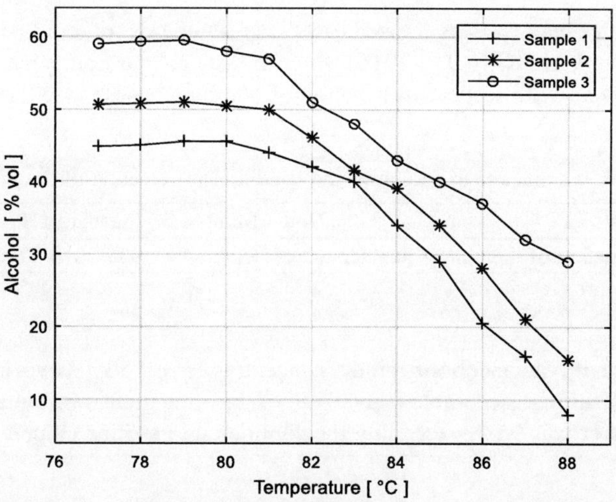

Fig. 6. Curve of obtained bioethanol in a period of 9 days of fermentation of the ripe fruit.

The data of Fig. 6 were taken for different ranges of distillation temperature and fermentation times. It can observed that sample 1 equivalent to the third and fourth fermentation days present alcoholic grades in the ethanol between 28° and 59.5°, sample 2 equivalent to the fifth, sixth and seventh day of fermentation present alcoholic grades in the ethanol between 16.3° and 51.5°, and sample 3 equivalent to the eighth and ninth day of fermentation present alcoholic grades in the ethanol between 8° and 45.5°. Therefore, increasing the distillation temperature impoverishes the alcoholic grades of the ethanol until obtaining water. Hence, the physical treatment should be controlled to obtain significant volumes of bioethanol to carry out the tests to transform it to biogasoline.

3.2 Stage for Obtaining Bioethanol

By means of experimental analysis, it was established a temperature of 340 °C in the interior of the reactor, because at temperatures below 310 [°C] and above 380 [°C] the oligomerization process in the catalyst (Zeolite HZSM-5) does not produce the expected reaction, generating as final product water and ethane; in addition, the amount of zeolite utilized in the reactor was 60 g, which resulted in a spatial velocity (WHSV) of 0.98 [h^{-1}], since at greater or smaller values the production of biogasoline impoverishes.

The conditions of temperature and spatial velocity established for the reaction of ethanol on the HZSM-5 catalyst are shown in Table 3.

Table 3. Condition of temperature and spatial velocity for obtaining the biofuel.

Temperature (T)	Spatial Velocity (WHSV)
340 [°C]	0.98 [h^{-1}]

The product obtained was treated over the standards of commercial gasoline according to standard INEN 935-2016, which evaluates and compares the characteristics of the biofuel with respect to gasoline of commercial use (see Table 4).

Table 4. Parameters to determine the conditions of reaction of the biofuel.

Analysis	Unit	Min	Max	Standard NTE-INEN
Determination of the octane number	RON	81	–	935-2016

Bioethanol with an alcoholic grade concentration of 50 [% vol] was utilized, obtaining as a result an acceptable conversion of gasoline greater or equal to 15%, and the octane number was 96.5, exceeding the commercial gasoline (super) of 92 octanes in Ecuador.

Table 5 shows the mass balance of biofuel from the ripe fruit of plantain, for which bioethanol of 50 alcoholic grades with a mass of 16447.86 g was used as base. These were injected to the reactor, obtaining water and hydrocarbon separately; these data were employed to calculate the non-condensable gases produced in the reaction, obtaining a mass of gasoline of 600 g.

Table 5. Mass balance of the biofuel.

Input	Products			
Ethanol 16447.86 [g]	HC Liquid Gasoline [g]	HC Gases [g]	Water Gases [g]	Total Water [g]
T = 340 [°C] WHSV = 0.98 $[h^{-1}]$	600	4410.76	3213.17	11437.1

4 Conclusions

During the process for obtaining the ethanol, it is necessary to control the distillation temperature and, overall, the fermentation time, because when the data of sample 1 taken during the third and fourth days were compared with the data of sample 3 taken during the fifth, sixth and seventh days, the alcoholic concentration decreased 14% of a total of 100 ml of ethanol. Similarly, it is mentioned the need of controlling the internal temperature of the reactor for the process of obtaining the fuel, since when the temperature is increased or decreased by 5% with respect to the established temperature, the reactor delivered water and hydrocarbons with molecular weights between C2 and C5 (gas). Due to this and following guidelines of the variables in the two processes (bioethanol and biofuel), C6 liquid hydrocarbon was obtained with a selective performance of 23.5% and with an octane rating of 96.5, which represents an increase of 4.7% in the anti-detonating quality with respect to the standard gasoline of commercial use in Ecuador.

References

1. Ruddy, D.A., Hensley, J.E., Nash, C.P., Tan, E.C.D., Christensen, E., Farberow, C.A., Schaidle, J.A.: Methanol to high-octane gasoline within a market-responsive biorefinery concept enabled by catalysis. Nature Catalysis **2**(7), 632–640 (2019)
2. Gomes, G.J., Dal Pozzo, D.M., Zalazar, M.F., Costa, M.B., Arroyo, P.A., Bittencourt, P.R. S.: Oleic Acid Esterification Catalyzed by Zeolite Y-Model of the Biomass Conversion. Topics in Catalysis (2019)
3. Popa, V.I.: Biomass for fuels and biomaterials. Biomass as renewable raw material to obtain bioproducts of high-tech value. In: 9th International Proceedings on Proceedings, pp. 1–33 (2018)

4. Palate, C., Olmedo, L., Ruiz, O.L.M.:. Diseño, construcción y pruebas de un sistema prototipo para la producción de etanol a partir de papa, zanahoria, remolacha y lacto suero. Riobamba: Escuela Superior Politécnica de Chimborazo: Facultad de Mecánica - Escuela de Ingeniería Mecánica (2010)
5. Usmani, M.A., Khan, I., Haque, A., Bhat, A.H., Mondal, D., Gazal, U.: Biomass-based composites from different sources. In: Lignocellulosic Fibre and Biomass-Based Composite Materials, pp. 45–76 (2017)
6. Carrión, M., Karina: Reutilización de residuos de la cáscara de banano (Musa Paradisiaca) y Plátanos (Musa Sapientum) para la producción de alimentos destinados al consumo humano. Guayaquil: Universidad de Guayaquil: Facultad de Ingeniería Química (2013)
7. Gualan, A., María, M.: Implementación de un destilador simple para la obtención de etanol a partir del penco azul (agave americano) para la planta de alimentos del Campus Juan Lunardi Paute. Cuenca: Universidad Politécnica Salesiana Sede Cuenca: Carrera de Ingeniería Agropecuaria Industrial (2014)
8. Vasquez, T., Anibal, S.: Oligomerization and catalytic ketonization in the Mixalco™ Process, p. 66. Texas A&M University, Texas (2013)
9. Amaya, F., Raúl: Evaluación de dos estados de madurez del plátano hartón Musa AAB utilizado en la elaboración de pan. Ibarra: Universidad Técnica del Norte: Facultad de Ingeniería en Ciencias - Agropecuarias y Ambientales: Carrera de Ingeniería Agroindustrial (2014)
10. Argote, V., Emilio, F., Cuervo, M., Alberto, R., Osorio, C., Esteban, D.-O., Johannes, Villada, H.S.: Evaluación De La Producción De Etanol a Partir De Melaza Con Cepas Nativas Saccharomyces cerevisiae. Biotecnología en el Sector Agropecuario y Agroindustrial **13**(2), 40–48 (2015)
11. Machín, C., Antonio, N., Carralero, G., Amarilys, C., Rodríguez, G.: Levadura Saccharomyces Cerevisiae Y La Producción De Alcohol. Revisión Bibliográfica. Yeast Saccharomyces Cerevisiae and The Production Of Alcohol. A Review. Revista ICIDCA. **50**, 20–29 (2016)
12. Alonso-Gómez, L.A., Bello-Pérez, L.A.: Materias primas usadas para la producción de etanol de cuatro generaciones: retos y oportunidades. Agrociencia **52**(7), 967–990 (2018). Recuperado en 10 de marzo de 2020, de http://www.scielo.org.mx/scielo.php?script=sci_arttext&pid=S1405-31952018000700967&lng=es&tlng=es
13. Vasquez, T., Sebastian: Trasnformation of acetona and isopropanol to hydrocarbons using HZSM-5 catalyst. Texas: Texas A&M University (2009)
14. Oregui-Bengoechea, M., Agirre, I., Iriondo, A., Lopez-Urionabarrenechea, A., Requies, J. M., Agirrezabal-Telleria, I., … Cambra, J.F.: Heterogeneous catalyzed thermochemical conversion of lignin model compounds: an overview. Topics in Current Chemistry **377**(6) (2019)
15. Goodarzi, F., Herrero, I.P., Kalantzopoulos, G.N., Svelle, S., Lazzarini, A., Beato, P., … Kegnæs, S.: Synthesis of mesoporous ZSM-5 zeolite encapsulated in an ultrathin protective shell of silicalite-1 for MTH conversion. Microporous and Mesoporous Materials, 109730 (2019)
16. Chen, J., Li, C., Zhou, Y., Sun, C., Sun, T.: An efficient, scalable and eco-friendly synthesis of 4,5-substituted pyrrole-3-carbonitriles by intramolecular annulation on Pd/C and HZSM-5. ChemCatChem (2019)
17. Salinas, M.: Obtención de Hidrocarburos de Alto Octanaje a Partir de Etanol. Quito: Escuela Politécnica Nacional: Facultad de Ingeniería Química y Agroindustria (2016)
18. Taco, S., Holtzapple, M.: Conversion of Isopropanol and Mixed - alcohols to Hydrocarbons Using HZSM-5 Catalyst in the MixAlco™. Texas (2013). https://doi.org/10.1002/aic.14008

Evaluation of the Performance of a Low Power Wind Turbine Applied to Isolated Communities of the Andean Region in Ecuador

Mauricio Carrillo(✉), Jesús Romero, and Alex Mayorga

Universidad Técnica de Ambato, Ambato, Ecuador
{cmcarrillo, jromero4338, asmayorga}@uta.edu.ec

Abstract. The study of the wind potential available on a small scale in Ecuador in different sectors applied to remote communities is the main motivation of the present investigation. In this sense, a low-power wind turbine has been designed and built, which was installed in the community of Rio Blanco, Tungurahua province, in order to satisfy basic energy needs of a rural house located at 4040 meters above sea level; with operational capacity for an annual wind speed of six meters per second that supplies one hundred and seventy watts of useful power. The main objective of this research was to observe the behavior of an installed wind turbine to determine its power curve, in addition to establishing the variation of the power coefficient according to the wind speeds registered at the site, the data was discretized for a better using base, the curves represent the samples taken every day that denote that the presence of gusts of wind does not remain constant, which dispersed their treatment due to extreme weather conditions in the area where the research was carried out. This will allow the improvement of the machine and the increase of its performance.

Keywords: Method of binning · Power curve · Wind turbine

1 Introduction

Electricity coverage in Ecuador, according to projected data from the Electricity Regulatory and Control Agency [1], stands at 85.74% for rural areas and 92.79% for urban areas. The rural sector is complex to analyze. The existence of dispersed population nuclei is a reality in Ecuadorian demography, which is magnified by the existence of a complex orography in the Sierra area as well as remote areas with limited access. This problem is shared with neighboring Andean countries such as Peru, Brazil and Bolivia. A side effect if you do not have access to the electricity system is poverty, according to the living conditions survey, of rural areas 41.8%, compared to urban areas 17.2% [2–5].

The use of renewable energy has been proposed as an alternative to provide electricity to isolated communities, as demonstrated by Ecuador's IDB/GED program [6]. However, projects with similar financing based on micro wind energy have not been implemented in Ecuador.

M. Botto-Tobar et al. (Eds.): CI3 2020, AISC 1277, pp. 515–530, 2021.
https://doi.org/10.1007/978-3-030-60467-7_42

1.1 Description of Micro Wind Systems Isolated from the Grid

Low-power wind turbines, as shown in Fig. 1, are small wind machines that are used to power loads that are far from conventional power grids, typically combined with battery banks to store electricity and subsequently distribute it, are designed to produce direct current between 12 V and 18 V that subsequently transforms alternating current to 110 V through an inverter. This class of wind turbines supplies less than 100 kW of power and is generally aimed at supplying rural homes with electricity [7, 8].

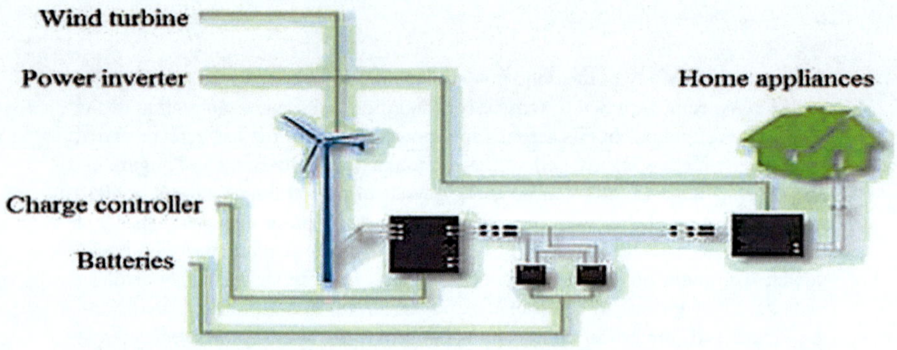

Fig. 1. Installation diagram of a micro wind system isolated from the grid [7]

Wind energy moves the blades of the generators, due to the shape they have, a pressure difference is generated that produces the force necessary to trigger the rotational movement on the main axis of the wind generator [4]. In an axial flow generator, the magnetic flux passes through the coils in a direction parallel to the axis of the machine. It consists of a front rotor, a stator and a rear rotor, both rotors are joined by a longitudinal axis. The stator contains copper wire coils, in which the voltage resulting from the variable flow produced by the magnetic field of the permanent magnets located in the front and rear rotors is induced. The stator is mounted on a fixed, stationary base. The permanent magnets are located in such a way that the magnetic flux passes from one rotor to another, crossing through the stator coils, while the rotor shaft rotates; the sine voltage is induced in the coil terminals [9] (Fig. 2).

1.2 Power Curve

The available electrical power versus wind speed is called the power curve which allows the energy delivered by the machine to be assumed [10, 11]. Power curves are obtained from measurements made in the field. The significant values of this curve are shown in Fig. 3:

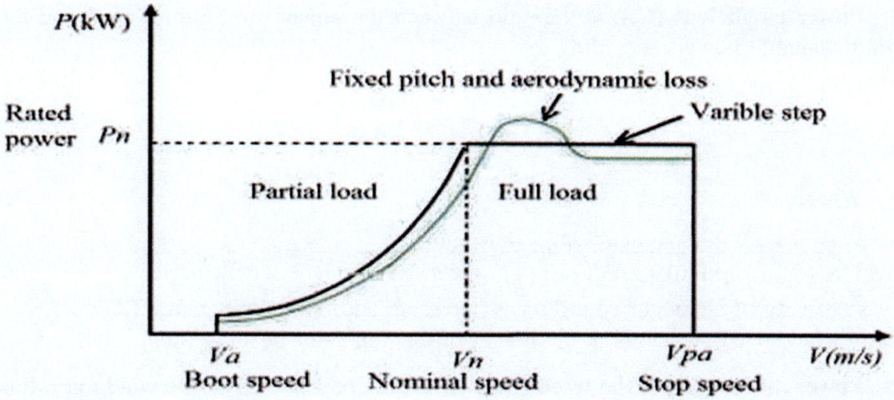

Fig. 2. Power curves of the same turbine, step change regulation and aerodynamic loss regulation [12]

Start speed V_a: also called cut in speed, it is determined when the wind turbine begins to deliver useful power. It is generally approximately 4 m/s in horizontal wind turbines that have three axes.

Rated speed V_n: also called rated speed is obtained when the wind speed reaches the nominal power of the turbine.

Disconnection or stop speed V_{pa} or V_d (cut speed): this is the wind speed at which the rotor stops due to the action of the regulation and control systems to prevent the risk of damage caused by high wind speed.

The ratio between the harnessed and the available power known as the power factor [13]. It is the fraction of the kinetic energy of the wind converted into kinetic energy of rotation in the wind turbine rotor. Its value ranges between 0 and 0.5925 [14]. The power coefficient indicates the efficiently with which the turbine transforms wind energy into electricity for use.

Wind speed is a vector magnitude. The main measuring instrument is the anemometer that records values by means of a sampling frequency and averages them over a set time [15]. Wind speed measurement should be done with a bowl anemometer with a speed range to be measured from 2 to 25 m/s [16].

1.3 Power Flow in a Wind Turbine

The power flow that starts at the available power (P_d) of the wind ends at the power (P_u).

The following powers and performances are distinguished in the power flow:

Wind power available (P_d): is the power of the wind, due to its kinetic energy.

Power coefficient (C_p): is the ratio between the power on the rotor shaft and the wind available power, according to:

$$C_p = \frac{P}{P_d} = \frac{P}{\frac{1}{2}\rho v^3 A} \tag{1}$$

Where:

P: power on the generator rotor shaft (*W*).
A: area swept by the wind turbine rotor (m^2).
ρ: density of air under conditions of pressure and wind temperature (Kg/m^2).
v: wind speed measured at the height of the generator bushing (m/s).

Power on the axis of the wind rotor (*P*): is the power used by the wind rotor. It is determined through the relationship:

$$P = C_p * P_d \tag{2}$$

Cp depends on the type of rotor, its turning speed and wind speed.

Drive power (P_{acc}): is the power on the output shaft of the multiplier box that is used for operating the operating machine.

Potencia útil final (P_u): Is the useful power supplied by the operating machine.

The overall performance (n_g) is also known as electric wind coefficient (C_e) defined according to:

$$n_g = \frac{P_u}{P_d} - n_m * n_c * C_p \tag{3}$$

$$n_g = C_e \tag{4}$$

In the case of an alternator the performance is designated as ($n_e = 1$) and the useful power (P_u) is the active electrical power (P_e) supplied by the alternator at its terminals is expressed as:

Three-phase generator:

$$P_e = \sqrt{3}U * I * cos\varphi \tag{5}$$

Single phase generator:

$$P_e = V * I * cos\varphi \tag{6}$$

where:

U, I: actual values of voltage and current in a three-phase connection.
V, I: actual values of voltage and current in a single-phase connection.
$Cos\varphi$: power factor [1].

1.4 Variables to Measure

Figure 3 shows how the data of the variables were taken: rotation speed, wind speed, current intensity, voltage, pressure and temperature.

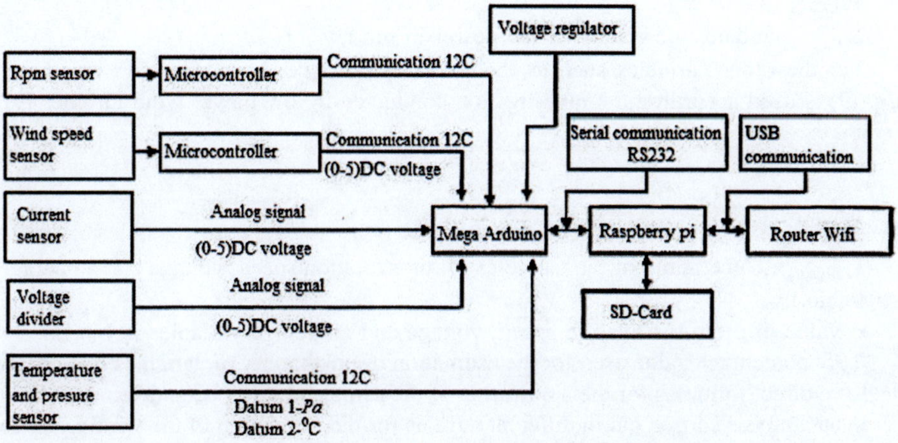

Fig. 3. Data acquisition system connection diagram

1.5 Data Analysis

The wind analysis requires data collection, measurement technique and data processing meth-od. For this experimental work, the IEC 61400-12-1 standard on the test of the power curve of low power wind turbines, it was taken as the basis [8, 16].

Data is acquired and stored in 10-minutes periods of contiguous data an internally, the acquisition system delivers the maximum, mean, minimum value, and standard deviation at the sample rate of 1 Hz [8].

The estimation of the measurement uncertainty is based on the application of the standard deviation to calculate the estimated error for the correction and validation of the results, in this case it will be applied to the power since the wind speed will be represented with the class mark of each interval [8, 17].

The uncertainty of the electric power will be realized by the following equations: Standard deviation of power:

$$\sigma_{P,i} = \sqrt{\frac{1}{N_i - 1}\sum_{j=1}^{N_i} \left(P_i - P_{n,i,j}\right)^2} \quad (7)$$

Where:

$\sigma_{(P,i)}$: standard deviation in bin i,
N_i: instantaneous or tenminutal distribution in bin i,
P_i: output power normalized and averaged in bin i,
$P_{(n,i,j)}$: normalized output power of data j in bin i.

Using Eq. (8) The standard uncertainty can be estimated:

$$S_i = S_{P,i} = \frac{\sigma_{P,i}}{\sqrt{N_i}} \tag{8}$$

Where:

$S_{(P,i)}$: standard uncertainty of the power in bin i,

For the other variables such as the rotor rotation speed, voltage, and amperage, Eq. (9) is used according the investigation conducted in low power wind turbines [9].

$$X_{a,b,c,d} = x*0.95 \tag{9}$$

Where:

$X_{(a,b,c,d)}$: Uncertainty of the variables of rotor rotation speed, voltage and amperage in bin i,

x: value of the rotor rotation speed, voltage and amperage variables in bin i,

0.95: percentage value used for the estimation of uncertainty of variables that do not have direct formulas for the application of the same and it has been taken according to an analysis carried out in different terrains for the estimation of the wind resource power curve at low scale [10].

1.6 Data Normalization

The first will be the air density above sea level according to the standard ISO atmosphere, and the second an average of the air density of the samples measured in situ during the test [8].

$$\rho_{10minutos} = \frac{B_{10minutos}}{R*T_{10minutos}} \tag{10}$$

Where:

$\rho_{10minutes}$: density of the air averaged in 10 min (kg/m^3)
$B_{10minutes}$: atmospheric pressure measured and averaged in 10 min (P_a)
R: specific constant of ideal gases (287.05 J/kg* ∘ K)
$T_{10minutes}$: absolute temperature of the air measured and averaged in 10 min (°K)

For the calculation of the normalized power we have the equation:

$$P_n = P_{10minutos} * \frac{\rho_o}{\rho_{10minutos}} \tag{11}$$

Where:

P_n: normalized power (W)
$P_{10minutes}$: measured and averaged power in 10 min (W)
ρ_o: air density (1,225 kg/m^3)

To determine the normalized wind, speed the equation is used:

$$V_n = V_{10minutos} * \left(\frac{\rho_{10minutos}}{\rho_o}\right)^{1/3} \tag{12}$$

Where:

V_n: normalized wind speed (m/s)
$V_{10minutes}$: wind speed measured and averaged (m/s)

1.7 Obtaining the Measured Power Curve

To determine the power curve, the method of the bines was used with the entire data set, using a range of 0.5 m/s, then the average values of power and normalized wind speed for each bin were calculated by means of the equations:

$$V_i = \frac{1}{N_i}\sum_{j=1}^{N_i} V_{n,i,j} \tag{13}$$

$$P_i = \frac{1}{N_i}\sum_{j=1}^{N_i} P_{n,i,j} \tag{14}$$

1.8 Power Coefficient

For the determination of the power coefficient Cp is detailed according to the following equation:

$$C_{p,i} = \frac{P_i}{\frac{1}{2}\rho_o A V_i^3} \tag{15}$$

Where:

$C_{(p,i)}$: power coefficient in bin i
V_i: normalized and averaged wind speed in bin i (m/s)

1.9 Annual Energy Production (AEP)

To estimates the AEP, measurements of the wind speed from 4 to 11 m/s should be made at the height of the center of the turbine using the equation:

$$AEP = N_h \sum_{i=1}^{N} [F(V_i) - F(V_{i-1})]\left(\frac{P_{i-1} + P_i}{2}\right) \tag{16}$$

Where:

N_h: number of hours in a year (approximately 8760)
N: number of bins

$$F(V) = 1 - e^{\left(-\frac{\pi}{4}\left(\frac{V}{V_{ave}}\right)^2\right)} \quad (17)$$

Where:

$F(V)$: Rayleigh cumulative probability distribution
V_{ave}: annual wind speed at the hub level
V: wind speed

2 Measured Power Curve

Table 1 shows the statistically obtained summary of the different parameters necessary for the interpretation of the machine's characteristic curves, as well as the representation of the aforementioned variables.

Table 1. Measured power curve

A	B	C	D	E	F	G	H
1	2.26	2.14	0.06	2794	0.02	0.0004	0.11
2	2.75	3.83	0.06	4365	0.03	0.0005	0.19
3	3.27	7.30	0.06	5022	0.06	0.0008	0.37
4	3.76	16.67	0.10	3993	0.15	0.0024	0.83
5	4.24	27.94	0.11	3530	0.27	0.0046	1.40
6	4.74	58.34	0.17	2765	0.65	0.0123	2.92
7	5.25	94.21	0.20	2939	1.02	0.0189	4.71
8	5.74	122.51	0.20	2070	1.61	0.0353	6.13
9	6.23	142.63	0.18	2090	1.87	0.0409	7.13
10	6.74	164.18	0.16	1948	2.25	0.0509	8.21
11	7.25	174.32	0.14	1338	2.89	0.0789	8.72
12	7.74	184.78	0.12	1229	3.21	0.0917	9.24
13	8.23	197.95	0.11	730	4.49	0.1662	9.90
14	8.73	204.82	0.09	563	5.33	0.2247	10.24
15	9.24	209.15	0.08	320	7.26	0.4056	10.46
16	9.72	215.06	0.07	211	9.17	0.6310	10.75
17	10.19	211.71	0.06	83	14.51	1.5930	10.59
18	10.62	207.70	0.05	28	24.63	4.6552	10.39

A: Bin number
B: Wind speed at bushing height (m/s)
C: Output power (W)
D: Power coefficient (Cp)
E: Data set number (10 min)
F: (Standard deviation of power)
G: Uncertainty A (Typical uncertainty for each Bin)
H: Uncertainty B (as measured in a small-scale wind turbine.)

Tablc 2 shows thc corrcctions.

Table 2. Corrected power curve

Bin N°	Corrected wind speed at hub height (m/s)	Output power corrected (W)	Corrected power coefficient (Cp)	Dataset No. (10 min)
1	2.25	2.09	0.056	2794
2	2.75	3.73	0.055	4365
3	3.25	7.12	0.064	5022
4	3.75	16.25	0.095	3993
5	4.25	27.24	0.109	3530
6	4.75	56.88	0.163	2765
7	5.25	91.84	0.195	2939
8	5.75	119.43	0.193	2070
9	6.25	139.05	0.175	2090
10	6.75	160.05	0.160	1948
11	7.25	169.92	0.137	1338
12	7.75	180.12	0.119	1229
13	8.25	192.92	0.106	730
14	8.75	199.59	0.092	563
15	9.25	203.72	0.079	320
16	9.75	209.37	0.069	211
17	10.25	205.62	0.059	83
18	10.75	200.18	0.050	28

Figure 4 shows the measurement of the speed of the wind turbine; it can be seen how the rotor begins to rotate when the wind overcomes the inertia of the generator with a speed of approximately 2.3 m/s; its curve shows an increasing behavior, they are directly proportional magnitudes.

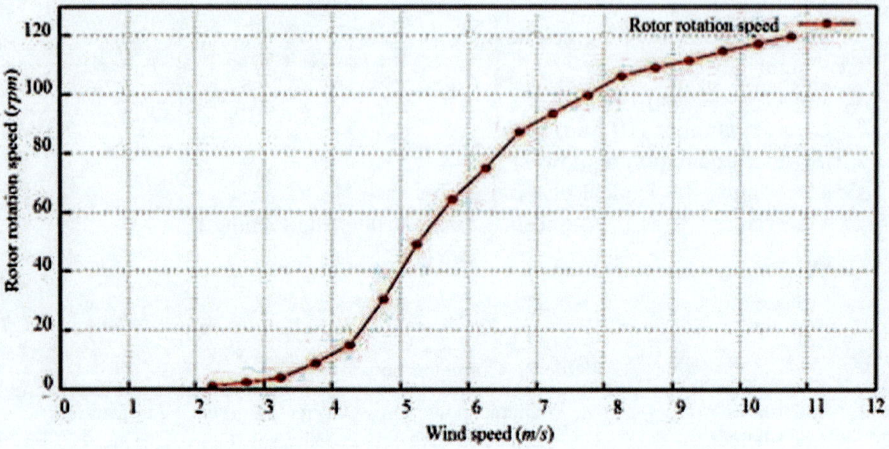

Fig. 4. Rotation speed of the rotor at the hub height.

Figure 5 shows the voltage delivered by the permanent magnet generator, its curve shows an increasing behavior, they are directly proportional magnitudes due to the higher wind speed the higher voltage delivered to the system.

The generator provides the necessary energy from 6.5 m/s when it reaches approximately 12 V required to charge the batteries.

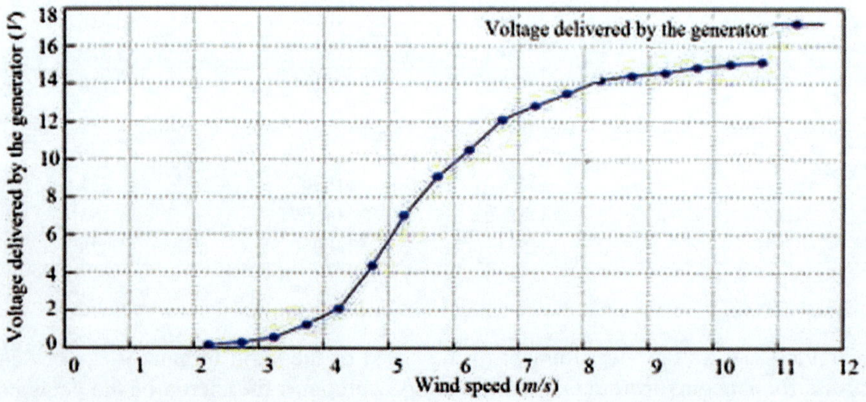

Fig. 5. Voltage delivered by the permanent magnet generator vs wind speed at bushing height.

Figure 6 shows the measurement of the amperage in the connection to the, its curve shows an increasing behavior, they are directly proportional magnitudes due to the higher wind speed greater amperage delivered to the system.

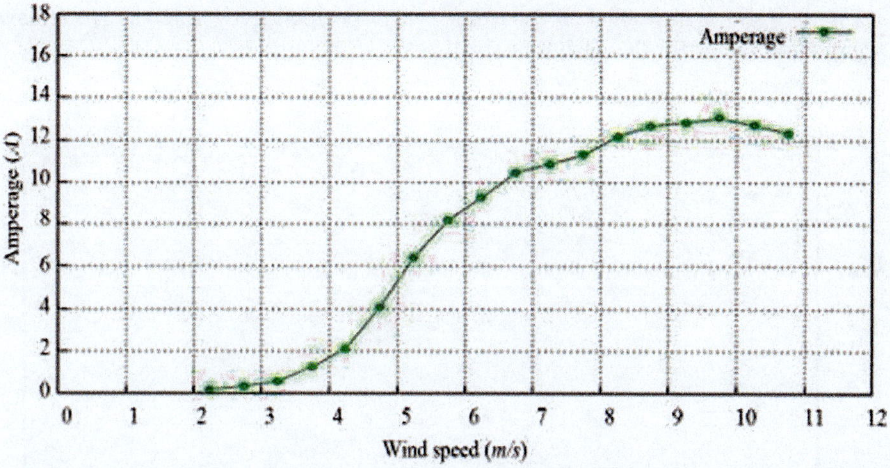

Fig. 6. Amperage generated in the connection to the load vs wind speed at the hub height.

Figure 7 shows a cloud of points classified in bins obtaining maximum, minimum, average and standard deviation values for each data set.

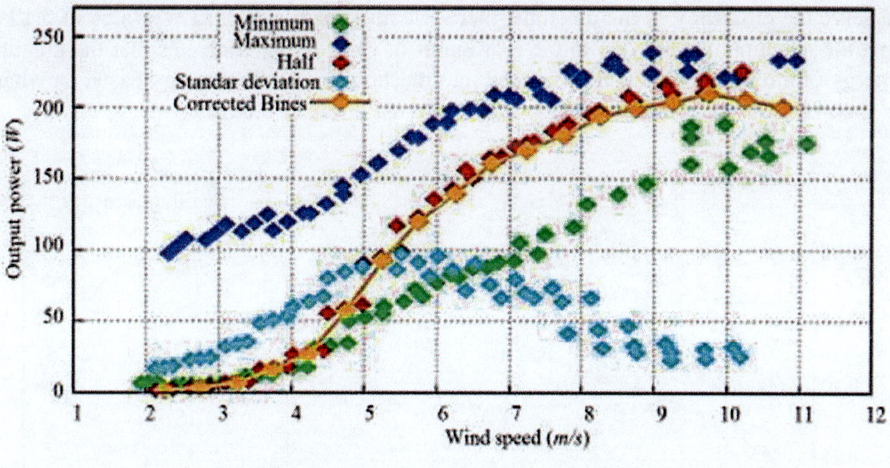

Fig. 7. Point cloud of the measured output power.

Figure 8 shows the characteristic curve of the machine that, as can be seen, begins to generate a power of 2.14 W from 2.3 m/s, when the inertia of the permanent magnet generator has been exceeded, at approximately 11 m/s it has a maximum power of 207.7 W, from this point the charge regulator does not allow the generator to produce more energy than necessary, which takes care of the batteries in the system after that, an increasing curve is observed due to the increase in power generation.

Figure 9 shows how efficient a machine is at a specific speed, which allows to have an important factor for the design of small-scale wind turbines.

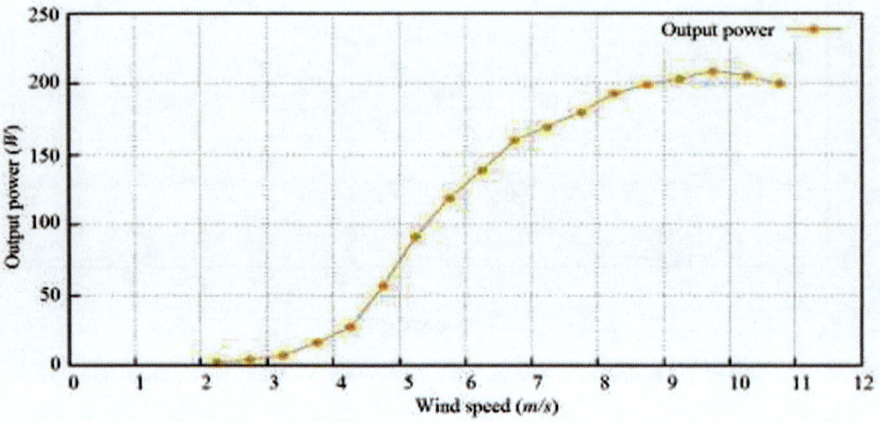

Fig. 8. Characteristic power curve of the permanent magnet wind turbine.

Figure 10 shows the coefficient curve and the wind speed, this curved allows us to observe the efficiency of the machine; there is a maximum power factor value of 0.195 that the machine reaches due to the rotation of its axis, which indicates that the kinetic energy of the wind is being transformed into mechanical rotation energy based on what is theoretically established with the maximum use limit of 0.59.

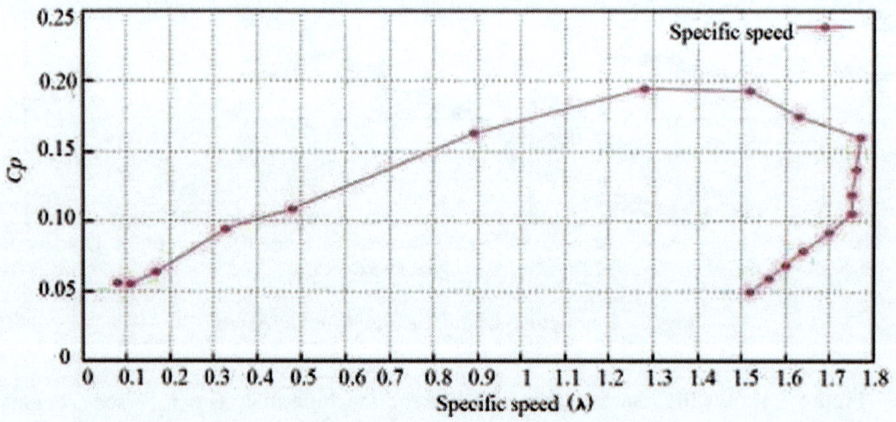

Fig. 9. Power coefficient vs Specific speed.

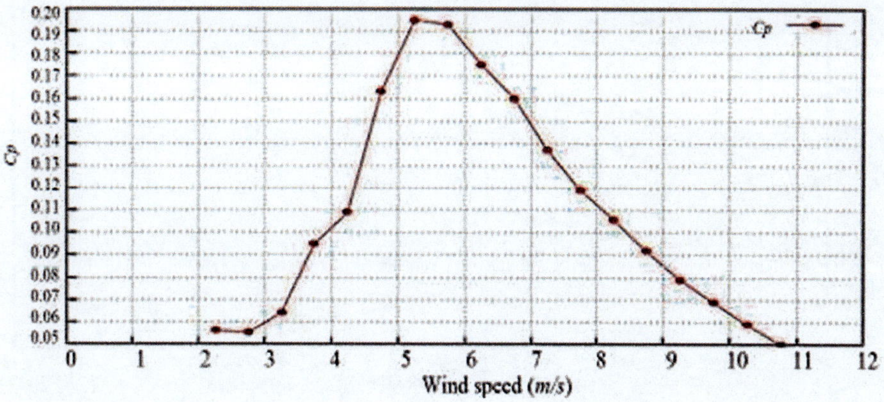

Fig. 10. Power coefficient vs Wind speed.

Figure 11 shows the overall coefficient of power; can be seen, there is a maximum value of 0.15 which represents the wind turbine's performance as a function of the power factor Cp and the electric generator's performance from the power curve.

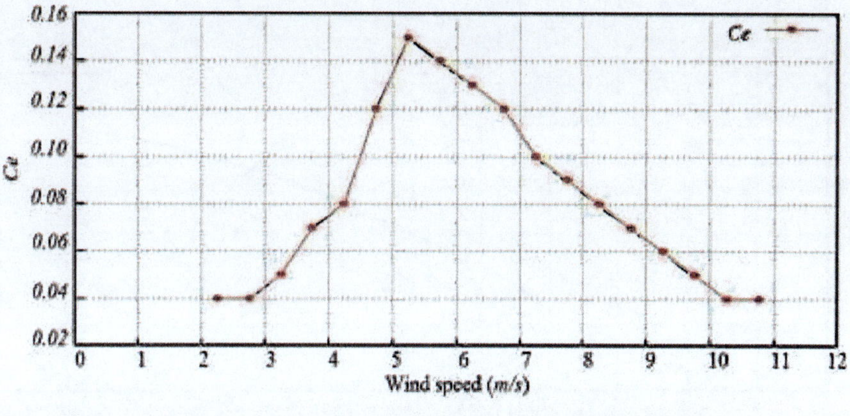

Fig. 11. Overall coefficient of electric power vs wind speed.

Figure 12 shows the torque coefficient curve, it can be seen that there is a maximum value of 0.075 at a speed of 2.3 m/s because if the wind turbine is stopped the wind needs a greater force to overcome the inertia and power spinning the blades, this curve is decreasing because after the inertia has been overcome the torque coefficient decreases, because the wind increases and the generator has gained kinetic energy for its rotation.

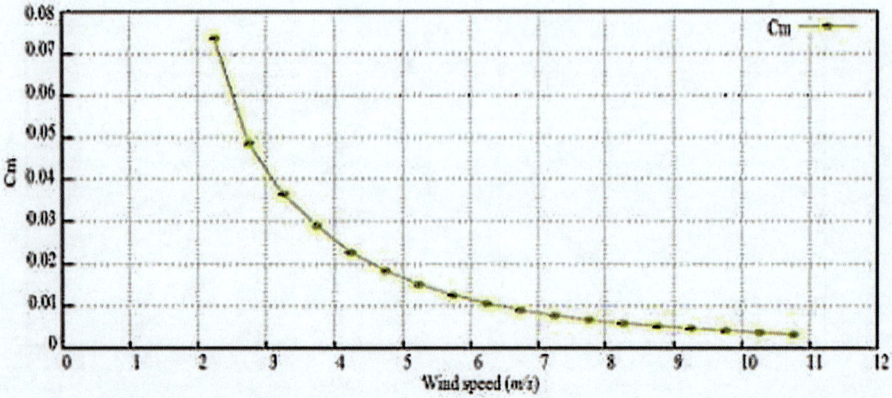

Fig. 12. Torque coefficient vs Wind speed.

Figure 13 shows the AEP, this calculation is based on a total of 8760 of annual energy production, it has been drawn from a wind speed of 3.75 m/s. Due to this amount of wind, representative energy is generated, which can be accumulated in the system for use.

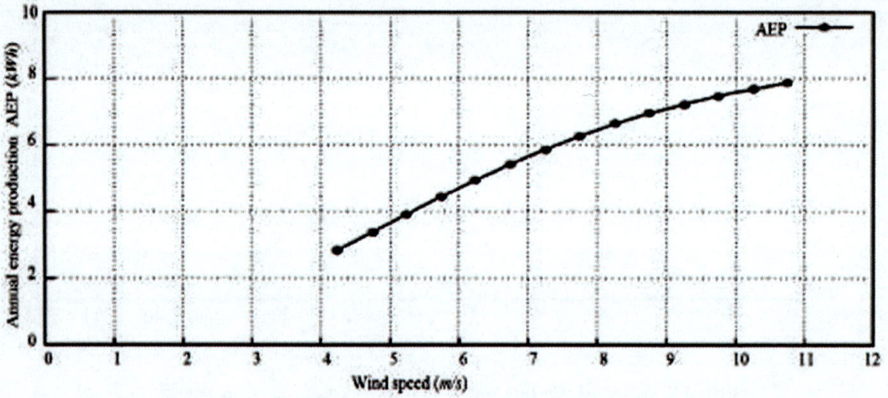

Fig. 13. AEP vs Wind speed.

3 Conclusions

It was observed that between 2 and 3 (m/s) the wind provides the kinetic energy necessary to overcome the inertia of the generator.

Wind speed between 2 to 12 (m/s) at bushing height, rotor rotation speed between 2 to 120 (rpm), voltage between 0.5 to 16 (V), and amperage was determined between 0.5 to 13 (A) delivered by the generator which describes the behavior of the machine during the time sampled in situ, it can be noted that as the wind speed increases the other variables show a proportional increase in their magnitudes, the ranges of the variables obtained are represented based on an average made by the method of the bins.

The wind turbine power coefficient curve was obtained, thus observing that it shows a Cp = 0.2 that represents 20% efficiency, a range below half the maximum energy limit that theoretically can be take advantage whose limit denotes Betz's theorem (Cp = 0.593) which represents 59.3% showing an acceptable efficiency in this type of machines.

The annual production obtained herein should not be taken into account to implement wind turbines of the same type as the one monitored one if the place is changed. The production of AEP would not be the same and would require a new study on the site that is going to perform the data collection for the subsequent installation and implementation of wind machines.

From the power coefficient obtained, changes can be made to the machine to increase the Cp value in order to obtain a better turbine efficiency. Among the different parameters that can be changed for improvement is to raise the height of the machine, increase the length of the paddles, design a mechanism that allows the angle to be varied on the blades so that air currents can be better used.

Despite the random behavior of the wind, it is shown that it is possible to evaluate the performance of small wind turbines in real conditions by applying the binning method without having to use a wind tunnel as the manufacturers of these machines do. This will allow the future to involve additional variables in the research and development of wind energy converters, improve their benefits and facilitate their commercialization in Andean populations of Ecuador.

Acknowledgements. This research Project has been funded by Dirección de Investigación y Desarrollo (DIDE), of Universidad Técnica de Ambato under the project: "Medios de conversión de energía eólica - solar a pequeña escala para satisfacer necesidades energéticas básicas en comunidades remotas de la provincia de Tungurahua", approved with resolution 1525-CU-P-2018 (Code: PFICM 16) is gratefully acknowledge.

References

1. ARCONEL: Plan Maestro de Electrificación 2009–2020, pp. 165-168 (2009)
2. Ferrer-Martí, L., Garwood, A., Chiroque, J., Ramirez, B., Marcelo, O., Garfí, M., Velo, E.: Evaluating and comparing three community small-scale wind electrification projects. Renew. Sustain. Energy Rev. **16**, 5379–5390 (2012)
3. Gómez, M.F., Silveira, S.: Rural electrification of the Brazilian Amazon-Achievements and lessons. Energy Policy **38**, 6251–6260 (2010)
4. Ferrer Martí, L., Domenech Léga, B., Canedo, W., Reza, C., Tellez, M., Domínguez, M.: Experiences of community wind electrification projects in Bolivia: evaluation and improvements for future projects. Vienna: In-Tech, pp. 85–106 (2011)
5. INEC: Encuesta Nacional de Empleo, Desempleo y Subempleo (ENEMDU), 1 edn., p. 13, Quito (2020)

6. Ministerio de Electricidad y Energía Renovable, Ecuador (MEER): Electrificación rural con energías renovables en zonas aisladas del Ecuador – proyecto BID/GEF (2014). Accessed: 12 julio 2019 [En línea]. http://historico.energia.gob.ec/electrificacion-rural-con-energias-renovables/
7. Ruíz, A.Y., López, D.E., Barberena, L.N., Sevilla, P.A.: Guía metodológica para desarrollar proyectos energéticos con fuentes renovables en la República de Nicaragua. Universidad Nacional Autónoma de Nicaragua, Managua, Tesis de Maestría (2012)
8. Atur, V.S., Goudappanavar, S.Y., Jangamshetti, S.H.: Novel control algorithm of storage system for small wind Turbine generator. In: Conf. International Conference on Current Trends in Computer, Electrical, Electronics and Communication (CTCEEC), pp. 383–388 (2017)
9. Carrillo, M., Medina, C., Mayorga, A.: Caracterización de un generador de flujo axial para aplicaciones en energía eólica. Ingenius **19**, 19–28 (2018)
10. Landberg, L.: Short-term prediction of the power production from wind farms. J. Wind Eng. Ind. Aerodyn. **80**, 207–220 (1999)
11. Bandi, M.M., Apt, J.: Variability of the wind turbine power curve. Appl. Sci. **6**, 262 (2016)
12. Lecuona Neumann, A.: La energía eólica: principios básicos y tecnología. Madrid, España: Leganés (2002)
13. Vahidzadeh, M., Markfort, C.D.: An induction curve model for prediction of power output of wind turbines in complex conditions. Energies **13**, 891 (2020)
14. Betz, A.: Schraubenpropeller mit geringstem Energieverlust. Gottinger Nachrichten. Nachrichten von der Gesellschaft der Wissenschaften zu Göttingen, Mathematisch-Physikalische Klasse **1919**, 193–217 (1919)
15. Villarrubia López, M.: Ingeniería de la energía eólica, vol. 5. Barcelona. España: Marcombo (2012)
16. IEC 61400-12-1, Wind turbines–Part 12-1: Power performance measurements of electricity producing wind turbines, Geneva, Switzerland: IEC (2005)
17. Hipp, C.H., Hall, M.: Evaluación de datos de medición Guía para la Expresión de la Incertidumbre de Medida. España. NIPO, (2007)

Design and Evaluation of the PID, SMC and MPC Controllers by State Estimation by Kalman Filter in the TRMS System

Byron Zapata(✉), Jaime Heredia, and Julio Proaño

Universidad Politécnica Salesiana, Cuenca, Ecuador
bzapata@ups.edu.ec

Abstract. The Twin Rotor Mimo System (TRMS) consists of two rotors namely tail rotor (∅) and main rotor (ψ). The main objective of this article is improving the stability of the TRMS system in rotation and elevation. To achieve that, a predictive controller based on the model (MPC), and a sliding mode controller (SMC) were implemented. The analytical modelling is based on the study of the system dynamics to obtain a linear model. Additionally, concepts of control involved in the design of the MPC, PID and SMC controllers are considered.

The controllers were evaluated on simulation and on the real TRMS system by performing several tests. In these tests the metrics considered were the response of the system and the control signal. The results show a comparison of settling times, rise times, integral of the absolute error (IAE) and integral of the squared error (ISE). We found that SMC controller have the best performance in comparison with PID and MPC controllers.

Keywords: Twin rotor mimo sistem · Controllers · Predictive controller based on the model · Sliding mode controller · Pitch · Yaw · Main · Tail · ISE · IAE

1 Introduction

The extensive range of applications of UAV systems have created a high interest in the control systems which seek to carry out different studies for the implementation of optimal and robust control techniques, to improve stability in these systems [1].

The TRMS is a laboratory equipment that simulates the behavior of a helicopter, the difference from a real helicopter is that in the TRMS system the angle of attack cannot be modified, so the aerodynamic force is obtained by varying the speed of its two motors. Despite this difference, the TRMS equipment helps to study the principles of MIMO coupled systems and the application of various controllers [2].

In [3], the authors use the technique PSO to estimate the parameters of the non-linear TRMS laboratory also they use the PID controllers as control methodology for the two rotors of TRMS. In [4] the authors propose to tune the PID controller using evolutionary computation methods to twin rotor mimo system (TRMS).

On the other hand, in [5]. The TRMS is decoupled into horizontal and vertical subsystem and two separate SMC's are designed for each subsystem. The propose was

M. Botto-Tobar et al. (Eds.): CI3 2020, AISC 1277, pp. 531–544, 2021.
https://doi.org/10.1007/978-3-030-60467-7_43

evaluated by simulation in compared with PID. In [6] SMC is used for the control of the TRMS and backstepping control is used to remove the Chattering and oscillating behaviors.

Finally, in [7] the authors propose the use of MPC controller because its results are better than LQR controller. In [8] the authors implement the MPC controller in Matlab.

All the previous works implement the controllers separately, however, in this work the PID, SMC and MPC controllers are evaluated all together in the elevation and rotation of the TRMS system. The evaluation is carried out in simulation and in a real TRMS system. The metrics considered for the evaluation are: ts (settling time), IAE (integral absolute error) and ISE (integral square error). Next the follow methodology is explained in detail

2 Methodology

Firs, in Fig. 1, the diagram of the free body and the forces acting on the system is shown.

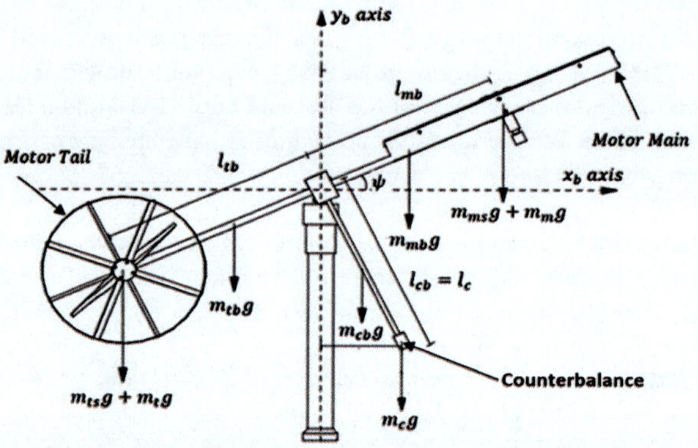

Fig. 1. Diagram of the free body TRMS [9]

Thus, the dynamics equations of the TRMS system for the pitch and yaw movements are shown in (1) [10].

$$I_1\ddot{\psi} = M_1 - M_G - M_{Gyr} - M_{B\psi}; \quad I_2\ddot{\phi} = M_2 - M_R - M_{B\phi} \tag{1}$$

Where:

M_1 y M_2 represent the static characteristics produced by the non-linearity of the motors (2) [9].

$$M_1 = a_1\tau_1^2 + b_1\tau_1; \quad M_2 = a_2\tau_2^2 + b_2\tau_2 \tag{2}$$

Table 1. TRMS model parameters [10]

Symbol	Description	Value
I_1	Moment of inertia of vertical rotor	6.8×10^{-2} kg.m^2
I_2	Moment of inertia of horizontal rotor	2.10^{-2} kg.m^2
a_1	Static characteristic parameter	0.0135
b_1	Static characteristic parameter	0.0924
a_2	Static characteristic parameter	0.02
b_2	Static characteristic parameter	0.09
M_g	Gravity momentum	0.32 N.m
$B_{1\psi}$	Friction momentum parameter	1.10^{-3} Nms^2/rad
$B_{2\phi}$	Friction momentum parameter	1.10^{-2} Nms^2/rad
K_{gy}	Gyroscopic momentum parameter	0.05 s/rad
k_1	Main motor gain	1
k_2	Tail motor gain	0.8
T_{10}	Main motor denominator parameters	1
T_{11}	Main motor denominator parameters	1.1
T_{20}, T_{21}	Tail motor denominator parameters	1
T_p	Cross reaction momentum parameters	2
T_o	Cross reaction momentum parameters	3.5
k_c	Cross reaction momentum gain	−0.2

M_G represent the gravity momentum and M_{Gyr} represent the Gyroscopic momentum. In Table 1 all the TRM model parameters are shown.

$$M_G = M_g \sin(\psi); \quad M_{Gyr} = K_{gy} M_1 \dot{\emptyset} \sin(\psi) \tag{3}$$

$M_{B\psi}$ y $M_{B\phi}$ represent the moments produced by frictional forces (4).

$$M_{B\psi} = B_{1\psi}\dot{\psi} + \tfrac{0.0326}{2}\sin(2\psi)\dot{\phi}^2; \quad M_{B\phi} = B_{2\phi}\dot{\phi} \tag{4}$$

M_R represent the cross reaction momentum.

$$M_R = \frac{k_c(T_o s + 1)}{(T_p + 1)}\tau_1 \tag{5}$$

The motors tail and main have an electrical equivalent transfer function (6). [10]

$$\frac{d\tau_1}{dt} = -\frac{T_{10}}{T_{11}}\tau_1 + \frac{k_1}{T_{11}}u_1; \quad \frac{d\tau_2}{dt} = -\frac{T_{20}}{T_{21}}\tau_2 + \frac{k_2}{T_{21}}u_2 \tag{6}$$

Using the before equations, we obtain the representation in states:

$$\psi = \dot{x}_1 = x_2 \tag{7}$$

$$\dot{\psi} = \dot{x}_2 = \frac{1}{I_1}\begin{pmatrix} a_1x_5^2 + b_1x_5 - M_g\sin(x_1) + 0.016\sin(2x_1)x_4^2 - B_{1\psi}x_2 \\ -k_{gy}\cos(x_1)x_4\left(a_1x_5^2 + b_1x_5\right) \end{pmatrix} \tag{8}$$

$$\phi = \dot{x}_3 = x_4 \tag{9}$$

$$\dot{\phi} = \dot{x}_4 = \frac{a_2}{I_2}x_5^2 + \frac{b_2}{I_2}x_6 - \frac{B_{2\phi}}{I_2}x_4 - 1.75\frac{k_c}{I_2}\left(a_1x_5^2 + b_1x_5\right) \tag{10}$$

$$\tau_1 = \dot{x}_5 = -\frac{T_{10}}{T_{11}}x_5 + \frac{k_1}{T_{11}}u_1; \quad \tau_2 = \dot{x}_6 = -\frac{T_{20}}{T_{21}}x_6 + \frac{k_2}{T_{21}}u_2 \tag{11}$$

Taylor's series method was used to obtain a linear approximation model of the system at equilibrium points: $x_0 = [0 \quad 0 \quad 0 \quad 0 \quad 0 \quad 0]^T; u_0 = [0 \quad 0]^T$. Obtaining the state space Eqs. (12) and (13).

$$\dot{x}(t) = \overbrace{\begin{bmatrix} 0 & 1 & 0 & 0 & 0 & 0 \\ -4.71 & -0.08 & 0 & 0 & 1.36 & 0 \\ 0 & 0 & 0 & 1 & 0 & 0 \\ 0 & 0 & 0 & -5 & 1.62 & 4.5 \\ 0 & 0 & 0 & 0 & -0.9 & 0 \\ 0 & 0 & 0 & 0 & 0 & -1 \end{bmatrix}}^{a} x(t) + \overbrace{\begin{bmatrix} 0 & 0 \\ 0 & 0 \\ 0 & 0 \\ 0 & 0 \\ 1 & 0 \\ 0 & 0.8 \end{bmatrix}}^{B} u(t) \tag{12}$$

$$y(t) = \overbrace{\begin{bmatrix} 1 & 0 & 0 & 0 & 0 & 0 \\ 0 & 0 & 1 & 0 & 0 & 0 \end{bmatrix}}^{C} x(t) + \overbrace{\begin{bmatrix} 0 & 0 \\ 0 & 0 \end{bmatrix}}^{D} u(t) \tag{13}$$

The Eqs. (12) and (13) have been represented in a transfer function matrix:

$$G(s) = \begin{bmatrix} \frac{2541}{(11s+10)(34s^2+3s+160)5} & 0 \\ \frac{17787}{s(11s+10)(s+5)1000} & \frac{18}{5s(s+1)(s+5)} \end{bmatrix} \tag{14}$$

The Relative Gain Array method is used to quantify degree of dynamic interactions between the inputs and outputs of the TRMS. Using $K = \lim_{s \to 0} G(s)$ in the Eq. (14), we obtain the terms of the relative gain array (15) [11].

$$\lambda = \frac{\frac{\partial yr}{\partial um}\Big|_{u_j}}{\frac{\partial yr}{\partial um}\Big|_{u_i}} = 1; \quad \Lambda = \begin{bmatrix} \lambda & 1-\lambda \\ 1-\lambda & \lambda \end{bmatrix} = \begin{bmatrix} 1 & 0 \\ 0 & 1 \end{bmatrix} \tag{15}$$

We use RGA's method to find the value of λ (λ = 1). The matrices (14) and (15) use hadamard product to obtain the transfer functions $G_p(s)$ and $G_y(s)$ [11].

$$G_p(s) = \frac{2541}{(11s+10)(34s^2+3s+160)5}; G_y(s) = \frac{18}{5s(s+1)(s+5)} \tag{16}$$

2.1 Sliding Mode Control (SMC)

SMC needs to define a sliding surface which is associated with a PID controller (17) [12].

$$S(t) = \frac{d^2}{dt^2}e(t) + \overbrace{3\lambda}^{K_d}\frac{d}{dt}e(t) + \overbrace{3\lambda^2}^{K_p}e(t) + \overbrace{\lambda^3}^{K_i}\int_0^t e(t)dt \tag{17}$$

To obtain the adjustment parameters, we use the Ziegler-Nichols method Table 2.

Table 2. PID control tuning parameters

	K_p	K_i	K_d
Yaw	4.9	1.7	3.5
Pitch	0.198	0.07	0.13

The control law U(t) combines the continuous (16) and discontinuous part of the system. Equations (18) and (19) represent the control law to movement yaw and pitch movement respectively [12].

$$U_y(t) = \frac{1}{3.6}\left(\ddot{X}(t)(6-\lambda_2) + \dot{X}(t)(5-\lambda_1) + \lambda_0 e(t) + K_D\frac{S(t)}{|S(t)|+\delta}\right) \tag{18}$$

$$U_p(t) = \frac{1}{1.35}\left(\ddot{X}(t)(0.997-\lambda_2) + \dot{X}(t)(4.786-\lambda_1) + 4.2780X(t) + \lambda_0 e(t) + K_D\frac{S(t)}{|S(t)|+\delta}\right) \tag{19}$$

The heuristic method by studying the quadratic error integral (ISE) allows defining the parameters K_D and δ of Eqs. (18) and (19) Table 3.

Table 3. Tuning parameters for Chattering reduction

Parameters	K_D	δ
Yaw	0.54	4.41
Pitch	101	0.2

2.2 Model Predictive Control (MPC)

The MPC is an optimal controller that uses an optimization function to incorporate restrictions owns of the system also it uses a model to predict future control actions.

The states space model with embedded integrator is used so that the error tends to zero. To incorporate the integrator, we use the Eqs. (20) [13].

$$\overbrace{\begin{bmatrix} \Delta x_{k+1} \\ y_{k+1} \end{bmatrix}}^{X_{k+1}} = \overbrace{\begin{bmatrix} A_d & o_d^T \\ C_d A_d & 1 \end{bmatrix}}^{A_e} \overbrace{\begin{bmatrix} \Delta x_k \\ y_k \end{bmatrix}}^{X_k} + \overbrace{\begin{bmatrix} B_d \\ C_d B_d \end{bmatrix}}^{B_e} \Delta u_k; y_k = \overbrace{\begin{bmatrix} o_d & 1 \end{bmatrix}}^{C_e} \overbrace{\begin{bmatrix} \Delta x_k \\ y_k \end{bmatrix}}^{X_k} \tag{20}$$

Equation (18) can be represented in states space, posteriorly it is discretized with a sampling time $T_s = 0.1$.These results are used in (20) to obtain the pitch movement (21).

$$Ap_e = \begin{bmatrix} 0.1567 & -1.9013 & -1.3331 & 0 \\ 0.3116 & 0.4671 & -0.4099 & 0 \\ 0.0958 & 0.4072 & 0.9257 & 0 \\ 0.1302 & 0.5533 & 1.2579 & 1 \end{bmatrix}; Bp_e = \begin{bmatrix} 0.3116 \\ 0.0958 \\ 0.0174 \\ 0.0236 \end{bmatrix}; Cp_e = \begin{bmatrix} 0 \\ 0 \\ 0 \\ 1 \end{bmatrix}^T \tag{21}$$

We use the same criterial for yaw movement.

$$Ay_e = \begin{bmatrix} 0.0016 & -0.6687 & 0 & 0 \\ 0.1337 & 0.8041 & 0 & 0 \\ 0.0392 & 0.3689 & 1 & 0 \\ 0.1411 & 1.32791 & 3.6 & 1 \end{bmatrix}; By_e = \begin{bmatrix} 0.1337 \\ 0.0392 \\ 0.0062 \\ 0.0224 \end{bmatrix}; Cy_e = \begin{bmatrix} 0 \\ 0 \\ 0 \\ 1 \end{bmatrix}^T \tag{22}$$

Output prediction equations and future control increment are represented in matrix (23) respectively.

$$Y = \begin{bmatrix} y_{k+1|ki} \, y_{k+2|ki} \cdots y_{k+N_p|ki} \end{bmatrix}^T; \Delta U = \begin{bmatrix} \Delta u_{k|ki} \, \Delta u_{k+1|ki} \cdots \Delta u_{k+N_c-1|ki} \end{bmatrix}^T \tag{23}$$

Grouping the products of the predictions and the control increments of (23), we obtain:

$$Y = \overbrace{\begin{bmatrix} C_eA_e \\ C_eA_e^2 \\ \vdots \\ C_eA_e^{N_p} \end{bmatrix}}^{F} X_k + \overbrace{\begin{bmatrix} C_eB_e & 0 & 0 & 0 \\ C_eA_eB_e & C_eB_e & \dots & 0 \\ \vdots & \vdots & \vdots & \vdots \\ C_eA_e^{N_p-1}B_e & C_eA_e^{N_p-2}B_e & \dots & C_eA_e^{N_p-N_c}B_e \end{bmatrix}}^{\Phi} \Delta U \quad (24)$$

MPC was designed using values of prediction horizon Np = 8 and control horizon Nc = 20 for pitch motion and Np = 4, Nc = 15 for yaw motion. These values are obtained by using ISE analysis.

Optimizer Design with Constraints

The main physical limitations of the TRMS system are two: motors voltage with a max value of $\pm5[V]$ and the value of the angles. For pitch movement −0.6 a 1.6 [rad] and for yaw movement −2.9 a 2.7 [rad]. The cost function with constraints is presented in (25) [14].

$$J = \frac{1}{2}\Delta U^T\left[\Phi^T\Phi + \bar{R}\right]\Delta \mathrm{U} + \left[FX_k - R_s\right]^T\Delta U \quad (25)$$

Associating the constraints of the control signal and the output of the system, we obtain the constraint matrix for pitch and yaw. They are presented in Eq. (26) [14].

$$\begin{bmatrix} T \\ \vdots \\ -T \\ \Phi \\ \vdots \\ -\Phi \end{bmatrix} \Delta \mathrm{U} \leq \begin{bmatrix} u_{max} - u_{k-1} \\ \vdots \\ -u_{min} + u_{k-1} \\ y_{max} - FX_k \\ \vdots \\ -y_{min} + FX_k \end{bmatrix} \quad (26)$$

2.3 Kalman Filter as States Observer

For the TRMS system, the Kalman filter is used to determine the torque of the pitch and yaw motors, for which the estimation of the states and the output of the system is necessary (27) [15].

$$\hat{x}_k = A_d\hat{x}_{k-1} + B_du_k + K_k\Delta y_k; \Delta y_k = y_k - \hat{y}_k \quad (27)$$

For the filter correction phase, the Kalman gain equation is used (28).

$$K_k = \left[P_kA_d^T + F_dQ_kF_d^T\right]C_d^T\left[C_d\left[A_dP_kA_d^T + F_dQ_kF_d^T\right]C_d^T + R_k\right]^{-1} \quad (28)$$

Where:

- A_d is state matrix in discrete time, F_d is a matrix of dimensions $(n, 1)$, Q_k: Covariance matrix of process noise y R_k: Covariance matrix of measured noise.

Matrices A_d, B_d y C_d for the movement of yaw and pitch are presented in (29) and (30) respectively.

$$A_{md} = \begin{bmatrix} 0.88 & -0.47 & -0.40 \\ 0.09 & 0.97 & -0.02 \\ 0.004 & 0.09 & 0.99 \end{bmatrix}; B_{md} = \begin{bmatrix} 0.094 \\ 0.004 \\ 0.002 \end{bmatrix}; C_{md} = \begin{bmatrix} 0 & 0 & 1.3 \end{bmatrix} \quad (29)$$

$$A_{td} = \begin{bmatrix} -0.04 & -0.65 & 0 \\ 0.13 & 0.73 & 0 \\ 0.05 & 0.44 & 1 \end{bmatrix}; B_{td} = \begin{bmatrix} 0.13 \\ 0.05 \\ 0.01 \end{bmatrix}; C_{td} = \begin{bmatrix} 0 & 0 & 3.6 \end{bmatrix} \quad (30)$$

The parameters of the covariance matrices of process noise and measurement noise are determined heuristically by reducing the ISE index Table 4.

Table 4. Tuning parameters for the Kalman filter

	R_k	Q_k
Yaw	10	0.025
Pitch	4	0.025

3 Results and Discussions

In this section MPC and the estimation of states using Kalman filter to stabilize the TRMS pitch and yaw responses are tested. Four scenarios are considered. Firstly, simulation model of TRMS scenario and secondly, a real implementation scenario. Both of them are evaluated with and without perturbations. Results are the response and control signal of both rotors. In addition, ISE and IAE indices for MPC, SMC and PID are considered (Table 5).

Table 5. Parameters table for simulation and real implementation

Parameter	Value
Simulation time [s]	100
Reference angle [rad]	0.5
Perturbation amplitude (Step Type) [rad]	0.2
Perturbation time [s]	50

The most relevant results and its discussions are presented below.

3.1 Simulated Scenario

The Fig. 2. presents the graphs of behavior of yaw movement of the TRMS system, using different controllers (Tables 6 and 7).

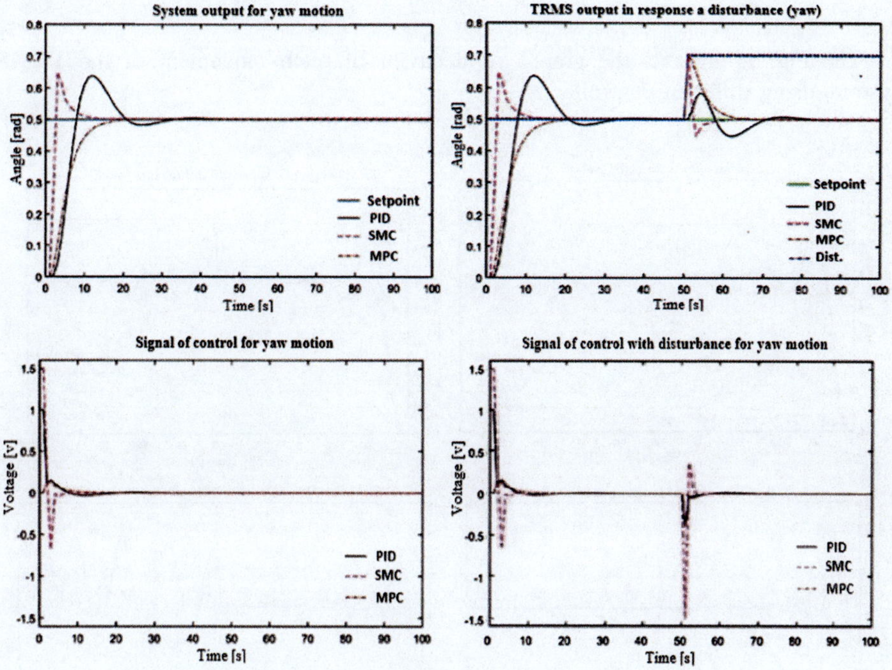

Fig. 2. Yaw movement control response using different controllers.

Table 6. Simulation results for yaw movement.

Scenario	Without perturbation				With perturbation		
	Ts [s]	Tr [s]	IAE	ISE	Ts [s]	IAE	ISE
MPC	15.7	8.1	1.82	0.42	11.8	2.5	0.3
SMC	9.8	1.2	0.92	0.21	6.5	1.3	0.2
PID	26.1	2.4	2.21	2.21	18.8	3.1	0.5

Table 7. Simulation results for pitch movement.

Scenario	Without perturbation				With perturbation		
	Ts [s]	Tr [s]	IAE	ISE	Ts [s]	IAE	ISE
MPC	20.0	10.1	3.4	1.1	12.2	2.8	0.17
SMC	9.98	5.2	0.06	0.00	4.3	0.3	0.04
PID	26.2	3.6	2.28	0.4	14.4	3.2	0.56

The Fig. 3. presents the graphs of behavior of pitch movement of the TRMS system, using different controllers.

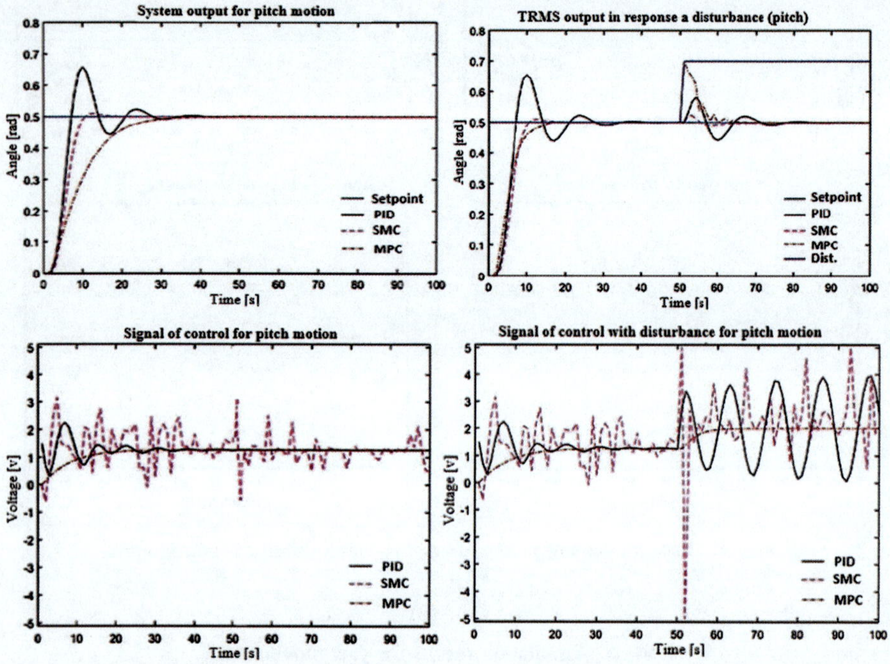

Fig. 3. Pitch movement control response using different controllers.

3.2 Real Implementation Scenario

The Fig. 4. presents the graphs of behavior of yaw movement of the TRMS real system, using different controllers (Tables 8 and 9).

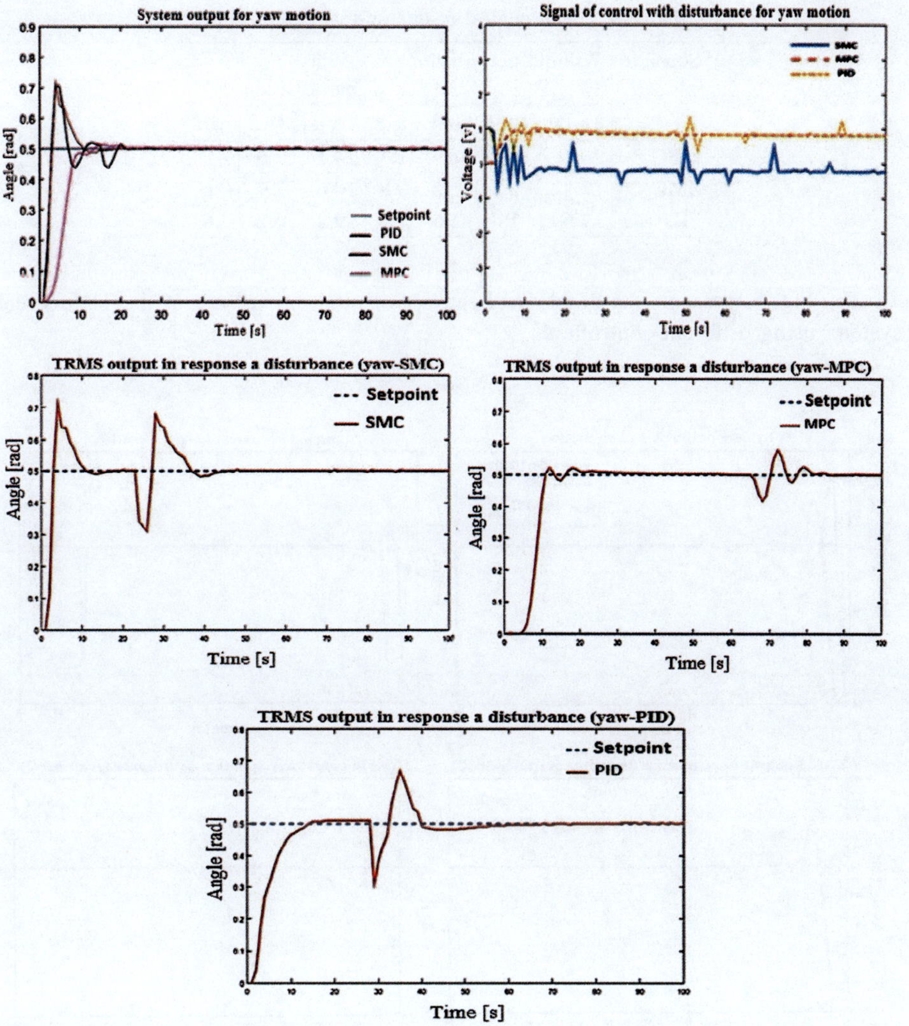

Fig. 4. Yaw movement control response using different controllers in the real system TRMS.

Table 8. Yaw control results on real equipment.

Scenario	Without perturbation				With perturbation		
	Ts [s]	Tr [s]	IAE	ISE	Ts [s]	IAE	ISE
MPC	24.4	4.7	2.9	1.5	28.5	2.5	1.9
SMC	14.2	1.4	1.7	0.8	23.2	1.9	1.0
PID	52.4	7.1	4.8	3.3	35.6	6.2	3.4

Table 9. Pitch control results on real equipment.

Scenario	Without perturbation				With perturbation		
	Ts [s]	Tr [s]	IAE	ISE	Ts [s]	IAE	ISE
MPC	26.7	11.2	1.9	1.3	24.1	3.6	2.5
SMC	17.8	2.5	1.2	0.8	10.6	2.9	0.6
PID	14.69	1.6	2.6	1.8	29.2	6.2	3.6

The Fig. 5. presents the graphs of behavior of pitch movement of the TRMS real system, using different controllers.

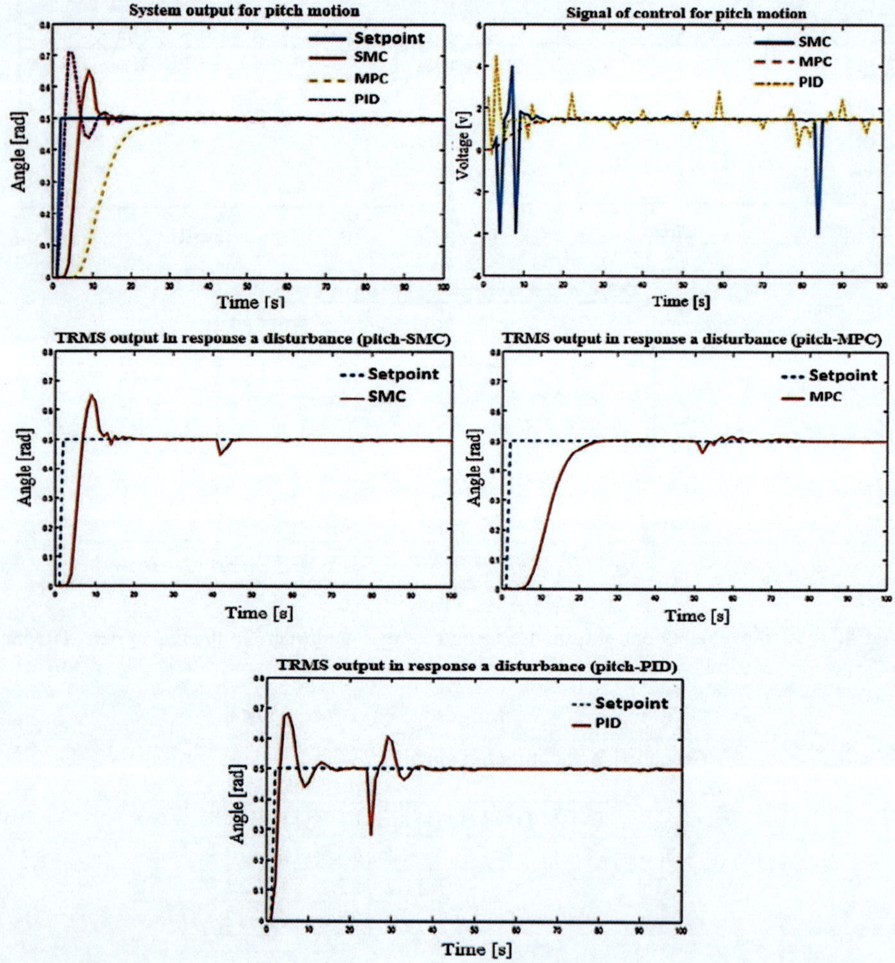

Fig. 5. Pitch movement control response using different controllers in the real TRMS.

4 Conclusions

Performance tests of the application of the SMC, MPC and PID controllers were performed to the yaw and pitch of the system, by applying a constant reference, by making a reference change in the process and by introducing a disturbance, thereby It was observed that the SMC controller has low performance rates and lower settling times than the other controllers.

The results obtained in the pitch and yaw control of the real system presented in test 3.3 and 3.4 were compared, where it can be observed that the SMC control has the performance rates, settling and lifting times less than the response of the MPC and PID controllers, in the same way it can be observed that when introducing a disturbance and a reference change in the process, the SMC control presents lower performance rates and settling times to the other controllers.

Through the tests carried out, it is verified that the application of the MPC control minimizes the control effort compared to the SMC and PID controllers, and that the SMC control has better performance rates and better response to disturbances.

References

1. Saska, M., Langr, J., Přeučil, L.: Plume Tracking by a Self-Stabilized Group of Micro Aerial Vehicles. Springer, Roma (2002)
2. Belmonte, L., Morales, R., Fernández, A., Somolinos, J.: Nonlinear cascade-based control for a twin rotor. In: Tech (2016)
3. Gautam, S., et al.: PSO based parameter estimation and PID controller tuning for 2-DOF nonlinear twin rotor MIMO System. Int. J. Autom. Control **12**(4), 582 (2018)
4. Tuning of Nonlinear PID Controller for TRMS Using Evolutionary Computation Methods. Tehnicki vjesnik - Technical Gazette, vol. 25, no. Supplement 1 (2018)
5. Shaikh, P. U., Karvekar, S.: Implementation of sliding mode controller for twin rotor multi input multi output system without state observer. In: 2016 IEEE 1st International Conference on Power Electronics, Intelligent Control and Energy Systems (ICPEICES), pp. 1–6. IEEE (2016)
6. Ilyas, M., et al.: Control law design for twin rotor MIMO system with nonlinear control strategy. Discrete Dyn. Nat. Soc. **2016**, 1–10 (2016)
7. Ulasyar, A., Zad, H.S.: Robust & optimal model predictive controller design for twin rotor MIMO system. In: 2015 9th International Conference on Electrical and Electronics Engineering (ELECO), pp. 854–858. IEEE (2015)
8. Tiwalkar, R.G., et al.: Model predictive controller for position control of twin rotor MIMO system. In: 2017 IEEE International Conference on Power, Control, Signals and Instrumentation Engineering (ICPCSI), pp. 952–957. IEEE (2017)
9. Christensen, R., Fogh, N., Hansen, R., Jensen, M., Larsen, S.A.: Paramanathan, Modelling and Control of a Twin-Rotor MIMO System. Aalborg University, Aalborg (2006)
10. TRMS 33-949S User Manual, East Sussex. Feedback Instruments, U.K
11. Chau, P.C.: Process Control: A First Course with MATLAB. Cambridge University Press, Cambridge (2002)
12. Lipták, B.G.: Instrument Engineers' Handbook. Vol. 2. CRC/Taylor & Francis, Boca Raton (2006). Open WorldCat, https://www.taylorfrancis.com/books/9781420064001

13. Wang, L.: Model Predictive Control System design and Implementation using MATLAB. Springer, Heidelberg (2009)
14. Camacho, E.F., Bordons, C.: Model Predictive Control. Springer, London (1999)
15. Kim, P., Huh, L.: Kalman Filter for Beginners: with MATLAB Examples. CreateSpace Independent Publishing Platform, Korea (2010)

Innovation in Biodegradable Textile Fibers for the Creation of Ecological Textiles

Ana C. Umaquinga-Criollo(✉), Cristina E. Molina-Martínez, Wendy M. Guerrero-Loyola, Tatyana K. Saltos-Echeverria, and Edgar D. Jaramillo-Vinueza

Universidad Técnica del Norte, 100105 Ibarra, Ecuador
acumaquinga@utn.edu.ec

Abstract. This research carries out a systematic, descriptive, documentary study in the textile area focus on biodegradability in textile fibers to know in this first approach the processes that have detected the material obtained by biological action to change and lose its original properties and chemical level decreasing in as much as possible the damage environment, benefic to the areas: textile, construction, agriculture, livestock, fish farming, among others. 40 relevant articles or review articles of interest of factor-impact Q1, Q2, Q3 are selected: 22 selected articles are Q1, 17 articles are Q2 and 1 articles Q3 of the last 10 years, which answered the research question asked on impact investigations in this area.

Keywords: Biodegradable · Fiber · Textile · Pollution · Ecology · Environment

1 Introduction

In recent years, according [1], environmental pollution has increased significantly all over the world. In this context, textile engineering has been the main cause of environmental damage, due to it produces a large amount of toxic waste to the planet. For this reason, environmental initiatives are being implemented in the textile industries to contribute to conservation of the environment.

The author [2], reveals that biodegradable products are currently very popular due to their ecological contribution. Based on this, the textile industry is innovating in creating biodegradable textiles to help reduce pollution rates. For this, synthetic fibers are being replaced by fibers of natural origin, which by its natural properties are easily degradable. Studies have been carried out in which some fibers such as jute, hemp, soybeans, cassava, among others, are modified and integrated, which mixed with other synthetic fibers manage to reduce their toxicity. These biodegradable fibers contribute to different areas to produce ecological and earth-friendly products.

Studies have been carried out in which some fibers such as jute, hemp, soybeans, cassava, among others, are modified and integrated, that mixed with other synthetic fibers reduce the toxicity. The investigation [3] reveals the physical textile properties of the fibers that are: tenacity of 100 MPa with an elongation at break of 30% and a modulus E of 0.5 GPa, While [4] talks about enzymes that can replace aggressive

M. Botto-Tobar et al. (Eds.): CI3 2020, AISC 1277, pp. 545–554, 2021.
https://doi.org/10.1007/978-3-030-60467-7_44

chemicals, catalyze the reaction, and operate under mild conditions [5]. It points out the many properties and versatility of cationic biopolymer chitosan which makes this polymer applicable to natural fibers [6]. Analyses the effect of light on bleached and dyed fabrics (jute/plain cotton and taladro, and doubts) and the dyeing properties of light-colored fabrics [7]. It states that special ether sulfonates allow to retain the antistatic effect on nylon fibers.

2 Materials and Methods

This study presents a descriptive and experimental analysis, beginning with a quantitative and qualitative observation from the fibrology area, biochemistry, weaving, dry cleaning, ecology, agribusiness, automotive, industrial, architecture, clothing and nanotechnology, thus it seeks to provide a first approach to the contribution of biodegradable textile fibers towards the fight against environmental pollution. The Scopus database is considered, followed by a systematic study of the information.

In the Scopus database, we selected 2 types of publications: articles and article review, considering articles from quartile Q1, Q2, Q3, due to the greater demand in their writing and in-depth information. With respect to the last 10 years, it is subsequently reflected in the search equation.

((TITLE-ABS-KEY (“biodegradable”) AND TITLE-ABS-KEY (“textile”))) AND (“fiber”) AND (LIMIT-TO (DOCTYPE, “ar”) OR LIMIT-TO (DOCTYPE, “re”)) AND (LIMIT-TO (SUBJAREA, “ENVI”)) AND (LIMIT-TO (PUBYEAR, 2020) OR LIMIT-TO (PUBYEAR, 2019) OR LIMIT-TO (PUBYEAR, 2018) OR LIMIT-TO (PUBYEAR, 2017) OR LIMIT-TO (PUBYEAR, 2016) OR LIMIT-TO (PUBYEAR, 2015) OR LIMIT-TO (PUBYEAR, 2014) OR LIMIT-TO (PUBYEAR, 2013) OR LIMIT-TO (PUBYEAR, 2012) OR LIMIT-TO (PUBYEAR, 2011) OR LIMIT-TO (PUBYEAR, 2010)).

In the results of this equation, 84 publications are obtained with a date cut January 16, 2020. 40 more relevant articles are selected for interest and quality of information, because this is a first approach to the study. Of the selected articles, 22 are Q1 while 17 are Q2 and 1 of Q3, which answer the research question.

In addition, articles with a historical sense are technically analyzed. In the Scopus database, 2 types of publications are selected: articles and review of historical articles, considering articles from quartile Q2 and Q4 due to the greater demand in their writing and deepening of the information, considering the years 1990s until 2001, it is later reflected in the search equation.

((TITLE-ABS-KEY (“biodegradable”) AND TITLE-ABS-KEY (“textile”))) AND (“fiber”) AND (LIMIT-TO (DOCTYPE, “ar”) OR LIMIT-TO (DOCTYPE, “re”)) AND (LIMIT-TO (SUBJAREA, “ENVI”)) AND (LIMIT-TO (PUBYEAR, 2001) OR LIMIT-TO (PUBYEAR, 2000) OR LIMIT-TO (PUBYEAR, 1999) OR LIMIT-TO (PUBYEAR, 1998) OR LIMIT-TO (PUBYEAR, 1997) OR LIMIT-TO (PUBYEAR, 1996) OR LIMIT-TO (PUBYEAR, 1995) OR LIMIT-TO (PUBYEAR, 1994) OR LIMIT-TO (PUBYEAR, 1993) OR LIMIT-TO (PUBYEAR, 1992) OR LIMIT-TO (PUBYEAR, 1991) OR LIMIT-TO (PUBYEAR, 1990)).

In the results of this equation, 18 publications were obtained, with court date of January 26, 2020. 5 most relevant historical articles are selected because of interest and quality of information, because this is a first advance of the study. Of the selected articles, 3 are Q4 and 2 are Q2, which answer the research question.

Below is a diagram that synthesizes the aforementioned information (Fig. 1).

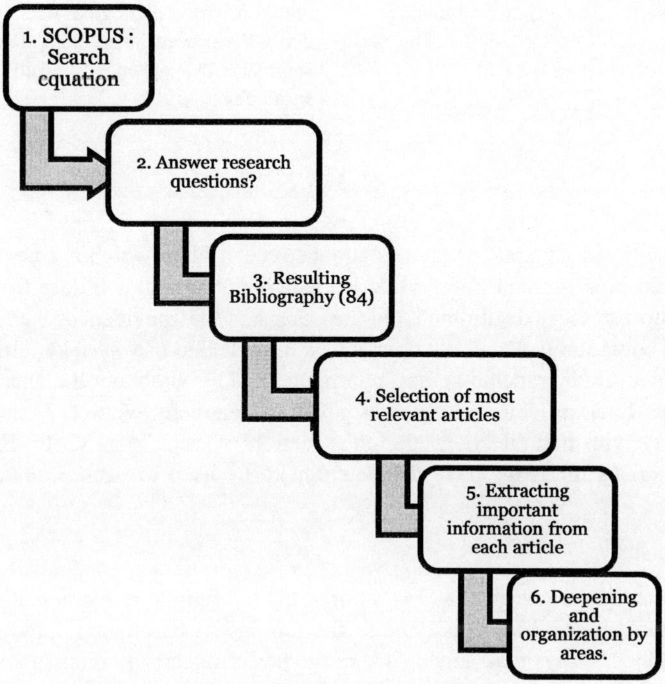

Fig. 1. Scopus search diagram

2.1 Research Questions

Two research questions are established to search for information in articles and reviews, referring to the research topic (Table 1).

3 Results and Discusion

A review is made of the 45 selected documents and their matrix. It has been noted that the most common problems performing research are: the platform in which we are working only works on the university campus, on the other hand, the documents in its entirety cannot be downloaded, and finally, certain articles do not have their DOI on the platform.

Initially, high-level bibliographies and research are reviewed and later a much more limited search is carried out to gain access only to research areas.

Table 1. Research questions

Research questions	Motivation
¿What research or literature of high impact there are in environmental area in the last 10 years?	Identify the articles that according to our criteria and their concepts are the most suitable to carry out the research
¿ what are the articles that research on biodegradable textiles and environmental benefits?	Identify the appropriate processes for obtaining good ecological fibers and their derivatives, used in multiple areas with semblants that are mostly positive, but also some negative

- Fibrology

Inside the fibrology area [8] reveals about soybean fiber, which is a new generation of regenerated protein plant fiber, while [9] shows the natural cellulose fiber extracted from the cotton stem by degummed. On the other hand [10] evidences the formation of total keratin compounds that could open a new approach to use wool keratin to produce sustainable and biodegradable keratin compounds. [11] points out the characterization of the banana fiber structure modified by alkaline treatment, while [12] shows how to reinforce clay with natural fibers, the construction of steep slopes, etc. Furthermore, [13] demonstrates that bags have the potential to be used to pack products with dry surfaces.

- Biochemistry

Within the biochemistry area [14] reports that the fatigue properties of compounds based on two biodegradable polymers: thermoplastic starch and filled polylactide mixture, while [15] says that satisfactory tissue performance and biodegradability from soy and glycerol proteins. On the other hand [16] demonstrates how soy protein sizing agents modified with cellulose nanocrystals are prepared. (CNC, [17] Informs about the in-situ synthesis of nanostructured MnO2 in natural fibers of fique, while [18] reports biodegradable sizing agents of triethanolamine modified soy protein (TEA) substitute for sizes of (PVA). [19] says that photodegradability is improved by adding titanium dioxide. [20] Reports on the growth of invasive and non-invasive macrophytes is determined by comparison with control transects, whereas [21] evidences that industrial tissue shows that TEA soy protein has a relative tissue efficiency 3% and 10% higher than PVA. Furthermore [22] shows that organic and plastic, the two main fractions of household waste vary considerably between different geographical areas, [23] Evidences that the cellulose fiber and the feed moisture of the mixture have a strong impact on the foam properties of cassava starch, although [24] reports that wheat gluten applied to polyester is necessary to replace poly (vinyl alcohol) (PVA). To the contrary, [25] concludes on the main bottlenecks that must be resolved to make the use of biocomposites more viable for food packaging applications.

- Fibrology in Marine Environment

Within the marine area [1] points out about the release of dangerous chemicals into an environment causing infertility on agricultural land and contamination of natural water sources, while [26] evidences that the environmental compartment of the water column was simulated under euphotic and apotic conditions. On the other hand, [27] developed a fast and non-invasive visualization monitoring platform for the hydrolytic degradation of biodegradable polymers.

- Weaving

Within the weaving area [28] reports on an enzyme-based process to achieve a surface pattern of the tissues as an alternative to conventional chemical processes, while [21] makes a series of optically transparent compounds using tunicated cellulose membranes.

- Dry Cleaner

Within the dry cleaning area [2] evidences that synthetic dyes are used intensively, a wide range of colors and shades, etc., while [29] says that indigo dyeing for cotton fibers achieves an ideal color depth in non-aqueous medium dyeing.

- Ecology

Within the area of ecology [30] shows that the traction, flexion and compression of 2# 3DWEC had better mechanical properties than 1# 3DWEC and 3# 3DWEC, while [31] points out the use of thermoplastics has caused an increase in pollutant emissions. Furthermore, [32] talks about polymer blending and curauá fibers are prepared by melt blending in an ex-twin screw extruder.

- Agroindustry

Within agribusiness, [33] reports on natural lignocellulosic biomass, which is a valuable raw material for soluble grade cellulose, while [34] talks about the polymerization through the formation of ester bonds occurred at elevated temperatures without any catalyst, and water was released as the only by-product.

- Automotive

Within the automotive area, [35] shows through the LCA method the possibility of using natural fibers replacing glass fibers, while [36] evidences the evolution of natural fiber compounds has a much stronger drive towards use of thermoplastic matrices.

- Industrial Engineering and Architecture

Within the industrial and architectural area, [37] evidences about geopolymers, which have become an ecological alternative to conventional construction materials, while [38] talks about the morphological analysis of the compounds revealed that the PBAT/PLA binder had more affinity with wood and textures.

On the other hand, [39] reports that cotton composite cellulose is successfully prepared from textile cotton, preforms woven using alkaline wash. [40] Evidence of contemporary and everyday composites with their valuable value as weight-saving materials, while [41] informs about the filling material for the pillows must be bulky, resistant, and comfortable.

- Clothing

Within the clothing, [42] informs about cotton threads, which are modified with pyrrole in the presence of ferric chloride.

- Nanotecnología

Within the nanotechnology area, [43] talks about soybean meal, it has a film flexibility of 30.1, 25.2 and 8.3% greater, thread adhesion and resistance to abrasion.

Finally, a consolidated statement of the above is presented. The 45 articles are divide in 8 area, the same ones are subdivided into different subareas (Table 2).

Table 2. Grouping of articles by areas

Área	Reference
Fibrology	[1, 8–13, 26, 27]
Biochemistry	[14–25]
Fibrology	[1, 26, 27]
Biochemistry	[21, 28]
Biochemistry	[2, 29]
Ecology	[30–32]
Agroindustry	[33]
Automotriz	[34, 35]
Engineering	[37–41]
Clothing	[42]
Nanotechnology	[43]

Therefore, it is concluded that each area and subarea have a certain number of articles:

- Fibrology (6)
- Marine environment (3)
- Biochemistry (13)
- Laundry (2)
- Weave (2)
- Agroindustry (2)
- Automotive (2)
- Ecology (3)
- Industrial engineering and architecture (5)
- Clothing (1)
- Nanotechnology (1)

4 Conclusions

This research makes a systematic study of information considering magazine articles from the first three quartiles, which ensures the quality of the information consulted. The contribution of this study is the identification of the alternatives proposed by the scientific community to reduce and mitigate the environmental impact from the area of biodegradable textile fibers.

The need to incorporate biodegradable products for the market has been stated, since it is increasing the amount of waste generated globally, and what is worse, that most of these cannot be eliminated.

The creation of biodegradable textile fibers helps with this problem because, as we know, biodegradation is a natural process that breaks down matter in a certain time. Therefore, by carrying out the correct and necessary biochemical processes that have been discovered, the manufacturing processes are faster and more efficient, making it possible to produce large quantities of biodegradable substitutes.

Awareness must be raised of the need to reduce pollution, for this reason it is being sought innovative and ecological methods in the textile area, which is the most polluting area. Especially in obtaining fibers and making products such as: clothing, plastic for food wraps, fishing nets, beverage containers, bags, among others. Not only must be awareness in the textile field, but also companies in general, must look for ways to create biodegradable products.

Future work

This study aims to be a first approach to the academic and business community on how to create ecological textiles. As future research work proposes the best alternatives already applied in real cases and achieving successful replication.

This first effort was the result of a classroom project that was obtained thanks to teamwork between teacher and students, which led to the development of knowledge and soft skills required in the personal and professional field. A key and priority aspect for achieving better knowledge and service to the community is to continue encouraging collaborative work among teachers - students and students – students, according to their level of study.

References

1. Gore, P.M., Naebe, M., Wang, X., Kandasubramanian, B.: Progress in silk materials for integrated water treatments: fabrication, modification and applications. Chem. Eng. J. **374**, 437–470 (2019). https://doi.org/10.1016/j.cej.2019.05.163
2. Beldean-Galea, M.S., Copaciu, F.M., Coman, M.V.: Chromatographic analysis of textile dyes. J. AOAC Int. **101**, 1353–1370 (2018)
3. Tendier, B.: Melt processing of a new biodegradable synthetic polymer in high-speed spinning and underpressure spunbonding process. J. Polym. Environ. **9**, 149–156 (2001). https://doi.org/10.1023/A:1020413321123
4. Doshi, R., Shelke, V.: Enzymes in textile industry - an environment-friendly approach. Indian J. Fibre Text. Res. **26**, 202–205 (2001)

5. Vulia, M.R., Canal, J.M., Jocic, D.: Green chemistry approaches to develop antimicrobial textiles based on sustainable biopolymers - a review. Ind. Eng. Chem. Res. **52**, 5245–5260 (1998). https://doi.org/10.1021/ie303627x
6. Bandyopadhyay, B.N., Salaskar, NR.: ECO-FRIENDLY TEXTILES
7. Wahle, B.: 1990 UNITED STATES PATENT, **12**, 2–4 (1999)
8. Yilmaz, D., Karaboyaci, M., Kiliç, H.: Comparison of selected properties of eco-friendly soybean and other fibres. Fibres Text East Eur. **23**, 14–24 (2015). https://doi.org/10.5604/12303666.1151694
9. Li, L., Zhao, L.: Nature cellulose fibre extracted from different cotton stalk sections by degumming. Fibres Text East Eur. **23**, 37–40 (2015). https://doi.org/10.5604/12303666.1167415
10. Fitz-Binder, C., Pham, T., Bechtold, T.: A second life for low-grade wool through formation of all-keratin composites in cystine reducing calcium chloride–water–ethanol solution. J. Chem. Technol. Bibotechnol. **94**, 3384–3392 (2019). https://doi.org/10.1002/jctb.6151
11. Barreto, A.C.H., Costa, M.M., Sombra, A.S.B.: Chemically modified banana fiber: structure, dielectrical properties and biodegradability. J. Polym. Environ. **18**, 523–531 (2010). https://doi.org/10.1007/s10924-010-0216-x
12. Abou Diab, A., Sadek, S., Najjar, S., Abou Daya, M.H.: Undrained shear strength characteristics of compacted clay reinforced with natural hemp fibers. Int. J. Geotech. Eng. **10**, 263–270 (2016). https://doi.org/10.1080/19386362.2015.1132122
13. Scheibe, A.S., De Moraes, J.O., Laurindo, J.B.: Production and characterization of bags from biocomposite films of starch-vegetal fibers prepared by tape casting. J. Food Process Eng **37**, 482–492 (2014). https://doi.org/10.1111/jfpe.12105
14. Liber-Kneć, A., Kuźniar, P., Kuciel, S.: Accelerated fatigue testing of biodegradable composites with flax fibers. J. Polym. Environ. **23**, 400–406 (2015). https://doi.org/10.1007/s10924-015-0719-6
15. Khiari, R., Dridi-Dhaouadi, S., Aguir, C., Mhenni, M.F.: Experimental evaluation of eco-friendly flocculants prepared from date palm rachis. J. Environ. Sci. **22**, 1539–1543 (2010). https://doi.org/10.1016/S1001-0742(09)60286-2
16. Xu, H., Yang, M., Hou, X., et al.: Industrial trial of high-quality all green sizes composed of soy-derived protein and glycerol. J. Clean. Prod. **135**, 1–8 (2016). https://doi.org/10.1016/j.jclepro.2016.06.041
17. Chacón-Patiño, M.L., Blanco-Tirado, C., Hinestroza, J.P., Combariza, M.Y.: Biocomposite of nanostructured MnO2 and fique fibers for efficient dye degradation. Green Chem. **15**, 2920–2928 (2013). https://doi.org/10.1039/c3gc40911b
18. Zhao, Y., Moser, C., Henriksson, G.: Transparent composites made from tunicate cellulose membranes and environmentally friendly polyester. Chemsuschem **11**, 1728–1735 (2018). https://doi.org/10.1002/cssc.201800627
19. Puls, J., Wilson, S.A., Hölter, D.: Degradation of cellulose acetate-based materials: a review. J. Polym. Environ. **19**, 152–165 (2011). https://doi.org/10.1007/s10924-010-0258-0
20. Hoffmann, M.A., Benavent González, A., Raeder, U., Melzer, A.: Experimental weed control of najas marina ssp. Intermedia and elodea nuttallii in lakes using biodegradable jute matting. J. Limnol. **72**, 485–493 (2013). https://doi.org/10.4081/jlimnol.2013.e39
21. Zhao, Y., Zhao, Y., Xu, H., Yang, Y.: A sustainable slashing industry using biodegradable sizes from modified soy protein to replace petro-based poly(vinyl alcohol). Environ. Sci. Technol. **49**, 2391–2397 (2015). https://doi.org/10.1021/es504988w
22. Miezah, K., Obiri-Danso, K., Kádár, Z., et al.: Municipal solid waste characterization and quantification as a measure towards effective waste management in Ghana. Waste Manag. **46**, 15–27 (2015). https://doi.org/10.1016/j.wasman.2015.09.009

23. Kaisangsri, N., Kowalski, R.J., Kerdchoechuen, O., et al.: Cellulose fiber enhances the physical characteristics of extruded biodegradable cassava starch foams. Ind. Crops Prod. **142**, 111810 (2019). https://doi.org/10.1016/j.indcrop.2019.111810
24. Chen, L., Reddy, N., Yang, Y.: Remediation of environmental pollution by substituting poly (vinyl alcohol) with biodegradable warp size from wheat gluten. Environ. Sci. Technol. **47**, 4505–4511 (2013). https://doi.org/10.1021/es304429s
25. Berthet, M.A., Angellier-Coussy, H., Guillard, V., Gontard, N.: Bio-based packaging. J. Appl. Polym. Sci. 133 (2016). https://doi.org/10.1002/app.42971
26. Beltrán-Sanahuja, A., Casado-Coy, N., Simó-Cabrera, L., Sanz-Lázaro, C.: Monitoring polymer degradation under different conditions in the marine environment. Environ. Pollut. **259**, 113836 (2020). https://doi.org/10.1016/j.envpol.2019.113836
27. Ma, H., Zhang, A., Zhang, X., et al.: Novel platform for visualization monitoring of hydrolytic degradation of bio-degradable polymers based on aggregation-induced emission (AIE) technique. Sens. Actuators B Chem. **304**, 127342 (2019). https://doi.org/10.1016/j.snb.2019.127342
28. Prajapati, C.D., Smith, E., Kane, F., Shen, J.: Selective enzymatic modification of wool/polyester blended fabrics for surface patterning. J. Clean. Prod. **211**, 909–921 (2019). https://doi.org/10.1016/j.jclepro.2018.11.079
29. Luo, Y., Pei, L., Wang, J.: Sustainable indigo dyeing and improvement of rubbing fastness of dyed cotton fiber using different fixing agents for obtaining eco-friendly cowboy products. J. Clean. Prod. **251**, 119728 (2020). https://doi.org/10.1016/j.jclepro.2019.119728
30. Meng, X., Fan, W., Ma, Y., et al.: Recycling of denim fabric wastes into high-performance composites using the needle-punching nonwoven fabrication route. Text. Res. J. (2019). https://doi.org/10.1177/0040517519870317
31. Nurul Fazita, M.R., Jayaraman, K., Bhattacharyya, D., et al.: Green composites made of bamboo fabric and poly (lactic) acid for packaging applications-a review. Materials **9**(6), 435 (2016). https://doi.org/10.3390/ma9060435
32. Harnnecker, F., dos Santos Rosa, D., Lenz, D.M.: Biodegradable polyester-based blend reinforced with curauá fiber: thermal, mechanical and biodegradation behaviour. J. Polym. Environ. **20**, 237–244 (2012). https://doi.org/10.1007/s10924-011-0382-5
33. Gallardo-Sánchez, M.A., Hernández, J.A., Casillas, R.R., et al.: Obtaining soluble-grade cellulose pulp from agave tequilana weber var. Azul bagasse. BioResources **14**, 9867–9881 (2019). https://doi.org/10.15376/biores.14.4.9867-9881
34. Zhao, Y., Xu, H., Mu, B., et al.: Functions of soymeal compositions in textile sizing. Ind. Crops Prod. **89**, 455–464 (2016). https://doi.org/10.1016/j.indcrop.2016.05.047
35. Alves, C., Ferrão, P.M.C., Silva, A.J., et al.: Ecodesign of automotive components making use of natural jute fiber composites. J. Clean. Prod. **18**, 313–327 (2010). https://doi.org/10.1016/j.jclepro.2009.10.022
36. Puglia, D., Sarasini, F., Santulli, C., Kenny, JM.: Challenges, potential and barriers for development. Green Biocomposites, pp. 13–29 (2017). https://doi.org/10.1007/978-3-319-46610-1
37. Silva, G., Kim, S., Aguilar, R., Nakamatsu, J.: Natural fibers as reinforcement additives for geopolymers – a review of potential eco-friendly applications to the construction industry. Sustain. Mater. Technol. **23**, e00132 (2020). https://doi.org/10.1016/j.susmat.2019.e00132
38. Muthuraj, R., Lacoste, C., Lacroix, P., Bergeret, A.: Sustainable thermal insulation biocomposites from rice husk, wheat husk, wood fibers and textile waste fibers: elaboration and performances evaluation. Ind. Crops Prod. **135**, 238–245 (2019). https://doi.org/10.1016/j.indcrop.2019.04.053

39. Grozdanov, A., Jordanov, I., Gentile, G., et al.: All-cellulose composites based on cotton textile woven preforms. Fibers Polym. **20**, 1243–1249 (2019). https://doi.org/10.1007/s12221-019-7970-0
40. Muhammad, A., Rahman, M.R., Hamdan, S., Sanaullah, K.: Recent developments in bamboo fiber-based composites: a review. Polym. Bull. **76**, 2655–2682 (2019). https://doi.org/10.1007/s00289-018-2493-9
41. Cafuta, D., Abu-Rous, M., Jary, S., et al.: Suitability of lyocell fiber for pillow fillings. Text. Res. J. **89**, 3722–3743 (2019). https://doi.org/10.1177/0040517518819844
42. Datta, M., Chaudhuri, A., Mitra, M., et al.: Development of biodegradable conductive cotton yarns by in-situ polymerisation of pyrrole. J. Text. Inst. **110**, 10–15 (2019). https://doi.org/10.1080/00405000.2018.1455312
43. Xu, X., Song, K., Xing, B., et al.: Thermal-tenacity-enhanced and biodegradable textile sizes from cellulose nanocrystals reinforced soy protein for effective yarn coating. Ind. Crops Prod. **140**, 111701 (2019). https://doi.org/10.1016/j.indcrop.2019.111701

Design in Exoskeleton Software for Lifting 50 Kg

Harry Arias Realpe(✉), Daniel Isaías Barzallo Núñez, David Trajano Basantes Montero, Daniel Andagoya Alba, and Lenin Merino

Instituto Superior Tecnológico Central Técnico, Quito, Ecuador
harias@istct.edu.ec

Abstract. The design of the exoskeleton HEXO is based on movements required by a per-son to walk and to lift heavy weight objects. This design contemplates just one required Degrees of Freedom (DOF) in each joint to complete these tasks. The design includes servo linear actuators to generate the translation of user and lift-ing required forces, because their nominal rate force is convenient in each action and were located to work without interfere with the normal performance of other limbs. Rotational servo actuators offered in the market were analyzed and includ-ed in a very first design, but they demonstrated to be non-convenient because their low generated torque and their overwhelming size that would increase the HEXO′s weight and interfere the desirable performance. Within de mechanical system design is included the charge analysis in main parts, those that would support all the structure and heavy object weight. This analysis is made into the most critical performed position. It also shows mass center for a stability analysis in the HEXO′s walking most critical positions. Control design of the HEXO is shown in a flow chart to follow by linear servo actuator activation in each limb, into walking and lifting tasks, it includes the conditional and final position to reach.

Keywords: Hexo · Servo motor · Mechanical effort

1 Introduction

Currently the field of robotics has also focused on the search for mechanisms orient-ed to the user's operation. It is for this reason that the study and analysis of robots that can be used directly by man to enhance movements, replace limbs, complement or replace basic human functions.

As the basic movement capacity of the human being is the weightlifting that by its own physiological characteristics is limited by physical damage that the person can suffer, both in back and in limbs, by moving objects that exceed the person's carrying capacity. Although it depends on the frame and size, there is no single load limit to lift; however, as suggested by occupational safety agencies, a weight between 20 and 25 kg is difficult for the vast majority of people to carry [30].

There are jobs that affect the conditions of the musculoskeletal system, due to continuous weightlifting, is one of the main work-related health problems. In addition,

M. Botto-Tobar et al. (Eds.): CI3 2020, AISC 1277, pp. 555–570, 2021.
https://doi.org/10.1007/978-3-030-60467-7_45

several of the industrial tasks require the use of forklifts for lifting, transport and placement of equipment, machinery elements, engines, materials, etc., which do not weigh even 5% of the working capacity of that vehicle and are would consider its use unnecessary, as it represented a waste of resources and would be allocated to more important tasks.

The design of an exoskeleton to lift loads up to 50 kg optimizes the execution time of this type of task, since they do not take up much space, help to execute functions without hindering other processes and without compromising the health of those who operate it, reducing the risk of back injuries, low back pain, or cumulative disorders due to deterioration of the skeletal muscle system. In addition, indirect contact of the exoskeleton operator with the load reduces the risk of trauma, cuts, fractures or burns due to accidents caused by destabilization of the load or poor grip of the load.

The same operator can raise objects several times in the same day without being exposed to so much occupational risk, favoring the acceleration and facilitating the fulfillment of certain tasks in the industrial field, in the military and domestic sphere. For these reasons the objective of the present work is to design, analyzing and simulating in software an exoskeleton to lift a maximum weight of 50 kg.

To achieve the general objective mentioned above, the specific objectives to be achieved this are:

- Determine the kinematics of movement and weightlifting in a person.
- Design a mechanism that allows to execute the most important degrees of freedom in movement and weightlifting.
- Design the electronic system that allows movement and control in the designed mechanical parts.
- Analyze mechanical components by simulating the exoskeleton.

1.1 Exoskeleton

Exoskeletons are described as orthotic robots that allows the user to boost strength, range, speed, etc. They are also characterized because the human-robot physical interface (pHRi) is fundamentally based on the application of internal forces directly from the user to the device. Exoskeletons whose kinematic chain is applied to upper limbs, lower extremities, or applied to the entire body are currently designed.

1.2 Legislation in Ecuador

As stated above, it depends on the person and their context, and you cannot set a single value as maximums allowed. In Ecuador, Decree No. 2393 applies, which is the "Regulation on the Safety and Health of Workers and Improvement of the Working Environment" to establish limits on the burden that a person can place to reduce or eliminate the risks involved in excess burden on a person.

According to Title IV, Chap. 5, art. 128, on material handling, the maximum weight to be lifted by a person is as follows:

Men up to 16 years 35 lb.

- Men aged 16 to 18 years 50 lb.
- Men over 18 years up to 175 lb.
- Women up to 18 years 20 lb.
- Women aged 18 to 21 years 25 lb.
- Women 21 years and older 50 lb.

In addition, it makes clear that a worker is not allowed to carry weights that may compromise his health or safety and must wear appropriate personal protective clothing to perform the activity. The design of an exoskeleton that allows to lift 50 kg of weight ensures that workers can perform physical demand tasks without being exposed to some type of injury, as this ensures that the maximum effort generated by each user does not over demand the physical ability. Prolonging the health and significantly decreasing the cumulative disorders that this task can involve. In addition, load lifting allows this task to be performed by a single person, reducing the risk of more workers being injured.

2 Methods and Materials

For the design of the exoskeleton, a methodology based on concurrent techniques applied in the design of the robot proposed by Méndez in 2003 was used, which allows returning from any design stage in order to correct, debug and improve the final product. Although this methodology is applied for manufacturing purposes up to the scale production stage, this work was only considered until the testing and refinement stage. The steps of the methodology are shown in Fig. 1.

Fig. 1. Project stages.

2.1 Planning

Within the planning are specified the requirements and restrictions to be considered in the design of the exoskeleton.

Design requirements

Requirements are identified in different groups (Table 1) :

Table 1. Design requirements.

Groups	Description
Functions:	Forward movement Flexion in lower extremities (knees, ankles and hips) Flexion in upper extremities (elbows and shoulders) Load lift up to 50 kg
Materials:	Impact resistant (tenacity) Corrosion resistance Cutting resistance Hardness Machinability
Design:	Ergonomic Restricted Mobility (At scroll angles) Linear and rotational actuators High strength
Energy:	Extended operating time
Control:	PLC programming Pushbuttons for movement start Continuous movements until the end of the subprogram
Security:	Stability Clamping elements for the user

Design Restrictions

Exoskeleton constraints need to be determined for design:

- Limitation in GDL's to avoid instability.
- Forward movement. It doesn't move laterally.
- Loads not greater than 50 kg.
- Low execution speed and translation to promote balance.
- Only for solid loads with CG in the center of the element.

2.2 Conceptual Development and General Diagram

Basically, the operation of the exoskeleton as a system is to obtain at its output the flexion of the joints that are designed and lift a load from an input force applied by the user. The diagram of the exoskeleton system is shown in Fig. 2.

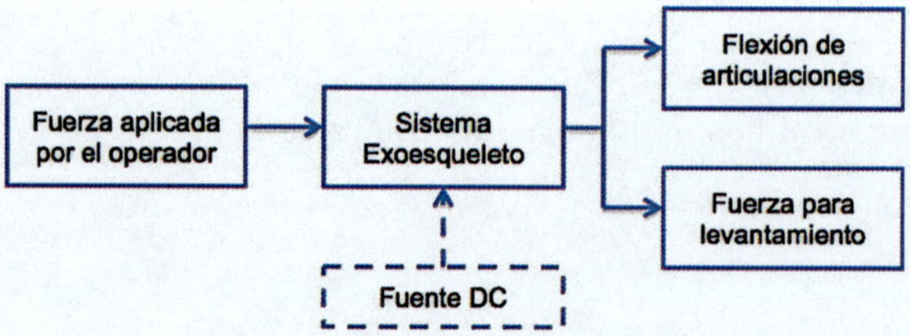

Fig. 2. General system diagram

As a resource, for the operation of this system, a power source is required to feed the mechanical, electronic and sensory subsystems of the exoskeleton.

2.3 Development at the System Level and Specific Diagram

Considering the general diagram applied to this design, the interrelation of the subsystems is presented in Fig. 3.

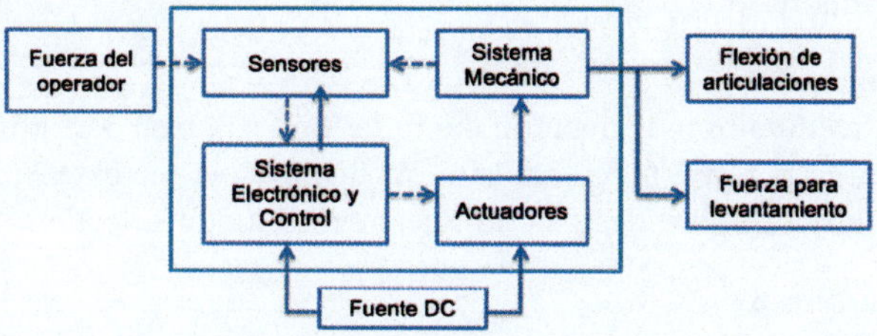

Fig. 3. General system diagram

The system input that is defined as a variable is the force applied by the operator. The basic system proposed in the structure of the Verein Deutscher, corresponds in this design to the mechanical subsystem that includes the support structure, limbs, joints and support devices. The processing of the information will be done on the selected driver. Being a mechatronic system, sensors and actuators will be interrelated through this controller.

Verification

The force generated by the user by pressing the button located on the end to make contact quickly and to send the signal to the PLC. In accordance with what is programmed in the controller, an activation signal is sent to the actuator to indicate its linear positioning, which generates the necessary force for the mechanical system to come into action, perform the required flexion in the joint and cause the movement.

To generate the walk, two synchronized signals are sent, one for the linear servos and the other for the rotational motors, so that the forward movement of the legs of the exoskeleton is generated.

For the lifting of the exoskeleton, the mechanical system will be detected by means of a linear encoder that receives the activation signal, increases the current in the servo and in the rotational movements of the thighs in this way greater force and torque will be obtained.

Validation

For the validation of the electronic system, the simulation was performed verifying the signals sent from the sensors to the controller. For the validation of the mechanical system, a simulation was performed by designed parts; which are the diagram of a limb, the support structure and the hip, obtaining the behavior and dynamics of these parts.

Within the design, the interrelationship between the main systems, the mechanical system and the electronics must be made. It is essential to check the behavior of the actuators according to the load to be lifted, including a safety factor to ensure at the time of operation.

2.4 Detailed Design

The detailed design is divided into two parts, the mechanical design and the exoskeleton assembly.

Mechanic system
The degrees of freedom of movement in the limbs that a person performs were reduced to those required to perform a vertical lift. For this reason, the extension and bending functions are selected in part design and linear servo motor selection.

Hexoskeleton assembly
The assembly of all the elements of mechanical design, allows to obtain an exoskeleton structure that does not present collisions between its parts when it performs some type of movement.

All the elements are joined to the corresponding parts using pins between the sections of the extremities, and by cavities with FAG 6008 bearings in the elements to rotate, as is the case of the servos and the extremities of the motor, see in Fig. 4.

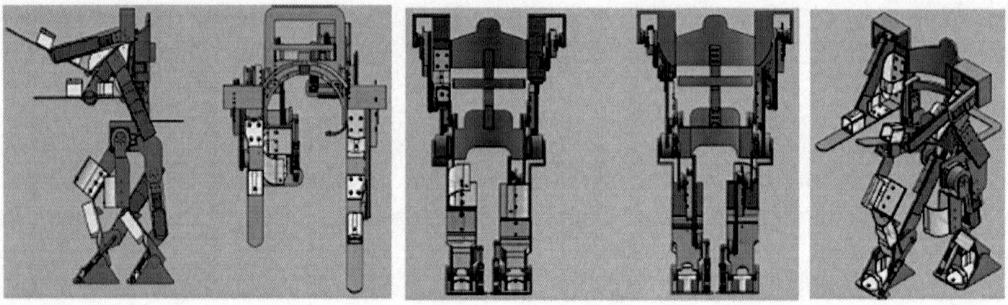

Fig. 4. View of the Hexo.

Electronic and control system
The electronic system uses buttons that are exposed to the user to activate the necessary movements and perform lifting and travel actions.

Selection of linear motor servo motors
According to the calculations performed, the maximum force required for the lifting of 50 kg is presented in section C of the back, with a total magnitude of 1425.51 N and an angle of action of 5°, see in Fig. 5.

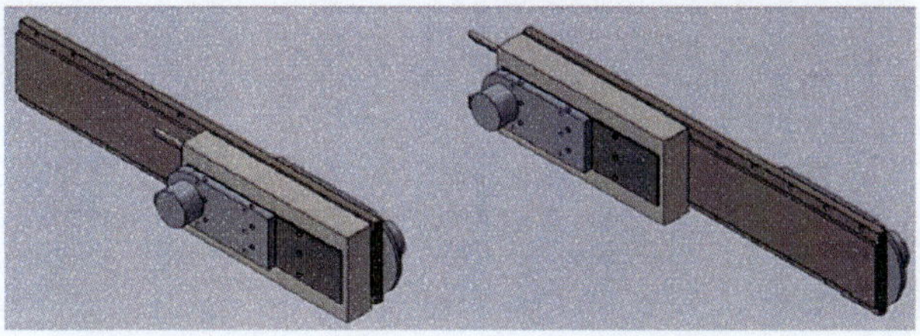

Fig. 5. Linear servo motors in limit positions

Selection of servo rotary motors

The thigh element includes a rotational motor servo that complements the rotation required for full movement. According to the calculations obtained, 19.42 N.m. is required to keep section C stable when performing weightlifting actions.

To determine the maximum weight to support of the rotational motor servo, according to the load time value, Eq. 1 is set: [32].

$$M_0 = F(L - A) \tag{1}$$

Where:
Mo: Time of Charging = 70
F: External force 3465.35 N
L: Load length = 0
A: Distance from the contact surface and bearing = 0.0202

Connections and algorithms

The amplifiers connect to each other via an optical cable with the SSCNET III/H network, with the master module being the Q172UDHCPU motion controller, which determines which amplifier is to command certain servos. Figure 6 shows the functional diagram of the Hexo, see in Fig. 6.

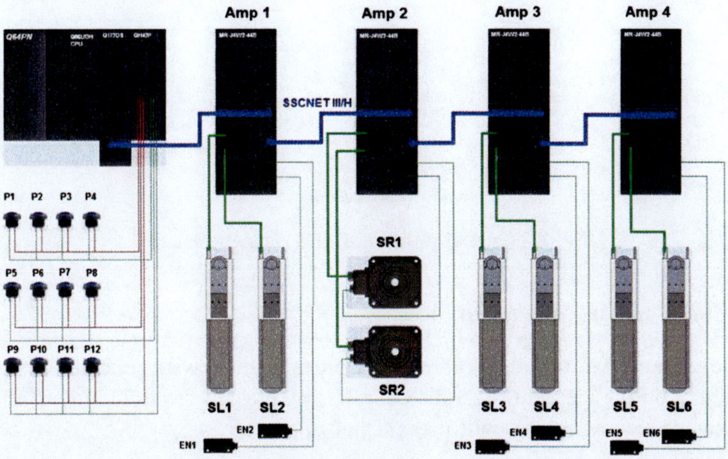

Fig. 6. Functional diagram of the Hexo.

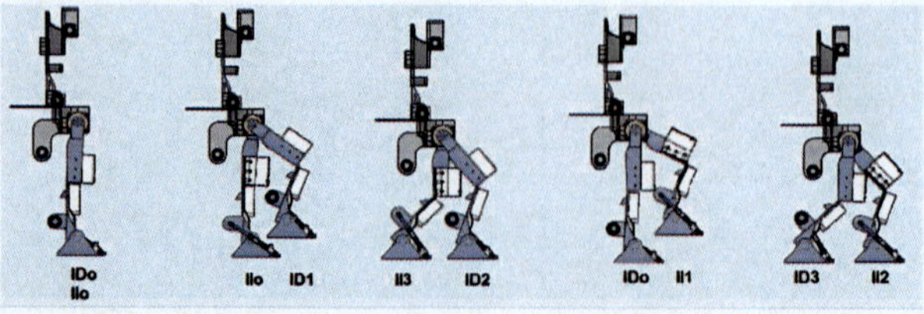

Fig. 7. Leg positions in exoskeleton displacement

A name is assigned to each of the positions that will be reached in the Hex, both in displacement and in rising. see in Fig. 7.

3 Analysis of Results

Using SolidWorks™ software analysis tools, each element of the structure is subjected to the different loads (forces) acting on it. The loads to be applied are calculated according to the most critical position to which the Exoskeleton will be subjected, including a weight to be loaded located at the end of the forearm.

The weight of the element itself is also considered, as a part of the user's body supported by this structure (The weight of each parts are calculated according to the percentage of mass corresponding to a body mass of the user of 80 kg) and the resulting load of weight linear motors servos if significant.

The results obtained, by applying the SimulationXpress™ tool, are displayed graphically with a color scheme indicating the stress variations in the element and the displacements that will be presented due to all loads. In addition, the expected deformation of the element is displayed.

The most critical positions of the Exoskeleton are considered those used for lifting the load, with the arms outstretched and legs collected, as shown in Fig. 8.

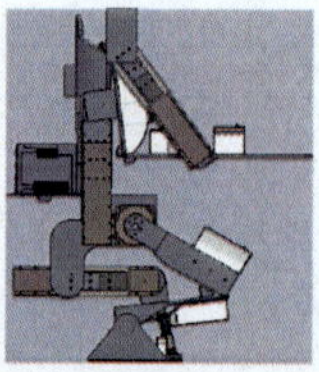

Fig. 8. Critical position of the Hexoeskeleton.

3.1 Analysis and Results on the Forearm

It is made an analysis of all participating forces, which will generate a stress and bending effect on the element. For this analysis, is selected the right side of the Hexo, since the results by symmetry will present the same values.

A weight of 25 kg is considered over the end of the forearm assuming a balanced distribution of the total weight of 50 kg between the two limbs. The following weights are presented for the analysis:

- Wab - Weight of forearm Structure - 1.2 kg.
- Wbh - Weight of the user's forearm - 1.84 kg.
- Wload - Load Weight - 25 kg.

Stress analysis

Properties of 6061-T6 aluminum throughout the skeleton:

- Volume: 0.000586292 m3.
- Density: 2700 kg/m3.
- Weight: 11.76 N.
- Elastic limit: 2.75and8 N/m2.
- Traction limit: 3.1and 8 N/m2 (Table 2).

Table 2. Forearm results

Simulation results	
Von Mises Stress Results: Minimum Effort: 3,93981 N/m2 in Node 888 Maximum Effort: 138526 N/m2 in Node 106056. Von Mises's efforts top view (a) and bottom (b) forearm	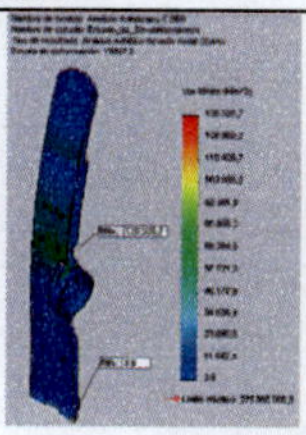 (a) 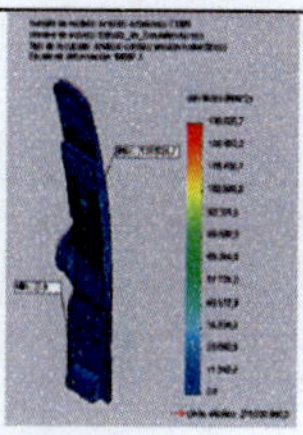 (b)
Displacement results: Minimum offset: 0 mm in 98 node; Maximum offset: 0.00313248 mm in node 102103.	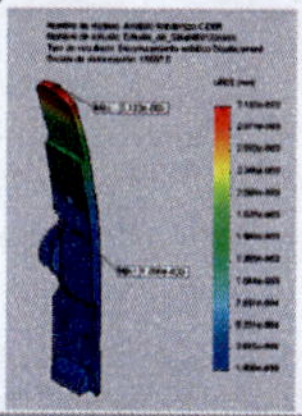
Factor of safety: Minimum factor: 1985.19 in node106056; Maximum factor: 6.98004e7 in node 888	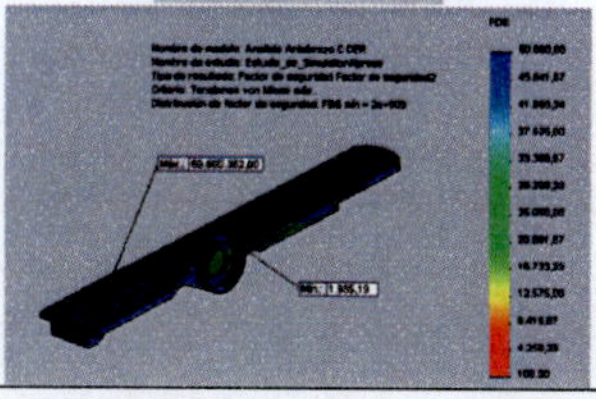

3.2 Analysis and Results in the Arm

Taking the values of forces obtained in the forearm joint with the arm, the acting forces that will generate bending on the element and the resulting force are deter-mined. The analysis is done again on the right side of the Hexo. The following weights are presented for the analysis:

- Wbr - Arm Structure Weight - 1.50 kg.
- Wbrh - Weight of the User's Arm - 2.16 kg (Table 3).

Table 3. Results in the arm

Simulation results	
Stress analysis. Applied forces: Wbrh - N 21.16 Wbh - 14.70 N Is ' = 135.13 N Ax - N543.98 F - 696.42 N	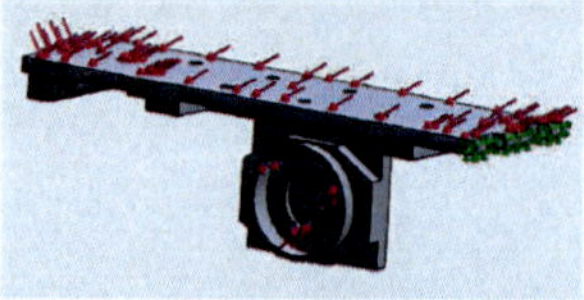
Stress Results: Minimum Effort: 3707.79 N/m^2 at Node 61721 Maximum Effort: 1.17533e8 N/m^2 in Node 99271	 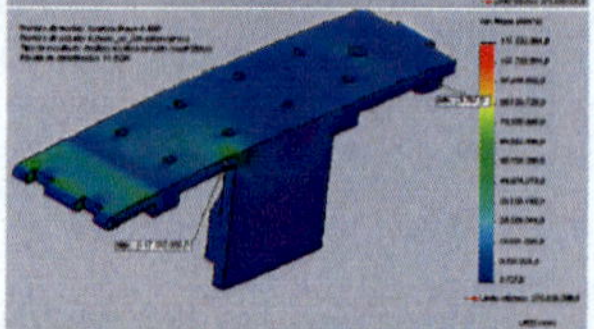
Displacement results: Minimum displacement: 0 mm in 309 nodes Maximum displacement: 3.17329 mm in node 57	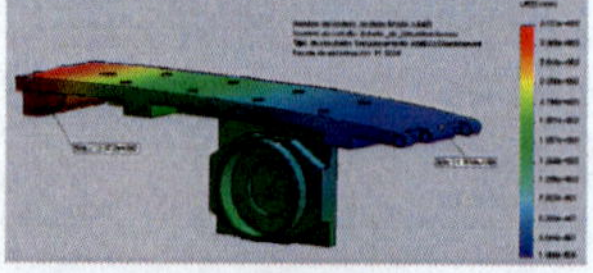
Safety factor: Minimum factor: 2.33977 in node 99271 Maximum factor: 74168.2 in node 61721	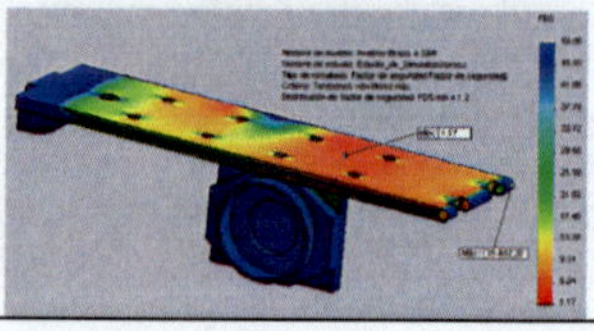

3.3 Force Analysis of the Linear Forearm Servo Motor

The actuating forces on this linear motor servo are analyzed to obtain the resulting ones that will act on a part of the back structure. These will be the forces that the servo will have to exert to achieve mobility in the forearm element. The analysis is continued with the elements belonging to the right side. The weight to act in this analysis is the same as the linear motor servo:

- Ws - Weight of the servo motor - 9.0 kg.

3.4 Analysis and Results in Section "A" of the Back

When considering the symmetry of the design, the values of the forces to be analyzed are the same on both the right and left of the Hexo. In addition, in this element converge the resulting forces in the linear motor servo that gives mobility to the forearm, and the support of the arm element. In the part of the support you get the resulting that will be the same force that acts on section B of the back.

This element supports the weight of the elements of the upper limbs and does not support any weight exerted by the user since this only serves as support of the upper back. The weight considered for this analysis is:

Wea - Weight of The Structure of Section A of the Back - 6.0 kg (Table 4).

Table 4. Results in section "A" of the back

Simulation results	
Applied forces: Wea = 58.80 N; Ay = 758.92 N; Ay2 = 758.92 N Ax - 320.69 N; Ax2 - 320.69 N Fx - 551.05 N; Fx2 - 551.05 N Fy = 488.56 N; Fy2 = 488.56 N	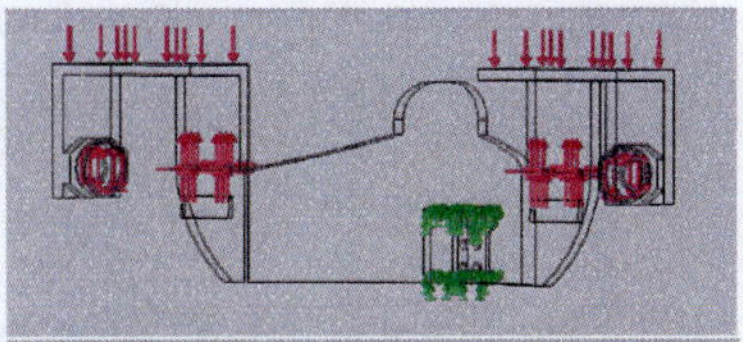
Stress Results: Minimum Effort: 21.35 N/m2 at Node 21598 Maximum Effort: 8,74642e7 N/m2 at Node 392	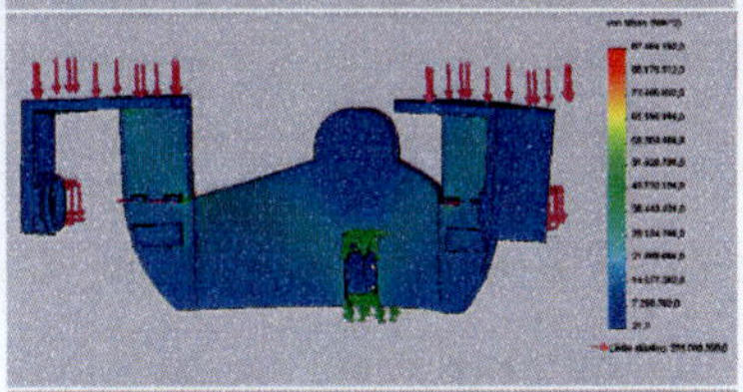
Displacement results: Minimum displacement: 0 mm in node 15 Maximum displacement: 3.26859 mm in node 18501	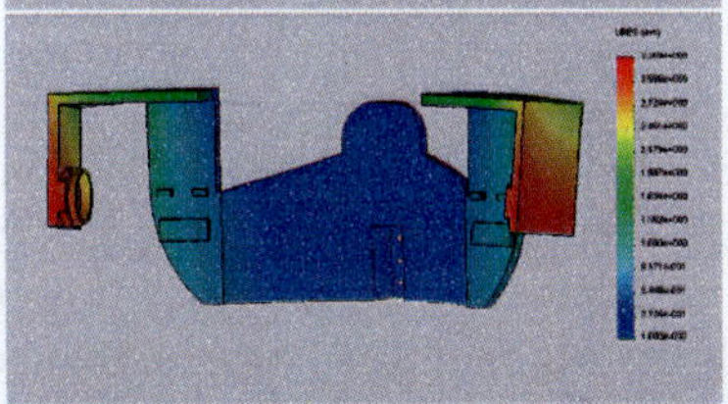
Safety factor: Minimum factor: 3,14414 in node 392 Maximum factor: 1.28806e7 in node 21598	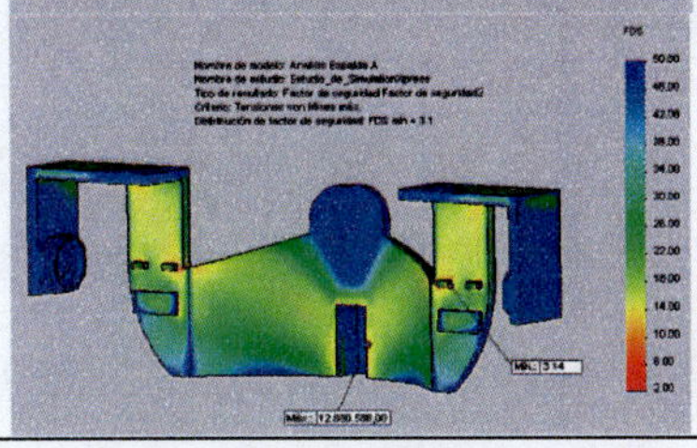

3.5 Analysis and Results in Section "B"– of the Back

This element represents the vertical support of the Hexo. The actuating forces are applied in the coupling zones with the other sections axially. The analysis tempers the loads generated in the previous elements and their consequent action on this element.

The lower part is considered as a clamp for analysis and contain the forces that will act on the next element of the back. The proper weight of this element is applied in the lumbar area of the design for the incidence in this analysis.

Web - Weight of The Structure of Section B of the Back - 2.0 kg.

3.6 Analysis and Results in Section "C"– of the Back

The driving forces due to the linear servo motors of the arms and pins are taken into account, and the servo motors applied to the thighs are presented in equal magnitude on both sides of the structure; Because of this, the analysis is performed with the acting forces on one side.

Also, the forces exerted by the weight of the entire electronic control system, which is applied in the center of mass of the designed tray, and by the weight of the same structure, which will be applied on top of this element for analysis in software.

In the part of the element where the rotating servo motor is assembled, the calculation of the time to be made to allow the rotation of the thigh element is carried out, in such a way that it complements the application of the force of the linear servo motor in the bolt to lift the Hexo.

This calculation of the resulting moment is previously presented in the x and y axes, in the z axis no additional moment will be needed because between the two sides they are counteracted by symmetry, allowing Hexo stability to occur only in the 2 main planes. The weights that act on this structure are the following:

Wcn - Weight of the electronic control system - 7.1 kg.

Wec - Weight of the structure of the C section of the back - 8.0 kg.

3.7 Analysis and Results in Thigh

At the top of this element act the forces obtained from the previous analysis in section C of the back. The analysis is performed on each of the perforations for the clamping bolts between this element and the rotational motor servo; these forces are applied in a distributed manner and in equal value on the inside of each drilling.

The analysis considers the weight of the thigh element, applied in the center of mass determined by the software. This element is designed in Aluminum 6061 T6. The weights acting on this structure are as follows:

Wm - Weight of the thigh structure - 1.5 kg.

3.8 Analysis and Results of the Support

In this element operates the forces generated by the linear servo motor and by the upper structure of the Exoskeleton. It is in this that there is supported to maintain the weight and position of the entire Exoskeleton, since it is closer to the floor and allows the user to bend with the entire structure.

When the Exoskeleton is straight, the design allows the base of the bobbin to rest on top of the foot.

The second form of support occurs when the Exoskeleton flexes to bend in lifting actions, in which case it is supported by two joints that fit into two cavities designed in the foot and which is also compatible with the additional mechanical resistance device when lifting actions are performed.

On the back, the FAG 6008 bearing cavity is designed, which allows the free rotation of the end of the linear motor servo that generates the movement of the lower leg. In this part the forces previously determined in the calculation section of the linear motor servo of the chin are acted.

A pin is designed as part of the piece in which the support element fits. In this element acts the normal force by the support generated against the floor. The weights acting on this structure are as follows:

Wc - Weight of the thigh structure 2.5 kg.

3.9 Analysis and Standing Results

In this element converge the forces generated by the linear motor servo and by the element of the quill; In addition, this element presents the normal reaction to the weight of the entire Hexo. Unlike other elements of the lower limbs, it is affected by the user's weight, the magnitude that is applied in stress analysis.

In the assembly part between the quill element and the foot, the Fx and Fy forces, determined in the previous section, act.

The forces to determine in this analysis are those of floor reaction on the floor of the element of the foot and the friction. The rest of the forces are those transferred by the previous elements. The weight of this is applied in the center of mass.

The weight of the user and the weight of the foot element are considered and the values of these acting forces on this structure are as follows:

Wp - Weight of foot structure - 3.0 kg.

Wh - User Weight - 80.0 kg.

3.10 Analysis of Results

In section D of the back, the value of the maximum effort is quite high, with a value of $6.17e^7$ N/m^2 and, due to its design, it appears in the area where it intersects with the lateral protection. Although it is a high value, it does not exceed the limit stress value of $2.75e^8$ N/m^2, so the support of this structure is ensured.

In the part of the foot, where the support element is assembled, although in the stress simulations a deformation of the element is shown, the value of the maximum stress is $1.43e^7$ N/m^2. The security of the support element is guaranteed, due to the characteristics of the material, providing a minimum safety factor of 19.14.

The back-section C is the main part of the Hexo because most actuators converge on it. In the analysis of the rotational moment, the required value is determined so that the structure does not rotate and cause Hex instability; This is why it is complemented by the use of rotational servo motors, which allow the thigh element to rotate when carrying out translational actions and complements the stability of the system when performing lifting actions. These rotational servo motors continuously generate a moment opposite to that caused by the forces acting on the entire Hexo.

In this same section C of the back, the tray that houses the electronic modules and amplifiers is complemented. The analysis considers the weight that this performs, and the value of the maximum effort of $2.97e^7$ N/m^2. Although it seems that it does not support the weight of the entire electronic system, there is no evidence of greater displacement of the tray, which ensures an adequate hold that keeps all the modules in place. For added security, the modules are bolted against the tray. The design of the structure allows the wiring to be simple and there are no tangles between cables or with the elements of the structure.

The location and displacement of the linear servo motor is designed considering the minimum positions to avoid element jams. In the critical lift position, when the Hexo is crouched, the bobbin linear servo motor is angled at 2° so that it can act on both axes of the lateral plane. If the motor becomes 0°, the system freezes and could not go up. To avoid reaching this position, safety stops are implemented in certain parts, in this way the maximum position of the extremities is protected so that it does not move beyond the limit and cause damage to the user, and so that the Hexo does not lock. In the linear actuators of the arms, this drawback is also treated with stops to protect the movement. The foot element requires that the surface on which it rests keep a coefficient of friction of 0.197 to ensure the stability of the Hexo. If the surface in which it is being used has a coefficient of friction less than the calculated value, the risk increases that the entire structure will destabilize and fall.

4 Conclusions

- For the purpose of the Hexo it is not necessary to design articulations and elements that emulate exactly the same degrees of freedom as a person, they are limited to selecting the degrees of freedom required to execute the necessary actions. It is concluded that the limbs only require flexion and extension to be able to move and lift weights, leaving aside for design adduction and abduction movements, and pronation and supination.
- Values thrown by the Simulation Xpress application show that each designed mechanical element performs well when subjected to extreme weight load forces. The displacements of each element are minimal, and although the displacement values of the arm element and section A of the back are around 3 mm, each ele-ment remains in place and has no representative deformations that involve mis-matches or hurdles in the Hexo system. Each of the elements maintains its shape and design characteristic without affecting the objective of the entire structure.
- The initial design undergoes many changes due to the characteristics of the first proposed linear motor servos.
- The maximum stress values obtained on each element ensure that the elements are not faulty when they are working.
- None of the elements have maximum stresses that exceed the yield limit.
- The safety factor on passes the value of 2 (being 2.33 in the element of the arm), which confirms the improbability of failures in the structure of the Hexo. As there are no faults, these elements are selected to be part of the hexo's final design.

- The electronic design presents conditionings in the programming of the execution of actions in the lower extremities. The posture is required to be safe before lifting and translation actions are carried out, because the stability of the entire Hexo can be compromised when trying to take actions under unprofitre conditions. The upper limbs do not need to be conditioned, because they can move freely and do not compromise the stability of the Hexo, nor does it jeopardize the execution of the actions of arms and forearms.

References

1. Aoki, E., Manabe, S., Konosu, H., Imaida, M., Nakashima, I., Elbihara, Y.: Leg assist device. United States patent, pp. 1–9 (2015)
2. Bolton, W.: Mecatrónica, Sistemas de Control Electrónico en la Ingeniería Mecánica y Eléctrica, 4ta edn. Alfaomega, México (2010)
3. Catálogo de rodamientos FAG WL 41 520/3 SB. FAG. Recuperado el 07 de abril de 2016 de (2000). http://ingemecanica.com/tutorialsemanal/objetos/figutut214/FAG.pdf
4. Chen, J., Liao, W.: A leg exoskeleton utilizing a magnetorheological actuator. In: Smart Materials and Structure Laboratory, PP. 824–828 (2006)
5. Chen, Y., Li, G., Zhu, Y., Zhao, J., Cai, H.: Design of a 6 DOF upper limb rehabilitation exoskeleton with parallel actuated joints. Bio-Med. Mater. Eng. **24**(6), 2527–2535 (2014)
6. Duncan, L.: Ergonomics and the Management of Musculoskeletal Disorders, 2dn edn. Elsevier, United States (1997)
7. Floyd, R.T.: Basic biomechanical factors & concepts. Strucutral kinesiology. Recuperado el 10 de marzo de 2016 de. http://www.kean.edu/~jeadams/docs/Kinesiology/Kines_Power_Points/Kines_PPT_PDF_Chap3.pdf. (s.f.)
8. Glowinski, S., Krzyzynski, T., Pecolt, S., Maciejewski, I.: Design of motion trajectory of an arm exoskeleton. Arch. Appl. Mech. **2015**, 76–85 (2014)
9. Guo, Q., Li, S., Jiang, D.: A lower extremity exoskeleton: human-machine coupled modeling, robust control design, simulation, and overload-carrying experiment. Math. Probl. Eng. **2015**, 2–9 (2015)
10. Hibbeler, R.C.: Mecánica de Materiales, 8va edn. Pearson Prentice Hall, México (2011)
11. Kazerooni, H.: Berkeley Lower Extremity Exoskeleton BLEEX. Recuperado el 15 de abril de 2015, de. http://bleex.me.berkeley.edu/wp-content/uploads/helmedia/images/CV/BLEEX-Summary.pdf. (s.f.)
12. Li, Y., Guan, X., Tong, Y., Xu, C.: Design and simulation study of the translational-knee lower extremity exoskeleton. Mechanika **21**, 207–212 (2015)
13. Linear shaft motor. Nippon pulse. Recuperado el 20 de enero de 2016, de (2016). http://nipponpulse.com/products/overview/linear-shaftservomotors
14. Louie, D.R., Eng, J.J., Lam, T.: Gait speed using powered robotic exoskeleton after spinal cord injury: a systematic review oand correlational study. J. Neuroengineering Rehabil. **2015**, 3–6 (2015)
15. Megatorque motor. NSK. Recuperado el 17 de enero de 2016 de (2009). http://onexia.com/nsk/pdf/NSK%211_PS_and_PN_Motor_Catalog.pdf
16. Megatorque motor system user´s manual. NSK. Recuperado el 19 de enero de 2016, de (2003). https://www.jp.nsk.com/services/download/manualpdf/C237–01.pdf

17. Melsec Q series motion controller programming manual. Mitsubishi. Recuperado el 20 de enero de 2016 de. http://dl.mitsubishielectric.com/dl/fa/document/manual/plc/sh080809eng/sh080809engs.pdf. (s.f.)
18. Melsec Q series motion controller users manual. Mitsubishi. Recuperado el 21 de enero de 2016 de. http://dl.mitsubishielectric.com/dl/fa/document/manual/plc/sh080007/sh080007o.pdf. (s.f.)
19. Melsec Q series users manual. Mitsubishi. Recuperado el 11 de enero de 2016 de (2004). https://www.plcsystems.ru/catalog/Mitsubishi/doc/SystemQ_MultipleCPU.pdf
20. Melsec system Q beginners manual. Mitsubishi. Recuperado el 12 de enero de 2016 de (2007). https://www.mitsubishi/mitsubishi-q-seriesbeginners-manual.pdf
21. System Q user's manual. Mitsubishi. Recuperado el 12 de enero de 2016 de (2007). https://www.plcsystems.ru/catalog/Mitsubishi/doc/SystemQ_DigitalIO_Manual.pdf
22. Micro actuator MA-35. Pi. Recuperado el 25 de enero de 2016, de (2014). http://www.pi-usa.us/products/Linearactuator-linearpusher/precision_actuator_overview.php
23. Mott, R.L.: Diseño de Elementos de Máquinas, 4ta edn. Pearson Prentice Hall, México (2006)
24. MXB rodless belt drive actuators. Tolomatic. Recuperado el 18 de enero de 2016, de (2015). http://www.tolomatic.com/products/productdetails/mxb-u-unguided-belt-driven-actuators#/
25. Navas, F.: Diseño mecatrónico aplicado a la robótica. UNAB, pp. 1–4 (2011)
26. Norton, R.: Diseño de Maquinaria, Síntesis y Análisis de Máquinas y Mecanismos, 3ra edn. Mc Graw Hill, México (2008)
27. Pons, J.: Wereable Robots, Biomechatronic Exoskeletons. Wiley, United States (2008)
28. Push button Switches. Multicomp. Recuperado el 06 de enero, de 2016, de (2014). http://uk.farnell.com/multicomp/mcpas6b2m1ce7/switch-pushbutton-spst-ip67-blue/dp/1638332
29. Resolute absolute optical encoder with Mitsubishi serial communications. Renishaw. Recuperado el 10 de abril de 2016, de (2016). http://resources.renishaw.com/en/details/–65959
30. Riesgos asociados a la manipulación manual de cargas en el lugar de trabajo. Facts. Recuperado el 15 de marzo de 2016, de (2007). https://osha.europa.eu/sites/default/files/publications/documents/es/publications/factsheets/73/Factsheet_73_Riesgos_asociados_a_la_manipulacion_manual_de_cargas_en_el_lugar_de_trabajo.pdf+&cd=1&hl=en&ct=clnk&gl=ec
31. Roncon, E., Moreno, J.C., Gallego, J.A., Pons, J.L.: Wereable Robots in Rehabilitation Engineering Tremor Suppression. Intech, Spain (2009)
32. Servo amplifier & motors Melservo-J4. Mitsubishi. Recuperado el 06 de marzo de 2016, de (2016). https://us.mitsubishielectric.com/fa/en/products/drive-products/ac_servos-melservo/melservo-j4
33. Servo geared motors KSG. Georgii Kobold. Recuperado el 15 de enero de 2016, de (2014). www.georgii-kobold.de
34. Servo system controller. Mitsubishi. Recuperado el 02 de marzo de 2016, de (2012). http://dl.mitsubishielectric.com/dl/fa/document/catalog/ssc/l03062/l03062f.pdf
35. System Q. Mitsubishi. Recuperado el 01 de marzo de 2016, de (2004). https://www.hiflexonline.nl/products/mitsubishi/plc/documentatie/Q_Automation_Book.pdf
36. Torque motors KTY-Q. Georgii Kobold. Recuperado el 15 de enero de 2016, de (2014). www.georgii-kobold.de
37. Walsh, C.J., Paluska, D., Pasch, K., Grand, W., Valiente, A., Herr, H.: Development of a lightweight, underactuated exoskeleton for load-carrying augmentation. MIT media lab, pp. 3485–3490 (2006)
38. Wolf, J., Parker, C., Borisoff, J., Mortenson, W.B., Mattie, J.: A survey at stakholder perspectives on exoskeleton technology. J. Neuroeng. Rehabil. pp. 2–12 (2014)

Author Index

M. Botto-Tobar et al. (Eds.): CI3 2020, AISC 1277, pp. 571–573, 2021.
https://doi.org/10.1007/978-3-030-60467-7